PATRICK LESTIENNE (Ed.)

Mitochondrial Diseases

Springer

Berlin
Heidelberg
New York
Barcelona
Hong Kong
London
Milan
Paris
Singapore
Tokyo

PATRICK LESTIENNE (Ed.)

Mitochondrial Diseases

Models and Methods

With 75 Figures

Springer

Dr. Patrick Lestienne
Directeur de Recherche INSERM
E 99-29 INSERM
Physiologie Mitochondriale
Université Victor Segalen Bordeaux 2
146, rue Léo Saignat
33076 Bordeaux Cedex, France
patrick.lestienne@phys-mito.u-bordeaux2.fr

The cover picture, taken by Barbara Stevens and Jane Sinclair, was donated by Björn Afzelius. It symbolizes heteroplasmy, which is frequently encountered in mitochondrial DNA diseases, suggesting their possible reversibility.

Library of Congress Cataloging-in-Publication Data

Mitochondrial diseases : models and methods / Patrick Lestienne (ed.). p. cm
Includes bibliographical references and index.

ISBN-13: 978-3-642-64166-4 e-ISBN-13: 978-3-642-59884-5
DOI: 10.1007/978-3-642-59884-5

1. Mitochondrial pathology–Molecular aspects. 2. Mitochondrial DNA. I. Lestienne, Patrick.
RB147.5.M58 1999 616'.042–dc21 99-25405 CIP

Softcover reprint of the hardcover 1st edition 1999

Production: PRO EDIT GmbH, 69126 Heidelberg, Germany
Cover Design: design & production GmbH, 69126 Heidelberg, Germany
Typesetting: STORCH GmbH, 97353 Wiesentheid, Germany
Computer to Film: Saladruck, Berlin, Germany

SPIN 10672061 31/3137 - 5 4 3 2 1 0 – Printed on acid-free paper

Foreword

For those like me who witnessed the beginning of the adventure of human mitochondrial pathology, one can only be astounded by the extent and unexpectedness of what the field has come to offer. Extent because nobody could have imagined the sheer size of the domain. Unexpectedness because hitherto it was impossible to imagine the clinical polymorphism that this pathology would represent.

The starting point was clear. Initially, there was the exceptional, and for a long time unique, observation of euthyroidian hypermetabolism that Luft and colleagues analyzed remarkably in biochemical and clinical terms. Thereafter, there was the support provided by the electron microscopy studies of Afzelius, and the very first visualization of mitochondrial abnormalities. That was way back in 1958. A few years later, progress in the cytology and cytochemistry of skeletal muscle tissue was to provide the means of detecting such abnormalities by examining sections with light microscopy. The colorful term "ragged red fibers", coined by W.K. Engel, became universally accepted, and this typical aspect with Gomori trichrome stain was to throw light on the frequency with which these mitochondrial abnormalities could occur under pathological conditions which, until then, had remained a total mystery regarding their mechanism: syndromes such as the ocular myopathies with their descending evolution and the oculocraniosomatic syndromes. We were at the beginning of the 1970s.

At about the same time, biochemical analysis of mitochondria-enriched subcellular fractions was becoming applicable to human muscle specimens, provided there was more material available than with a simple biopsy. The analysis of the various respiratory chain complexes then paved the way for understanding this pathology. Almost concomitantly, A. G. Engel and C. Angelini next demonstrated the possible correction of abnormalities in long-chain fatty acid degradation by the administration of l-carnitine, a cofactor with a very simple structure. In that same year of 1973, Di Mauro showed how a deficiency in mitochondrial carnitine palmityl transferase could result in certain lipidosis and paroxystic myoglobinurias.

A watershed was reached in the middle of the 1980s with the discovery of tools to explore the mitochondrial genome, which was then sequenced. The first demonstrations of these gene alterations were made almost simultaneously in London by the team headed by J. Morgan-Hughes and in Angers by P. Lestienne and G. Ponsot. This heralded an explosion in studies on the mitochondrial cytopathies. It was already known that double affection of the central nervous and muscle was frequent in this pathology. The abbreviations MERRF and MELAS, coined by Di Mauro‘s team, then came to designate syndromes in which there was both central nervous and muscle involvement. In this way, it became clear that Leigh's syndrome, Leber's disease, and a range of cardiac, digestive, immune and endocrinal pathologies were all related to mitochondrial abnormalities. Neonatal medicine moved forward with the discovery of cytochrome oxidase deficiencies, which were frequently severe, sometimes benign and regressive. Although this list is far from complete regarding the various aspects involved in mitochondrial pathology, it is the schematic overview of the question as seen through the eyes of one pathologist.

All this progress could never have been so rapid and dramatic without breakthroughs in pure mitochondrial biology, especially the fine structure of mitochondrial crests, the way the respiratory chain functions, the production of ATP, and the structure and evolution of mitochondrial DNA. Without doubt, in the future we will need to look not only at how all this progress in pure biology has had an impact on human pathology, but also at the questions that these issues raise. This sort of symbiosis is to be found in almost all sectors of neuromuscular pathology.

In my opinion, such collaboration between pathologists and biologists is one of the keys to the vast progress which has been made in what is indeed a short period of time. Regarding France, special mention should be given to the driving force represented by the AFM and the 5-year INSERM network "Biologie Moléculaire et Pathologie des Mitochondries Humaines". Our thanks go to those who, with the inspiration by Cécile Marsac and under the guidance of Patrick Lestienne, have collaborated on this network. Most of the authors contributing to this volume have been associated with this project. The result is testimony to the richness and depth of this network. Their combined efforts represent a further step in the understanding of mitochondrial biology, its dysfunctioning and the ways in which man may, in the near future, be able to correct these serious pathological disorders.

Michel Fardeau
Medical and Scientific Director
of the Institute of Myology
Salpêtrière Hospital

Preface

The aim of this book is to gather several aspects of mitochondrial diseases put forward by various research teams who exchanged knowledge over the 5-year INSERM network, "Biologie Moléculaire et Pathologie des Mitochondries Humaines". We hope that this work will contribute to further progress in the understanding of the basic aspects underlying human diseases.

I wish to express my thanks to the Association Française pour la lutte contre les Myopathies (AFM), the Institut National de la Recherche et de la Santé Médicale (INSERM), the Ministère de l'Enseignement Supérieur et de la Recherche, and the Fondation pour la Recherche Médicale for their support.

Finally, I would like to acknowledge the contributing authors for accepting my invitation to collaborate on this project. Special thanks are extended to Drs. C. Godinot, C. Marsac, J. P. Mazat and G. Ponsot for their advice, and the editorial team of Springer-Verlag for their help in the realization of this enterprise.

June 1999 PATRICK LESTIENNE

Contents

Introduction

P. Lestienne

Mitochondria, Witness to Our Origins, and Localization of Respiration and Energy Production

The origins of life probably came from the formation of membranes allowing the accumulation and concentration of interactive macromolecules able to perpetuate a genetic message. The independence of these systems was made possible by the assimilation and the stocking of chemical energy. The genes encoding for the enzymatic protosystems were then progressively appended to produce intracellular organelles: the mitochondria. Although mitochondrial DNA displays size and informational content variations during its evolution, organization and expression are well conserved, indicating a monophyletic origin.

In 1789 Antoine Lavoisier demonstrated that the molecular oxygen of air allowed the combustion of organic compounds, and the first mitochondria were observed around 1840 by cytologists.

The first experiments on respiration mechanisms and characterization of cytochrome oxidase were reported in the early 1930s by Warburg. In 1949, Lehninger and Kennedy demonstrated that mitochondria are the very place where the Krebs cycle occurs, as well as the site of fatty acid oxidation (Lynen's helix) and oxidative phosphorylation. Subcellular fractionation by differential centrifugation was developed at the same period by Claude and deDuve, who showed with Hogeboom and Hotchkiss that respiration takes place within the mitochondria.

The dichotomic nature of the genetic information responsible for respiration in yeast was established by Ephrussi in 1949. In 1952 Palade and Sjöstrand presented the first pictures of mitochondria at high resolution, showing that there was an outer and an internal membrane folded into cristae, the very place of energy production and respiration.

During the following years, research was oriented towards the study of respiration with submitochondrial particles formed by fragments of inner vesicular membranes, or intact mitochondria. This led to new information on the composition, kinetics, and coupling sites of the respiratory chain. Most results could be explained by the presence of highly energetic non-phosphorylated compounds as intermediates between electron transport and ATP synthesis. A chemical coupling was envisaged by different laboratories and demonstrated by Slater. Studies on beef heart mitochondria were then started in Green's laboratory, showing that ubiquinone and iron-sulfur proteins, besides the cytochromes already characterized by Keilin in 1925, represent the main components of electron transport. In 1960, Racker and his associates succeeded in isolating the ATPase from submitochondrial particles, and demonstrated that it was used as a coupling factor able to reconstitute oxidative phosphorylation in the particles lacking in it.

Meanwhile, Luft and collaborators described the first patient presenting a mitochondrial disease due to a defect in respiratory chain coupling.

As time went by, it became more and more likely that oxidative phosphorylation required an intact membrane structure, allowing interaction between the electron transporters and ATP ase. However, the importance of the membrane in the conversion of energy was far from acceptance. A breakthrough came in 1961 when Peter Mitchell proposed his chimiosmotic theory according to which the energy transfer between electron transport and ATP synthesis could be mediated through an electrochemical transmembrane proton gradient. In effect, the complexes of electron transfer and the reconstituted ATPase in the membrane were demonstrated to generate a transmembrane proton gradient which could be used for ATP synthesis coupled with electron transfer. Molecules like uncouplers that dissipate the proton gradient were found to abolish the coupling between electron transport and phosphorylation.

In parallel with these enzymatic studies, research on mitochondrial DNA continued. A new breakthrough occurred in 1963 when Nass, then Schatz, characterized vertebrate and yeast mitochondrial DNA. Using electron microscopy, Vinograd showed the presence of circular molecules, sometimes triple-stranded, corresponding to replication intermediates. One of them of about 600 nucleotides, was called the D loop or displacement loop. The characterization of a thymidine kinase (–) mutant enabled to suggest the asymmetric replication mode of mammalian mitochondrial DNA.

In 1979, Barrell demonstrated some differences between the human mitochondrial genetic code, and the universal one. Two years later, the entire sequence of a human mitochondrial DNA from placenta was established by Sanger and collaborators. A number of open reading frames with unknown functions (URFs) was then ascribed to respiratory subunits by Attardi, who also demonstrated the polycistronic nature of some of the mitochondrial mRNA. mRNA editing was discovered in the mitochondria of trypanosomes by Benne and Brakenhoff towards 1985.

The first genetic basis of mitochondrial DNA diseases was established in 1988 with the first description of heteroplasmic mitochondrial DNA deletions in some neuromuscular diseases, and the main mutation associated with Leber's hereditary optic neuropathy.

Although ATP may be formed by anaerobic glycolysis by enzymatic reactions from glucose to pyruvate, mitochondrial respiration from pyruvate to CO_2 allows a 20-fold higher yield in ATP production, such that an individual may synthesize and use his own body weight of ATP per day. The general reactions of oxidative phosphorylation are outlined in Fig. 1 which indicates the genetic origin, either mitochondrial or nuclear, of the different subunits forming the respiratory chain complexes.

Understanding the atomic and electronic mechanisms allowing the reduction of O_2 into H_2O coupled with ATP synthesis has been a challenge for many laboratories since the elucidation of the three-dimensional structure of bovine heart cytochrome c oxidase and ATPase. Besides the fundamental knowledge which will emerge, it may be possible, as for cytochrome oxidase, to recreate these basic reactions in vitro with synthetic molecules, in order to alleviate some of the energetic metabolism dysfunctions whose origins and consequences are dealt with in this book.

The first chapters describe several aspects of mitochondrial DNA organization and expression, distinctive mitochondrial genetics, clinical diagnosis of mitochondrial diseases and mitochondrial DNA alterations. Special emphasis is given to the unusual

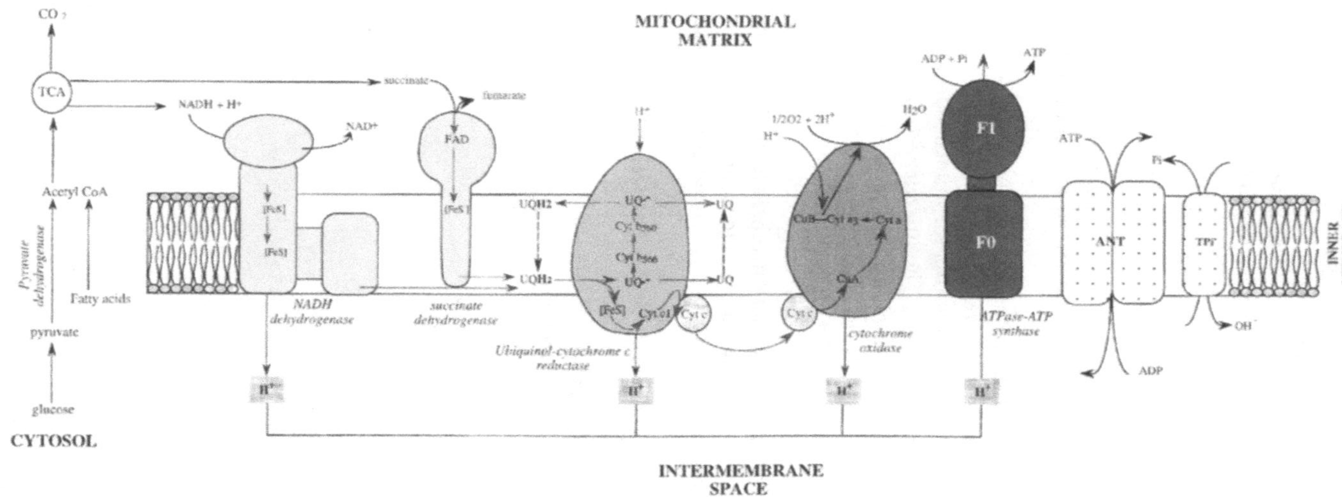

Complexes	I	II	III	IV	V
mitochondrial DNA	7	0	1	3	2
Nuclear DNA	36	4	10	10	14
Inhibitor	Rotenone	Malonate	Antimycin	Cyanide	Oligomycin

Fig. 1. Schematic representation of respiratory chain complexes involved in oxidative phosphorylation. *Complex I* NADH-ubiquinone oxido-redcutase; *complex II* succinate-ubiquinol oxidoreductase; *complex III* ubiquinol-cytochrome c reductase or bc1 complex; *complex IV* cytochrome c oxidase; *complex V* ATPase-ATP synthase; *ANT* adenine nucleotide translocator or ATP carrier: *TPi* phosphate carrier; FeS iron-sulfur protein; *UQ* ubiquinone; *Cyt* cytochrome; *TCA* tricaboxylic acid or Krebs cycle:*Pi* inorganic phosphate. The table indicates the number of subunits coded by the mitochondrial or nuclear DNA in each complex and the reagents specifically inhibiting each complex

structure of human mitochondrial tRNA transcripts. Each enzymatic complex is thereafter described, together with its known deficiencies in human diseases. Some key nuclear-encoded proteins are also detailed as the participate either or in the regulation of mitochondrial/cytosolic ATP/ADP concentration as the ADP/ATP carrier, in the conversion of ATP into energetic compounds such as creatine phosphate by the creatine kinase isoforms in the production of substrate for the Krebs cycle as pyruvate dehydrogenase. Their use in the diagnosis of some diseases as well as their potential targets in specific human therapies are highlighted.

The next section presents studies on the conformation of mitochondrial DNA in living cells, on heteroplasmic deletions in Drosophila, and on the control mitochondrial genome structure during senescence in fungi.

Due to their central role in cell bioenergetics, mitochondria play fundamental roles in programmes cell death, termed apoptosis, as well as in ageing. Recent data indicate the presence of hormone receptors such as T3 located within the mitochondria, thus emphasizing their potential role in cell differentiation. The uneven distribution of mitochondria within cells, as well as their mobility within neurons, result from specific interactions with the cytoskeleton. These various facets of mitochondrial function in cell biology and the in vivo measurement of NADH using laser fluorescence constitute the third part of this volume.

As possible new methods of mitochondrial manipulation, other studies report the characterization of yeast mutants which favours the intra-mitochondrial import of highly hydrophobic proteins that are normally mitochondrially encoded, and the precise mechanism of tRNA import which displays several common features as in plants; cytoplasmic plant male sterility can also be induced by mitochondrial gene engineering.

In the last section, several methodological approaches to the diagnosis of mitochondrial disorders are described, i.e. histologic, enzymatic and genetic methods.

CHAPTER 1

Organization and Expression of the Mitochondrial Genome 1

J. Veziers and P. Lestienne

Contents

1 Introduction

The first studies which suggested that mitochondria contain part of the genetic information responsible for respiration were carried out on yeast by Ephrussi and collaborators (Ephrussi et al. 1949). They showed that the genetic material which produced few "petites colonies" on a glucose medium, which were unable to respire, did not segregate according to usual Mendel's laws when the "petite colonie" was crossed with the wild-type strain. Instead, genetic material segregated according to new rules independent of nuclear chromosomes. Indeed, the proportion of "petite colonie" could be randomly found in the descendants, instead of the usual Mendelian repartition of 25/25/50.

The mitochondrial DNA of vertebrates was indirectly characterized by Nass and Nass (1963) who showed that the specific coloration of DNA by osmium tetroxide disappeared after the treatment of mitochondrial sections with DNAse. Then, Schatz et al. (1964) purified the association of mitochondrial DNA with mitochondria by using a sucrose gradient. The conclusive result was found by Sanger, who sequenced the entire mitochondrial DNA from the human placenta (Anderson et al. 1981).

E 99-29 INSERM physiologie mitochondriale Université Bordeaux 2, 146 rue Léo Saignat, 33076 Bordeaux, France

Meanwhile, various laboratories reported that the formation of "petites colonies" could also segregate according to Mendel's laws. This indicates the importance of nuclear gene products in biogenesis and mitochondrial function, which results from the concerted expression of both genomes, implying autosomal dominant or recessive transmissions to certain mitochondrial phenotypes. As the following chapter deals with mitochondrial genetics in mammals, we will briefly outline the main established characteristics of human mitochondrial DNA.

2 Mitochondrial DNA Heredity

Analysis of polymorphism in restriction of mitochondrial DNA (mtDNA) showed their exclusively maternal origin (Hutchinson et al. 1974). This observation has recently been refined by polymerase chain reaction (PCR) analysis in mice which unexpectedly revealed that about one paternal mtDNA could be present among 10 000 maternal molecules (Gyllensten et al. 1991). However, the persistence and transmission of mtDNA from spermatozoa seem to be restricted to interspecies crosses, while their elimination occurs in intraspecies crosses (Kaneda et al. 1995).

3 Mitochondrial DNA Polymorphism

The analysis of mtDNA polymorphism rapidly disclosed a much more important polymorphism than that of the nuclear DNA. The fixation rate of mitochondrial mutations is about 10 times higher than that of the nucleus (Whittman et al. 1986). The most polymorphic region is located in the D loop, or displacement loop, a non-coding region that controls the expression of mtDNA (Parsons et al. 1997).

4 Mitochondrial DNA Heteroplasmy

As evidenced in 1983 by Solignac and coworkers in the non-coding region of mtDNA from *Drosophila,* heteroplasmy is defined as the presence of different sequences of mtDNA within a single individual. This observation was further extended to various species, including humans, indicating that the sequence identity of the 10^{16} mtDNA molecules in humans may slightly vary, although only one difference has been reported after sequencing of different mtDNA clones representing 49 kilobases (Monnat and Loeb 1985).

5 Mitochondrial Genes and Pseudogenes Localized in the Nucleus

The evidence for mitochondrial pseudogenes localized in the nucleus was brought by hybridizing the nuclear DNA from a yeast strain devoid of mtDNA (rho°) with mtDNA probes (Farelly and Butow 1983). This observation was then extended to humans (Fukada et al. 1985). By comparing the respective mutation rates, is has been deduced that this event occured 25 million years ago for yeast, and between 400 000 years (Hu and Thilly 1995) and 770 000 years ago for humans (Wallace et al. 1997). They have evoled at the same rate as the nuclear DNA. The analysis of these mitochondrial sequences inserted into the chromosomes indicates that these events would result from recombinations involving direct or inverted repeats (Blanchard and Smith 1996). Some mutations in nuclear genes controlling the emigration of the mtDNA towards the nucleus have been identified in yeast (Thorsness and Weber 1996).

6 Human Mitochondrial DNA Organization

Human mitochondrial DNAs are covalently double-standed closed circular molecules. So far, about 20 mtDNAs from various animal species have been sequenced. Differing from yeast and plants, vertebrate mtDNA does not contain introns (Attardi 1984). Besides the non-coding sequence, "D loop", this genome is very compact, since structural genes are generally separated by tRNA genes. Two mRNAs are translated in two different reading frames, yielding four proteins (ATPase 6 and 8, subunits 4 and 4L). Furthermore, the mitochondrial genetic code differs from the cytosolic one (Barrell et al. 1979).

Human mitochondrial DNA is composed of 16 569 base pairs according to the "Cambridge" reference sequence (Anderson et al. 1981). The light strand (31.2% C, 30.9% A, 24.7% T, 13.1% G) encodes for subunit 6 of NADH dehydrogenase (complex I) and 8 tRNAs. The heavy strand encodes for the 12 other subunits of the respiratory chain complexes, 2 ribosomal RNAs (12 S and 16 S) and 14 tRNAs (Fig. 1). The control region for transcription and replication of the heavy strand DNA, called by extension the "D loop", is located between the genes of $tRNA^{Pro}$ and $tRNA^{Phe}$.

6.1 Transcription

The two main promoters (HSP, heavy strand promoter) and LSP (light strand promoter) allows a bidirectional and independent transcription of each strand (Hixson and Clayton 1985) with a non-specific RNA polymerase (145 kDa) which presents in yeast homologies with phages T3 and T7 RNA polymerase. mtTFA (Transcription Factor Activator) (25 kDa) allows specific transcription from each promoter by binding and bending the DNA close to the promoters (Fischer et al. 1992). This protein presents functional homologies with the HMG "high mobility group" proteins (Parisi and Clayton 1991) and is formed of two tandem domains of fixation to the DNA, and one C terminal end activator of transcription: the distance between fixation of the mtTFA and the promoter is critical for transcription activation (from −10 to −40 base pairs of the promoters). mtTFA also binds to CSB I "conserved sequence box I", and with a lower affinity to CSB II and CSB III. Although the mouse and human mtTFA may be interchangeable in vitro, the identity of the contact points with DNA is below 50% (Fischer et al. 1989); however, yeast mtTFA cannot replace the human one, although it

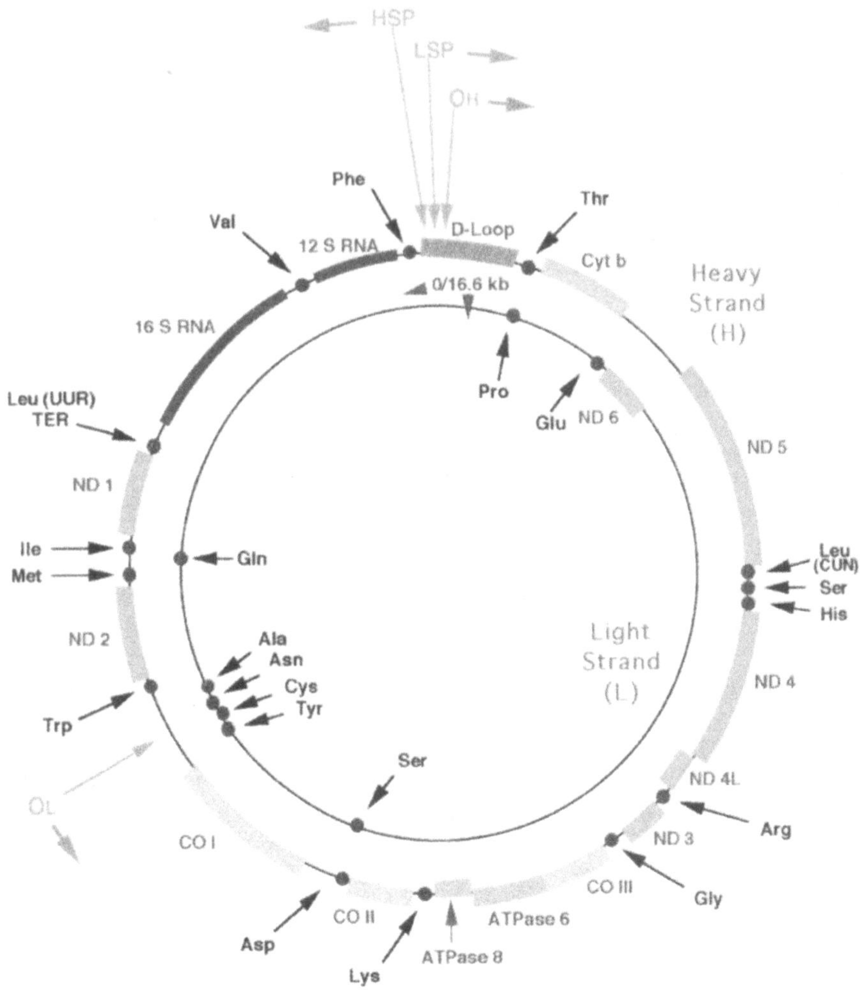

Fig. 1. Organization of human mitochondrial DNA. The numbering of the L strand is conterlockwise (1 → 16.569; Anderson et al. 1981). *Circles* indicate tRNA genes. The D-loop, delineated by tRNA Phe and Pro genes, contains the H strand origin of replication (OH) and the light and heavy strand promoters (LSP and HSP). *TER* designates the terminator of transcription located within the $tRNA^{Leu(UUR)}$ gene. co = cytochrome; c oxidase; ND = NADH dehydrogenase

can be bound to the DNA. At low concentrations, mtTFA first activates the transcription from LSP, allowing formation of the primer which will be required for further replication. The increase in mtTFA concentration (1 molecule for 35 base pairs) triggers transcription from the HSP (Dairaghi et al. 1995), then further increase in mtTFA, in vitro, induces a decrease of transcription from both promoters. mtTFA is not only located in the mitochondria, but also specifically in the nucleus of spermatozoa and spermatides; the two proteins encoded by the same nuclear gene differ by alternative splicing, yielding different cell compartments (Larsson et al. 1996). This may induce a reduction of intramitochondrial mtTFA and thus a decreased mtDNA copy number, which could partly account for the monoparental origin of mtDNA. The function of mtTFA in the structure and regulation of expression of nuclear genes remains to be established. The transcription is polycistronic from each HSP and LSP. The attenuation of transcription from the HSP occurs at the level of a tridecamer sequence located within the $tRNA^{Leu(UUR)}$ (Christianson and Clayton 1988). A protein complex composed of two polypeptides of 34 kDa and of one of 31 kDa (Daga et al. 1993) binds on this sequence, and can function bidirectionally, allowing a 15- to 50-fold higher synthesis of 16 S and 12 S rRNA than the genes located downstream. Endonuclease-like RNAse P (Doersen et al. 1985), will then cleave the polycistronic transcripts, yielding tRNA which will be modified (Martin 1995) and maturated by the addition of the 3′ CCA by tRNA-nucleotidyl transferase. Some stop codons will be formed by the 3′ polyadenylation of premessenger RNA with polyA polymerase. The 12 S ribosomal RNA is also monoadenylated while the 16 S rRNA may be polyadenylated by up to eight residues.

6.2
Protein Synthesis

As in bacteria and chloroplasts, protein synthesis starts with an N formyl methionine-tRNA, and is inhibited by chloramphenicol, which binds the 16 S rRNA. Mitoribosomes have a sedimentation constant of 55S (28S + 39S), and contain 85 nuclear-encoded proteins. There is a higher flexibility in the pairing of the third base of mitochondrial anticodons, compared with cytosolic ones. The main difference from the universal genetic code results from the possibility of pairing the uracil of the third base (wobble) of the anticodons with A, U, G, C in eight classes of codons designating Thr, Pro, Arg, Leu, Ala, Gly, Val, Ser, and by the pairing of that uracil with purines (G, A) in six classes of codons (Lys, Met, Glu, Gln, Trp, Leu). On the other hand, the guanosine at the third position of anticodons base pairs with pyrimidines (C, U) of eight classes of codons (Asn, Ser, Ile, His, Asp, Tyr, Cys, Phe). Therefore, there are two kinds of codons specifying Leu (CUN, UUPur) and Ser (AGpyr, UCN), and four corresponding mitochondrial tRNA genes.

The reading of mitochondrial codons differs from that of the cytosol in five kinds of amino acids: four Arg codons (instead of six), four Ter codons (instead of three), two Met codons (instead of one), two Ile codons (instead of three) and two Trp codons (instead of one). The universal termination codon UGA is translated as Trp in mitochondria, while AGA/AGG, which normally encode Arg, serve as termination codons, and the AUA codon is recognized as Met instead of Ile (Table 1). This suggests that the

Table 1. Differences between universal and mitochondrial genetic codes

Codons	Universal	Mammalian mitochondria	Yeast mitochondria
UGA	STOP	TRP	TRP
AUA	ILE	MET	MET
CUA	LEU	LEU	THR
AGA-AGG	ARG	STOP	ARG

decoding mechanisms have diverged, most likely to the benefit of each system, providing a frontier preventing lethal exchanges between the organelle and the nucleus. The constitution of the mitochondrially encoded proteins located within the inner membrane is a perequisite to maintain respiration and ATP production. The weaker translational constraints associated with the presence of only 22 tRNA, able to base pair essentially with the first two bases of the anticodons, also allowed a wider sequence divergence of the genome, which could account for the more rapid evolution of mtDNA compaired with the nuclear DNA. Furthermore, these peculiarities allow the translation of the whole genome with a restricted number of tRNA genes without the necessity of reported cytoplasmic import. Protein synthesis appears to be associated with the inner membrane (Escaig-Hayes et al. 1991).

6.3 Replication

The half-life of mtDNA has been estimated at 1 week in the heart, and 1 month in the brain (Gross et al. 1969).

Each mtDNA molecule contains two distinct origins of replication, one for each strand (Clayton 1982). The replication cycle starts with the synthesis of the heavy strand at OH (at position 191); the synthesis progresses until the light strand origin of replication (OL), located two thirds along the molecule, is reached. The displaced heavy strand is thus single stranded and adopts a stem and loop structure recognized by a DNA primase (Wong and Clayton 1985) associated with the cytoplasmic 5.8 S rRNA. The DNA primase then initiates the synthesis of a short primer RNA which is elongated by the DNA polymerase gamma made up of two subunits of 140 and 54 kDa (Gray and Wong 1992). The cDNA of the large subunit has been characterized (Ropp and Copeland 1996). Interestingly, the first exon corresponding to the N-terminal domain contains repeated triplets (10 repeats of CAG) and CAA, CAG encoding for 12 Gln.

Contrary to mtTFA, whose amount and stability depends on the presence of mtDNA, the DNA polymerase gamma is stably expressed in cell culture devoid of mtDNA (Davis et al. 1996).

A DNA ligase then ligates the daughter molecules in the presence of ATP, and a DNA gyrase will increase the linking number with ATP. The whole cycle of replication lasts about 2 hours in vivo (Clayton 1982), between the S and G2 cell phases (Fig. 2).

The initiation of the heavy strand DNA synthesis results from the cleavage of the primary transcript synthesized from the LSP by a ribonucleoprotein, the MRP RNAse (mitochondrial RNA processing enzyme) which requires a nuclear-encoded RNA of 265 nucleotides (Chang and Clayton 1987). This RNA displays some complementary sequences to the LSP transcript at the level of CSB II. The elongation of the primer RNA by the DNA polymerase gamma yields the "7S DNA", or D loop, which ends about 50 nucleotides further than the TAS (termination associated sequence), recognized by a termination protein of 48 kDa (Madsen et al. 1993). An antitermination, or reinitiation of DNA synthesis at the TAS level allows further elongation of the D loop, providing the heavy strand daughter DNA.

The RNA cofactor of MRP RNase is not only located within mitochondria, but also in the nucleolus where it intervenes in 5,8 S rRNA processing (Lygerou et al. 1996). The

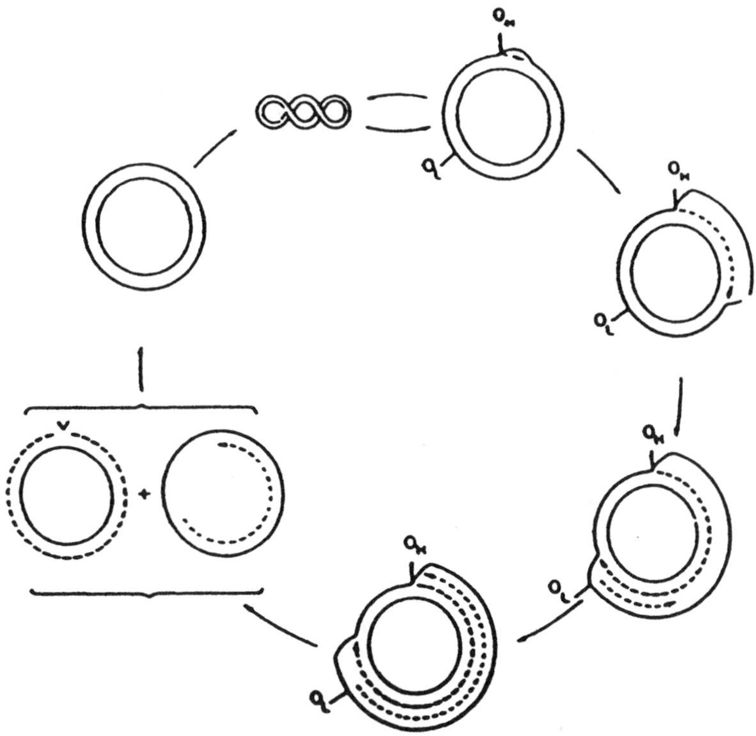

Fig. 2. Replication cycle of mammalian mitochondrial DNA. (According to Clayton 1982)

secondary structure presents a high homology with different eucaryotes, even with yeast (Schmitt et al. 1993). The deletion between nucleotides 118 and 175 of this RNA would prevent its intra-mitochondrial import (Li et al. 1994). The gene of the RNA of MRP RNase has been deleted in Saccharomyces cerevisiae, leading to its death (Schmitt and Clayton 1992).

The characterization of the POP1 gene, encoding a part of the MRP RNase of Saccharomyces cerevisiae allowed demonstration of its necessity for the survival of the strain, as well as showing a number of homologies with RNase P (Lygerou et al. 1996). The secondary structure of the RNA cofactor of MRP RNase and that of human RNase P are very similar, suggesting their shared origin and functional relationships (Gold et al. 1989). Some auto-antibodies (anti-th-40) immunoprecipiate both enzymatic activities, and notably a 40-kDa polypeptide (Lygerou et al. 1994). These observations suggest that the MRP RNase and the RNase P constitute a macromolecular complex intervening in both cytoplasmic ribosome biogenesis (5,8 S rRNA) (Lee et al. 1996) and mitochondrial biogenesis.

Another protein from yeast, SNM1, composed of 198 amino acids presenting some potential RNA binding sites, would be a specific and essential component of MRP RNase. The ribonucleic system of MRP RNase has been very well conserved during evolution, since the enzyme of Schizosaccharomyces pombe can specifically cleave the human LSP transcript at the level of CSB II (Paluth and Clayton 1995).

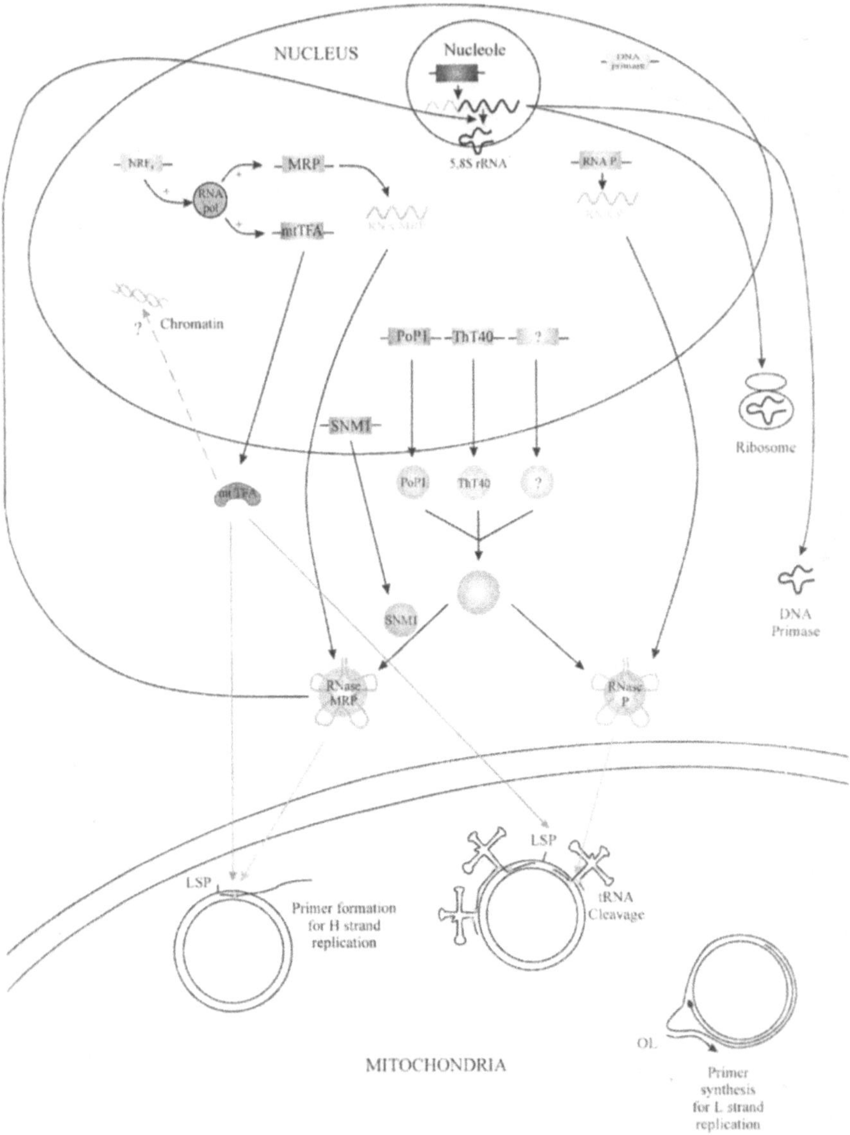

Fig. 3. Relationships between the nuclear genes encoding the main ribonucleoproteins involved in the replication and translation of mitochondrial DNA. RNase MRP is localized in mitochondria playing a role in primer formation for H strand DNA replication, and processing in the nucleole the 5,8 S rRNA, a cofactor of DNA primase that initiates L-strand DNA replication. The product of ThT40 constitutes a subunit of RNase MRP and RNase P. The POP1 gene has been characterized in the yeast Saccharomyces cerevisiae

The regulation of mtDNA replication may thus be controlled by the nucleus at a number of levels (1) synthesis of RNA polymerase and mtTFA, allowing LSP transcription, (2) cleavage of the primary transcript at the level of CSB II by MRP RNase whose RNA cofactor biosynthesis is under the control of NRF (Nuclear Respiratory Factor) 1; (3) elongation of the D loop until TAS; (4) regulation of the termination or antitermination of the D loop; and (5) regulation of DNA polymerase and primase subunits, as well as that of the 5,8 S rRNA. On the other hand, mitochondria could play the role of indicator or effector to regulate these nuclear steps by the oxido-redox status of the nucleotide pools, or by still other unrevealed molecules triggering their biogenesis (Nunnari and Walter 1996).

7 Conclusion

In conclusion, more and more data point out the tiny relationships in the so far understated relationships of nucleus-mitochondria coordinated expression, as well as the conservation of the basic mechanisms during the evolution of these organelles. These puzzling elements are schematically drawn in Fig. 3, which also suggests the subtle imbalance in the molecules playing a fundamental role in mitochondrial DNA expression and replication. Further studies on the feedback of mitochondrial products on nuclear gene expression, notably on NRF, will shed further light on mitochondrial biogenesis, as well as on the evolution of the endosymbiosis of these organelles.

The determination of the mtDNA sequence of the heterotrophic flagellate Reclinomonas americana again supports the endosymbiotic theory of evolution of these eubacterial-like organisms (Lang et al. 1997). It presents a standard genetic code as well as several genes involved in its own transcription and translation; notably the bacteria-like RNA polymerase which has now escaped to the nucleus and has been replaced by simplified nuclear viral-like RNA polymerase.

References

Anderson S, Bankier AT, Barrell BG, Debruijn MHL, Coulson AR, Drouin J, Eperon IC, Nierlich DP, Roe BA, Sanger F, Schreier PH, Smith AJH, Staden R, Young IG (1981) Sequence and organization of the human mitochondrial DNA. Nature 290:457–465

Attardi G (1985) Animal mitochondrial DNA: an extreme example of genetic economy. Int Rev Cytol 93:93–135

Barrell BG, Bankier AT, Drouin J (1979) A different genetic code in human mitochondria. Nature 282:189–194

Blanchard JL, Smith GW (1996) Mitochondrial DNA migration events in yeast and humans: integration by a common end-joining mechanism and alternative perspectives on nucleotide substitution patterns. Mol Biol Evol 13:537–548

Chang D, Clayton DA (1987) A mammalian mitochondrial RNA processing activity contains nuclear-encoded RNA. Science 235:1178–1184

Christianson T, Clayton DA (1988) A tridecamer DNA sequence supports human mitochondrial RNA 3′-end formation in vitro. Mol Cell Biol 8:4502–4509

Clayton DA (1982) Replication of animal mitochondrial DNA. Cell 28:693–705

Daga A, Micol V, Hess D, Aebersold R, Attardi G (1993) Molecular characterization of the transcription terminator factor from human mitochondria. J. Biol Chem 268:8123–8130

Dairaghi DJ, Shadel GS, Clayton DA (1995) Human mitochondrial transcription factor A and promoter spacing integrity are required for transcription initiation. Biochim Biophys Acta. 1271:127–134

Davis AF, Ropp PA, Clayton DA, Copeland WC (1996) Mitochondrial DNA polymerase gamma is expressed and translated in the absence of mitochondrial DNA maintenance and replication. Nucleic Acids Res 24:2753–2759

Doersen CJ, Guerrier-Takeda C, Altman S, Attardi G (1985) Characterization of an RNase P activity from HeLa mitochondria. J Biol Chem 260:5942–5949

Ephrussi B, Hottinguer H, Tavlitzki J (1949) Action de l'acriflavine sur les levures. Etude génétique du mutant "Petite colonie". Ann Inst Pasteur 76: 419–451

Escaig-Hayes F, Grigoriev V, Peranzi G, Lestienne P, Fournier J-G (1991) Analysis of human mitochondrial transcripts using electron microscopic in situ hybridization. J Cell Sci 100:851–862

Farelly F, Butow RA (1983) Rearranged mitochondrial genes in the yeast nuclear genome. Nature 301:296–301

Fischer RP, Parisi M, Clayton DA (1989) Flexible recognition of rapidly evolving promoter sequences by mitochondrial transcription factor 1. Genes Dev 3:2202–2217

Fischer RP, Lisowski T, Parisi M, Clayton DA (1992) DNA wrapping and bending by a mitochondrial high mobility group-like transcriptional activator protein. J Biol Chem 267:3358–3367

Fukada M, Wakasugi S, Tsuzuki T, Nomiyama H, Shimada K (1985) Mitochondrial DNA-like sequences in the human nuclear genome. J Mol Biol 186:257–266

Gold HA, Topper JN, Clayton DA, Craft J (1989) The RNA processing enzyme RNase MRP is identical to the Th RNP and related to RNase P. Science 245:1377–1380

Gray H, Wong TW (1992) Purification and identification of subunit structure of the human mitochondrial DNA polymerase. J Biol Chem 267:5835–5841

Gross NJ, Getz GS, Rabinowitz M (1969) Apparent turnover of mitochondrial deoxyribonucleic acid and mitochondrial phospholipids in the tissues of the rat. J Biol Chem 244:1552–1562

Gyllensten U, Wharton D, Josephson A, Wilson A (1991) Paternal inheritance of mitochondrial DNA in mice. Nature 352: 255–257

Hixson J, Clayton DA (1985) Initiation of transcription from each of the two human mitochondrial promoters requires unique nucleotides at the transcriptional start sites. Proc Natl Acad Sci USA 82:2660–2664

Hu G, Thilly WG (1995) Multicopy nuclear pseudogenes of mitochondrial DNA reveal recent acute genetic change in the human genome. Curr Genet 28:401–404

Hutchinson CA, Newbold JE, Potter SS, Edgell MH (1974) Maternal inheritance of mammalian mitochondrial DNA. Nature 251:536–538

Kaneda H, Hayashi JI, Takahama S, Taya S, Fischer-Lindahl K, Yonekawa H (1995) Elimination of paternal mitochondrial DNA in intraspecific crosses during early mouse embryogenesis. Proc Natl Acad Sci USA 92:4542–4546

Lang F, Burger G, O'Kelly CJ, Cedergren R, Golding GB, Lemieux C, Sankoff D, Turnel M, Gray MW (1997) An ancestral mitochondrial DNA resembling a eubacterial genome in miniature. Nature 387:493–497

Larsson NG, Garman JD, Oldorfs A, Barsh GS, Clayton DA (1996) A single mouse gene encodes the mitochondrial transcription factor A and a testis-specific nuclear HMG-box protein. Nat Genet 13:296–302

Lee B, Matera G, Ward DC, Craft J (1996) Association of RNase mitochondrial RNA processing enzyme with ribonuclease P in higher ordered structures in the nucleolus: a possible coordinate role in ribosome biogenesis. Proc Natl Acad Sci USA 93:11471–11476

Li K, Smagula CS, Parsons WJ, Richardson JA, Gonzales M, Hagler HK, Williams RS (1994) Subcellular partitioning of MRP RNA assessed by ultrastructural and biochemical analysis. J Cell Biol 124:871–882

Lygerou Z, Mitchell P, Petfalski B, Tollervey D (1994) The POP1 gene encodes a protein component common to the RNase MRP and RNase P ribonucleoproteins. Genes Dev 8:1423–1433

Lygerou Z, Allmang C, Tollervey D, Séraphin B (1996) Accurate processing of a eucaryotic precursor ribosomal RNA by ribonuclease MRP in vitro. Science 272:268–270

Madsen CS, Ghivizzani SC, Hauswirth WW (1993) Protein binding to a single termination-associated sequence in the mitochondrial DNA D-Loop region. Mol Cell Biol 13:2162–2171

Martin NC (1995) Organellar tRNAs: biosynthesis and function. In: Söll D, Rajbandhary UL (eds) tRNA structure, biosynthesis, and function. ASM Press, Washington, DC, pp 127–140

Monnat RJ, Loeb LA (1985) Nucleotide sequence preservation of human mitochondrial DNA. Proc Natl Acad Sci USA 82:2895–2899
Nass S, Nass MMK (1963) Intramitochondrial fibers with DNA characteristics. II. Enzymatic and other hydrolytic treatments. J Cell Biol 19:613–629
Nunnari J, Walter P (1996) Regulation of organelle biogenesis. Cell 84:389–394
Paluth JL, Clayton DA (1995) Schizosaccharomyces pombe RNase MRP RNA is homologous to metazoan RNase MRP RNAs and may provide clues to interrelationships between RNase MRP and RNase P. Yeast 11:1249–1264
Parisi MA, Clayton DA (1991) Similarity of human mitochondrial transcription factor 1 to high mobility group proteins. Science 252:965–969
Parsons TJ, Muniec DS, Sullivan K, Woodyatt N, Alliston-Greiner R, Wilson M, Berry DL, Holland KA, Weedn V, Gill P, Holland M (1997) A high observed substitution rate in the human mitochondrial DNA control region. Nat Genet 15:363–367
Ropp PA, Copeland WC (1996) Cloning and characterization of the human mitochondrial DNA polymerase. Genomics 36:449–458
Schatz G, Haslbrunner E, Tuppy H (1964) Deoxyribonucleic acid associated with yeast mitochondria. Biochem Biophys Res Commun 15:127–132
Schmitt ME, Clayton DA (1992) Yeast site-specific ribonucleoprotein endoribonuclease MRP contains an RNA component homologous to mammalian RNase MRP RNA and essential for cell viability. Genes Dev 6:1975–1985
Schmitt ME, Bennett JL, Dairaghi D, Clayton DA (1993) Secondary structure of RNase MRP RNA as predicted by phylogenic comparison. FASEB J 7:208–213
Solignac M, Monnerot M, Mounolou JC (1983) Mitochondrial DNA heteroplasmy in Drosophila mauritania. Proc Natl Acad Sci USA 80:6942–6946
Thorsness PE, Weber ER (1996) Escape and migration of nucleic acids between chloroplasts, mitochondria, and the nucleus. Int Rev Cytol 165:207–234
Wallace DC, Stugard C, Murdock D, Schurr T, Brown MD (1997) Ancient mtDNA sequences in the human nuclear genome: a potential source of errors in identifying pathogenic mutations. Proc Natl Acad Sci USA 94:14900–14905
Whittmann TS, Clark AG, Stoneking M, Cann RL, Wilson AC (1986) Allelic variation in human mitochondrial genes based on pattern of restriction site polymorphism. Proc Natl Acad Sci USA 83:9611–9615
Wong TW, Clayton DA (1985) DNA primase of human mitochondria is associated with structural RNA that is essential for enzymatic activity. Cell 45:817–825

CHAPTER 2

Mitochondrial DNA Inheritance in Mammals

2

D. CASANE[1] and M. GUÉRIDE[2]

Contents

1 Introduction

In animal species, including humans, mitochondrial DNA is highly polymorphic at the population level, but a unique mitochondrial genotype is generally found in an individual (Wilson et al. 1985; Avise et al. 1987; Harrison 1989). mtDNA polymorphism in individuals or populations and its evolution depend on the mechanism determining the segregation of mtDNA molecules in somatic and germ cell lines.

In mammals, the contribution of paternal cytoplasm to the formation of the zygote is negligible if any, therefore mtDNA is maternally inherited (Hutchinson et al. 1974). In addition, specific properties of mitochondrial and nuclear genomes have to be taken into account for their transmission to daughter cells:

---

[1] Laboratoire de Biologie Cellulaire, Bat 444, Université Paris Sud, 91405 Orsay Cedex, France
[2] UPRESA 8087, CNRS, Batiment Fermat, Université Versailles/St. Quentin, 45 Avenue des Etats-Unis, 78035 Versailles, France

1. Multiple mtDNA copies are present in each cell, from a few hundred to several thousand depending on the cell line (Robin and Wong 1988). The number of mitochondria and mtDNA molecules may change during the cell differentiation process, while there are two and only two copies of the nuclear genome. Due to the high copy number of mtDNA molecules, the occurrence of several mitochondrial alleles within a cell is quite possible, whereas there is only a maximum of two alleles at a chromosomal locus.
2. During the cell cycle, nuclear DNA is replicated once and only once; then duplicated chromosomes are accurately distributed to each daughter cell. Allelic segregation occurs only through meiosis. At the same time mtDNA molecules can replicate more than once or not at all, and mitochondria can divide and be distributed to daughter cells. If some polymorphism exists in a cell lineage, random partitioning of mitochondria at each cell division can lead to the evolution of the relative frequencies of mtDNA types. It follows from this that an intracellular genetic drift of mitochondrial genomes may occur, in contrast to the nuclear genome.

At the population level, the evolution of mitochondrial polymorphism through generations depends on the selective value of the alleles, and on random variations of their frequencies. Selection pressure and genetic drift can occur at various levels: (1) inside mitochondria between mtDNA molecules, (2) in cells between mitochondria, (3) in organisms between cells, and (4) in populations between individuals (female gametes sampling).

This chapter summarizes recent data on mtDNA transmission in the germ cell line and during somatic cell divisions. Our current knowledge comes from analyses of mtDNA genotype transmission in humans with hereditary mitochondrial diseases, as well as analyses of inheritance of mitochondrial heteroplasmy in animal models or in cell cultures. The recent discovery of a relationship between mtDNA mutations and severe human pathologies gave rise to numerous studies on the transmission of mutated DNA during cell divisions, during development and through generations. These studies reveal how difficult it is to predict the distribution of mutant mtDNA in the progeny from the known polymorphism in the mother's tissues.

2
Distribution of Mitochondria at Mitosis

Our knowledge about mitochondrial doubling and partitioning between daughter cells is very limited. For example, it is not precisely known when the doubling of the organelles takes place in relation to the mitosis process. A morphometric study (Posakony et al. 1977) using HeLa cells suggests that mitochondria divide during the cell cycle independently of mitosis. It is generally accepted that mtDNA synthesis takes place all through the cell cycle, but it is not known if the doubling of DNA molecules precedes or follows the division of the organelle. These processes are likely to be under nuclear control. Furthermore, physical and spatial constraints can be predicted at various levels: DNA protein association in a nucleoid structure and probably association of this structure with the membrane (Barat et al. 1985), relations of mitochondria with the cytoskeleton etc. In situ localization of mitochondrial DNA replication in mammalian cells suggests that mtDNA replication first occurs in a perinuclear position. Newly replicated mtDNA then appears to rapidly distribute throughout the dynamic cellular mitochondrial network (Davis and Clayton 1996).

3
Mitochondrial Heteroplasmy

Analysis of the transmission of mtDNA molecules is based on the detection of distinct mitochondrial genotypes within an individual or a cell (heteroplasmy).

Two kinds of mutations create mtDNA polymorphism, and consequently heteroplasmy, with various frequencies:

1. *Point Mutations.* Polymorphism at a single nucleotide position is rare within an individual whereas it is common between individuals in populations. Heteroplasmy appears to be a transient state resulting in the loss or a rapid fixation of mutations.
2. *Length Variations:* Length variation can be generated by the repetition of a single base (for example, a cytosine track in humans; Bendall and Sykes 1995), but generally it is due to tandem repeats (Rand 1993). Tandem repeats have been often observed in the non-coding region of mammalian mtDNA, and length polymorphism of mtDNA molecules arises from variable copy numbers of the repeat unit. In some species, mtDNA is polymorphic in all the individuals. A high mutation rate seems to account for such a high level of polymorphism within individuals and for its maintenance through generations. Large deletions or duplications can also generate length polymorphism; as these mutations are very likely to be lethal, mutated molecules are always found associated with intact mtDNA molecules.

4
Variations of the Mitochondrial DNA Polymorphism During Cell Divisions

4.1
Theoretical Approach and Methods of Analysis

The evolution of polymorphism will be discussed for the germline and then for somatic cell lines. A schematic representation of the evolution of mitochondrial polymorphism through generations is given in Fig. 1. Assuming that an unfertilized oocyte contains two mtDNA molecular types with different frequencies, as indicated by the white and grey areas, these frequencies will not be modified by fertilization. The germline is rapidly individualized from the somatic cell lines during the first stages of embryogenesis, the fundamental difference between the two being that the former will survive through generations while the latter disappears with the death of the individual.

The variations in the distribution of molecular polymorphism depends on three parameters: (1) the selective value σ_i, (2) the mutation rate μ_i, and (3) the effective number Ne_i.

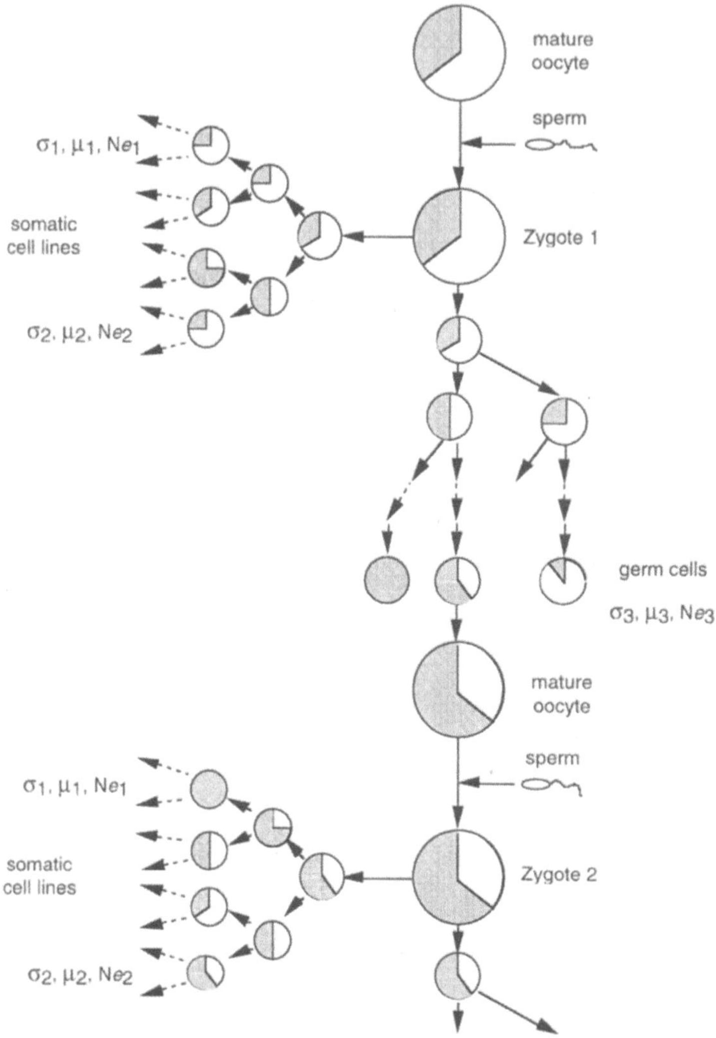

Fig. 1. Mitochondrial transmission in germ and somatic cell lines. (Adapted from Chapman et al. 1982). Mutation rates (μ_1, μ_2, μ_3), selective values ($\sigma_1, \sigma_2, \sigma_3$) and effective numbers (Ne_1, Ne_2, Ne_3) of a molecular type can be different in the various somatic cell lines and in the germline. The *white* and *grey* areas correspond to the relative frequencies of two mtDNA types in heteroplasmic cells

1. Differences in the selective values of two mtDNA molecular types can be expressed at several levels and modify their frequencies. At the level of the organelle, a replicative advantage for one molecular type can produce its accumulation and the loss of the other type. At the level of the cell, the shift toward mitochondria containing a particular mtDNA can be responsible for mitochondrial polymorphism between cells conferring different selective values on them. Finally, a mitochondrial polymorphism between individuals can confer different selective values to the indivi-

duals, genetic mitochondrial pathologies being an extreme case of a potentially low selective value.
2. Mutations modify the frequency of an mtDNA type; with the exception of a very high mutation rate, this variation is usually low, on a short time scale.
3. The variation of the relative frequencies of genotypes, resulting from random partitioning of a finite number of mtDNA molecules to daughter cells at each division, depends, in a large part, on the effective number of mtDNA molecules, Ne_i, in a particular cell lineage (i).

Ne corresponds to the number of molecules in an "ideal" population of mtDNA with the following properties: (1) the size of the population does not vary during cell divisions or cell differentiation; (2) at each cell cycle all the molecules undergo one replication cycle and only one, and are subsequently partitioned at random to daughter cells; (3) the evolution of the distribution of frequencies of mtDNA molecules corresponds to the evolution observed in the true population. Generally, the estimated value obtained for Ne in a particular cell line is lower than the true number of mtDNA molecules for several reasons: mtDNA can be replicated more than once or not at all during a cell cycle (Bogenhagen et al. 1977), possibly depending on its position in the cell; the number of molecules can change considerably during the differentiation process; the partitioning can be asymmetric. As Ne is equal to the harmonic mean of the real number of molecules (Crow and Kimura 1970), it is often close to the lowest number of molecules present in the cell line. The lower Ne is, the more important the expected variation in frequencies will be.

In the absence of selection, and if mutation rates are low, a good estimation of Ne will allow tracking of the evolution of polymorphism. The mean p_n and the variance V_n of the frequency of a mitochondrial genotype after n cell divisions are given by the formula (Solignac et al. 1984):

$$p_n = p_0 \quad (p_0 = \text{initial frequency}),$$
$$V_n = p_0 (1 - p_0) [1 - (1 - 1/Ne)^n]. \tag{1}$$

When n increases, the variance increases up to the limit $p_0 (1 - p_0)$, and the mean frequency remains unchanged. The prediction of the evolution of polymorphism requires accurate estimations of these parameters, but few quantitative analyses have yet been done in mammals.

4.2 Transmission of Mitochondrial DNA in the Germline

To understand the transmission of polymorphism through successive generations it is necessary to get some knowledge about the dynamics of mtDNA molecules during the germ cell divisions. A schematic representation of the evolution of frequencies of two mtDNA molecular types is given in Fig. 1. The female grown from zygote 1 contains a population of germ cells with various distributions of mtDNA polymorphism. The importance of the variation depends on the parameters mentioned above. In this example, one cell contains a pure mtDNA genotype while this molecular type is only weakly represented in another cell. The polymorphism is then different between the progeny and the mother and among siblings. In humans, the number of cell divisions in the female germline has been estimated to be about 30. A theoretical example of the evolution of the distribution of an mtDNA molecular type, assuming 30 cell cycles, is given in Fig. 2. Taking the initial percentage to be 33%, with an effective number Ne = 100, a flat distribution is observed, predicting possible important differences between the mother and the progeny. In contrast, with Ne equal to or over 1000, only slight variations between frequencies are expected. Assuming that mutation rates are

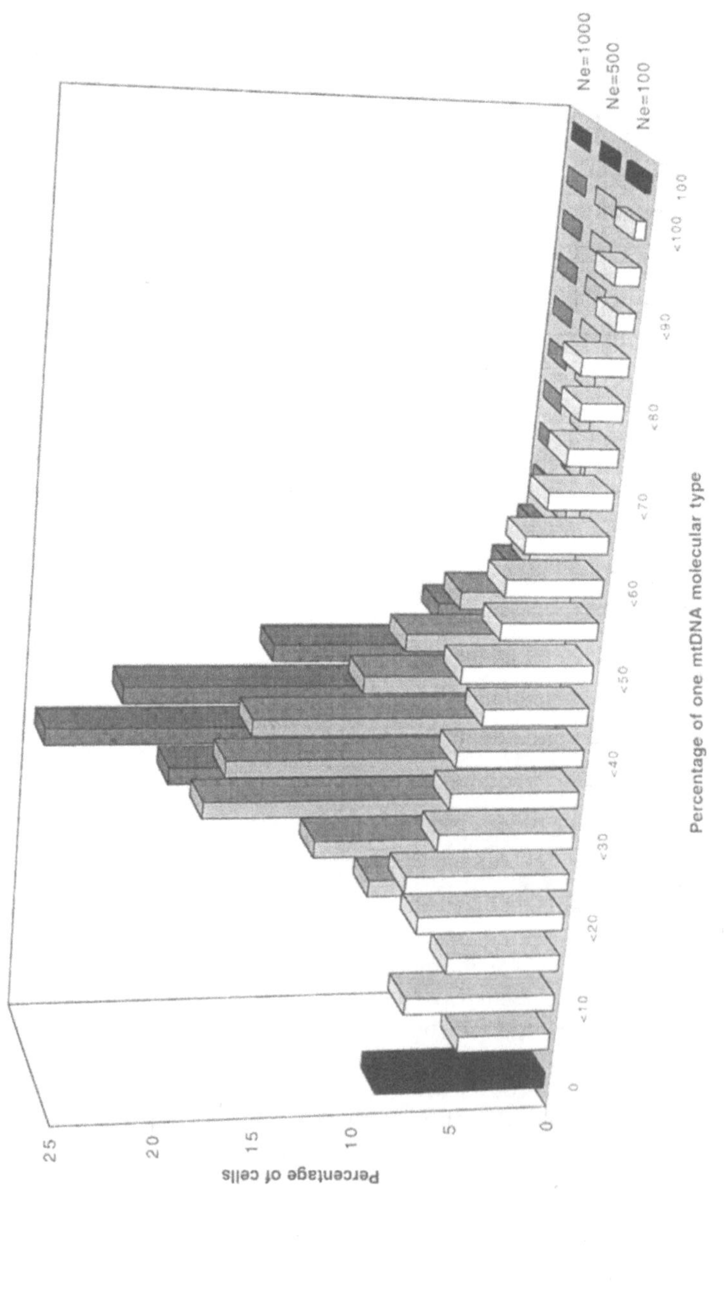

Fig. 2. Distribution of the frequencies of an mtDNA type: effect of genetic drift. Simulations were made to determine the distribution of a mitochondrial allele in germ cells, after 30 replication cycles, using three different values for Ne: 100, 500 and 1000, respectively. The initial frequency in the zygote was taken to be 33%

low and that there is no selective advantage for one molecular type, the frequencies observed in an individual should be close to frequencies present in the original zygote. The genetic drift in the germline can thus be estimated from the comparison of the distribution of frequencies between the mother and the offspring.

4.2.1
In Mice

In a recent work (Jenutz et al. 1996) an estimation of the effective number of mtDNA molecules has been made in mice germlines. Females heteroplasmic at the restriction site *Rsa* 1 in ND2 gene, nt 3961, were produced by electrofusing cytoplasts from NZB or BALB zygotes to one-cell embryos of the other type (BALB or NZB). These embryos were transplanted, at two-cell stage, into pseudopregnant females to complete development. Five heteroplasmic females carrying from 3% to 7% of the donor mtDNA were obtained. These were then mated to BALB males, and mtDNA was studied in the resulting progeny. The mean proportion of the donor mtDNA genotype in the offspring was not significantly different from the frequency in the founder female, suggesting that the mutation is neutral. The effective number Ne calculated for the five females using Eq. (1) is 185 (range 76-867). From cytological observations, the number of mitochondria in oogonia had been estimated to be approximately 40, which would correspond to about 200 mtDNA molecules. In this case, the calculated effective number is close to the real number. These results suggest a random partitioning of mitochondria during cell divisions in the germline.

4.2.2
In Cows

In Holstein cows it has been reported (Ashley et al. 1989) that heteroplasmy at position 364 in mtDNA shifted toward homoplasmy within a few generations, maybe only one. The authors proposed that an important reduction in the number of mtDNA molecules, "a bottleneck", could occur in the female germline. A random partitioning of a limited number of mitochondria could account for a rapid segregation and a shift toward homoplasmy within very few generations.

4.2.3
In Humans

Pedigree analysis of families with mtDNA polymorphism have not allowed a precise calculation of the effective number; only rough estimations are available.

Example 1. In a study of a neutral polymorphism at nt 14560 in ND6 gene, using Eq. (1), Ne was estimated to be in the range 36-180 (Howell et al. 1992a). Point mutations in NADH-dehydrogenase subunits ND1, ND4 and ND6 are associated with Leber disease (LHON). The mutation is essentially homoplasmic, but has been found to be heteroplasmic in about 15% cases. An analysis of some family pedigrees shows an evolution toward homoplasmy. In one case the frequency of the mutation at position 11778 in ND1 evolved from 32% to 95% in two generations (Howell et al. 1994); in another family heteroplasmic for a mutation at nt 3460 in ND1 (Howell et al. 1992b), the analysis of mtDNA of seven people over two generations reveals that mtDNA segregation has led to the three possible situations: homoplasmy for the mutation, homoplasmy for the wild-type allele and heteroplasmy.

Example 2. The tRNA lys A $\rightarrow$ G mutation, nt 8334, is found in patients with MERRF syndrome. The frequency of mutant DNA was determined in lymphocytes of relatives of two patients, over four generations in one family (Larsson et al. 1992). The following observations could be made:

1. The mutation was not necessarily found in the offspring of heteroplasmic women; four mothers with 10% to 33% mutant DNA in lymphocytes had transmitted the mutation to only some of their children (7 out of 14 investigated children).
2. The frequency of the mutation was highly variable among siblings: four of the six children of a woman with 28% mutant DNA in lymphocytes had 0%, 15%, 49% and 72% mutant DNA respectively.
3. There was a correlation between the level of the mutation in the mother's lymphocytes and the risk of transmission of the mutation to the offspring; above a threshold of about 43% the mutation is very likely to be transmitted to all the children.

The rapid fixation of the mutated or wild type allele and the high variability of the proportions of mutant DNA among the offspring agree with the hypothesis of a rather low effective number of mtDNA molecules. The observed distribution is quite similar to the simulated distribution setting Ne = 100 (Fig. 2). It shows how difficult it is to predict the evolution of polymorphism through generations. It is noteworthy that the value of Ne, about 100, which apparently agrees with the observations, is much lower than the estimated number of mtDNA molecules in women's ova, about 100 000 (Chen et al. 1995). This number is probably lower in oogonia; a dramatic reduction in the number of mtDNA molecules, early during oogenesis, with a subsequent expansion up to thousands of molecules, "the bottleneck model", is generally suggested, although not proved. Additional factors could contribute to a low value of Ne, such as heterogeneous divisions of mitochondria inside the cell or their asymmetric partitioning to daughter cells.

4.2.4
In Rabbits

Length heterogeneity of mtDNA molecules has been observed in rabbits (Ennafaa et al. 1987; Biju-Duval et al. 1991) and in all the other species of the order Lagomorpha studied so far. Rabbit strains of well defined lineage are available, making this species a valuable material for studies of the mtDNA transmission through generations or during somatic cell divisions.

Two sets of repeats have been characterized at the 5′ end of the non-coding region (Mignotte et al. 1990; Fig. 3). A first set consists of a 153 bp sequence repeated in tandem (long repeats, LR), in which case the variability in copy number may be not neu-

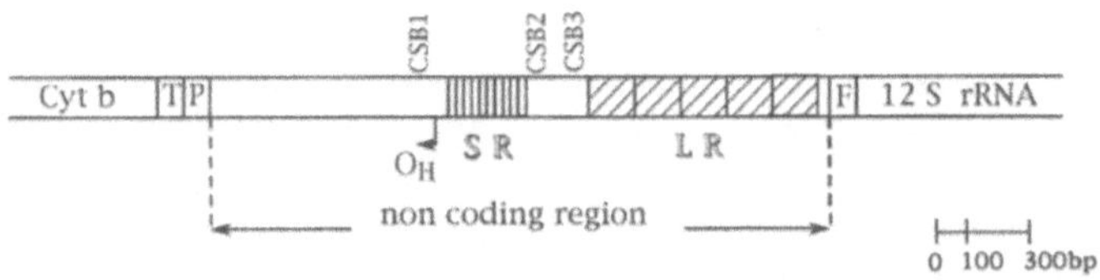

Fig. 3. Localization of tandem repeats in the non-coding region of rabbit mitochondrial DNA. *LR* 153-bp repeat, *SR* 20-bp repeat, O_H replication origin for the H strand, *CSB* conserved sequence block in vertebrates

tral, as it has been shown that this amplification can produce multiple initiation transcription sites (Dufresne et al. 1996). The second set of repeats consists of a stretch of 20 bp elements (short repeats, SR). Both sets are involved in the generation of heteroplasmy.

In a study of length polymorphism in a female rabbit lineage we have made a rough estimate of the effective number of mtDNA molecules in the germline, using the variation in frequencies of molecules with four or five long repeats, LR (Casane et al. 1994). We have found Ne = 500, a number which agrees with the number of mitochondria observed. This result is consistent with the hypothesis that, in this species, mitochondria are independent segregating units which are partitioned at random.

Estimates of Ne have been made for four kinds of mammal so far, giving different values; very low for bovines, low for humans and mice and rather high for rabbits. The reason for this apparent variability could just be due to imprecision in the estimations, moreover, there is no available estimation of selective values for the mtDNA types.

4.3 Transmission of Mitochondrial DNA in Somatic Cell Lines

The segregation behaviour of mtDNA during somatic dell divisions is not well known. From the few studies available it is clear that several parameters will have to be taken into account in the analyses of the effects of mtDNA mutations.

In the absence of mitotic segregation, nuclear DNA polymorphism is, with rare exceptions (genomic rearrangements), identical in all somatic cells of an individual. This is obviously not the case in mtDNA polymorphism (Fig. 1). As previously mentioned, the evolution of mtDNA polymorphism during cell divisions depends on three parameters (σ_i, μ_i, Ne_i) which can vary from one cell line to the next. For example, the effective number Ne is expected to be lower in cell lines containing few mitochondria, such as chondrocytes, than in hepatocytes rich in mitochondria. This would mean that variability in the polymorphism would be more important in the former than in the latter. These variations being random, a same initial polymorphism in two zygotes can lead to different distributions in the somatic cell lines of the corresponding individuals. A mutation could have been eliminated in a cell line of an individual and could remain in the same cell line of one of his siblings (Fig. 1).

From studies of human mitochondrial disorders it appears that the deleterious effect of a mutation is expressed when a threshold level of mutant DNA has been reached. The higher the frequency is in the zygote, the higher will be the risk of the thre-shold level being reached in a cell line. The mitochondrial genetic drift in somatic cell lines could explain, at least in part, the various effects of the same mutation observed in a given tissue of different individuals. This genetic drift would not be responsible for a systematic bias of an mtDNA type frequency according to the tissue. In other words, variable mutation frequencies are expected according to cell lines, but they are not systematically higher or lower in one cell line compared with another. However, such discrepancies among tissues have been observed in humans and in rabbits.

4.3.1
In Humans

Larsson et al. (1992) showed that, in patients with MERRF syndrome and even in asymptomatic relatives, the frequency of the mutation tRNA lys, at position 8334, was different in the various tissues: muscle > white blood cells > fibroblasts. In the same way, Howell et al. (1994) reported in a study of heteroplasmic LHON patients that the percentage of the mutation in ND1, nt 11778, was found to be always lower in blood than in any other tissue.

How can this bias be explained? Is the threshold level for the expression of the dysfunction variable according to the tissue? Does the intracellular selective value of the mutated allele depend on the cellular environment?

4.3.2
In Rabbits

In each individual a variable copy number of long (LR) and short (SR) repeats is responsible for the mtDNA polymorphism We have analyzed the polymorphism distribution due to the long repeats in various tissues (Casane et al. 1994). In the domestic rabbit lineage studied, mtDNA molecules contained two to ten LRs with a major mtDNA population having five LRs (mtDNA-5 LR) in each individual. As well as the predominant mtDNA-5 LR, the frequency of molecules with four LRs (mtDNA-4 LR) was high enough to allow an accurate estimation. The proportions of these two molecular types were analyzed in liver and kidneys in order to detect a possible bias in frequencies, and to identify the potential causes (Fig. 4). Two observations were made: a strong correlation between the frequencies of mtDNA-4 LR in the two organs, and a systematically higher frequency of this molecular type in liver. Two non-exclusive hypotheses could account for these observations:

1. *Variation in Mutation Rates.* The mutation rate, per time unit, of mtDNA-5 LR to mtDNA-4 LR (deletion of one unit) is higher in liver than in kidneys for two possible reasons: (1) the rate of mtDNA turnover is similar in both organs but the mutation rate, per replication cycle, is higher in liver; (2) the mutation rate per replication cycle is similar in both organs but the rate of mtDNA turnover is higher in liver.
2. *Variation in Selective Values:* The selective values of molecules with 5 or 4 LRs are different in different tissues, for example molecules with 4 LRs could have a replicative advantage in liver.

Fig. 4. Comparison of the frequencies of mtDNA-4LR in kidneys and liver. R^2 Correlation coefficient

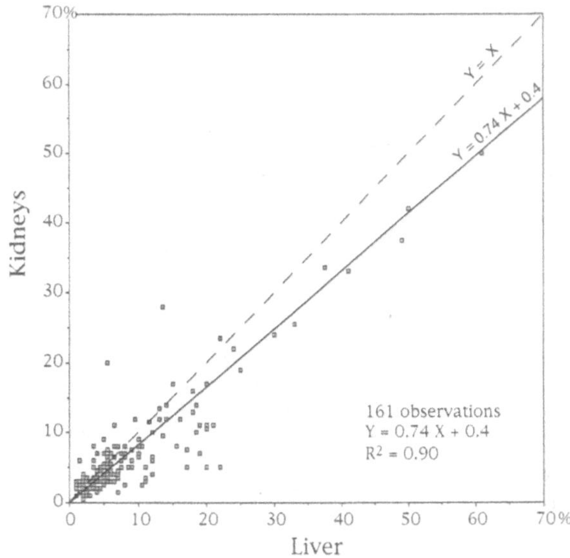

The experimental data were compared with theoretical curves generated under selective or mutational models. We came to the conclusion that the higher frequency of mtDNA-4 LR in the liver could result from a higher selective advantage. In addition, the simulation showed that a little variation of selective values for a molecular type between various tissues can produce important differences in the pattern of mtDNA polymorphism.

An analysis of mtDNA polymorphism in more tissues (Casane et al. 1997) clearly showed that mtDNA molecules with more than five long repeats accumulate with time in gonads when mtDNA molecules with less than five long repeats accumulate in the other tissues (Table 1). It is currently impossible to know if these divergent evolutions of polymorphism are the consequences of: (1) selection for longer molecules in gonads and shorter ones in the other tissues; or (2) mutational mechanisms generating mainly longer molecules in the gonads and shorter ones in the other tissues.

In agreement with our observations, in a recent study of heteroplasmic mice carrying two mtDNA genotypes, Jenuth et al. (1997) concluded that tissue specific selective pressures should exist for different mtDNA molecular types in the same animal.

From data obtained with rabbits and mice and from observations in humans, it is obvious that the proportions of mtDNA molecular types having passed through the bottleneck can be subsequently modified by factors which influence the segregation in somatic cells during development. Predictions of polymorphism in one tissue on the basis of the knowledge of polymorphism in another tissue of the mother proves to be more hazardous than previously thought.

Table 1. Tissue-specific polymorphism in adult rabbits

Animals	Organ						Sex		
	Kidneys Σ% mtDNA types		Brain Σ% mtDNA types		Liver Σ% mtDNA types		Ovaries/testicles Σ% mtDNA types		
	<5LR	>5LR	<5LR	>5LR	<5LR	>5LR	<5LR	>5LR	
449	4	<1					<1	9.5	F
549	5	<2			8	0	1	10	F
2.8.3	9	<1			12.5	0	0	1	F
539	13	<1			23	0	4	5	F
2.6.8	4	5			14	<1	1	11	F
JoN.7.1	4.5	<2	3	0	12	<1	1.5	7.5	F
D	12	<2			12	<1	<2	<4	M
E	7.5	4			6.5	<1	<1	4	M
G	3.5	2	2.5	<1	13.5	<1	2	5.5	M
JoM	11	3			20.5	4	1	25	F
JoN	3.5	<1	3	<1	23.5	<2	<1	20.5	F
JoO	4	3	4	<1	29	<1	2	7	F
F	7	2	5	<1	26.5	<2	<1	16.5	M

Sum of the frequencies of the mtDNA types with less than five repeats and sum of the frequencies of the mtDNA types with more than five repeats in several organs of adult rabbits.

4.4 In Vitro Cell Cultures

Quantitative analyses have been made possible with in vitro culture of cells derived from tissues heteroplasmic for mtDNA mutations and with the availability of a series of cybrids (fusion of cells depleted of mtDNA, ϱ0, with the cytoplasm of heteroplasmic cells).

The percentage of mtDNA with the transition A → G, at position 3243, in the tRNA Leu gene (MELAS) was determined in fibroblast-derived cell clones of four individuals, three belonging to the same family (Matthews et al. 1995); 16 to 32 clones isolated from cell lines containing 39%, 58%, 64% and 67% mutant mtDNA respectively were analyzed. The mean percentage of mutant mtDNA in the whole set of clones was not significantly different from the percentage in the cell line from which they were derived but there was an asymmetry in the distribution of heteroplasmy for this mutation. It must be noted that, in this experiment, cell cultures were made in a medium requiring a functionl respiratory chain in the cells. Serial passages were done to assess the evolution of the variance in the distribution of heteroplasmy: after 15 successive generations the proportions of mutant mtDNA did not change. This absence of increase in the variance does not favour the hypothesis of a random partitioning of mitochondria. Similar results were obtained with cybrids. To interpret their observations, the authors suggested that mtDNA must segregate into progeny cells as heteroplasmic units, and they suggest the existence in higher eukaryotic cells of a structure similar to the mitochondrial network described in yeast (Stevens 1981).

In another study, with cybrids obtained by the fusion of ϱ0 cells with cytoplasm of myoblasts heteroplasmic for the same tRNA Leu mutation (77% mutant mtDNA), different observations were made (Shoubridge 1995). When grown in a selective medium (requiring functional mitochondria) 56% of the cybrids evolved toward homoplasmy for the wild type allele. In long-term cell cultures grown in a non-selective medium (supplemented with pyruvate and uridine) the mean level of heteroplasmy remained stable but the variance was highly increased, which is consistent with a stochastic distribution.

In a study using 13 cybrid clones (ϱ0 cells fused with cytoplasm of myoblasts heteroplasmic for the same MELAS mutation), grown in a non-selective medium, a rapid shift toward homoplasmy for the mutation was observed in five clones, the proportion of mutant DNA remaining stable in the eight other clones, whatever the original level (Yoneda et al. 1992). The authors suggested that a marked replicative advantage for the mutant mtDNA was probably responsible for the shift toward homoplasmy, selective mtDNA replication taking place in defective mitochondria.

In these three studies the authors excluded a possible selective advantage of cells growing faster depending on their mitochondrial genotypes.

In vitro analyses did not provide clear-cut conclusions; the results are sometimes contradictory. It shows the difficulty in unravelling the effects of the mutation and the mechanism of partitioning itself. However, this approach is promising and should be carried on with a rigorous statistical analysis

5 Conclusion

Data on mitochondrial partitioning and on the evolution of mtDNA polymorphism obtained so far are not conclusive. An accurate estimation of the effective number of segregation units requires an analysis of a neutral mtDNA polymorphism on large samples. An estimation of selective values and expression thresholds of mutations, particularly mutations involved in human pathologies, should be done using various cell lines. Present data are too limited to allow an estimation of these two parameters.

Mitochondrial genotype segregation is a complex and dynamic process; more in vivo and in vitro analyses are needed to get a comprehensive overview of mtDNA inheritance. Experimental conditions and quantitative studies should be improved to get a good estimate of the parameters required for an understanding of the transmission of a deleterious mtDNA mutation.

References

Ashley M, Laipis PJ, Hauswirth WW (1989) Rapid segregation of heteroplasmic bovine mitochondria. Nucleic Acids Res 18:7325–7331

Avise JC, Arnold J, Ball RM, Bermingham E, Lamb T, Neigel JE, Reeb CA, Saunders NC (1987) Intraspecific phylogeography: the mitochondrial DNA bridge between population genetics and systematics. Annu Rev Ecol Syst 18:489–522

Barat M, Rickwood D, Dufresne C, Mounolou JC (1985) Characterization of DNA-protein complexes from the mitochondria of *Xenopus laevis* oocytes. Exp Cell Res 157:207–217

Bendall KE, Sykes BC (1995) Length heteroplasmy in the first hypervariable segment of the human mtDNA control region. Am J Hum Genet 57:248–256

Biju-Duval C, Ennafaa H, Monnerot M, Mignotte F, Sorigeur RC, El Gaaïed A, El Hili A, Mounolou JC (1991) Mitochondrial DNA evolution in lagomorphs: origin of systematic heteroplasmy, organisation of diversity in European rabbits. J Mol Evol 33:92–102

Bogenhagen D, Clayton DA, Mouse L (1977) Cell mitochondrial DNA molecules are selected randomly for replication throughout the cell cycle. Cell 11:719–727
Casane D, Dennebouy N, de Rochambeau H, Mounolou JC, Monnerot M (1994) Genetic analysis of systematic mitochondrial heteroplasmy in rabbits. Genetics 138:471–480
Casane D, Dennebouy N, de Rochambeau H, Mounolou JC, Monnerot M (1997) Non neutral evolution of tandem repeats in the mitochondrial DNA control region of lagomorphs. Mol Biol Evol 14:779–789
Chapman RW, Stephens JC, Lansman RA, Avisec JC (1982) Models of mitochondrial DNA transmission genetics and evolution in higher eukaryotes. Genet Res (Camb) 40:41–57
Chen X, Prosser R, Simonetti S, Sadlock J, Jagiello G, Schon EA (1995) Rearranged mitochondrial genomes are present in human oocytes. Am J Hum Genet 57:239–247
Crow JF, Kimura M (1970) An introduction to population genetics theory. Harper and Row, New York
Davis FD, Clayton DA (1996) In situ localization of mitochondrial DNA replication in intact mammalian cells. J Cell Biol 135:883–893
Dufresne C, Mignotte F, Guéride M (1996) The presence of tandem repeats and the initiation of replication in rabbit mitochondrial DNA. Eur J Biochem 235:593–600
Ennafaa H, Monnerot M, El Gaaïed A, Mounolou JC (1987) Rabbit mitochondrial DNA: preliminary comparisons between some domestic and wild animals. Genet Sel Evol 19:279–288
Harrison RG (1989) Mitochondrial DNA as a genetic marker in population and evolutionary biology. Trends Ecol Evol 4:6–11
Howell N, Halvorson S, Kubacka, I, McCullough DA, Bindoff LA, Turnbull DM (1992a) Mitochondrial gene segregation in mammals: is the bottleneck always narrow? Hum Genet 90:117–120
Howell N, McCullough D, Bodis WI (1992b) Molecular genetic analysis of a sporadic case of Leber hereditary optic neuropathy (Letter). Am J Hum Genet 50:443–446
Howell N, Xu M, Halvorson S, Bodis WI, Sherman J (1994) A heteroplasmic LHON family: tissue distribution and transmission of the 11778 mutation (Letter). Am J Hum Genet 55:203–206
Hutchinson CA, Newbold JE, Potter SS, Edgell MH (1974) Maternal inheritance of mammalian mitochondrial DNA. Nature 251:536–538
Jenuth JP, Peterson AC, Fu K, Shoubridge EA (1996) Random genetic drift in the female germline explains the rapid segregation of mammalian mitochondrial DNA. Nat Genet 13:146–151
Jenuth JP, Peterson AC, Shoubridge EA (1997) Tissue-specific selection for different mtDNA genotypes in heteroplasmic mice. Nat Genet 16:93–95
Larsson N-G, Tulinius MH, Holme E, Oldfors A, Andersen O, Wahlström J, Aasly J (1992) Segregation and manifestations of the mtDNA $tRNA^{Lys}$ A → $G^{(8344)}$ mutation of myoclonus epilepsy and ragged-red fibers (MERRF) syndrome. Am J Hum Genet 51:1201–1212
Matthews PM, Brown RM, Morten K, Marchington D, Poulton J, Brown G (1995) Intracellular heteroplasmy for disease-associated point mutations in mtDNA: implications for disease expression and evidence for mitotic segregation of heteroplasmic units of mtDNA. Hum Genet 96:261–268
Mignotte F, Gueride M, Champagne AM, Mounolou JC (1990) Direct repeats in the non-coding region of rabbit mitochondrial DNA: involvement in the generation of intra and inter-individual heterogeneity. Eur J Biochem 194:561–571
Posakony JW, England JM, Attardi G (1977) Mitochondrial growth and division during the cell cycle in HeLa cells. J Cell Biol 74:468–491
Rand DM (1993) Endotherms, ectotherms, and mitochondrial genome-size variation. J Mol Evol 37:281–295
Robin ED, Wong R (1988) Mitochondrial DNA molecules and virtual number of mitochondria per cell in mammalian cells. J Cell Physiol 136:507–513
Shoubridge EA (1995) Segregation of mitochondrial DNAs carrying a pathogenic point mutation ($tRNA^{leu3243}$) in cybrid cells. Biochem Biophys Res Commun 213:189–195
Solignac M, Genermont J, Monnerot M, Mounolou JC (1984) Genetics of mitochondria in *Drosophila:* mtDNA inheritance in hjeteroplasmic strains of *D. mauritiana.* Mol Gen Genet 197:183–188

Stevens B (1981) Mitochondrial structure. In: Strathern JN, Johns EW, Broach JR (eds) The molecular biology of the yeast *Saccaromyces.* Cold Spring Harbor Laboratory Press, Cold Spring Harbor 471–504

Wilson AC, Cann RL, Carr SM, Georege M, Gyllensten UB, Helm-Bychowski KM, Higushi RG, Palumbi SR, Prager EM, Sage RD, Stoneking M (1985) Mitochondrial DNA and two perspectives on evolutionary gegnetics. Biol J Linn Soc 26:375–400

Yoneda M, Chomyn A, Martinuzzi A, Hurko O, Attardi G (1992) Marked replicative advantage of human mtDNA carrying a point mutation that causes the MELAS encephalomyopathy. Proc Natl Acad Sci USA 89:11164–11168

CHAPTER 3

Molecular Basis of Mitochondrial DNA Diseases

3

P. Lestienne[1], M. F. Bouzidi[2], Isabelle Desguerre[3], and Gérard Ponsot[3]

Contents

[1] E 99-29 INSERM-Physiologie Mitochondriale, Université de Bordeaux 2, 146 rue Léo Saignat, 33076 Bordeaux, France
[2] CGMC CNRS, MR 5534-CNRS Université Lyon 1, 69622 Villeurbanne, France
[3] Service de Neuropédiatrie, Hôpital Saint Vincent de Paul, 75674 Paris Cedex, France

1 Introduction

The concept of mitochondrial disease was first described in 1962, in a patient with hypermetabolism of nonthyroid origin associated with defective mitochondrial coupling[1] (reviewed by Luft 1994). Between 1960 and 1970, several clinical and histological descriptions of mitochondrial myopathies were published in the litterature. Using modified Gomori trichrome staining, Engel and Cunninghan (1963) identified irregular mitochondrial deposits on the muscle fibers of patient with myopathy, and called them ragged red fibers (RRF). After 1970, mitochondrial diseases were observed which affected tissues other than muscles, especially the central nervous system. Shapira et al. (1977) introduced the term of mitochondrial encephalomyopathy. From 1980, research made faster progress, and in 1987, a biochemical classification of the mitochondrial cytopathies was proposed (DiMauro et al. 1987), as follows: defects in transport, defects in substrate utilisation, defects in the Krebs cycle, defects in the respiratory chain, and defects in oxidation/phosphorylation coupling. The year 1988 saw the first descriptions of mtDNA deletions in mitochondrial myopathies (Holt et al. 1988), in the Kearns-Sayre syndrome (Lestienne and Ponsot 1988), and the first mtDNA point mutation in Leber's hereditary optic neuropathy (Wallace et al. 1988). Almost all mtDNA - caused disorders depend on mutant gene dosage which is generally heteroplasmic with the normal mtDNA and functionally recessive. We will review here the main aspects of mtDNA alterations leading to respiratory chain impairments responsible for diseases.

2 Diagnosis

2.1 Clinical

Identification of the clinical phenotype provides important information for the diagnosis. The most common clinical symptoms are cerebral, muscular, cardiac and neurosensory disorders. The may be associated in well individualized syndromes like

[1] "Defective mitochondrial coupling" is defined as the loss of the mechanism that allows the energy released by electron transfer across the respiratory chain (oxidation) to be recovered for the synthesis of ATP molecules from ADP (phosphorylation).

myopathies with external ophthalmoplegia, the Kearns-Sayre or the Leigh syndrome, MELAS or MERRF. Although clinical signs mainly occur in the nervous system and muscle, other organs may be concomitantly or predominantly affected, such as the osteoarticular system, the kidney, liver, hematopoietic system, digestive tract and cutaneous tissue. The progressive injury to different organs during the evolution of disease is fairly characteristic of mitochondrial cytopathies, regardless of age at onset and the initial symptom. The central nervous system is very often affected in the late phase of evolution. Consequently, some forms of mitochondrial cytopathy should be considered in cases of an unexplained association of signs involving several organs with different embryologic origins.

In children, mitochondrial cytopathies are often pluriviseral and very frequently affect the CNS, whereas in adults, they are more often monovisceral myopathies. Mitochondrial diseases can occur at any age. Their evolution varies: often, it is severe, leading to death in the first months of life or in childhood, but sometimes the disease lasts for many years.

2.2 Genetics

Maternal inheritance is an initial criteria to postulate a mtDNA disease, as in MERRF, MELAS, NARP, and in mtDNA duplications. Nevertheless, spontaneous occurrences of mtDNA deletions were reported in Kearns-Sayre and Pearson's syndrome, as well as some point mtDNA mutations. Autosomal dominant or recessive forms were reported in diseases affecting the integrity of mtDNA, and of nuclear encoded proteins involved in the constitution of the respiratory chain, as in complex I (van den Heuvel et al. 1998).

2.3 Diagnostic Tests

Hyperlactacidemia is the most common biochemical abnormality suggestive of respiratory chain dysfunction. When this test is combined with measurement of the lactate/pyruvate ratio (i.e. the state of cytoplasmic oxido-reduction) and of the β-hydroxy-butyrate/aceto-acetate ratio (state of mitochondrial oxido-reduction), in both the fasting and fed states, it can be definitely related to a mitochondrial pathway deficiency. The presence of hyperlactacidemia with an increased lactate-to-pyruvate ratio and paradoxical cetosis under fed conditions with an increased β-hydroxy-butyrate-to-aceto-acetate ratio is highly suggestive of a respiratory chain deficiency.

Biochemical laboratory tests must be done on blood samples obtained without general or local anoxia. They must be deproteinized at the patient's bedside and immediately taken to the laboratory in a mixture of water and ice. Measurement of cerebrospinal fluid lactate is the best test for orientating the diagnosis in neurological mitochondrial cytopathies. Provocative tests should be carried out when basal screening tests are inconclusive. The glucose loading test (2 g/kg orally), and ergonometer bicycle exercise with determination of blood glucose, lactate, pyruvate, and ketone bodies should be done.

Among the neurological tests, magnetic resonance imaging (MRI) provides important information for diagnosis by showing symetrical lesions in the thalamus and subthalamic nuclei and in the putamen and caudate. The presence of infarcts, chiefly in the posterior regions, is suggestive of the MELAS syndrome. Magnetic resonance spectroscopy (mRS) of protons (^{1}H) and phosphorus (^{31}P) permits in vivo study of, for instance, glucose, ATP, PCr and lactate metabolisms in muscle and the central nervous system, and gives a picture of their metabolic state at a given time in these tissues (Bresolin et al. 1991; Grod et al. 1991; Mattews et al. 1991). Muscular metabolism can thus be explored by measuring the phosphocreatine/anorganic phosphate ratio (PCr/PI) at rest, during exercise and during recovery. In mitochondrial myopathies, the PCr/Pi ratio is lower than normal at rest and diminishes strikingly during exercise, even when the effort required is moderate; during recovery, this ratio returns to normal too slowly. Similarly, phosphorus (^{31}P) and proton (^{1}H) magnetic resonance spectroscopy (mRS) combined with brain MRI can be used to evaluate the brain metabolism of these different metabolites (Grod et al. 1991). This type of noninvasive investigation is destined to become increasingly important for the study of mitochondrial diseases whose expression is confirmed to the brain.

The study of muscle histology (see Chap. 25, this Vol.) may show the presence of ragged red fibers. Electron microscopy confirms intramitochondrial paracrystalline inclusions which are mainly seen in mitochondrial cytopathies with mtDNA mutations. Immuno-histochemical techniques allow the study of succinate dehydrogenase and cytochrome oxidase (cox) activity. Most RRF are cox-negative but all cox-negative fibers are not RRF.

The study of the respiratory chain in different necessitates specialized laboratories (see Chap. 26, this Vol.). Polarographic studies measure oxygen consumption by mitochondria - enriched fractions with a Clark electrode in the presence of various oxidative substrates (malate + pyruvate, malate + glutamate, succinate, palmitate, etc.) and ADP. The choice of substrates and complex inhibitors enables the deficiency to be localized in the different respiratory chain complexes. Spectrophotometric study which measures respiratory enzyme activities separately or in groups, using specific electron acceptors and donors, is an indispensable complement to polarography. It is important to be able to test several tissues, because of the mitotic segregation process. Polarographic and spectrophotometric studies detect respiratory chain abnormalities in only 30% of cases, even when the indications are correctly formulated (Rustin et al. 1994). The subunits of the different respiratory chain complexes can be studied in muscle samples, by immunoblotting. This technique shows the absence or alteration of the mobility of a particular subunit in one or several complexes.

Analysis of mitochondrial DNA (mtDNA) (see Chap. 27, this Vol.) is essential for the diagnosis of mitochondrial cytopathies. Besides identification of the molecular defects (deletion, duplication, point mutation, depletion), it allows estimation of the percentages of normal and mutant mtDNAs (heteroplasmy) in the tissue under examination. The choice of tissue for analysis depends on the organ sustaining clinical injury. The tissue mainly used are lymphocytes, fibroblasts and muscle. It is indispensable to obtain immortalized lymphocytes and a fibroblast culture (in the presence of uridine and pyruvate) for future enzymological or molecular investigations. When the clinical picture is suggestive, for instance, of Leber's disease, the Kearns-Sayre syndrome, MELAS, MERRF or the Leigh syndrome, a search for mtDNA abnormalities may constitute the initial examination. Additional tests with HeLa cells devoid of mtDNA by long-term treatment with ethidium bromide, termed Rho°, which are fused with the patient's platelets containing only mitochondria, make it possible to determine whether a case of mitochondrial cytopathy is of nuclear or mitochondrial origin (Attardi and Chomyn 1996).

3 Main mtDNA Mutations

Due to extensive mtDNA polymorphism and to nuclear pseudogenes (Hirano et al. 1997), several criteria are required to assess whether an mtDNA mutation is pathogenic: (1) it changes a nucleotide and/or amino acid well conserved during evolution, (2) it induces an amino acid or a nucleotide change important for the protein or for the rRNA or tRNA structure and function or mtDNA metabolism, (3) there is a significant proportion of affected patients presenting the mutation regardless of the unaffected population from the same ethnic group, (4) the mutation is heteroplasmic and its amount is correlated with the severity of the disease, and (5) the introduction of mitochondria from the patient into recipient ϱ° cells devoid of mtDNA is correlated with a decrease of oxygen consumption, as compared with normal cells. A compilation of the reported mtDNA mutations is presented in Table 1.

3.1 Mutations in Genes Coding Proteins

3.1.1 *Leber's Disease or LHON (Leber's Hereditary Optic Neuropathy)*

Leber's disease is characterized by a sudden bilateral loss of vision due to optic nerve atrophy which begins at about 20 years of age, and preferentially affects men with a bias of 70%. Although homogeneous at the clinical level, LHON has so far been associated with more than 20 different mtDNA mutations. Only three of them are considered as primary, and the others are considered as secondary and may be associated together or have either synergistic or suppressor effects (Brown et al. 1997), and are often associated with different haplotypes. They are generally located within gene encoding subunits of complex I which represent about half of the mtDNA coding capacity. Other main mutations are located within cytochrome oxidase subunit III, and cytochrome b genes. The first primary mutation, detected in 60–70% of cases, was described by Wallace as a G to A transition at position 11778, converting a highly conserved arginine into histidine of codon 340 of subunit 4 of NADH dehydrogenase (Wallace et al. 1988).

The second primary mutation, found in 10–20% of the patients, is a G to A transition at nucleotide 3460, converting a well - conserved alanine into threonine at codon 52 of subunit 1 of complex I. This mutation induces a reduction of 80% of sensitivity to rotenone, and of the electron transfer dependent on ubiquinone (Majander et al. 1996).

A third primary mutation is a G to A transition at position 14484, substituting a methionine into valine in subunit 6 of NADH dehydrogenase. These three primary

Table 1. Reported mitochondrial DNA base substitution and diseases. (Modified from Wallace et al. 1996)

Mutation	Gene	Disease	Mutation	Gene	Disease
G1243A	rRNA 12S	NIDDM	A8906G[a]	ATPase 6	Cardiomyopathy
A1555G	rRNA 12S	Deafness	T8993G	ATPase 6	NARP/Leigh
G1642A	$tRNA^{Val}$	MELAS	T8993C	ATPase 6	Leigh
G3196A	rRNA 16S	Alzheimer/Parkinson	T9101C	ATPase 6	LHON
A3243G	$tRNA^{Leu(UUR)}$	MELAS/CPEO/diabetes	T9176C	ATPase 6	FBSN
T3250C	$tRNA^{Leu(UUR)}$	Mt myopathy	G9438A	COIII	LHON
A3251G	$tRNA^{Leu(UUR)}$	Mt myopathy	T9540C[a]	COIII	MELAS
A3252G	$tRNA^{Leu(UUR)}$	MELAS/cardiomyopathy	G9559C[a]	COIII	MELAS
A3254G[a]	$tRNA^{Leu(UUR)}$	Cardiomyopathy	G9738T	COIII	LHON
C3256T	$tRNA^{Leu(UUR)}$	MELAS/cardiomyopathy	G9894A	COIII	LHON
A3260G	$tRNA^{Leu(UUR)}$	MELAS/cardiomyopathy	T9957C	COIII	MELAS
T3271C	$tRNA^{Leu(UUR)}$	MELAS	T9997C	$tRNA^{Gly}$	Cardiomyopathy
T3291C	$tRNA^{Leu(UUR)}$	MELAS	A10006G	$tRNA^{Gly}$	CIPO
A3302G	$tRNA^{Leu(UUR)}$	Mt myopathy	A10044G	$tRNA^{Gly}$	Encephalopathy
C3303T	$tRNA^{Leu(UUR)}$	Cardiomyopathy	C10400T[a]	ND3	MELAS
T3394C	ND1	LHON	T10873C[a]	ND4	MELAS
A3397G	ND1	Alzheimer/Parkinson	A11084G[a]	ND4	MELAS
G3460A	ND1	LHON	A11251G[a]	ND4	MELAS
A3505G	ND1	NIDDM	A11696	ND4	LHON
A4136G	ND1	LHON	G11778A	ND4	LHON
T4160C	ND1	LHON	C12246G	$tRNA^{Ser(AGY)}$	CIPO
T4216C	ND1	LHON	A12308G[a]	$tRNA^{Leu(CUN)}$	CPEO
A4269G	$tRNA^{Ile}$	MELAS/cardiomyopathy	T12311C	$tRNA^{Leu(CUN)}$	CPEO
T4285C	$tRNA^{Ile}$	CPEO	G12315A	$tRNA^{Leu(CUN)}$	CPEO
A4295G	$tRNA^{Ile}$	Cardiomyopathy	A12320G	$tRNA^{Leu(CUN)}$	Mt myopathy
A4300G	$tRNA^{Ile}$	Cardiomyopathy	C12562T	ND5	NIDDM
A4317G	$tRNA^{Ile}$	MELAS/cardiomyopathy	A12612G[a]	ND5	LHON
C4320T	$tRNA^{Ile}$	Cardiomyopathy	C12705T[a]	ND5	MELAS
T4336C	$tRNA^{Gln}$	Alzheimer/Parkinson	T13143C[a]	ND5	MERRF
C4883T[a]	ND2	MELAS	A13258T[a]	ND5	Cardiomyopathy
A4917G	ND2	LHON	G13708A	ND5	LHON
G5046A	ND2	NIDDM	G13730A	ND5	LHON
C5178A[a]	ND2	MELAS	G14368C[a]	ND6	LHON
G5244A	ND2	LHON	G14459A	ND6	LHON
G5460A[a]	ND2	Alzheimer	T14484C	ND6	LHON
G5460T[a]	ND2	Alzheimer	T14596A	ND6	LHON
C5554T[a]	$tRNA^{Trp}$	Cardiomyopathy	T14709C	$tRNA^{Glu}$	Mt myopathy/diabetes
A5692G	$tRNA^{Asn}$	CPEO	T14783C[a]	Cyt b	MELAS
G5703A	$tRNA^{Asn}$	CPEO	T14798C[a]	Cyt b	LHON
A5814G	$tRNA^{Cys}$	MELAS	A14927G[a]	Cyt b	Cardiomyopathy
G5821A[a]	$tRNA^{Cys}$	Cardiomyopathy	G15110A[a]	Cyt b	Deafness
G7444A	COI	LHON	G15257A	Cyt b	LHON
A7445G	$tRNA^{Ser(UCN)}$	Deafness	G15301A[a]	Cyt b	LHON
T7512C	$tRNA^{Ser(UCN)}$	MERRF/MELAS	C15452A[a]	Cyt b	LHON
A7673G[a]	COII	Cardiomyopathy	G15615A[b]	Cyt b	PEI
A8308G[a]	$tRNA^{Lys}$	Cardiomyopathy	G15812A	Cyt b	LHON
G8334A	$tRNA^{Lys}$	NIDDM	G15884C[a]	Cyt b	NIDDM
A8344G	$tRNA^{Lys}$	MERRF	A15923G	$tRNA^{Thr}$	Mt myopathy
T8356C	$tRNA^{Lys}$	MERRF/MELAS	A15924G[a]	$tRNA^{Thr}$	Mt myopathy
G8363A	$tRNA^{Lys}$	MERRF/cardiomyopathy	G15927A[a]	$tRNA^{Thr}$	Cardiomyopathy

Table 1. (Continue)

Mutation	Gene	Disease	Mutation	Gene	Disease
A8701G[a]	ATPase 6	MELAS	A15951G[a]	$tRNA^{Thr}$	Cardiomyopathy
T8851C	ATPase 6	FBSN	C15990T	$tRNA^{Pro}$	Mt myopathy

Mt myopathy: mitochondrial myopathy; PEI: progressive exercise intolerence; LHON: Leber hereditary optic neuropathy; CIPO: chronic intestinal pseudo-obstruction with myopathy and ophthalmoplegia; CPEO: chronic progressive external ophthalmoplegia; FBSN: familial bilateral strial necrosis; MELAS: mitochondrial encephalopathy, lactic acidosis, and stroke-like episodes; MERRF: myoclonic epilepsy and ragged red fibers; NARP: neurogenic muscle weakness, ataxia, and retinis pigmentosa; NIDDM: non-insulin-dependent diabetes mellitus.

[a] Polymorphism.

[b] Bouzidi et al. (1996).

LHON mutations are detected in about 90% of cases (Brown et al. 1997). Even though all the maternal descendants may present the mutations, some of them will develop the disease for still unknown reasons (triggering factors?).

3.1.2
Multiple Sclerosis

An association between LHON mutations and multiple sclerosis has been reported (Harding et al. 1992). Additional data revealed that secondary LHON mutations may indeed contribute to the development of multiple sclerosis (Kalman et al. 1996; Chalmers et al. 1996) as well as mutations in the $tRNA^{Thr}$ gene (Mayr-Wohlfart et al. 1996).

3.1.3
NARP (Neuropathy, Ataxia, Retinis, Pigmentosa)

A maternally inherited T to G transversion at nucleotide 8993 was found in family members presenting blindness, ataxia, dementia, seizures, muscle weakness and sensory neuropathy (NARP; Holt et al. 1990). Symptoms appear at various ages, usually after the age of 5, and include a motor deficiency due to axonal sensorimotor neuropathy, ataxia, retinitis pigmentosa, often an extra-pyramidal and pyramidal syndrome, progressive dementia and epileptic seizures. Hyperlactatorachia is constant but polarography and spectrophotometric studies do not show any respiratory chain deficiency. The mutation converts a highly conserved leucine to arginine at position 156 of ATPase subunit 6. It was found de novo in a Leigh's patient (Takahashi et al. 1998). Other mutations in this gene were found associated with Leigh's disease (cf. Chap. 9, this Vol.) which starts in the first years of life with clinical signs of brain stem dysfunction, including cranial nerve palsies, especially oculomotor palsies (ptosis and ophthalmoplegia), nystagmus and pyramidal and extra-pyramidal signs. Respiratory disorders (apnea and polypnea) are fairly suggestive of Leigh syndrome. Peripheral neuropathy (axonal sensorimotor neuropathy) is observed in 10 to 20% of cases. Leigh's syndrome frequently pursues a remitting course. Death occurs during respiratory failure sometimes brought on by intercurrent infections. Levels of lactate and pyruvate are elevated in the CSF and, less constantly, in the blood. MRI or CT scans usually show bilateral hypodensity of the basal ganglia nuclei (putamen and globus pallidus). Neuropathological abnormalities are very characteristic and predominate in the basal ganglia (putamen), brain stem, and the posterior columns of the spinal cord. They include vascular proliferation, demyelinization and astrocytosis; the neurons are little affected. Several biochemical abnormalities have been found to be associated with these disorders, particularly pyruvate dehydrogenase complex deficiency (cf. Chap. 12, this Vol.), cytochrome-c-oxidase deficiencies, and mitochondrial ATPase-6 gene point mutations.

3.2 Mutation in tRNA Genes

Although the 22 tRNA genes present low mitochondrial information content (about 10%), they represent the most frequent place for pathogenic mutation reported so far, since at least 36 of mutations meet the criteria for pathogenicity. The positions of the numerous mutations in the cloverleaf structure are reported in the following chapter.

3.2.1 *tRNA$^{Leu(UUR)}$ Gene*

This gene is the most important target of mtDNA mutations, since ten of them are transitions which have been associated with various clinical phenotypes including MELAS, insulino-independent diabetes, myopathies and cardiomyopathies. It is worthwhile noting that this gene and the D loop are preferentially relaxed by the mitochondrial DNA topoisomerase I (Topcù and Castora 1995).

MELAS (Mitochondrial Encephalopathy, Lactic Acidosis, and Stroke-Like Episodes). The stroke-like episodes suggestive of this syndrome begin between the ages of 5 and 15 years. It is often accompanied by headaches and vomiting (Pavliakis et al. 1984; Lestienne and Bataillé 1994). The severity of the signs vary, ranging from moderate localized deficiencies that resolve rapidly, to serious migraine with hemiplegia. During the evolution of the disease, digestive signs of the relapsing subocclusion type appear fairly frequently. The infarcts seen on CT scan and MRI are commonly observed in the posterior temporal, parietal and occipital lobes. The may be associated with other neurological abnormalities such as ventricular dilatation, cortical atrophy and basal ganglia calcifications. Hyperlactacidemia is very often observed. Isolated mitochondria in muscle tissue exhibit complex I deficiency.

In about 80% of MELAS cases, a heteroplasmic A to G transition at position 3243 has been reported (Goto et al. 1990). It is located in the DHU loop of the tRNA^{Leu} and is well conserved from sea urchins to humans. This same mutation has been found to be associated with other phenotypes, including myopathy, cardiomyopathy, myoclonic epilepsy, deafness associated with ataxia, endocrine signs and renal tubulopathy (Hu et al. 1991). In a family presenting maternal inheritance of insulino-independent diabetes mellitus associated with deafness (Van den Ouweland et al. 1992), the same mutation was discovered. Apart from a reduction of ATP level in pancreatic β cells leading to possible impairment in insulin secretion, an auto-immune origin was proposed, as an association between HLA markers and the frequency of patients presenting diabetes associated with this mutation was reported (Kobayashi et al. 1996). The mutation also induces a reduction of the mitochondrial membrane potential, inducing a reduction of mitochondrial calcium uptake, yielding a cell calcium overload (Moudy et al. 1995).

A second T to C substitution at position 3271 has been reported in 3 out of 40 MELAS patients (Goto et al. 1991).

3.2.2 *tRNA*[Lys] *Gene*

Five pathogenic mutations were described in this gene, the first one being associated with MERRF (myoclonic epilepsy with ragged ref fibers). The onset of MERRF syndrome may occur in children or adults. It combines progressive myoclonic epilepsy, ataxia, deafness and myopathy. The severity of intellectual degradation varies. In a few cases, lipomas were found in the cervical region. Muscle biopsy showed the presence of RRF and in many cases deficiency of complexes I and IV. In certain cases, evolution of the disease may be accompanied by relapsing hemiparesia, suggesting MELAS syndrome. Cases characterized by non insulin-dependent diabetes beginning at adult age, combined with perceptive deafness, myoclonic epilepsy and myopathy have been described in subjects with the 8344 mutation (Shoffner and Wallace 1995). In 80 to 90% of MERRF cases, a G to A substitution at nucleotide 8344 is located in the TUCG

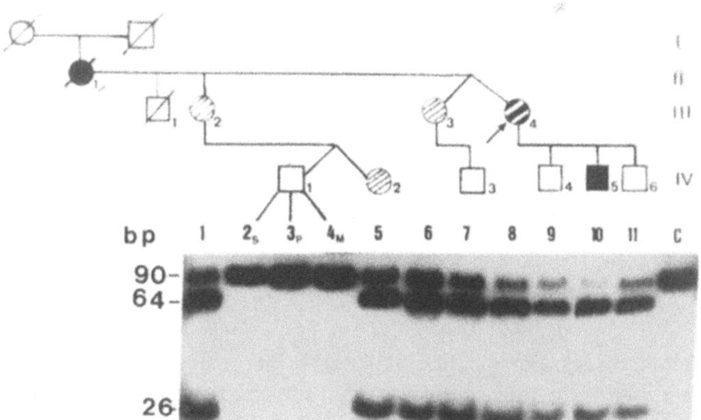

Fig. 1. Variations in the MERRF mutation in family members. The *upper band* on the gel represents the wild-type mtDNA, while the two shorter ones result from digestion by NaeI of the mutant mtDNA. Affected patients are shown by *black symbols*, mildly affected by *striped circles*; strongly affected by *heavily striped circle*. *Squares* Males; *circles* females. The quantification of heteroplasmy was done by densitometric scanning, or by counting radioactivity *(last line)*. Note the slight variations of the wild-type mtDNA associated with the disease, as well as the lack of detectable mutant in the dizygotic twins (IV 1) (*2s* blood, *3P* skin, *4m* muscle). (Penisson-Besnier et al. 1992)

loop of the $tRNA^{Lys}$. Biochemically, the mutation predominantly induces a reduction of complex I and IV activities. Measurement of cytochrome oxidase activity in several heteroplasmic myotube clones displaying various proportions of mutant mtDNA showed normal activities up to 85% of mutant, indicating its functional recessivity (Boulet et al. 1992; Fig. 1), which also induces a decrease in mitochondrial membrane potential (James et al. 1996).

3.2.3
$tRNA^{Met}$ Gene

More recently, mutations in the $tRNA^{Met}$ gene have been found associated either with splenic lymphoma (G4450A) (Lombès et al. 1998) or with exercise intolerance and muscle dystrophy (T4409C; Vissing et al. 1998). It is surprising that the same gene may be associated with so different pathological phenotypes.

3.3
Mutations in Ribosomal RNA Genes

3.3.1
12S rRNA Gene

A transition A1555G has been described in family members presenting deafness associated with aminoglycosid treatment (Prezant et al. 1993). It affects a well - conserved domain of the 12S rRNA. Aminoglycosides alter translation by binding to this nucleotide, and by stabilizing a mispaired tRNA. Aminoglycosid treatment therefore should follow the diagnosis for the absence of this mutation in families presenting variable hearing losses (Fischel-Ghodsian 1998).

An insertion of about five cytosine residues located between nucleotide positions 956–965 was reported in an Alzheimer's/Parkinson's patient (Shoffner et al. 1993).

3.3.2
16S rRNA Gene

A heteroplasmic variant G3196A was described in an Alzheimer's patient (Shoffner et al. 1993). In Rett syndrome, characterized by psychomotor deterioration and autistic behavior which preferentially affects females, a transition C $\rightarrow$ T at position 2835 was described in 7 out of 15 patients (Tang et al. 1997).

4
Heteroplasmic mtDNA Deletions

Kearns-Sayre syndrome is characterized by ptosis, ophthalmoplegia, pigmentary retinopathy, a complete cardiac bloc branch and elevated proteins in the cerebrospinal fluid. At the beginning, the ophthalmoplegia is absent, and muscular injury develops during the second stage.

The analysis of the restriction length polymorphism with single-site cutting restriction enzymes, followed by Southern blot probed with mitochondrial DNA from human placenta, showed the presence of the normal 16.7-kb band in the blood and grown fibroblasts from the affected child presenting Kearns-Sayre syndrome, and in her mother's lymphocytes (Lestienne and Ponsot 1988). However, the same analysis performed on the child's skeletal muscle DNA revealed the presence of one additional but shorter band of 12 kb. Using various restriction enzymes and mitochondrial DNA probes, it was then concluded that the shorter molecule corresponds to an mtDNA presenting a large 4.7-kb deletion. Deleted genes encoded for four subunits of complex I, three subunits of complex IV and the two ATPase subunits, as well as $tRNA^{Glu, Arg, His, Ser, Leu}$ genes (Lestienne and Ponsot 1988; Nelson et al. 1989a). Heteroplasmy with the normal mtDNA appears to prevent lethality. The analysis of mtDNA from various tissues in different cases showed that mtDNA heteroplasmy is a general feature, but that the amount of dmtDNA (deleted mtDNA) varies from one organ to another (i.e.: heart 24%, kidney 65%, brain 69%, diaphragm 73%, skeletal muscle 74%, liver 86%; Nelson et al. 1992).

4.1 Genetic Localization of the Deletions

The analysis of more than 200 cases has revealed that deletions are generally localized in the large arc spanning between OH from OL, the heavy and light strand origins of replication. During H-strand DNA replication, the parental displaced H strand DNA is single-stranded, suggesting that deletions occur during the first step of the mtDNA replication cycle (Nelson et al. 1989b; Fig. 2). Their length and position vary from one patient to another (Lestienne 1989). Some deletions may be as long as 8.4 kb (Degoul et al. 1991b) and 10.4 kb, eliminating OL (Ballinger et al. 1994), while others may be as short as 15 base pairs (Keightley et al. 1996). In about 40% of cases, the "common deletion" spans nucleotides 8.482 to 13.460 (Schon et al. 1989). In most of the cases, they induce the loss of the essential tRNA genes, thus preventing any translation of the transcripts of the structural genes of the respiratory chain complexes unless complemented by tRNA from the normal mtDNA. Rarely, however, deletions occur between OL and OH, eliminating the 12S and 16S ribosomal RNA genes.

4.2 Heredity of Deletions

In almost all of the cases, heteroplasmic deletions appear sporadically. A maternal transmission has been reported in one case (Poulton et al. 1991), but not in others (Larsson et al. 1992). It may be hypothesized that the expansion of the dmtDNA occurs during oogenesis. If one among the about 2000 mtDNA molecules of the immature oocyte is deleted by half, and assuming that its replication rate is two-fold faster than the normal one, it is kinetically possible that after the 8–9 replication rounds during oocyte maturation, an equimolar amount of both molecules may be present (Lestienne 1992). The presence of 0.1% of deleted mtDNA has been detected in normal oocytes (Chen et al. 1995).

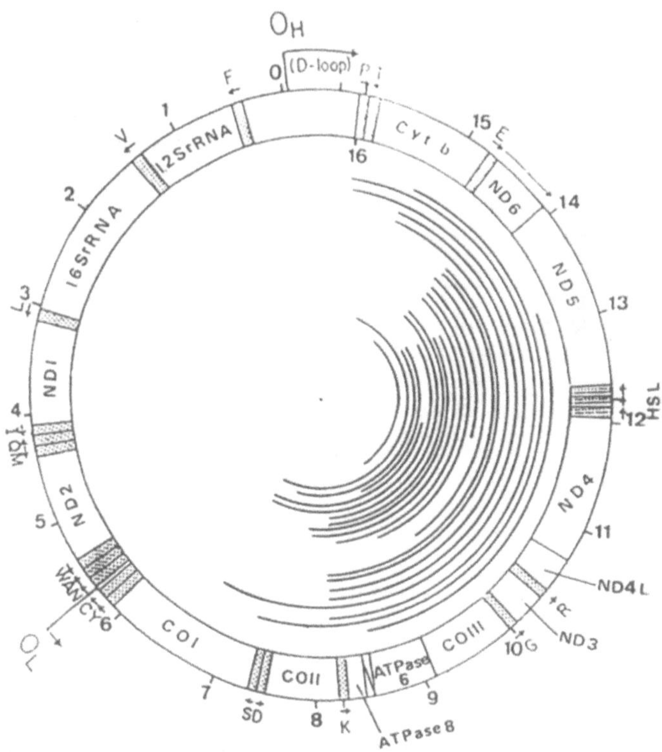

Fig. 2. Localization of the heteroplasmic mtDNA deletions. The deleted regions are indicated by the inner lines

In Wolfram syndrome, characterized by optical atrophy, deafness and early-onset diabetes mellitus, a low amount of deletion of 8.5 kb was detected in non-affected parents, but up to 23% in the lymphocytes of the affected child, indicating that an autosomal recessive nuclear gene product is involved in the expansion of the deleted molecule of this infant (Barrientos et al. 1996a).

4.3
Multiple Deletions

Several multiple mtDNA deletions have been detected in family members presenting autosomal dominant progressive external ophathalmoplegia (adPEO; Zeviani et al. 1989). Candidate nuclear genes such as mtTFA, single-stranded DNA binding protein, and endonuclease G have not been found associated with these pleioplasmic mtDNA deletions. However, several loci at position 10 q 23.3–24.3 (Suomalainen et al. 1995), 3 p 14.1–21.2 (Kaukonen et al. 1996) and 4 p 16 (Barrientos et al. 1996b) have been ascribed to the occurrence of some multiple deletions. Severall sporadic cases have been reported including cardiomyopathy, male hypofertility, periodic attacks of paralysis, and encephalomyopathy. The various phenotypes associated with mtDNA arrangements are shown in Table 2.

Table 2. Association between different diseases, mtDNA alterations and their heredities

Disease	mtDNA alteration	Heredity	Reference[a]
Mitochondrial myopathy	Single deletion	Sporadic	1
	Tandem duplication	Sporadic	2
Kearns-Sayre syndrome	Single deletion	Sporadic	3
	Tandem duplication	Sporadic	4
Pearson's syndrome	Single deletion	Sporadic	5
	Deletion/Duplication	Sporadic	6
Wolfram syndrome	Single deletion	Sporadic	7
	Single deletion	Autosomal recessive	8
	Multiple deletions	Autosomal recessive	9
Symmetric lipomatosis	Single deletion	Sporadic	10
	Multiple deletions	Sporadic	11
Diabetes mellitus, deafness	Deletion/duplication	Maternal	12
Diabetes mellitus	Tandem duplication	Maternal	13
Progressive ophthalmoplegia	Multiple deletions	Autosomal dominant	14
Progressive encephalomyopathy	Multiple deletions	Autosomal dominant	15
Cardiomyopathy	Multiple deletions	Sporadic	16
Dilated cardiomyopathy	Multiple deletions	Maternal	17
Encephalomyopathy	Multiple deletions	Sporadic	18
MNGIE[b]	Multiple deletions	Autosomal recessive	19
Periodic attacks of paralysis	Multiple deletions	Sporadic	20
Polymyalgia rheumatica	Multiple deletions	Sporadic	21
Male hypofertility	Multiple deletions	Sporadic	22

[a] 1, Holt et al. (1988); 2, Poulton et al. (1989); 3, Lestienne and Ponsot (1988); 4, Poulton et al. (1995); 5, Rötig et al. (1991); 6, Superti-Fuga et al. (1993); 7, Rötig et al. (1993); 8, Barrientos et al. (1996); 9, Barrientos et al. (1996b); 10, Campos et al. (1996); 11, Klopstock et al. (1994); 12, Ballinger et al. (1994); 13, Dunbar et al. (1993); 14, Zeviani et al. (1989); 15, Cormier et al. (1991); 16, Ozawa et al. (1990); 17, Soumalainen et al. (1992); 18, Checcarelli et al. (1994); 19, Hirano et al. (1994); 20, Prelle et al. (1993); 21, Reynier et al. (1994); 22, Lestienne et al. (1997)
[b] Mitochondrial neurogastrointestinal encephalomyopathy

4.4 Origins and Mechanism Intervening in Deletion Formation

The determination of the nucleotidic sequence at the deletion boundaries revealed that in the majority of cases they occur between two perfect direct repeats of the normal mtDNA, one of them being eliminated during the deletion process (Johns et al.

Table 3. Sequences of direct repeats involved in mtDNA rearrangements

Sequence	R[a]	Reference
-ACCTCCCTCACCA	10/13	Schon et al. (1989)
-CCCCTCCCCA	9/10	Martin-Négrier et al. (1998)
-CACCCCATCC	8/10	Ballinger et al. (1994)
-ACCCCCC	7/ 8	Brockington et al. (1993)
-CCCCCTCTA	8/ 9	Thomas et al. (1998)
-TTTTTCT	7/ 7	Keightley et al. (1996)

[a] Pyrimidines.

1989; Mita et al. 1990; Degoul et al. 1991b); in about 30% of cases, there are partial direct repeats, and in 10% of the remaining cases, no repeats can be seen suggesting that different mechanisms occur (Degoul et al. 1991a). The common deletion of 4.977 base pairs (bp) which accounts for about 40% of cases with perfect direct repeats occurs at the level of a 13-bp direct repeat, and the deletion of 15 bp in Cox III gene is located at a 7-bp, direct repeat. The mechanisms inducing the deletions are still unclear as they are sporadic. By analogy with the observations on E. coli and with the yeast rho$^-$ (containing short tandem amplified mtDNA sequences), the deletions might occur by recombination mechanisms involving homologous direct repeats. Such strand-transfer activity has been detected in mammalian mitochondria (Thyagarajan et al. 1996), buts its function remains to be established (Howell 1997).

According to a "slippage mispairing" model (Shoffner et al. 1989), the first direct repeat (located near OH) of the displaced H-strand DNA would base pair with the second direct repeat. The 3′ part of the H strand DNA might then be digested and ligated to produce a deleted molecule after completion of replication of the light strand. In vitro and in vivo experiments point to the importance of slippage mispairing in deletion and duplication formations (Madsen et al. 1993), as well as the sequencing after cloning of an apparent single deletion, which has revealed two species differing by four nucleotides (Degoul et al. 1991b). The conformation of some sequences close to the deletion boundaries (Hou and Wei 1996), the oxidative damage of the mtDNA (Hayakawa et al. 1992), and the specific DNA binding proteins could favor the formation of deletions if correction mechanisms are deficient. It is worthwhile noticing that the direct repeats involved in mtDNA rearrangements display a higher polypyrimidine rich sequence (Table 3), compared with the light strand direct repeat overall pyrimidine contents (mean value of 85% against 61%). The neo-synthesized polypurine – rich sequence could hypothetically favour the strand transfer of the replication complex to the second direct repeat.

5 Duplications

Partial duplications have been associated with mitochondrial myopathies (Poulton et al. 1989), diabetes mellitus and deafness (Ballinger et al. 1994) and other syndromes such as proximal tubulopathy and cerebellar ataxia (Rötig et al. 1992). They are generally maternally inherited (Dunbar et al. 1993; Martin-Négrier et al. 1998). The partial tandem duplicated sequence involves the shorter region located between OL and OH.

A short tandem duplication of 260 base pairs containing the two promoters of transcription is also associated with large – scale deletions (Brockington et al. 1993). The reported amount of heteroplasmy (5%) is, however, about ten times lower than that of the deletion. In a MELAS syndrome (see Sect. 3.2.1) associated with insulino-independent diabetes and hyperthyroidism, the same tandem duplication of 260 bp was detected in carriers of the A3243G mutation (Li et al. 1996).

A similar tandem duplication of 200 bp containing the heavy and light strand promoters was also recently characterized in a patient presenting a mutation in the cytochrome b gene (Bouzidi et al. 1998). Contrasting with previous reports, the amount of the two lesions is similar, suggesting that they are located within the same mtDNA molecule. The junction of the duplication displays a direct repeat of 12 bp. The short tandem D-loop duplication apparently does not affect the mtDNA copy number or the transcription rates of the rearranged homoplasmic molecule in cell culture (Hao et al. 1997).

6 Triplications

The analysis of the mtDNA restriction length polymorphism of the members of a family presenting a maternally inherited diabetes mellitus and deafness over four generations revealed the presence of tandem duplication, as well as a triplication in the muscle in one of the patients (Martin-Négrier et al. 1998). The mtDNA rearrangement also occured at the level of a direct repeat of 10 bp.

7 Depletions

Some reductions, down to 90%, of the mtDNA have been detected in the neonatal period of some children, with a rapid and generally fatal result (Moraes et al. 1991). Studies of primary grown myoblasts from the muscle biopsy of an affected infant unexpectedly revealed that the amount of mtDNA was normal during the first passages (while mtDNA content was decreased by 95% in the muscle biopsy), but then gradually diminished in several myoblast clones (Taanman et al. 1997). Repopulation of ρ^o cells with the patient's mitochondria could be achieved, indicating their normal mtDNA structure and the time-dependent inactivation (or inhibition) of nuclear gene(s). Sequencing of the origins of replication did not reveal abnormalities, except the so-called polymorphisms.

Some reversible mtDNA depletions were reported upon treatment of AIDS patients with AZT (Arnaudo et al. 1991), a powerful inhibitor of DNA polymerase gamma.

8
Amplifications

An isolated cDNA, obtained by subtractive hybridization from glioblastoma, was found to correspond to an amplified mtDNA sequence (Liang and Hays 1996). Three-to-ten-fold mtDNA amplifications were also detected in acute myeloid leukaemia (Boultwood et al. 1996) and in thyroid and oncocytic tumors (Tallini et al. 1994).

9
Experiments with Cell Culture

9.1
MELAS

A reduction in mitochondrial protein synthesis and a decrease in respiratory chain activities were measured on clonal cell cultures repopulated with various ratios of normal and mutant mtDNA (King et al. 1992). The polycistronic transcript RNA 19, corresponding to the 16S rRNA + $tRNA^{Leu(UUR)}$ + ND 1, significantly increases in cells harboring the mutant tRNA. This unprocessed abnormal transcript could induce the formation of abnormal ribosomes that might prevent normal translation (Schon et al. 1992).

The impairment of mitochondrial translation due to the A3243G mutation has been restored in cell culture by the characterization of a $tRNA^{Leu(CUN)}$ suppressor with an anticodon UAA instead of UAG, which can compensate the mutation borne by the $tRNA^{Leu(UUR)}$ mutant (Meziane et al. 1998).

9.2
MERRF

Studies with cybrid cells repopulated with various amounts of mutant mtDNA have indicated a decreased steady-state level of the mutant $tRNA^{Lys}$ by about 35% as compared with the wild type, as well as a decrease in the charged mutant $tRNA^{Lys}$ (48% against 66%; Enriquez et al. 1995). These observations indicate that the G8344A mutation impairs the tRNA aminoacylation reaction, thereby inducing a reduced protein synthesis as well as the formation of abnormal shorter polypeptides due to premature translation stops at lysine codons. They also demonstrated that about 10% of the normal genome allows normal cell respiration.

9.3
Large-Scale Rearrangements

Experiments carried out by Hayashi et al. (1991) showed that the rearranged mtDNA transcript could be translated in up to 60% of the deleted mtDNA. At higher ratios, of mutant mtDNA, mitochondrial protein synthesis abruptly diminishes, confirming the functional dominance of the deleted genome (Shoubridge et al. 1990). In this cell system, irrespective of the nuclear context, a translational complementation, and thus the cohabitation of both kinds of mtDNA within a mitochondrion, occurs in up to 60% of deleted mtDNA and 40% of normal mtDNA. Mitochondrial complementation was also reported with chloramphenicol-resistant cybrids repopulated with deleted mtDNA (Takai et al. 1997) whose chimeric transcript could be translated in the presence of chloramphenicol, unambiguously indicating interorganellar interactions, and trans complementation by the 16S rRNA which confers chloramphenicol resistance. Nevertheless, no data have so far reported the detection of the rearranged polypeptide from cells or tissues of patients presenting deletions. The frequency of appearance of the common deletion in ϱ° HeLa cells repopulated with mitochondria from a 25-year-old female has been estimated to be about 6×10^{-8} per replication cycle and per mtDNA (Shenker et al. 1996).

9.4 Role of the Nuclear Background

Interestingly, transfer of mitochondria from myoblasts of the heteroplasmic MELAS-carrying mutation to the ϱ° 206 cell line, followed by cloning, showed various outcomes of the mutant mtDNA; upon 2–3 months of growth, cells tended to homoplasmy in favour of the mutant genome. In addition, mutant cybrids had a marked replicative advantage over the wild-type ones (Yoneda et al. 1992). Further studies carried out with ϱ° cells from different origins (osteosarcoma and lung carcinoma) revealed different outcomes of both genomes, notably either a stable amount for each mtDNA, or a reduction of the mutant one (Dunbar et al. 1995).

The cell-specific nuclear gene product(s) has been shown to play a major role in the outcome of mtDNA tandem heteroplasmic duplication in cell culture experiments. In several subcloning experiments carried out with recipient ϱ° osteosarcoma cells repopulated with partially duplicated mtDNA, a clone containing 25% of partial duplication and 75% partial triplication was shown, while no wild-type mtDNA was found; in contrast, similar experiments carried outh with lung carcinoma cybrids, showed that the partial duplication was systematically lost to the benefit of the wild-type one (Holt et al. 1997), a similar observation to that found with heteroplasmic deletions (Bourgeron et al. 1993).

10 Treatment

The clinical and biochemical heterogeneity of mitochondrial cytopathies, as well as the unpredictable character of their spontaneous evolution, make it difficult to evaluate the different approaches to their treatment. The lack of homogeneity of the therapeutic evaluation criteria used, and the frequent administration of several types of medical drug to the same patient, further complicate assessment of the results of treatment.

The objective of mitochondrial cytopathy treatment is to increase ATP production by the mitochondria. The main drugs currently used are antioxidants which transport electrons from one respiratory chain complex to another. They also frequently stabi-

Table 4. Main drugs used to treat mitochondrial cytopathies

Name of drug	Paediatric dose (day)	Adult dose (day)
Co-enzyme Q-10	5–10 mg/kg	100–300 mg
Idebenone	3–5 mg/kg	80–200 mg
Menadione (Vit. K)	1–1.5 mg/kg	25–35 mg
Ascorbate	50 mg/kg	1–2 g
Thiamine (Vit. B1)	200 mg	400–800 mg
Riboflavin (Vit. B2)	50–100 mg	100 mg
Succinate	Not used	6 g

lize the complexes in the internal mitochondrial membrane and act on free radicals as antioxidants. The main oxidative reduction agents used (Table 4) act on fairly well localized parts of the respiratory chain (Calvani et al. 1993): thus, the co-enzyme Q_{10} and idebenone respectively transfer the electrons from I to III, and from complexes II to III; succinic acid transfers electrons straight to the co-enzyme Q (ubiquinone), thus bypassing complex I; the K vitamins (phylloquinone [K_1] and menadione [K_2] and ascorbic acid (vitamin C) transfer electrons to complex IV, thus bypassing complex III; thiamine stimulates the production of NADH, whose electrons go to complex I; riboflavine acts as a cofactor of electron transfer to complexes I and II. At a dose of 6 g/day, succinic acid was found to reduce the number of neurological accidents in a case of MELAS syndrome (Calvani et al. 1993). Riboflavine combined with L-carnitine led to improvement in a case of myopathy with complex I deficiency, and also in another case of myopathy associated with neuropathy and again with complex I deficiency (Calvani et al. 1993). Vitamin C monotherapy may have prolonged survival by 5 years in a case of cytochrome-C deficiency. Corticoids combined with thiamine and L-carnitine induced a favorable response in four cases of myopathy, one of which involved cardiomyopathy. At doses ranging from 12.5 to 100 mg/kg/day, dichloro-acetate combined with thiamine improved motor performance and reduced lactic acidosis in a case of MELAS.

11 Prenatal Diagnosis and Genetic Counselling

(Degoul et al. 1991a; Shoffner and Wallace 1995)

The items of information that are indispensable at the preliminary stage are the pedigree tree of patients, the clinical phenotype of the index case and of other members of the same family, and the determination of all the biochemical and genetic abnormalities in the subject affected. In particular, in mitochondrial cytopathies, the mitochondrial genome must always be studied.

Mitochondrial Cytopathies with mtDNA Abnormality. If there is a point mutation with a clinical picture indicative of Leber's disease, MELAS, MERRF, NARP, or Leigh syndrome, or a large-scale duplication, maternal inheritance can be assumed. In such cases, the risk for men is nil, but is high for the descendants of women, both girls and boys. When there is a heteroplasmic mtDNA deletion in the index case, the risk of recurrence is very small, but cannot be completely ruled out because of possible gonadal mosaicism. The presence of mtDNA depletion or pleioplasmic deletion involves the risk of autosomal recessive or dominant transmission.

Mitochondrial Cytopathies Without mtDNA Abnormality. In these cases, nuclear DNA mutation must be suspected. Measurement of respiratory enzyme activities in cultured fibroblasts in the proband is the indispensable preliminary stage for prenatal diagnosis. In case of subsequent pregnancy, the measurement of respiratory enzyme activities in cultured amniocytes or choriocytes represents the unique possibility of prenatal diagnosis, but conclusions, when possible, must be drawn with extreme caution (only 40 to 50% of enzyme deficiencies are expressed in cultured fibroblasts of the proband).

12
Conclusions and Prospects

Since the first description of several mtDNA rearrangements associated with human diseases a decado ago, we are still facing several problems concerning their mechanistic origin, as well as the exact relationship between these genetic lesions and the expression of a phenotype.

Indeed, the same disease may be caused either by large - scale mtDNA rearrangements or by point mutations, as shown notably in CPEO, in diabetes mellitus and deafness, and in cardiomyopathy. Conversely, the same mutation may induce different phenotypes as shown for A3243G in the $tRNA^{Leu(UUR)}$ gene associated with MELAS or CPEO or diabetes, the A3252G, the C3256T, the A3260G in $tRNA^{Leu(UUR)}$ the A4269G and A4317G in the $tRNA^{Ile}$ gene associated with MELAS or cardiomyopathy, and for deafness associated either with mutations A1555G in the 12S rRNA or A7445G near the $tRNA^{Ser(UCN)}$ gene. Cell culture experiments revealed severe respiratory chain impairments due to defects of mitochondrial protein synthesis as shown on deletions and in cases of tRNA mutations. Furthermore, abnormal polypeptides due to either intramitochondrial complementation in cases of deletions, or premature translation termination in cases of tRNA mutations, have been detected which may interfere with the overall mitochondrial metabolism and hence bioenergetics. Additional data pointing to imbalances in nuclear transcripts (Heddi et al. 1993) may give further insights into the in vivo effects of these mtDNA alterations.

Animal models mimicking these disorders may soon be available; homozygous mice knocked out for the mtTFA gene (whose protein triggers LSP transcription and thus primer for H strand DNA replication) do not undergo significant embryogenesis due to their loss of mtDNA (Larsson et al. 1998). On the other hand, heterozygous mice display a two-fold reduction in mtDNA copy number. While a decrease in activity from an mtDNA - encoded protein - forming complex was detected in the heart, no significant reduction in enzymatic activities could be shown in the liver and the kidney, in comparison with normal mice.

Since mtDNA mutations and deletions are generally heteroplasmic, approaches to gene therapy by inhibition of replication of the altered mtDNA with peptide nucleic acids (Taylor et al. 1997) may provide rational tools to reverse the expression and amount of the defective genome to the advantage of the normal one.

The scope of mitochondrial diseases is now extended with the finding of a mitochondrial localization for the nuclear gene product associated with Friedreich ataxia (Koutnikova et al. 1997) and Wilson's disease (Lutsenk et al. 1998). Moreover, abnormal Paraplegin, a mitochondrial metalloprotease, induces mitochondrial dysfunctions in hereditary spastic paraplegia (Casari et al. 1998), and mutations in a nuclear gene involved in the biogenesis of cytochrome c oxidase may cause Leigh syndrome (Zhu et al. 1998).

A forward line of research in the mammalian mtDNA genetics will undoubtedly come from a better knowledge of the rules governing their repartition within mitochondria and within cells, as well as from the mechanisms involved in the maintenance of their structure (Anson et al. 1998; Takao et al. 1998). Further studies will also shed light onto the numerous facets of genotype-phenotype relationships. Conversely, the accumulation of pathogenic mtDNA mutations may disclose new mitochondrial-nuclear interactions that were definitely unexpected.

References

Anson RM, Croteau DL, Stierum RH, Filburn C, Parsell R, Bohr WA (1998) Homogenous repair of singlet oxygen-induced DNA damage in differentially transcribed regions and strands of human mitochondrial DNA. Nucleic Acids Res 26:662–668

Arnaudo E, Dalakas M, Shanske S, Moraes CT, DiMauro S, Schon EA (1991) Depletion of muscle mitochondrial DNA in AIDS patients with Zidovudine-induced myopathy. Lancet 337:508–510

Attardi G, Chomyn A (eds) (1996) Mitochondrial biogenesis and genetics. Methods in enzymology, vol 264. Academic Press, New York

Ballinger SW, Shoffner JM, Gebhart S, Koontz DA, Wallace DC (1994) Mitochondrial diabetes revisited. Nat Genet 7:458–459

Barrientos A, Casademont J, Saiz A, Cardellach F, Volpini V, Solans A, Tolosa E, Urbano-Marquez A, Estivill X, Nunes V (1996a) Autosomal recessive Wolfram syndrome associated with an 8.5 kb mtDNA single deletion. Am J Hum Genet 58:963–970

Barrientos A, Volpini V, Casademont J, Genis D, Mansanares J-P, Ferrer I, Carral J, Cardellach F, Urbano-Marquez A, Estevill X, Nunes V (1996b) A nuclear defect in the 4p16 region predisposes to multiple mitochondrial DNA deletions in families with Wolfram syndrome. J Clin Invest 97:1570–1576

Boulet L, Karpati G, Shoubridge EA (1992) Distribution and threshold expression of the $tRNA^{Lys}$ mutation in skeletal muscle of patients with myoclonic epilepsy and ragged red fibers (MERRF). Am J Hum Genet 51:1187–1200

Boultwood J, Fidler C, Mills I (1996) Amplification of mitochondrial DNA in acute myeloid leukaemia. Br J Haematol 95:426–431

Bourgeron T, Chrétien D, Rötig A, Munnich A, Rustin P (1993) Fate and expression of the deleted mitochondrial DNA differs between human heteroplasmic skin fibroblast and Epstein-Barr virus-transformed lymphocyte cultures. J Biol Chem 268:19369–19376

Bouzidi MF, Carrier H, Godinot C (1996) Antimycin resistance and ubiquinol cytochrome c reductase instability associated with a human cytochrome b mutation. Biochim Biophys Acta 1317:199–209

Bouzidi MF, Poyau A, Godinot C (1998) Coexistence of high level of a cytochrome b mutation and of a tandem 200 base pairs duplication in human muscle mtDNA. Hum Mol Genet 7:385–391

Bresolin N, Martinelli P, Barbiroli B, Zaniol P, Ausendenda C, Montagna P, Gallanti A, Comi GP, Scarlato G, Lugaresi E (1991) Muscle mitochondrial DNA deletion and 31P-NMR spectroscopy alterations in a migraine patient. J Neurol Sci 104:182–184

Brockington M, Sweeney MG, Hammans SR, Morgan-Hughes JA, Harding AE (1993) A tandem duplication in the D loop of human mitochondrial DNA associated with deletions in mitochondrial myopathies. Nat Genet 4:67–71

Brown MD, Sun F, Wallace DC (1997) Clustering of caucasian Leber hereditary optic neuropathy patients containing the 11778 or 14484 mutations on an mtDNA lineage. Am J Hum Genet 60:381–387

Calvani M, Koverech A, Caruso G (1993) Treatment of mitochondrial diseases. In: DiMauro S, Wallace DC (eds) Mitochondrial DNA in human pathology. Raven Press, New York, pp 173–198

Campos Y, Martin MA, Navarro C, Gordo P, Arenas J (1996) Single large-scale mitochondrial DNA deletion in a patient with mitochondrial myopathy associated with multiple symmetric lipomatosis. Neurology 47:1012–1014

Casar G, De Fusco M, Ciarmatori S, Zeviani M, Mora M, Fernandez P, De Michele G, Filla A, Cocozza S, Marconi R, Dürr A, Fontaine B, Ballabio A (1998) Spastic paraplegia and OXPHOS impairment caused by mutations in paraplegin, a nuclear-encoded mitochondrial metalloprotease. Cell 93:973–983

Chalmers RM, Robertson N, Compston DAS, Harding AE (1996) Sequence of mitochondrial DNA in patients with multiple sclerosis. Ann Neurol 40:239–243

Checcarelli N, Prelle A, Moggio M, Comi G, Bresolin N, Papadimitriou A, Faiolari G, Bordoni A, Scarlato G (1994) Multiple deletions of mitochondrial DNA in sporadic and atypical cases of encephalomyopathy. J Neurol Sci 123:74–79

Chen X, Prosser R, Simonetti S, Sadlock J, Jagiello G, Schon EA (1995) Rearranged mitochondrial genomes are present in human oocytes. Am J Hum Genet 57:239–247

Cormier V, Rötig A, Tardieu M, Colonna M, Saudubray J-M, Munnich A (1991) Autosomal dominant deletions of the mitochondrial genomse in a case of progressive encephalomyopathy. Am J Hum Genet 48:643–648

Degoul F, Nelson I, Lestienne P, François D, Romero N, Duboc D, Eymard B, Fardeau M, Ponsot G, Paturneau-Jouas M, Chaussain M, Leroux JP, Marsac C (1991a) Deletions of mitochondrial DNA in Kearns-Sayre syndrome and ocular myopathies: genetic, biochemical and morphological studies. J. Neurol Sci 101:168–170

Degoul F, Nelson I, Amselem S, Romero N, Obermaier-Kusser B, Ponsot G, Marsac C, Lestienne P (1991b) Different mechanisms inferred from sequences of human mitochondrial DNA deletions in ocular myopathies. Nucleic Acids Res 19:493–496

DiMauro S, Bonilla E, Zeviani M (1987) Mitochondrial myopathies. J Inher Metab Dis 10 (Suppl 1):113–128

Dunbar DR, Moonie PA, Swingler RJ, Davidson D, Oberts R, Holt IJ (1993) Maternally transmitted partial direct tandem duplication of mitochondrial DNA associated with diabetes mellitus. Hum Mol Genet 2:1619–1620

Dunbar DR, Moonie PA, Jacobs HT, Holt IJ (1995) Different cellular backgrounds confer a marked advantage to either mutant or wild-type mitochondrial genomes. Proc Natl Acad Sci USA 92:6562–6566

Engel WK, Cunningham CG (1963) Rapid examination of muscle tissue: an improved trichrome stain method for fresh-frozen biopsy sections. Neurology 13:919–923

Enriquez JA, Chomyn A, Attardi G (1995) mtDNA mutation in MERRF syndrome causes defective aminoacylation of $tRNA^{Lys}$ and premature translation termination. Nat Genet 10:47–55

Fischel-Ghodsian N (1998) Mitochondrial mutations and hearing loss: paradigm for mitochondrial genetics. Am J Hum Genet 62:15–19

Goto Y-I, Nonaka I, Horai S (1990) A mutation in the $tRNA^{Leu(UUR)}$ gene in MELAS subgroup of mitochondrial encephalomyopathies. Nature 348:651–653

Goto Y-I, Nonaka I, Horai S (1991) A new mutation associated with mitochondrial myopathy, encephalopathy, lactic acidosis and stroke-like episodes (MELAS). Biochim Biophys Acta 1097:238–240

Grod W, Krageloh-Mann I, Klox V (1991) Metabolic and destructive brain disorders in children: findings with localised proton MR-spectroscopy. Radiology 181:173–181

Hao H, Manfredi G, Moraes CT (1997) Functional and structural features of a tandem duplication of the human mtDNA promoter region. Am J Hum Genet 60:1363–1372

Harding AE, Sweeney MG, Miller DH, Mumford CJ, Kellar-Wood H, Menard D, McDonald WI, Compston DAS (1992) Occurrence of a multiple sclerosis-like illness in women who have a Leber's hereditary optic neuropathy mitochondrial DNA mutation. Brain 115:979–989

Hayakawa M, Hattori K, Sugiyama S, Ozawa S (1992) Age-associated oxygen damage and mutations in mitochondrial DNA in human hearts. Biochem Biophys Res Commun 189:979–985

Hayashii JI, Ohta S, Kiruchi A, Takemitsu M, Goto Y, Nonaka I (1991 Introduction of disease-related mitochondrial DNA deletions into Hela cells lacking mitochondrial DNA results in mitochondrial dysfunction. Proc Natl Acad Sci USA 88:10614–10618

Heddi AE, Lestienne P, Wallace DC, Stepien G (1993) Mitochondrial DNA expression in mitochondrial myopathies and coordinated expression of nuclear genes involved in ATP production. J Biol Chem 268:12156–12163

Hirano M, Silvestri G, Blake DM, Lombes A, Minetti C, Bonilla E, Hays AP, Lovelace RE, Butler I, Bertorini TE (1994) Mitochondrial neurogastrointestinal encephalomyopathy (MNGIE): clinical biochemical, and genetic features of an autosomal recessive mitochondrial disorder. Neurology 44:721–727

Hirano M, Shtilbans A, Mayeux R, Davidson MM, DiMauro S, Knowles JA, Schon EA (1997) Apparent mtDNA heteroplasmy in Alzheimer's disease patients and in normal due to PCR amplification of nucleus-enbedded mtDNA pseudogenes. Proc Natl Acad Sci USA 94:14894–14899

Holt IJ, Harding AE, Morgan-Hughes JA (1988) Deletions of muscle mitochondrial DNA in patients with mitochondrial myopathies. Nature 331:717–719

Holt IJ, Harding AE, Petty RKH, Morgan-Hughes JA (1990) A new mitochondrial disease associated with mitochondrial DNA heteroplasmy. Am J Hum Genet 46:428–433

Holt IJ, Dunbar DR, Jacobs HT (1997) Behaviour of a population of partially duplicated mitochondrial DNA molecules in cell culture: segregation, maintenance and recombination dependent upon nuclear background. Hum Mol Genet 6:1251–1260

Hou J-H, Wei Y-H (1996) The unusual structures of the hot-regions flanking large-scale deletions in human mitochondria. Biochem J 318:1065–1070

Howell N (1997) mtDNA recombination: what do in vitro data mean? Am J Hum Genet 16:18–22

Hu DN, Qiu WQ, Wi BT, Fang LZ, Zhou F, Gu YP, Zhang WH, Yan JH, Ding YO (1991) Genetic aspects of antibiotic induced deafness: mitochondrial inheritance. J Med Genet 28:79–84

James AM, Wei Y-H, Pang C-Y, Murphy MP (1996) Altered mitochondrial function in fibroblasts containing MELAS or MERRF mitochondrial DNA mutations. Biochem J 318:401–407

Johns DR, Rutlege SL, Stine OC, Hurko O (1989) Directly repeated sequences associated with pathogenetic mitochondrial DNA deletions. Proc Natl Acad Sci USA 86:8059–8062

Kalman B, Lublin FD, Alder H (1996) Characterization of the mitochondrial DNA in patients with multiple sclerosis. J Neurol Sci 140:75–84

Kaukonen JA, Amati P, Suomalainen A, Rötig A, Piscaglia M-G, Salvi F, Weissenbach J, Fratta G, Comi G, Peltonen L, Zeviani M (1996) An autosomal locus predisposing to multiple deletions of mtDNA on chromosome 3p. Am J Hum Genet 58:763–769

Keightley JA, Hoffbuhr K, Burton MD, Salas V, Johnston WSW, Penn AMW, Buist NRM, Kennaway NG (1996) A microdeletion in cytochrome c oxidase (COX) subunit III associated with COX deficiency and recurrent myoglobinuria. Nat Genet 12:410–416

King GMP, Koga Y, Davidson M, Schon EA (1992) Defect in mitochondrial protein synthesis and respiratory chain activity segregates with the $tRNA^{Leu(UUR)}$ mutation associated with mitochondrial myopathy, encephalopathy, lactic acidosis and stroke-like episodes. Mol Cell Biol 12:480–490

Klopstock T, Naumann M, Schlke B, Bischof F, Seibel P, Kottlors M, Eckert P, Reiners K, Toyka KV, Reichmann H (1994) Multiple symmetric lipomatosis: abnormalities in complex IV and multiple deletions in mitochondrial DNA. Neurology 44:862–866

Kobayashi T, Oka Y, Katagiri H, Falorni A, Kasuga A, Takei I, Nakanishi K, Murase T, Kosaka K, Lernmark A (1996) Association between HLA and islet cell antibodies in diabetic patients with a mitochondrial DNA mutation at base pair 3243. Diabetologia 39:1196–1200

Koutnikova H, Campuzano V, Foury F, Dollé P, Cazzalini O, Koenig M (1997) Studies of human, mouse and yeast homologues indicate a mitochondrial function for frataxin. Nat Genet 16:345–351

Larsson NG, Eiken HG, Boman H, Holme E, Oldorfs A, Tulinius MH (1992) Lack of transmission of deleted mtDNA from a women with Kearns-Sayre syndrome to her child. Am J Hum Genet 50:360–363
Larsson NG, Wang J, Wilhelmsson H, Oldorfs A, Rustin P, Lewandowski M, Barsh GS, Clayton DA (1998) Mitochondrial transcription factor A is necessary for mtDNA maintenance and embryogenesis in mice. Nat Genet 18:231–236
Lestienne P (1989) Mitochondrial and nuclear DNA complementation in the respiratory chain function and defects. Biochimie 71:1115–1123
Lestienne P (1992) Mitochondrial DNA mutations in human diseases: a review. Biochimie 74:123–130
Lestienne P, Bataillé N (1994) Mitochondrial DNA alterations and genetic diseases: a review. Biomed Pharmacother 48:199–214
Lestienne P, Ponsot G (1988) Kearns-Sayre syndrome with muscle mitochondrial DNA deletion. Lancet 1:885
Lestienne P, Reynier P. Chrétien M-F, Penisson-Besnier I, Malthiéry Y, Rohmer V (1997) Oligo-asthenospermia associated with multiple mitochondrial DNA rearrangements. Mol Hum Reprod 3:811–814
Li J-Y, Kong K-W, Chang M-H, Cheung S-C, Lee H-C, Pang C-Y, Wei Y-H (1996) MELAS syndrome associated with a tandem duplication in the D-loop of mitochondrial DNA. Acta Neurol Scand 93:450–455
Liang BC, Hays L (1996) Mitochondrial DNA copy number changes in human gliomas. Cancer Lett 105:167–173
Lombès A, Bories D, Girodon E, Frachon P, Ngo MM, Breton-Gorius J, Tulliez M, Amselem S, Goossens M (1998) The first pathogenic mitochondrial methionine tRNA point mutation is discovered in splenic lymphoma. Hum Mutat (Suppl) 1:S173–S180
Luft R (1994) The development of mitochondrial medicine. Proc Natl Acad Sci USA 91:8731–8738
Lutsenk S, Cooper MJ (1998) Localization of the Wilson's disease protein product to mitochondria. Proc Natl Acad Sci USA 95:6004–6009
Madsen CS, Ghivizzani SC, Hauswirth WW (1993) In vitro and in vivo evidence for slipped mispairing in mammalian mitochondria. Proc Natl Acad Sci USA 90:7671–7675
Majander A, Finel M, Savontaus M-L, Nikoskelainen E, Wikstöm M (1996) Catalytic activity of complex I in cell lines that possess replacement mutations in the ND genes in Leber's hereditary optic neuropathy. Eur J Biochem 239:201–207
Martin-Négrier ML, Coquet M, Teissier-Moretto B, Lacut J-Y, Dupon M, Bloch B, Lestienne P, Vital C (1998) Partial triplication of the mtDNA in maternally transmitted diabetes and deafness. Am J Human Genet 63:1227–1232
Mattews PM, Allaire C, Shoubridge EA, Karpati G, Carpenter S, Arnould DL (1991) In vivo muscle magnetic resonance spectroscopy in the clinical investigation of mitochondrial disease. Neurology 41:114–120
Mayr-Wohlfart U, Paulus C, Henneberg A, Rödel G (1996) Mitochondrial mutation in multiple sclerosis patients with severe optic involvement. Acta Neurol Scand 94:167–171
Meziane AE, Letinen SK, Hance N, Nijtmans LGJ, Dunbar D, Holt IJ, Jacobs HT (1998) A tRNA suppressor mutation in human mitochondria. Nat Genet 18:350–353
Mita S, Rizzuto R, Moraes CT, Shanske S, Arnaudo E, Fabrizi GM, Koga Y, DiMauro S, Schon EA (1990) Recombination via flanking direct repeats is a major cause of large-scale deletions of human mitochondrial DNA. Nucleic Acids Res 18:561–567
Moraes CT, Shanske S, Tritschler H-J, Aprille JR, Andreetta F, Bonilla E, Schon EA, DiMauro S (1991) mtDNA depletion with variable tissue expression: a novel genetic abnormality in mitochondrial diseases. Am J Hum Genet 48:492–501
Moudy AM, Handran SD, Goldberg M, Ruffin N, Karl I, Kranz-Eble P, DeVivo DC, Rothman SM (1995) Abnormal calcium homeostasis and mitochondrial polarization in a human encephalomyopathy. Proc Natl Acad Sci USA 92:729–733
Nelson I, d'Auriol L, Galibert F, Ponsot G, Lestienne P (1989a) Nucleotidic mapping and kinetic model for a 4666 base pairs heteroplasmic mitochondrial DNA deletion in Kearns-Sayre syndrome. C R Acad Sci III 309:403–407

Nelson I, Degoul F, Obermaier-Kusser B, Romero N, Borrone C, Marsac C, Vayssiere JL, Gerbitz K, Fardeau M, Ponsot G, Lestienne P (1989b) Mapping of heteroplasmic mitochondrial DNA deletions in Kearns-Sayre syndrome. Nucleic Acids Res 17:8117–8124

Nelson I, Bonne G, Degoul F, Marsac C, Ponsot G, Lestienne P (1992) Kearns-Sayre syndrome with sideroblastic anaemia: molecular investigatons. Neuropediatrics 23:199–205

Ozawa T, Tanaka M, Sugiyama S, Hattori K, Ito T, Takahashi A, Sato W, Takada G, Mayumi B, Yamamoto K, Adachi K, Koga Y, Toshima H (1990) Multiple mitochondrial DNA deletions exist in cardiomyocytes of patients with hypertrophic or dilated cardiomyopathy. Biochem Biophys Res Commun 170:830–836

Pavlakis SG; Phillips PC, DiMauro S, DeVivo DC, Rowland LP (1984) Mitochondrial myopathy, encephalopathy, lactic acidosis, and stroke-like episode: a distinctive clinical syndrome. Ann Neurol 16:481–488

Penisson-Besnier I, Degoul F, Desnuelle C, Dubas F, Josi K, Emile J, Lestienne P (1992) Uneven distribution of mitochondrial DNA mutation in MERRF dizygotic twins. J Neurol Sci 110:144–148

Poulton J, Gardiner RM, Deadman ME (1989) Tandem direct duplications of mitochondrial DNA in mitochondrial myopathy: analysis of nucleotide sequence and tissue distribution. Nucleic Acids Res 17:10223–10229

Poulton J, Deadman ME, Ramacharan S, Gardiner RM (1991) Germ-line deletions of mtDNA in mitochondrial myopathy. Am J Hum Genet 48:649–653

Poulton J, Morten KJ, Marchington D, Weber K, Brown GK, Rötig A, Bindoff L (1995) Duplications of mitochondrial DNA in Kearns-Sayre syndrome. Muscle Nerve (Suppl) 3:S154–S158

Prelle A, Moggio M, Checcarelli N, Comi G, Bresolin N, Battistel A, Bordoni A, Scarlato G (1993) Multiple deletions of mitochondrial DNA in a patient with periodic attacks of paralysis. J Neurol Sci 117:24–27

Prezant TR, Agapian JV, Bohlman MC (1993) Mitochondrial ribosomal RNA mutation associated with both antibiotic-induced and non-syndromic deafness. Nat Genet 4:289–294

Reynier P, Pellissier J-F, Harle J-R, Malthiéry Y (1994) Multiple deletions of the mitochondrial DNA in polymyalgia rheumatica. Biochem Biophys Res Commun 205:375–380

Rötig A, Cormier V, Koll F, Mize CE, Saudubray J-M, Veerman A, Pearson HA, Munnich A (1991) Site-specific deletions of the mitochondrial genome in the Pearson marrow-pancreas syndrome. Genomics 10:502–504

Rötig A, Bessi J-L, Romero N, Cormier V, Saudubray J-M, Narcy P, Lenoir G, Rustin P, Munich A (1992) Maternally inherited duplication of the mitochondrial genome in a syndrome of proximal tubulopathy, diabetes mellitus, and cerebellar ataxia. Am J Hum Genet 50:364–370

Rötig A, Cormier V, Chatelain R, François J, Saudubray J-M, Rustin P, Munnich A (1993) Deletion of mitochondrial DNA in a case of early-onset diabetes mellitus, optic atrophy, and deafness (Wolfram syndrome, MIM 222300). J Clin Invest 91:1095–1098

Rustin P, Chrétien D, Gérard B, Bourgeron T, Rötig A, Saudubray JM, Munnich A (1994) Biochemical and molecular investigations in respiratory chain deficiencies. Clin Chim Acta 228:35–51

Schon EA, Rizzuto R, Moraes CT, Nakase H, Zeviani M, DiMauro S (1989) A direct repeat is a hotspot for large-scale deletions of human mitochondrial DNA. Nature 244:346–349

Schon EA, Koga Y, Davidson M, Moraes CT, King MP (1992) The mitochondrial $tRNA^{Leu(UUR)}$ mutation in MELAS: a model for pathogenesis. Biochim Biophys Acta 1101:206–209

Shapira Y, Harel S, Russel A (1977) Mitochondrial encephalomyopathies: a group of neuromuscular disorders with defects in oxidative metabolism. Isr J Med 13:161–164

Shenker R, Navidi W, Tavaré S, Dang MH, Chomyn A, Attardi G, Cortopassi G, Arnheim N (1996) The mutation rate of the human mitochondrial DNA deletion $mtDNA^{4977}$. Am J Hum Genet 59:772–780

Shoffner JM, Wallace DC (1995) Oxidative phosphorylation diseases. In: Schriver XR, Beaudet AL, Sly WS, Valle D (eds) The metabolic and molecular bases of inherited disease, 7th edn. McGraw-Hill, New York, pp 1535–1609

Shoffner JM, Lott MT, Voljavec AS, Souedan SA, Costigan DA, Wallace DC (1989) Spontenaous Kearns-Sayre/chronic external ophthalmoplegia plus syndrome associated with a mitochondrial DNA deletion: a slip-replication model and metabolic therapy. Proc Natl Acad Sci USA 86:7952–7956

Shoffner J, Brown M, Torroni A, Lott MT, Cabell MF, Mirra SS, Beal MF, Yang C-C, Gearing M, Salvo R, Watts RL, Juncos JL, Hansen LA, Crain BJ, Fayard M, Reckord CL, Wallace DC (1993) Mitochondrial DNA variants observed in Alzheimer disease and Parkinson disease patients. Genomics 17:171–184

Shoubridge EA, Karpati G, Hasting KE (1990) Deletion mutants are functionally dominant over wild-type mitochondrial genomes in skeletal muscle fiber segments in mitochondrial disease. Cell 62:43–49

Suomalainen A, Paetau A, Leinonen H, Majander A, Peltonen L, Somer H (1992) Inherited idiopathic dilated cardiomyopathy with multiple deletions of mitochondrial DNA. Lancet 340:1319–1320

Suomalainen A, Kaukonen J, Amati P, Timonen R, Haltia M, Weissenbach J, Zeviani M, Somer H, Peltonen L (1995) An autosomal locus predisposing to deletions of the mitochondrial DNA. Nat Genet 9:146–151

Superti-Fuga A, Schoenle E, Tuchschmid P, Caduff R, Sabato V, DeMattia D, Gitzelmann R, Steinmann B (1993) Pearson bone marrow-pancreas syndrome with insulin-dependent diabetes, progressive renal tubulopathy, organic aciduria and elevated fetal haemoglobin caused by deletion and duplication of mitochondrial DNA. Eur J Paediatr 152:44–50

Taanman J-W, Bodnar AG, Cooper JM, Morris AAM, Clayton PM, Leonard JV, Shapira AHV (1997) Molecular mechanisms in mitochondrial DNA depletion syndrome. Hum Mol Genet 6:935–942

Takahashi S, Makita Y, Oki J, Miyamoto A, Yanagawa J, Naito E, Goto Y-I, Okumo A (1998) De novo mtDNA nt 8993 (T $\rightarrow$ G) mutation resulting in Leigh syndrome. Am J Hum Genet 62:717–719

Takai D, Inoue K, Goto Y-I, Nonaka I, Hayashi J-I (1997) The interorganellar interaction between distinct human mitochondria with deletion mutant mtDNA from a patient with mitochondrial disease and HeLa mtDNA. J Biol Chem 272:6028–6033

Takao M, Aburatani H, Kobayashi K, Yasui A (1998) Mitochondrial targeting of human DNA glycosylases for repair of oxidative DNA damage. Nucleic Acids Res 26:2917–2922

Tallini G, Ladanyi M, Rosai J, Jhawar SC (1994) Analysis of nuclear and mitochondrial DNA alterations in thyroid and renal oncocytic tumors. Cytogenet Cell Genet 66:253–259

Tang J, Qi Y, Bao X-H, Wu X-R (1997) Mutational analysis of mitochondrial DNA of children with Rett syndrome. Pediatr Neurol 17:327–330

Taylor RW, Chinnery PF, Turnbull DM, Lightowlers RN (1997) Selective inhibition of mutant mitochondrial DNA replication in vitro by peptide nucleic acids. Nat Genet 15:212–215

Thomas MG, Cook CE, Miller KWP, Waring MJ, Hagelberg E (1998) Molecular instability in the COII-tRNA^{Lys} intergenomic region of the human mitochondrial genome: multiple origins of the 9-bp deletion and heteroplasmy for expanded repeats. Philos Trans R Soc Lond B 353:955–965

Thyagarajan B, Padua RA, Campbell C (1996) Mammalian mitochondria possess homologous recombination activity. J Biol Chem 271:27536–27543

Topcù Z, Castora FJ (1995) Mammalian mitochondrial DNA topoisomerase I preferentially relaxes supercoils in plasmids containing specific mitochondrial DNA sequences. Biochim Biophys Acta 1264:377–384

Van den Heuvel L, Ruitenbeek W, Smeets R, Gelman-Kohan Z, Elpeleg O, Loeffen J, Trijbels F, Mariman E, de Bruijn D, Smeitink J (1998) Demonstration of a new pathogenic mutation in human complex I deficiency: a 5-bp duplication in the nuclear gene encoding the 18-kD (AQDQ) subunit. Am J Hum Genet 62:262–268

Van den Ouweland JMW, Lemkes H, Ruitenbeek W, Sandkuijl LA, de Vijdler MF, Struyvenberg PAA, Van de Kamp JJP, Maasse JA (1992) Mutation in mitochondrial tRNA$^{Leu(UUR)}$ gene in a large pedigree with maternally transmitted type II diabetes mellitus and deafness. Nat Genet 1:368–371

Vissing J, Salamon MB, Arlien-Soborg P, Norby S, Manta P, DiMauro S, Schmalbruch H (1998) A new mitochondrial tRNA^{Met} gene mutation in a patient with dystrophic muscle and exercise intolerance. Neurology 50:1875–1878

Wallace DC, Singh G, Lott MT, Hodge JA, Schurr TG, Lezza AMS, Elsas LJ II, Nikoskelainen Ek (1988) Mitochondrial DNA mutation associated with Leber's hereditary optic neuropathy. Science 242:1427–1430

Wallace DC, Lott MT, Brown MD, Huoponen K, Torroni A (1996) Report of the committee on human mitochondrial DNA. In: Cuticchia AJ (ed) Human gene mapping 1995: a compendium. Johns Hopkins University Press, Baltimore, pp 910–954

Yoneda M, Chomyn A, Martinuzzi A, Hurko O, Attardi G (1992) Marked replicative advantage of human mtDNA carrying a point mutation that causes the MELAS encephalomyopathy. Proc Natl Acad Sci USA 89:11164–11168

Zeviani M, Servidei S, Gellera C, Bertini C, DiMauro S, Didonato S (1989) An autosomal dominant disorder with multiple deletions of mitochondrial DNA starting at the D-loop region. Nature 339:309–311

Zhu Z, Yao J, Fu K, DeBie I, Macmillan C, Cuthbert AP, Newbold RF, Wang J-C, Chevrette M, Brown GK, Brown RM, Shoubridge EA (1998) SURF1, encoding a factor involved in the biogenesis of cytochrome c oxidase, is mutated in Leigh syndrome. Nat Genet 20:337–343

CHAPTER 4

Structural Studies on Human Mitochondrial tRNAs

4

C. Florentz[1], H. Brulé[1,2], M. Helm[1] and R. Giegé[1]

Contents

1 Introduction

Transfer ribonucleic acids (tRNAs) are the key elements in protein biosynthesis. Indeed, these RNAs, charged with a specific amino acid, bind to messenger RNA and transfer their amino acid to the growing peptide chain. In order to fulfill this role, the tRNA has to interact with numerous macromolecular partners such as aminoacyl-tRNA synthetases (enzymes which catalyze binding of the specific amino acid), elongation factors, messenger RNA and different sites on the ribosome. The molecular mechanisms governing these various interactions, largely studied for cytosolic tRNAs, are based on the recognition of the three-dimensional structure of tRNA and on the recognition of signals on the surface of this structure (Söll and RajBhandary 1995). This has been best studied in the aminoacylation reaction, the key reaction of the whole process of protein synthesis (e.g. Saks et al. 1994; McClain 1995; Giegé et al. 1998). In particular, it is well established that any perturbation affecting either the structure an/or the recognition signals has the potential to lead to a functional deficiency in these molecules.

In recent years, numerous correlations have been established between the presence of point mutations in the mitochondrial genes encoding tRNAs, and maternally inherited neuromuscular diseases (Table 1 and reviewed by e.g. Wallace 1994; Lesti-

[1] UPR 9002 du CNRS, Institut de Biologie Moléculaire et Cellulaire, 15, rue René Descartes, 67084 Strasbourg Cedex
[2] U75 de l'INSERM, Faculté de Médecine Necker, 156 rue de Vaugirard, 75730 Paris Cedex 15, France

Table 1. Mutations in human mitochondrial tRNA genes and pathologies

tRNA	Mutation[a]	Pathology[b]	Reference
Asn	G5703A	CPEO	Moraes et al. (1993b)
	T5692C	CPEO	Seibel et al. (1994)
Asp	A7543G		El-Schahawi et al. (1995)
Cys	T5814C	Encephalopathy	Manfredi et al. (1996)
Glu	T14709C	Myopathy, diabetes mellitus	Hao et al. (1995)
			Hanna et al. (1995)
Gly	T9997C	Cardiopathy	Merante et al. (1994)
	A10006G	CIPO	Lauber et al. (1991)
	A10044G	Sudden death	Santorelli et al. (1996b)
Ile	A4269G	Cardio-, encephalopathy, deafness	Taniike et al. (1992)
	T4274C	CPEO	Chinnery et al. (1997)
	T4285C	PEO	Silvestri et al. (1996)
	A4295G	Cardiopathy	Merante et al. (1996)
	A4300G	Cardiopathy	Casali et al. (1995)
	A4317G	Cardiopathy, MELAS	Tanaka et al. (1990)
	C4320T	Cardiopathy, encephalopathy	Santorelli et al. (1995)
$Leu_{(CUN)}$	G12301A	Myelodysplasia	Gattermann et al. (1996)
	T12311C	CPEO	Hattori et al. (1994)
	G12315AS	Encephalomyopathy	Fu et al. (1996)
	A12320G	Myopathy	Weber et al. (1997)
$Leu_{(UUR)}$	A3243G	MELAS, diabetes mellitus	Goto et al. (1990)
			Kobayashi et al. (1990)
			van den Ouweland et al. (1992)
	A3243T	Encephalomyopathy	Shaag et al. (1997)
	T3250C	Myopathy	Goto et al. (1992)
	A3251G	PEO, myopathy, sudden death	Sweeney et al. (1993)
	A3252G	Encephalo-, retinopathy, diabetes	Morten et al. (1993)
	C3254G	Myopathy	Kawarai et al. (1997)
	C3256T	MERRF, retinopathy, diabetes mellitus	Moraes et al. (1993b)
	C3256T	MELAS	Sato et al. (1994)
	A3260G	Myopathy, cardiopathy	Zeviani et al. (1991)
	T3264C	Diabetes	Suzuki et al. (1997)
	T3271C	MELAS	Goto et al. (1991)
	del.T3271	Encephalomyopathy	Shoffner et al. (1995)
	T3291C	MELAS	Goto et al. (1994)
	A3302G	Myopathy	Bindoff et al. (1993)
	C3303T	Cardiopathy, myopathy	Silvestri et al. (1994)
Lys	A8344G	MERRF	Shoffner et al. (1990)
	T8356C	MERRF	Silvestri et al. (1992)
	G8363A	Encephalo-, cardiopathy, deafness	Santorelli et al. (1996a)
Pro	G15990T	Myopathy	Moraes et al. (1993a)
$Ser_{(AGY)}$	C12246A	CIPO	Lauber et al. (1991)
$Ser_{(UCN)}$	T7512C	MERRF-MELAS	Nakamura et al. (1995)
	ins.C7466	Hearing loss, ataxia, myoclonus	Tiranti et al. (1995)
Thr	G15915A	Encephalomyopathy	Nishino et al. (1996)
	A15923G	Lethal multisystemic disorders	Yoon et al. (1991)
Trp	G5549A	Encephalopathy, deafness	Nelson et al. (1995)
Tyr	C5877T	CPEO	Sahashi et al. (1994)
Val	G1642A	MELAS	Taylor et al. (1996)
	G1644T	Leigh syndrome	Chalmers et al. (1997)

[a] Numbering of the human mitochondrial genome according to Anderson et al. (1981). The letter preceding the number corresponds to the wild-type nucleotide in the L-strand and the letter following the number corresponds to the mutated nucleotide.
[b] CPEO and PEO: (chronic) progressive external ophthalmoplegia. MELAS: mitochondrial myopathy, encephalopathy with lactic acidosis and stroke-like episodes. MERRF: myoclonic epilepsy and ragged-red fibers syndrome. CIPO: chronic intestinal pseudo-obstruction with myopathy and ophthalmoplegia.

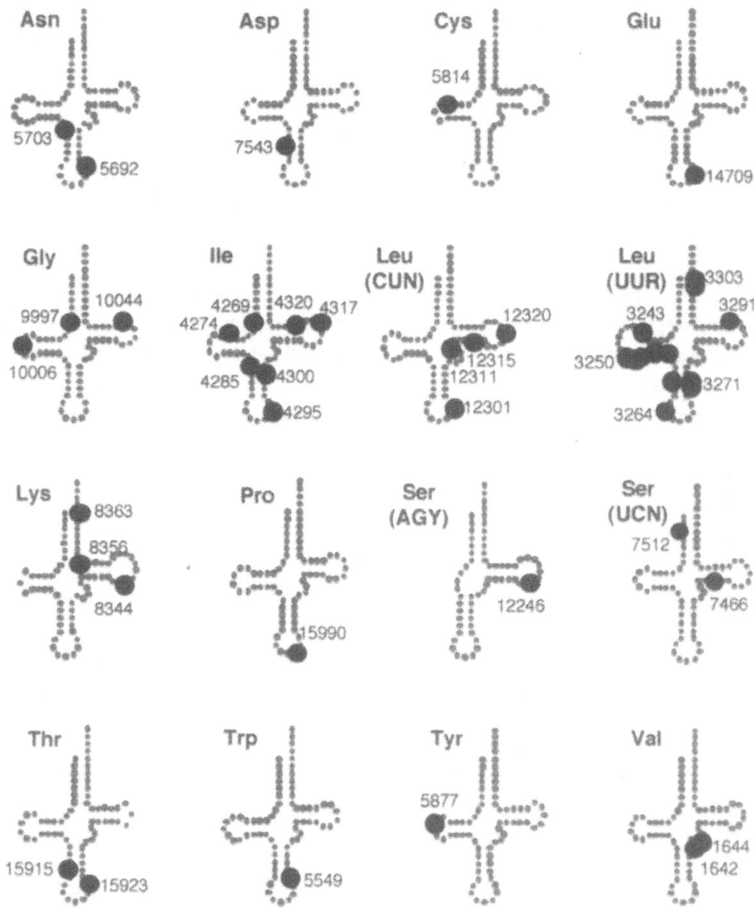

Fig. 1. Theoretical secondary structures of human mitochondrial tRNAs involved in neuromuscular diseases, and location within these structures of point mutations discovered in the corresponding genes. The *small dots* depict wild-type nucleotides whereas *larger dots* represent known mutated positions. Numbering of mutations is that of the mitochondrial genome. Although drawn on the same structure, the different mutations affecting one tRNA gene occur in different patients. In the case of $tRNA^{Leu(UUR)}$, only some mutations are numbered for more clarity

enne and Bataillé 1994; Larsson and Clayton 1995). The molecular mechanisms by which the mutations lead to pathologies are presently being investigated, but the diversity of postitions of mutated nucleotides within the genes (Fig. 1), as well as the variability of the mutation/phenotype relationships, suggest that several mechanisms of action exist at the level of the tRNA, product of the gene.

Point mutations may interfere at several levels. They may directly affect (1) the production of tRNA with a correct three-dimensional structure (maturation of the 3′ and 5′ ends, addition of a CCA sequence at the 3′ end of the tRNA [a sequence which is not encoded by the gene but is fundamental for the functional properties of the tRNA], post-transcriptional modification of several nucleotides, secondary and/or tertiary misfolding), or (2) the various functions of the mature tRNA. Since all these events are interrelated, any mutation may have an indirect role by a cascade event. For example, a mutation hindering the occurrence of a particular post-transcriptional modification could lead to the loss of an identity signal for aminoacylation.

In most cases already analyzed, it has been shown that mutations affect protein biosynthesis (e.g. Seibel et al. 1991; King et al. 1992; Hayashi et al. 1993; Hao and Moraes 1996). Following on from this, the steps leading to maturation and functional activity of the tRNAs are potential sites of action of the mutations. Since the three-dimensional structure of tRNAs is highly important for both or these steps, it becomes clear that knowledge of structural characteristics of tRNA and the study of the impact of "pathologic" mutations on these characteristics, is a fundamental prerequisite for the understanding of the mechanisms leading to pathologies.

In this chapter we will first recall the principal approaches available for structural analysis of RNA, then summarize the present knowledge on mitochondrial tRNA structures, and finally describe the particular studies on human mitochondrial $tRNA^{Lys}$. A mutation at postition 8344 of the corresponding nucleotide leads to the MERRF syndrome (Shoffner et al. 1990).

2
Approaches for the Structural Investigation of RNAs

Numerous approaches allowing access to structural understanding of RNAs are available and have been reviewed several times (e.g. Puglisi and Tinoco 1989; Gesteland and Atkins 1993; Nagai and Mattaj 1994; Kolchanov et al. 1996). These approaches, which will be briefly summarized here, fall into three classes, some allowing access to global features on conformation or on thermodynamic characeristics of the molecule, others leading to very detailed structural information.

The first type of approach combines knowledge of primary structures with the use of computer programs leading to prediction of secondary structures. These studies include phylogenetic comparisons and site directed mutagenesis. A second class includes biophysical investigations such as UV spectrophotometry, allowing the establishment of melting curves, and thus melting temperatures of the RNA, electrophoresis on non–denaturing gels allowing the detection of alternative conformers, light diffusion for the establishment of radius of gyration, etc.

The third category includes the use of structural probes in solution, combined with graphic modeling, and techniques such as NMR and X-ray crystallography. Both these last approaches lead to atomic structures. Thus, the structure of yeast $tRNA^{Phe}$ (Kim et al. 1974; Robertus et al. 1974) and $tRNA^{Asp}$ (Moras et al. 1980) have been establishment by crystallography, whereas the structure of RNA fragments has been obtained by NMR (Limmer et al. 1993; Viani-Puglisi et al. 1994). Both of these approaches require large amounts of material (several milligrams of pure RNA) and have intrinsic limitations, such as the size of the molecules to be studied or the difficultes in getting crystals (Ducruix and Giegé, 1992). Use of structural probes in solution allows, with knowledge of primary sequences and with low amounts of material, establishment of the secondary structures of large RNA fragments and modeling of 3-D structures (e.g. Krol et al. 1990; Isel et al. 1993; Felden et al. 1996). This approach is based on the differential reactivity of atoms within an RNA, depending on their location and their contribution to the nucleic acid structure. With tRNAs of known crystallographic structures, it has been demonstrated that these reactivities reflect the geometry of the RNA (Romby et al. 1987). Thus, an atom which is buried in the heart of the molecule or which takes part in Watson-Crick interactions, will not be reactive to probes when the RNA is in native conformation. In contrast, atoms located near the surface and not involved in chemical interactions will be reactive. Probes are used under conditions where they introduce less than one cut or modification per RNA molecule. The numbers and types of probes used are continuously increasing (Kolchanov et al. 1996; Giegé et al. 1999). In Table 2, the most widely used probes are listed according to the type of information they can yield. Some of them are enzymes, others are chemical reagents. For instance, nuclease S1 hydrolyses phosphodiester linkages in single stranded regions only. In contrast, ribonuclease V1 is specific for phosphodiester bonds in double-stranded or highly structured regions. Among chemical reagents, some will split accessible phosphodiester linkages (imidazole and lead acetate) whereas others will react with specific atoms of bases (e.g. diethylpyrocarbonate, dimethylsulfate, kethoxal). Finally, the reactivities of phosphates and riboses can be precisely tested with specific reagents (ENU and Fe-EDTA respectively).

When applied to the study of a particular RNA molecule in its native conformation, the use of a full set of probes gives information concerning any nucleotide and phosphodiester bond. These data usually are sufficient to decipher the secondary structure of the molecule. This approach also gives information about tertiary interactions between atoms, which will be taken into account for the establishment of three-dimensional models. These models are built up on graphic screens with the help of newly developed softwares (Westhof and Michel 1995).

Table 2. Chemical and enzymatic probes in current use in the determination of solution structure of RNA. The different probes are classified in regard to the type and precision of the information yielded

Probes	Molecular weight	Targets
Global outlook on the structure:		
RNAse T2	36000	Single stranded regions
Nuclease S1	32000	Single stranded regions
RNAse T1	11000	Guanosines in single stranded regions
RNase U2	12500	Adenosines in single stranded regions
Rnase V1	15900	Double stranded and structured regions
Lead (acetate)	207	Ribose-phosphate linkages
Imidazole	68	Ribose-phosphate linkages
Secondary structure determination (reactivity of Watson-Crick positions):		
DMS	126	N3 of cytidines, N1 of adenosines
Kethoxal	148	N1 and N2 of guanosines
CMCT	424	N3 of uridines, N1 of guanosines
Complementary information on tertiary structure:		
DEPC	174	N7 of adenosines
DMS	126	N7 of guanosines
ENU	117	Phosphates
OH° (Fe-EDTA)	17	Riboses

DMS: dimethylsulfate; CMCT: carbodiimide; DEPC: diethylpyrocarbonate; ENU: ethylnitrosaurea

3
Present Knowledge on Mitochondrial tRNA Structure

Structural analyses of mitochondrial tRNAs are limited, especially for humans. These tRNAs often display odd secondary structures when compared with classical cloverleaves (Dirheimer et al. 1995). The mayor differences concern the D and T stem-loop domains, which are sometimes completely missing. When present, the loops are of unusual size and sequence, and often do not contain the conserved nucleotides

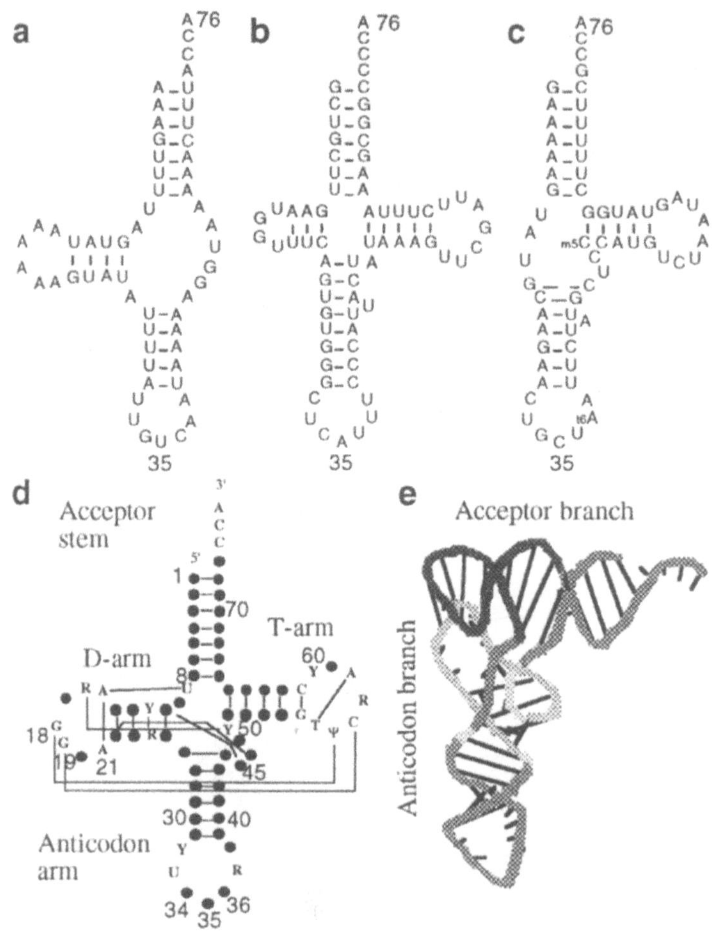

Fig. 2. Examples of mitochondrial tRNA secondary structures (**a–c**): Nematode $tRNA^{Asp}$ (**a**), *Tetrahymena pyriformis* $tRNA^{Met}$ (**b**) and bovine $tRNA^{Ser(AGY)}$ (**c**). Sequences (**a**) and (**b**) are deduced from their corresponding genes, whereas (**c**) was obtained by sequencing the tRNA. These sequences are taken from Sprinzl et al. (1996). Numbering of nucleotides is according to Sprinzl et al. (1996), the central anticodon nucleotide is always numbered 35, the 3′ terminal adenosine is always numbered 76. The "cloverleaf" secondary structure (**d**) and the "L-shaped" tertiary structure (**e**) of canonical tRNAs are shown. In (**d**), nucleotides conserved in cytosolic tRNAs are indicated as well as the tertiary interaction network leading to the three-dimensional folding. *R* Purine, *Y* pyrimidine, *T* ribosylthymine, *Ψ* pseudouridine

involved in tertiary interactions (see examples in Fig. 2). Since mitochondrial tRNAs are effectors of mitochondrial protein synthesis, they should adopt the three-dimensional L-shape characteristic for cytosolic tRNAs (Fig. 2e). Indeed, experimental data confirm that this is the case at least for some tRNAs (deBruin and Klug 1983; Yokogawa et al. 1991; Ueda et al. 1992; Wakita et al. 1994). In addition, theoretical considerations indicate that, although differing widely from one to an other with respect to their secondary structures, mitochondrial tRNAs may share the same three-dimensional structure as canonical tRNAs (Steinberg and Cedergren 1994).

4 Structural Study of Human Mitochondrial $tRNA^{Lys}$ In Vitro Transcript

We studied the solution structure of the human mitochondrial lysine-specific tRNA. Due to obvious difficulties in obtaining human biological material, especially from patients, $tRNA^{Lys}$ of a wild-type sequence and its mutated MERRF variant (position 8344) were prepared by cloning the corresponding synthetic genes, followed by in vitro transcription. This method has the advantage of being quick and efficient. It yields RNA molecules deprived of post-transcriptional modifications. Used for the study of the classical cytosolic tRNAs, it showed that transcripts deprived of modified nucleotides had globally the same behavior as natural molecules, both concerning their structural and functional properties (Giegé et al. 1993; Saks et al. 1994; McClain 1995). In only a few cases, post-transcriptional modifications were shown to be essential for functionality (see Giegé et al. 1998, for references).

Structural properties of wild-type and mutated in vitro transcribed $tRNA^{Lys}$ were compared using a thermodynamic approach and solution structure probing. Comparative analysis of melting curves showed that the melting points of the two transcripts differ by 1 °C, the MERRF variant being more stable than the wild-type transcript (Fig. 3b). This differential behavior was also confirmed using denaturing gradient gel electrophoresis (Fig. 3a). In this method, molecules migrate in parallel to a temperature gradient. The more the temperature is elevated, the more the molecule unfolds, which in turn slows its migration in the gel (Riesner et al. 1991).

The solution structure of the wild-type $tRNA^{Lys}$ transcript was investigated using structural probes (Helm et al. 1998). Typical autoradiograms showing reactivities of adenines towards DEPC and reactivities of the ribose-phosphate backbone towards imidazole are displayed in Fig. 3c, d respectively. For instance, one can see that nucleotides A7, A8 and A9 are fully protected against DEPC under both native and semi-denaturing conditions, and that the 3′ end of the molecule is not sensitive to imidazole. Other probes such as DMS, kethoxal and the enzymes listed in Table 1 were also used. Altogether, these data suggested that the secondary structure was an atypical and extended hairpin, totally differing from a cloverleaf (Fig. 4a). This quite unexpected result, corroborated by transient electric birefringence measurements (Leehey et al. 1995), lead us to postulate a role for modified nucleotides in the correct folding of the natural $tRNA^{Lys}$. In particular, the role of the modification of nucleotide 9, which impairs the establishment of a Watson-Crick base pair with nucleotide 64 was investigated through cloning and in vitro transcription of $tRNA^{Lys}$ variants mimicking the effect of the presence of this modified nucleotide. Structural analysis of these

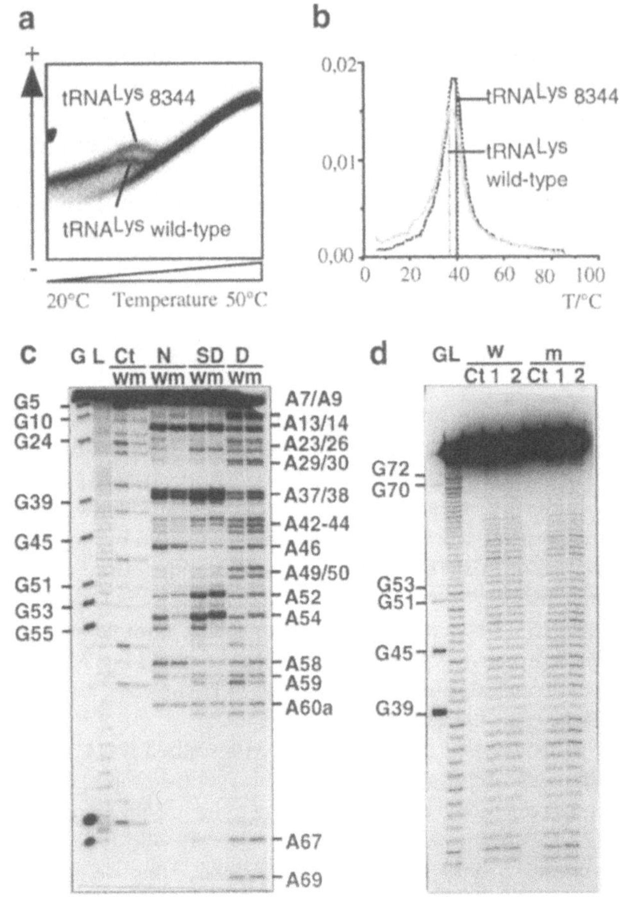

Fig. 3. Example of structural analysis of tRNAs: the case of human mitochondrial $tRNA^{Lys}$ prepared by in vitro transcription of synthetic genes of wild-type and 8344-mutated sequences. **(A)** Thermodynamic analysis using temperature gradient gel electrophoresis (TGGE): autoradiography of a 12% polyacrylamide gel. **(B)** Thermodynamic analysis using optical melting curves: the first order derivatives are presented. **(C, D)** Solution structure analysis: reactivities of adenines towards diethylpyrocarbonate **(C)** and reactivities of phosphodiester bonds towards imidazole **(D)**. Autoradiographies of 12% polyacrylamide gels. The different lanes corresponds to the product of statistical hydrolysis of 5′ end labeled transcripts. Both autoradiographs are annotated as follows: *G* ribonuclease T1 hydrolysis; *L* alcaline ladder; *Ct* untreated control; *N* native conditions; *SD* semi-denaturing conditions; *D* denaturing conditions; *w* wild-type transcript; *m* mutated transcript (mutation 8344); *1* and *2* two different concentrations in imidazole. Annotations at *left* and *right* of the autoradiographs allow positioning of nucleotides

variants showed that the newly discovered hairpin structure was a particularity of the in vitro transcribed wild-type molecule only. Indeed, the variants adopted a cloverleaf structure (Fig. 4b), as does the native and fully modified $tRNA^{Lys}$ (Helm et al. 1998). Thus, the correct folding of the $tRNA^{Lys}$ molecule is highly dependent on the presence of modified nucleotides, especially on the methylation of nitrogen 1 of adenosine 9.

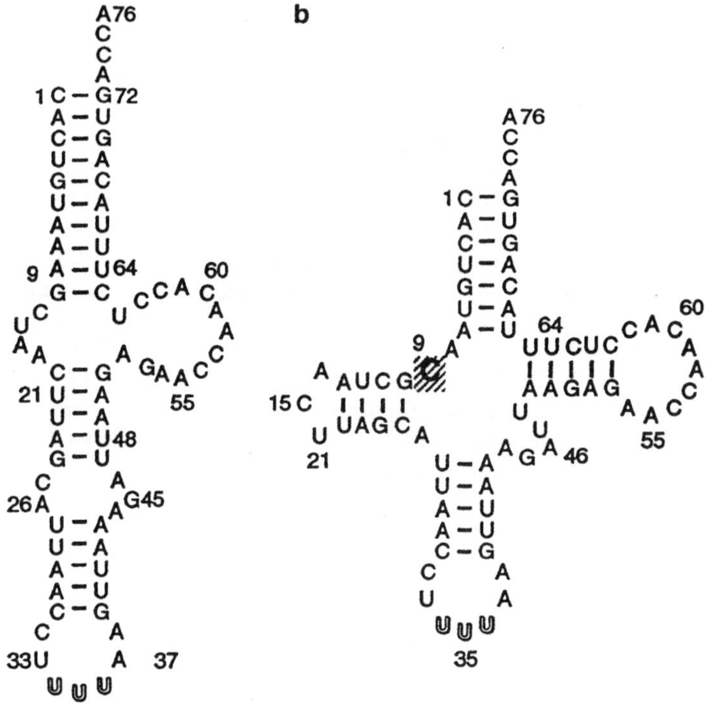

Fig. 4. (**a**) Experimentally determined hairpin structure of the wild-type tRNA^{Lys} in vitro transcript, as deduced from reactivities towards enzymatic and chemical probes (Helm et al. 1998). (**b**) Experimentally determined cloverleaf structure of variant (A9 → C9) mimicking the effects of the modified nucleotide m^{1}A at position 9 *(hatched nucleotide)*. The anticodon triplets *-UUU-* is emphasized in both structures. Numbering is the same as in Fig. 2

This methylation, actually present in the native tRNA as checked by sequencing of the native molecule, hinders base pairing of adenosine 9 with nucleotide 64. This situation is possible in the transcripts and leads to lengthening of the acceptor stem by three base pairs (compared with the situation found in a regular cloverleaf structure), which in turn induces a new type of folding for the whole molecule.

5 Conclusion and Perspectives

The determination of the solution structure of the human mitochondrial tRNA^{Lys} transcripts allowed the discovery of a very unusual tRNA structure, even for a mitochondrial tRNA. In addition, it led to the discovery of a particular and original potential role for a modified nucleotide, namely that of a "chaperone". This finding of a new role for modified nucleotides is of major importance in understanding structure-function relationships in RNAs in general, and also in understanding genotype-phenotype relationships in mitochondrial neuromuscular disorders linked to mutations in tRNA genes. Indeed, two conjectures can be proposed on the basis of this work: first, since simple nucleotide substitutions can lead to a folding quite different from that of the original molecule, such changes will have disastrous consequences on the functionality of the RNA. Secondly, a nucleotide substitution in a tRNA may impair its recognition by a modifying enzyme, and thus may impair the formation of some post-transcriptional modifications. Their absence may lead to profound structural changes, with the same disastrous consequences on function.

A systematic structural study of human mitochondrial tRNA couples (wild-type and disease-related variants) should provide a global outlook on the structural impact of mutations. Through the construction of structural variants mimicking the presence of modified nucleotides, is should be possible on one hand to compare the structure of wild-type and pathogenic tRNAs, and on the other hand to investigate the importance and mode of action of modified nucleotides in mitochondrial tRNAs. A systematic sequencing of native human mitochondrial tRNAs would complement these structural and functional studies.

Acknowledgements. The work was supported by the Association Française contre les Myopathies (AFM).

References

Anderson S, Bankier AT, Barrel BG, deBruijn MHL, Coulson AR, Drouin J, Eperon JC, Nierlich DP, Roe BA, Sanger F, Schreier PH, Smith AJH, Staden R, Young IG (1981) Sequence and organization of the human mitochondrial genome. Nature 290:457–465

Bindoff LA, Howell N, Poulton J, McCullough DA, Morten KJ, Lightowlers RN, Turnbull DM, Weber K (1993) Abnormal RNA processing associated with a novel tRNA mutation in mitochondrial DNA. A potential disease mechanism. J Biol Chem 268:19559–19564

Casali C, Santorelli FM, D'Amati G, Bernucci P, DeBiase L, DiMauro S (1995) A novel mtDNA point mutation in maternally inherited cardiomypopathy. Biochem Biophys Res Commun 213:588–593

Chalmers RM, Lamont PJ, Nelson I, Ellison DW, Thomas NH, Harding AE, Hammans SR (1997) A mitochondrial DNA $tRNA^{Val}$ point mutation associated with adult-onset Leigh syndrome. Neurology 49:589–592

Chinnery PF, Johmson MA, Taylor RW, Durward WF, Turnbull DM (1997) A novel mitochondrial tRNA isoleucine gene mutation causing chronic progressive external ophthalmoplegia. Neurology 49:1166–1168

DeBruijn MHL, Klug A (1983) A model for the tertiary structure of mammalian mitochondrial transfer RNAs lacking the entire "dihydrouridine" loop and stem. EMBO J 2:1309–1321

Dirheimer G, Keith G, Dumas P, Westhof E (1995) Primary, secondary and tertiary structures of tRNAs. In: Söll D, RajBhandary U (eds) tRNA: structure, biosynthesis, and function. ASM Press, Washington DC, pp 93–126

Ducruix A, Giegé R (1992) Crystallization of nucleic acids and proteins: a practical approach. IRL Press, Oxford

El-Schahawi M, Santorelli FM, Malkin E, Shanske S, DiMauro S (1995) A new mitochondrial DNA mutation in the $tRNA^{Asp}$ at np 7543 is associated with myoclonic seizures and developmental delay. J Mol Med 74:B37

Felden B, Florentz C, Giegé R, Westhof E (1966) A central pseudoknotted three-way junction imposes tRNA-like mimicy and the orientation of three 5′ upstream pseudoknots in the 3′ terminus of tobacco mosaic virus RNA. RNA 2:201–212

Fu K, Hartlen R, Johns T, Genge A, Karpati G, Shoubridge EA (1996) A novel heteroplasmic $tRNA^{Leu(CUN)}$ mt DNA point mutation in a sporadic patient with mitochondrial encephalomyopathy segregates rapidly in skeletal muscle and suggests an approach to therapy. Hum Mol Genet 5:1835–1840

Gattermann N, Retzlaff S, Wang Y-L, Berneburg M, Heinisch J, Wlaschek M, Aul C, Schneider W (1996) A heteroplasmic point mutation of mitochondrial $tRNA^{Leu(CUN)}$ in non-lymphoid haemopoietic cell lineages from a patient with acquired idiopathic sideroblastic anaemia. Br J Haematol 93:845–855

Gesteland RF, Atkins JF (1993) The RNA world. Cold Spring Harbor Laboratory Press, Cold Spring Harbor

Giegé R, Puglisi JD, Florentz C (1993) tRNA structure and aminoacylation efficiency. Prog Nucleic Acid Res Mol Biol 45:129–206

Giegé R, Sissler M, Florentz C (1998) Universal rules and idiosyncratic features in tRNA identity. Nucleic Acids Res 26:5017–5035

Giegé R, Helm M, Florentz C (1999) Chemical and enzymatic probing of RNA structure. In: Söll D, Nishinuva S (eds) Prebiotic chemistry, molecular fossils, nucleosides and RNA. Pergamon Press, Oxford (in press)

Goto Y, Nonaka I, Horai S (1990) A mutation in the $tRNA^{Leu(UUR)}$ gene associated with the MELAS subgroup of mitochondrial encephalomyopathies. Nature 348:651–653

Goto Y, Nonaka I, Horai S (1991) A new mtDNA mutation associated with mitochondrial myopathy, encephalopathy, lactic acidosis and stroke-like episodes (MELAS). Biochim Biophys Acta 1097:239–240

Goto Y, Tojo M, Tohyama J, Horai S, Nonaka I (1992) A novel point mutation in the mitochondrial $tRNA^{Leu(UUR)}$ gene in a family with mitochondrial myopathy. Ann Neurol 31:672–675

Goto Y, Tsugane K, Tanabe Y, Nonaka I, Horai S (1994) A new point mutation at nucleotide pair 3291 of the mitochondrial $tRNA^{Leu(UUR)}$ gene in a patient with mitochondrial myopathy, encephalopathy, lactic acidosis, and stroke-like episodes (MELAS). Biochem Biophys Res Commun 202:1624–1630

Hanna MG, Nelson I, Sweeney MG, Cooper JM, Watkins PJ, Morgan-Hughes JA, Harding AE (1995) Congenital encephalomyopathy and adult-onset myopathy and diabetes mellitus: different phenotypic associations of a new heteroplasmic mtDNA tRNA glutamic acid mutation. Am J Hum Genet 56:1026–1033

Hao H, Moraes CT (1996) Functional and molecular mitochondrial abnormalities associated with a C → T transition at position 3256 of the human mitochondrial genome. The effects of a pathogenic mitochondrial tRNA point mutation in organelle translation and RNA processing. J Biol Chem 271:2347–2352

Hao H, Bonilla E, Manfredi G, DiMauro S, Moraes CT (1995) Segregation patterns of a novel mutation in the mitochondrial tRNA glutamic acid gene associated with myopathy and diabetes mellitus. Am J Hum Genet 56:1017–1025

Hattori Y, Goto YI, Sakuta R, Nonaka I, Mizuno Y, Horai S, (1994) Point mutations in mitochondrial tRNA genes: sequence analysis of chronic progressive external ophthalmoplegia (CPEO). J Neurol Sci 125:50–55

Hayashi J-I, Ohta S, Takai D, Miyabayashi S, Sakuta R, Goto Y-I, Nonaka I (1993) Accumulation of mtDNA with a mutation at position 3271 in $tRNA^{Leu(UUR)}$ gene introduced from a MELAS patient to HeLa cells lacking mtDNA results in progressive inhibition of mitochondrial respiratory function. Biochem Biophys Res Commun 197:1049–1055

Helm M, Brulé H, Degoul F, Cepance C, Leroux JP, Giegé R, Florentz C (1998) The presence of a modified nucleotide is required for the cloverleaf folding of a human mitochondrial tRNA. Nucleic Acids Res 26:1636–1643

Isel C, Marquet R, Keith G, Ehresmann C, Ehresmann B (1993) Modified nucleotides of $tRNA_3^{Lys}$ modulate primer/template loop-loop interaction in the initiation complex of HIV-1 reverse transcription. J Biol Chem 268:25269–25272

Kawarai T, Kawakami H, Kozuka K, Izumi Y, Matsuyama Z, Watanabe C, Kohriyama T, Nakamura S (1997) A new mitochondrial DNA mutation associated with mitochondrial myopathy: $tRNA^{Leu(UUR)}$ 3254C-to-G. Neurology 49:598–600

Kim SH, Suddath FL, Quigley GJ, McPherson A, Sussman JL, Wang AHJ, Seeman NC, Rich A (1974) Three-dimensional tertiary structure of yeast phenylalanine transfer RNA. Science 185:435–440

King MP, Koga Y, Davidson M, Schon EA (1992) Defects in mitochondrial protein synthesis and respiratory chain activity segregate with the $tRNA^{Leu(UUR)}$ mutation associated with mitochondrial myopathy, encephalopathy, lactic acidosis and strokelike episodes. Mol Cell Biol 12:480–490

Kobayashi Y, Momoi MY, Tominaga K, Momoi T, Nihei K, Yanagisawa M, Kagawa Y, Ohta S (1990) A point mutation in the mitochondrial $tRNA^{Leu(UUR)}$ gene in MELAS (mitochondrial myopathy, encephalopathy, lactic acidosis and stroke-like episodes). Biochem Biophys Res Commun 173:816–822

Kolchanov NA, Titov II, Vlassova IE, Vlassov VV (1996) Chemical and computer probing of RNA structure. Prog Nucleic Acid Res Mol Biol 53:131–197

Krol A, Westhof E, Bach M, Lührmann R, Ebel J-P, Carbon P (1990) Solution structure of human U1 snRNA. Derivation of a possible three-dimensional model. Nucleic Acids Res 18:3803–3811

Larsson N-G, Clayton DA (1995) Molecular genetic aspects of human mitochondrial disorders. Annu Rev Genet 29:151–178

Lauber J, Marsac C, Kadenbach B, Seibel P (1991) Mutations in mitochondrial tRNA genes: a frequent cause of neuromuscular diseases. Nucleic Acids Res 19:1393–1397

Leehey MA, Squassoni CA, Friederich MW, Mills JB, Hagerman PJ (1995) A noncanonical tertiary conformation of a human mitochondrial transfer RNA. Biochemistry 34:16235–16239

Lestienne P, Bataillé N (1994) Mitochondrial DNA alterations and genetic diseases: a review. Biomed Pharmacother 48:199–214

Limmer S, Hofmann HP, Ott G, Sprinzl M (1993) The 3′-terminal end (NCCA) of tRNA determines the structure and stability of the aminoacyl acceptor stem. Proc Natl Acad Sci USA 90:6199–6202

Manfredi G, Schon EA, Baonilla E, Moraes CT, Shanske S, DiMauro S (1996) Identification of a mutation in the mitochondrial $tRNA^{Cys}$ gene associated with mitochondrial encephalopathy. Hum Mutat 7:158–163

McClain WH (1995) The tRNA identity problem: past, present and future. In: Söll D, RajBhandary U (eds) tRNA: structure, biosynthesis, and function. ASM Press, pp 335–347

Merante FM, Tein I, Benson L, Robinson BH (1994) Maternally inherited hypertrophic cardiomyopathy due to a novel T-to-C transition at nucleotide 9997 in the mitochondrial $tRNA^{Gly}$ gene. Am J Hum Genet 55:437–446

Merante FM, Myint T, Tein I, Benson L, Robinson BH (1996) An additional mitochondrial $tRNA^{Ile}$ point mutation (A-to-G at nucleotide 4295) causing hypertrophic cardiomyopathy. Hum Mutat 8:216–222

Moraes CT, Ciacci F, Bonilla E, Ionasescu V, Schon EA, DiMauro S (1993a) A mitochondrial tRNA anticodon swap associated with a muscle disease. Nat Genet 4:284–288

Moraes CT, Ciacci F, Bonilla E, Jansen C, Hirano M, Rao N, Lovelace RE, Rowland LP, Schon EA, DiMauro S (1993b) Two novel pathogenic mitochondrial DNA mutations affecting organelle number and protein synthesis. Is the $tRNA^{Leu(UUR)}$ gene an etiologic hot spot? J Clin Invest 192:2906–2915

Moras D, Comarmond MB, Fischer J, Weiss R, Thierry JC, Ebel JP, Giegé R (1980) Crystal structure of yeast $tRNA^{Asp}$. Nature 288:669–674

Morten KJ, Cooper JM, Brown GK, Lake BD, Pike D, Poulton J (1993) A new point mutation associated with mitochondrial encephalomyopathy. Hum Mol Genet 2:2081–2087

Nagai K, Mattaj IW (1994) RNA-protein interactions. Frontiers in molecular biology. IRL Press, Oxford

Nakamura M, Nakano S, Goto Y-I, Ozawa M, Nagahama Y, Fukuyama H, Akiguchi I, Kaji R, Kimura J (1995) A novel point mutation in the mitochondrial $tRNA^{Ser(UCN)}$ gene detected in a family with MERRF/MELAS overlap syndrome. Biochem Biophys Res Commun 214:86–93

Nelson I, Hanna MG, Alsanjari N, Scaravilli F, Morgan-Hugues JA, Harding AE (1995) A new mitochondrial DNA mutation associated with progressive dementia and chorea: a clinical, pathological, and molecular genetic study. Ann Neurol 37:400–403

Nishino I, Seki A, Maegaki Y, Takeshita K, Horai S, Nonaka I, Goto YI (1996) A novel mutation in the mitochondrial $tRNA^{Thr}$ gene associated with a mitochondrial encephalomyopathy. Biochem Biophys Res Commun 225:180–185

Puglisi JD, Tinoco IJ (1989) Absorbance melting curves of RNA. Methods Enzymol 180:304–325

Riesner D, Henco K, Steger G (1991) Temperature-gradient gel electrophoresis: a method for the analysis of conformational transitions and mutations in nucleic acids and proteins. In: Chrambach A, Dunn MJ, Radolas BJ (eds) Advances in electrophoresis, VCH, vol 4. Weinheim, pp 169–249

Robertus JD, Ladner JE, Finch JT, Rhodes D, Brown RS, Clark BFC, Klug A (1974) Structure of yeast phenylalanine tRNA at 3Å resolution. Nature 250:546–551

Romby P, Moras D, Dumas P, Ebel J-P, Giegé R (1987) Comparison of the tertiary structure of yeast $tRNA^{Asp}$ and $tRNA^{Phe}$ in solution. Chemical modification study of the bases. J Mol Biol 195:193–204

Sahashi K, Ibi T, Ohno K, Tanaka M, Tashiro M, Marui K, Nakao N, Ozawa T (1994) Two types of newly recognized familial mitochondrial mypoathies. Muscle Nerve Suppl 1:S139

Saks ME, Sampson JR, Abelson JN (1994) The transfer RNA identiy problem: a search for rules. Science 263:191–197

Santorelli FM, Mak SC, Vazquez-Acevedo M, Gonzalez-Astiazaran A, Ridaura-Sanz C, Gonzalez-Halphen D, DiMauro S (1995) A novel mitochondrial DNA point mutation associated with mitochondrial encephalo-cardiomyopathy. Biochem Biophys Res Commun 216:835–840

Santorelli FM, Mak S-C, El-Schahawi M, Casali C, Shanske S, Baram TZ, Madrid RE, DiMauro S (1996a) Maternally inherited cardiomyopathy and hearing loss associated with a novel mutation in the mitochondrial $tRNA^{Lys}$ gene G8363A. Am J Hum Genet 58:933–939

Santorelli FM, Schlessel JS, Slonim AE, DiMauro A (1996b) Novel mutation in the mitochondrial DNA tRNA glycine gene associated with sudden unexpected death. Pediatr Neurol 15:145–149

Sato W, Hayasaka K, Shoji Y, Takahashi T, Takada G, Saito M, Fukawa O, Wachi E (1994) A mitochondrial $tRNA^{Leu(UUR)}$ mutation at 3256 associated with mitochondrial myopathy, encephalopathy, lactic acidosis, and stroke-like episodes (MELAS). Biochem Mol Biol Int 33:1055–1061

Seibel P, Degoul F, Bonne G, Romero N, François D, Paturneau-Jouas M, Ziegler F, Eymard B, Fardeau M, Marsac C, Kadenbach B (1991) Genetic biochemical and pathophysiological characterization of a familial mitochondrial encephalomyopathy (MERRF). J Neurol Sci 105:217–224

Seibel P, Lauber J, Klopstock T, Marsac C, Kadenbach B, Reichmann H (1994) Chronic progressive external ophthalmoplegia is associated with a novel mutation in the mitochondrial $tRNA^{Asn}$ gene. Biochem Biophys Res Commun 204:482–489

Shòffner JM, Bialer MG, Pavlakis SG, Lott M, Kaufman A, Dixon J, Teichberg S, Wallace DC (1995) Mitochondrial enxephalomyopathy associated with a single nucleotide pair deletion in the mitochondrial $tRNA^{Leu(UUR)}$ gene. Neurology 45:286–292

Shoffner JM, Lott M, Lezza AMS, Seibel P, Ballinger SW, Wallace DC (1990) Myoclonic epilepsy and ragged red fiber disease (MERRF) is associated with mitochondrial DNA $tRNA^{Lys}$ mutation. Cell 61:931–937

Shaag A, Saada A, Steinberg A, Navon P, Elpeleg ON (1997) Mitochondrial encephalomyopathy associated with a novel mutation in the mitochondrial $tRNA^{Leu(UUR)}$ gene (A3243T). Biochem Biophys Res Commun 233:637–639

Silvestri G, Moraes CT, Shanske S, Oh SJ, DiMauro S (1992) A new mtDNA mutation in the $tRNA^{Lys}$ gene associated with myoclonic epilepsy and ragged-red fibers (MERRF). Am J Hum Genet 51:1213–1217

Silvestri G, Santorelli FM, Shanske S, Whitley CB, Schimmenti LA, Smith SA, DiMauro S (1994) A new mtDNA mutation in the $tRNA^{Leu(UUR)}$ gene associated with maternally inherited cardiomyopathy. Hum Mutat 3:37–43

Silvestri G, Servidei S, Rana M, Ricci E, Spinazzola A, Paris E, Tonali P (1996) A novel mitochondrial DNA point mutation in the $tRNA^{Ile}$ gene is associated with progressive external ophthalmoplegia. Biochem Biophys Res Commun 220:623–627

Söll D, RajBhandary UL (eds) (1995) tRNA: structure, biosynthesis, and function. ASM Press, Washington, DC

Sprinzl M, Steegborn C, Hübel F, Steinberg S (1996) Compilation of tRNA sequences and sequences of tRNA genes. Nucleic Acids Res 24:68–72

Steinberg S, Cedergren R (1994) Structural compensation in atypical mitochondrial tRNAs. Nat Struct Biol 1:507–510

Suzuki Y, Suzuki S, Hinokio Y, Chiba M, Atsumi Y, Hosokawa K, Shimada A, Ashina T, Matsuoka K (1997) Diabetes associated with a novel 3264 mitochondrial $tRNA^{Leu(UUR)}$ mutation. Diabetes Care 20:1138–1140

Sweeney MG, Bundey S, Brockington M, Poulton KR, Winer JB, Harding AE (1993) Mitochondrial myopathy associated with sudden death in young adults and a novel mutation in the mitochondrial DNA leucine transfer $RNA^{(UUR)}$ gene. J Med 86:709–713

Tanaka M, Ino H, Ohno K, Hattori K, Sato W, Ozawa T, Tanaka T, Itoyama S (1990) Mitochondrial mutation in fatal infantile cardiomyopathy. Lancet 336:1452

Taniike M, Fukushima H, Yanaghihara I, Tsukamoto H, Tanaka J, Fujimura H, Nagai T, Sano T, Yamaoka K, Inui K, Okada S (1992) Mitochondrial $tRNA^{Ile}$ mutation in fatal cardiomyopathy. Biochem Biophys Res Commun 186:47–53

Taylor RW, Chinnery PF, Haldane F, Moris AAM, Bindoff LA, Wilson J, Turnbull DM (1996) MELAS associated with a mutation in the valine transfer RNA gene of mitochondrial DNA. Ann Neurol 40:459–462

Tiranti V, Chariot P, Carella F, Toscano A, Soliveri P, Girlanda P, Carrara F, Fratta GM, Reid FM, Mariotti C, Zeviani M (1995) Maternally inherited hearing loss, ataxia and myoclonus associated with a novel point mutation in mitochondrial $tRNA^{Ser(UCN)}$ gene. J Biol Chem 270:6298–6307

Ueda T, Yotsumoto Y, Ikeda K, Watanabe K (1992) The T-loop region of animal mitochondrial $tRNA^{Ser(AGY)}$ is a main recognition site for homologous seryl-tRNA synthetase. Nucleic Acids Res 20:2217–2222

Van den Ouweland JMW, Lemkes HHPJ, Ruitenbeek W, Sandkuijl LA, de Vijlder MF (1992) Mutation in mitochondrial $tRNA^{Leu(UUR)}$ gene in a large pedigree with maternally transmitted type II diabetes mellitus and deafness. Nat Genet 1:368–371

Viani-Puglisi E, Puglisi JD, Williamson JR, RajBhandary U (1994) NMR analysis of tRNA acceptor stem microhelices: discriminator base change affects tRNA conformation at the 3′ end. Proc Natl Acad Sci USA 91:11467–11471

Wakita K, Watanabe Y-I, Yogogawa T, Kumazawa Y, Nakamura S, Ueda T, Watanabe K, Nishikawa K (1994) Higher-order structure of bovine mitochondrial $tRNA^{Phe}$ lacking the "conserved" GG and TΨCG sequences as inferred by enzymatic and chemical probing. Nucleic Acids Res 22:347–353

Wallace DC (1994) Mitochondrial DNA sequence variation in human evolution and disease. Proc Natl Acad Sci USA 91:8739–8746

Weber K, Wilson JN, Taylor L, Brierley E, Johnson MA, Turnbull DM, Bindoff LA (1997) A new mtDNA mutation showing accumulation with time and restriction to skeletal muscle. Am J Hum Genet 60:373–380

Westhof E, Michel F (1995) Prediction and experimental investigation of RNA secondary and tertiary foldings. In: Nagay K, Mattaj IW (eds) RNA – protein interactions. IRL Press, Oxford, pp 25–51

Yokogawa T, Watanabe Y, Kumazawa Y, Ueda T, Hirao I, Miura K, Watanabe K (1991) A novel cloverleaf structure found in mammalian mitochondrial $tRNA^{Ser(UCN)}$. Nucleic Acids Res 19:6101–6105

Yoon KL, Aprille JR, Ernst SG (1991) Mitochondrial $tRNA^{Thr}$ mutation in fatal infantile respiratory enzyme deficiency. Biochem Biophys Res Commun 176:1112–1115

Zeviani M, Gellera C, Antozzi C, Rimoldi M, Morandi L, Villani F, Tiranti V, DiDonato S (1991) Maternally inherited myopathy and cardiomyopathy: association with mutation in mitochondrial DNA $tRNA^{Leu(UUR)}$. Lancet 338:143–147

CHAPTER 5

Structure, Function and Pathology of Complex I 5

H. Duborjal[1], R. Beugnot[1], V. Procaccio[1,2], J. P. Issartel[1] and J. Lunardi[1,2]

Contents

1 Introduction

In eukaryote cells, substrates are terminally oxidized by enzymatic complexes of the respiratory chain located in the inner membrane of mitochondria. Oxidation of NADH or $FADH_2$ by these enzymes is coupled to an ejection of protons into the intermembrane space. The resulting electrochemical gradient drives ATP synthesis which is achieved by re-entry of protons through the ATP synthetase.

The NADH-ubiquinone oxidoreductase (complex I, EC 1.6.5.3) catalyzes the oxidation of NADH and reduction of ubiquinone. Its activity combines the transfer of two electrons with the ejection of four protons, and is energetically coupled to ATP synthesis. First assays of purification of the bovine enzyme were reported as early as 1962 by Hatefi et al. An energetically coupled NADH-ubiquinone oxidoreductase is also present in cytoplasmic membranes of a large number of bacteria. Despite a simpler

[1] Laboratoire de BioEnergétique Cellulaire et Pathologique, EA UJF 2411, DBMS/CEA Grenoble, 17 rue des Martyrs, 38054 Grenoble Cedex 9, France
[2] Biochimie de l'ADN, CHU Grenoble, 38043 Grenoble Cedex, France

structure, the prokaryotic enzyme displays biochemical properties very similar to that of its mitochondrial counterpart. It must be noted, however, that complex I is not found in some bacterial strains such as *Bacillus subtilis* or in some eukaryote organisms like *Saccharomyces cerevisiae* or *Schizosaccharomyces pombe.*

2 General Features of Complex I

2.1 Structural Organization

Complex I is a multimetric enzyme which contains a large number of cofactors, including FMN, ubiquinone(s) and [Fe-S] clusters. Detailed information regarding the structure and subunit composition of complex I results from studies carried out on bovine and *Neurospora crassa* enzymes. The tridimensional structure, reconstructed from high resolution microscopy analysis by Guénebaut et al. (1997), shows a general L shape for the *N. crassa* enzyme (Fig. 1). The base of the L corresponds to the membranous domain while the long arm points out into the mitochondrial hydrophilic matrix compartment. The NADH substrate binding site should be located in this extrinsic region. Beef heart mitochondrial complex I is a 900 kDa assembly of at least 43 different subunits. Primary amino acid sequences of complex I subunits have been obtained from protein and nucleic acid sequencing. Sequencing of the human mitochondrial DNA (mtDNA) has revealed the presence of open reading frames coding for seven complex I subunits: ND1 to ND6 and ND4L (Chomyn et al. 1985). In addition to

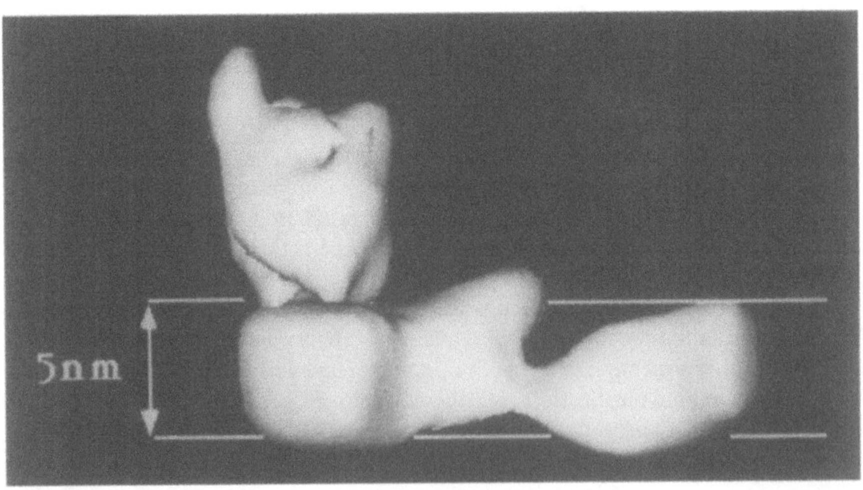

Fig. 1. Tridimensional reconstruction of the *Neurospora crassa* mitochondrial complex I structure. The hydrophobic domain is embedded in the mitochondrial inner membrane indicated by *lines.* The peripheral region points out into the mitochondrial matrix (Guénebaut et al. 1997). This reconstruction was deduced from microscopy data. (Reproduced with the kind authorization of the Journal of Molecular Biology, Academic Press)

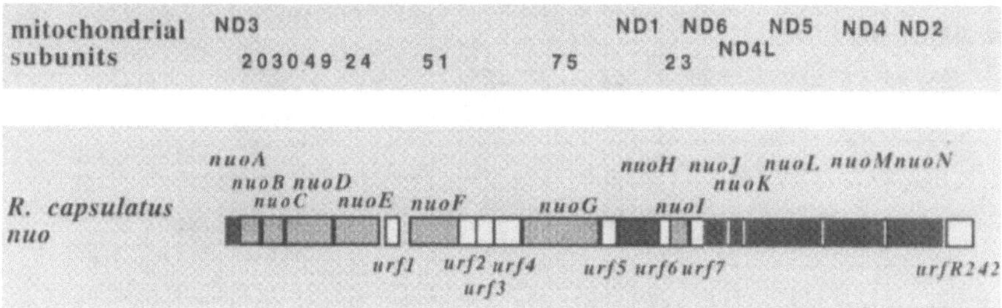

Fig. 2. Structure of the *nuo* operon which encodes the NADH-ubiquinone oxidoreductase in *Rhodobacter capsulatus. Lower part nuoA-nuoN,* structural genes encoding subunits homologous to subunits of the mammalian mitochondrial complex I; *urf* unidentified reading frame. *Upper part* Mitochondrial subunits homologous to the NUO subunits

these ND subunits, 36 subunits of the beef heart mitochondrial complex I have been extensively reviewed by Walker (1992).

As energy coupled NADH-ubiquinone oxidoreductase activity has been described in bacteria, prokaryotic models of complex I were developped to overcome problems due to the structural complexity of the mitochondrial enzyme. A *nuo* operon encoding the bacterial complex I has been identified in *Paracoccus denitrificans* (Yagi et al. 1993), *Escherichia coli* (Weidner et al. 1993), *Rhodobacter capsulatus* (Dupuis 1992; Dupuis et al. 1995) and *Thermus thermophilus* (Yano et al. 1997). The genes clustered in these *nuo* operons code for 14 proteins all homologous to subunits of the mitochondrial enzyme (Fig. 2). The bacterial enzyme therefore appears as a minimal system compared with mitochondrial complex I. On the other hand, the larger number of subunits identified in the mitochondrial enzyme raises the question of the role of these additional subunits. It must be noted that such an increase in complexity is also encountered when comparing bacterial and mitochondrial cytochrome oxidases (complex IV) and ATP synthetases (complex V).

2.2
Roles of the Subunits and Cofactors

All seven subunits encoded by mitochondrial DNA in eukaryotes have their homologues in bacterial complex I. According to secondary structure predictions, the hydrophobic subunits encoded by mtDNA are essential for the organisation of the membranous domain of the enzyme. In respect to the catalytic activity of complex I, these subunits were shown to be involved in the interaction with quinones and inhibitors such as rotenone and piericidin. Due to their membranous localization, these subunits are also likely to participate in the proton-pump mechanism which is coupled to NADH oxidation. The localization of the other subunits is a more difficult task. The 20 kDa (or PSST), 23 kDA (or TYKY) and 24 kDa subunits harbour domains which have a more or less pronounced hydrophobic character, depending on the species, and may be embedded in the membrane or at least be close by. In agreement with this assumption, the NUOI subunit of the *R. capsulatus* complex I, which is homologous to the 23 kDa subunit, was shown to play a critical role in the assembly of the membranous and

```
                                        10        20        30        40        50
R. capsulatus   nuo I          MAFDYVRAAKYFVLWDFIKGFALGMKYFVAPKPTLNYPHEKGPLSPRFRGE
B. taurus    TYKYVNLREPSMD.KSVTD...QTLLWTEL.R.LGMTLS.LFREPA.I...F...........

      60        70        80        90        100       110       120
HALRRYPSGEERCIACKLCEAICPAQAITIDAEPRDDGSRRTTRYDIDMTKCIYCGYCQEACPVDAIVEGPN
...................V........E....A..................F...............

  130       140       150       160
FEYATETREELFYTKEKLLENGARWEAEIARNIEMDAPYR
..FS...H...L.N.....N..DK......A..QA.YL..
```

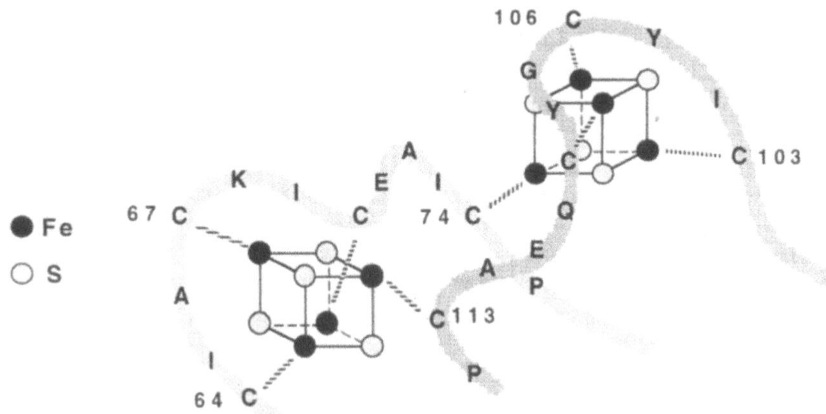

Fig. 3. Sequence comparison between the *R. capsulatus* NUOI subunit and the mitochondrial bovine 23-kDa subunit. Numbering is based on the *R. capsulatus* NUOI sequence. *Dots* correspond with identical amino acids. *Underlined sequences* indicate the two cystein motifs involved in the organization of the [4Fe-4S] clusters in the NUOI subunit. The *lower part* of the figure illustrates a schematic view of the potential structure of these two clusters

extrinsic parts of the enzyme (Chevallet et al. 1997). Using immunomicroscopy observations obtained after immunoreaction, the 49-kDa subunit was localized close to the membrane in the *N. crassa* complex I (Guénebaut et al. 1997). Based on photolabelling experiments, an NADH binding site was assigned to the 51 kDa subunit. This subunit also carries a well conserved motif for an FMN binding site as well as a motif of four cystein residues which is possibly involved in the binding of a [4Fe-4S] cluster. From these data, it was infered that the first catalytic step of the oxidation of NADH takes place at the 51 kDa subunit.

Complex I contains 22 to 24 atoms of iron and of labile sulfur which are organized in [Fe-S] clusters. These centres are characterized by a specific electronic paramagnetic resonance (EPR) signature. Complex I contains at least five EPR detectable [Fe-S] clusters: two binuclear clusters, known as N1a and N1b, and three tetranuclear clusters called N2, N3 and N4. Interestingly, the [Fe-S] centres present in *P. denitrificans* or *R. capsulatus* complex I display EPR spectroscopic features remarkably

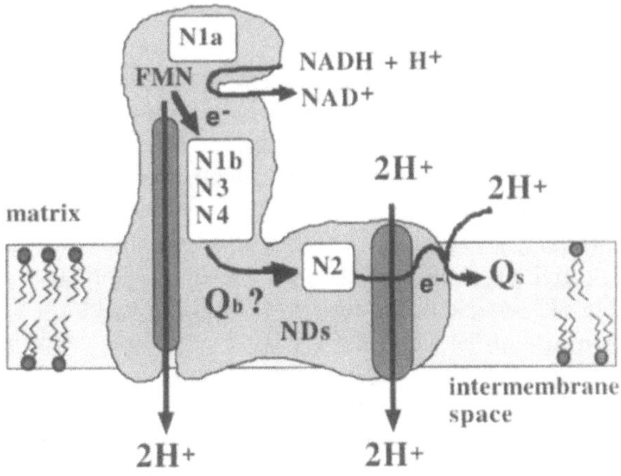

Fig. 4. Schematic view of the structure and mechanism of complex I. *Qb* Bound quinone, *Qs* soluble quinone, *N1a, N1b, N2, N3, N4* [Fe-S] clusters, *NDs* subunits encoded by the mtDNA

similar to those of the mammalian enzyme (Sled' et al. 1993). However, EPR studies allow only a partial analysis of the [Fe-S] cluster content: clusters with a mere structural role may not be detected by this approach (Yagi et al. 1993). From EPR data obtained with isolated subunits or entire complex I and from amino acid sequence analysis, a tentative assignment of the different iron-sulfur clusters has been proposed. In the bovine mitochondrial complex I, a number of nuclear DNA-encoded subunits: the 20-, 23-, 24-, 51- and 75-kDa subunits are believed to contain Fe-S clusters (Albracht and de Jong 1997). Sequence comparisons have shown that the 23-kDa subunit contains two motifs of four cystein residues which are very similar to those described for bacterial ferredoxins (Dupuis 1992). These motifs could accommodate two [4Fe-4S] clusters organized as presented in Fig. 3. A [2Fe-2S] cluster could be present on the 24-kDa subunit while the 75-kDa subunit could possibly organize two [4Fe-4S] clusters and one [2Fe-2S] cluster. However, assigning the different Fe-S clusters and defining their role in the electron transfer process remain tentative. Figure 4 presents a very schematic view of the general structure and mechanism of complex I (for a comprehensive review see Brandt 1997).

3 Inhibitors of Complex I

Until recently, the list of complex I inhibitors was limited to a small number of inhibitors such as rotenone, piericidin or amytal. During the past few years, a large number of new molecules showing complex I inhibitory properties have been described: methylphenylpyridinium derivatives, oxycarbocyanine laser dyes, quinazoline ethers such as fenazaquine, quinazolamines, pyridinamines and various natural compounds such as A and B aurachines, aureothine, capsaicin, acetogenins (1- and 2-rolliniastatines, otivarine, cherimolline), PI and PL myxalamides, thiangazole, phenoxan and phenylamide A (for review, see Hollingworth and Ahammadsahib 1995; Degli-Esposti 1998). Most of these molecules have insecticide and acaricide properties. This explains the strong interest which is shown by agrochemical companies in the development of such new molecules. These molecules generally display a strong hydrophobic character and their formulae are often characterized by the presence of a polycyclic structure related to quinones. At least two quinone binding sites are thought to be present in complex I, a situation which has already been described for complex III (bc_1 complex) of the mitochondrial respiratory chain. All these inhibitors would thus interact with complex I at the quinone binding sites. They have been grouped into at least two classes by a mutual exclusion criterion.

Rotenone is a flavonoid derivative which is extracted from roots of *Derris*-type shrubs. It was historically renowned for its toxicity toward fishes. Despite its poor selectivity and reduced stability, rotenone is still largeley commercialized as an "ecological" pesticide. However, this labelling should not hide the potential risk of long, term use of rotenone. On the basis of photolabelling experiments carried out with dihydro-rotenone, and analog of rotenone, this inhibitor was reported to interact with the ND1 subunit of complex I (Earley et al. 1987). Binding of rotenone to complex I leads to an almost complete inhibition of the NADH dependent respiration of mitochondrial membranes (I_{50} = 20 nM). Besides this major effect, other toxic manifestations of complex I inhibition have been described: (1) rotenone at a concentration of 5 μM induces apoptosis of HL60 cells (Matsunaga et al. 1996), (2) intracerebral injection of high doses of rotenone in mice causes a necrosis of the nigrostriatum associated with a syndrome similar to Parkinson's disease. This effect resembled that reported for MPTP (methyl-phenyl-tetrahydropyridine). MPTP is a drug contaminant which induces a Parkinson's-like syndrome in drug addicts (Ramsay et al. 1991). MPTP is metabolized by mono-amine-oxidase into MPP^+ (methyl-phenyl-pyridinium) in dopaminergic neurons. Due to both a positive charge and an amphoteric character, MPP^+ specifically accumulates in mitochondria, where despite a lower affinity for complex I than rotenone, MPP^+ inhibits complex I activity. Finally, piericidin (isolated from *Streptoverticillium mobaraensis)* is a powerful inhibitor of complex I (I_{50} = 1 nM) but, like rotenone, displays a rather low species selectivity. The characterization of piericidin-resistant mutants of *R. capsulatus* has recently shown that the piericidin binding site maps to the 49-kDa subunit of complex I (Darrouzet et al. 1998). Efforts have thus been engaged to produce new complex I inhibitors with increased species specificity, like fenazaquin, which is now commercialized as a pesticide. Surprisingly, some potent complex I inhibitors have been found in food, such as capsaicin in hot pepper and acetogenins in papaya fruit.

Some inhibitors of complex I such as acetogenins have also exhibited antitumoral properties. However, the antiproliferative effect which has been observed *in vitro* as well as in vivo (xenografts of ovarian tumors in mice) has been shown not to be attributed to the inhibitory effect on complex I. In fact, these compounds might act by chelating calcium. An antipromoter activity of rotenone responsible for its antitumoral activity has also been reported in mice. Noticeably, 15 out of 35 compounds showing antitumoral activity and identified during a survey by the National Cancer Institute (USA) proved to have inhibititory effects on mitochondrial respiration. From a theoretical point of view, such molecules can be of potential clinical use if associated with other classical antiproliferative drugs. As they target the cellular energetic machinery, they are believed to decrease the efficiency of ATP-dependent multidrug resistance (MDR) mechanisms.

Physiopathological consequences of the inhibition of human mitochondrial complex I can be very severe. Pathological symptoms are observed as soon as a tissue is damaged by a toxic molecule which targets complex I or as a direct consequence of genetic mutations affecting the genes encoding complex I subunits. The biogenesis of mitochondrial complex I depends on both the nuclear and mitochondrial genomes. This situation, which is also shared by mitochondrial respiratory complexes III, IV and V, increases the risk of functional deficiencies in the respiratory complexes. A dramatic example which illustrates the importance of mitochondrial function is that of blindness caused by instillation of chloramphenicol (an inhibitor of mitochondrial translation) into the ocular globe.

On the other hand, functional anatomy of mitochondria can lead to disorders whose consequences have to be evaluated. In this respect, the physiological activity of the respiratory chain leads to the synthesis of semi-quinones which are a potential source of reactive oxygen species (ROS; Papa and Skulachev 1997). Any dysfunction in complex I may thus produce deleterious chemical species for the mitochondria (Robinson 1998). This deleterious effect is amplified by mitochondrial features such as: the absence of protection of mitochondrial DNA (mtDNA) by proteins such as histones, the vicinity of mtDNA to the respiratory chain, a replication mechanism involving single-stranded DNA, limited mtDNA repair mechanisms and the low efficiency of γ DNA polymerase. For all these reasons, mtDNA is a privileged target for ROS. This phenomenon also seems to take part in the physiological process of ageing (Shinenaga et al. 1994).

4 Localization and Structure of the Human Complex I Genes

In mammals, the seven ND subunits are encoded by the mitochondrial genome. It must be noted, however, that in other eukaryote organisms such as in plants, para-

Table 1. Chromosomal localization of human complex I genes. Detail information on these genes is available at http://www-dsv.cea.fr/MitoPick/Default.html

Subunit name	Other names	Chromosomal localization	References
ND1 to ND6 ND4L		mtDNA	Chomyn et al. (1985)
10 kDa	NDUFV3, NUOM	21q22.3	De Coo et al. (1997)
20 kDa	NDUFS7, PSST, NUKM	19p13.3	Hyslop et al. (1996)
23 kDa	NDUFS8, NUIM	11q13	Procaccio et al. (1997)
24 kDa	NDUFV2, NUHM	18p11.3	De Coo et al. (1995);
	NDUFV2P1 (pseudogene)	19q13.3–13.4	Hattori et al. (1995)
39 kDa	NDUFS2L, NUEM	12p13	Baens et al. (1993)
49 kDa	NDUFS2, NUCM	1q23	Procaccio et al. (1998)
51 kDa	NDUFV1, NUBM	11q13	Spencer et al. (1992); Ali et al. (1993)
75 kDa	NDUFS1, NUAM	2q33–34	Duncan et al. (1992)
B8	NDUFA2, NI8M	5q31	Dunbar et al. (1997)
B13	NDUFA5, NUFM	7q32	Pata et al. (1997);
	NUFM pseudogene	11p15	Russel et al. (1997)
B22	NDUFB9, NI2M	8q13.3	Gu et al. (1996)
MWFE	NDUFA1, NIMM	Xq24	Zhuchenko et al. (1996); Frattini et al. (1997)

mecia, trypanosomes and in the mold *Dictyostelium discoideum*, the mitochondrial genomes encode a different number of complex I subunits. On the other hand, the bovine nuclear genome encodes at least 36 different subunits. To date, if the cDNA sequences of the human genes are known (see http://www-dsv.cea.fr/MitoPick/Default.html), only a fraction of the genes of the human complex I have been assigned to chromosomes and in most cases their structure is yet to be defined (Table 1). The gene encoding the 51-kDa subunit has been localized to chromosome 11q13 at 30 kbp from the P1-1 gene encoding glutathione S-transferase. Interestingly, the gene encoding the 23-kDa subunit has also been localized to the same region. However, the estimated distance between the two genes definitely excludes the presence of a complex I gene cluster. All genes studied so far are present in one copy and no isoforms have been described. It must be noted however, that two pseudogenes localized to chromosomes 19q13 and 11p15 respectively, have been characterized for subunits 24 kDa and B13.

5 Molecular Basis of Pathologies Associated with Complex I Deficiencies

5.1 Investigation of Mitochondrial Human Pathologies

Defects in the mitochondrial respiratory chain are responsible for various human pathologies grouped under the general name of mitochondrial cytopathies. These pathologies are mostly but not exclusively neuro-muscular diseases. Genetic studies of these diseases are hampered by the fact that both nuclear and mitochondrial genomes can be involved. Another general problem often encountered when studying mitochondrial pathologies is the large gap between available molecular and cellular data and the clinical observations of patients. In typical cases, a clear maternal inheritance of the pathology is a strong argument for associating the disease with an alteration of the mitochondrial genome. In most situations, however, the picture is not as clear, and more investigations are required to associate the mitochondrial phenotype with a specific molecular defect. The analysis of the molecular basis of pathologies characterized by a complex I deficiency uses various investigative approaches: measurement of the cellular redox-state, histological studies, immunochemistry analysis, mitochondrial or permeabilized fiber oxygraphy, determination of isolated enzymatic activities, spectroscopic measurements. NMR investigations and molecular genetic analyses.

Early immunological studies showed a modification of the polypeptidic composition of mitochondria isolated from biopsies from patients presenting complex I deficiencies. A selective absence of 13- and 75-kDa subunits has been reported in patients showing a severe lactic acidosis (Moreadith et al. 1987), while absence of the 20- or 24-kDa subunits has also been described in other patients. Genetic studies aimed at the characterization of molecular defects associated with mitochondrial cytopathies in general, and with complex I deficiencies in particular, have largely been developed since the first report of a mitochondrial DNA mutation in 1988 (Holt et al. 1988; Lestienne and Ponsot 1988). As it could be observed in the case of nuclear mutations (1) distinct mutations in the same mitochondrial gene could be responsible for distinct pathologies. (2) distinct mutations in different mitochondrial genes could be associated with the same disease phenotype, and (3) the same mutation could lead to various pathological pictures which differ by variations in disease severity and age of onset. For example, two mutations in the *ndI* gene have been associated with non insulino dependent diabetes (NIDDM) and with a predisposition to develop Alzheimer or Parkinson's diseases, whereas mutations at nucleotides 3460 *(ndI)* and 11778 *(nd4)* are associated with Leber's disease. On the other hand, mutation of nucleotides 5460 *(nd2),* 11084 *(nd4)* and 14459 *(nd6)* have been identified in patients suffering respectively from Alzheimer's disease, from myo-encephalopathy with lactic acidosis and strokes (MELAS) or from a complex pathology combining Leber's disease and dystonia.

5.2 Biological Models for Studying Mutations

5.2.1 The Transmitochondrial Cell Line Model

LHON (Leber's hereditary optical neuropathy) is one of the most documented diseases at the DNA level. No less than 18 different mutations have been associated with Leber's disease (Table 2). All mutations affect subunits of respiratory complexes encoded by mtDNA and 11 of them specifically affect complex I genes. When analyzing mtDNA mutations, one of the main problems is to demonstrate a relationship between the mutation and a direct implication in the pathological process. Different indirect criteria may be used to establish the causality of a mutation: (1) does the mutation segregate with the pathology? (2) does the mutation affect a well conserved amino acid? (3) does the mutation induce a drastic amino acid change? (4) does the mutation appear as a rare polymorphism? However, in the absence of a definite functional proof, one has to be cautious when considering the direct causality of a mutation. To further complicate the problem, in some families pathological pictures and analysis of the mode of inheritance have suggested that a nuclear genetic component might be involved.

Nevertheless, a tentative classification of the mutations associated with LHON as primary or secondary mutations has been proposed. Mutations *ndI-3460, nd4-11778* and *nd6-14484* are considered to be primary mutations i.e. mutations whose presence is sufficient *per se* to cause the disease. These mutations are found in 80% of caucasian patients suffering from the disease. In order to analyse the respiratory and biochemical phenotypes associated with these mutations, an approach based on the analysis of transmitochondrial cell lines has been developed (King and Attardi 1989). The functional impact of the *nd4-11778* mutation was studied in cell lines bearing the mutation at a homoplasmic state. The cell lines were isolated after injection of mitochondria harbouring the *nd4-11778* mutation purified from patients' muscle biopsies into a ϱ° cell line devoid of mtDNA. The digitonin-permeabilized cells showed a significant decrease in NADH supported oxygen consumption. However, the rotenone sensitive NADH dehydrogenase activity of this *nd4-11778* transmitochondrial cell line remained unchanged compared with that of normal cell lines when measurements were carried out on purified mitochondria membranes using decylubiquinone as the

Table 2. Mutations of mitochondrial genes encoding complex I subunits and associated with a pathology

Mutation position	Affected subunit	Associated pathology	Mutation nature	Amino acid change	*R. capsulatus* conservation	Primary mutation
3316	ND1	NIDDM	G → A	A → T	no	?
3394	ND1	Leber	T → C	Y → H	no	?
3397	ND1	AD/PD	A → G	M → V	no	no
3460	ND1	Leber	G → A	A → T	yes	yes
4136	ND1	Leber	A → G	Y → C	no	?
4160	ND1	Leber	T → C	L → P	yes	yes
4216	ND1	Leber	T → C	Y → H	no	no
4917	ND2	Leber	A → G	D → N	no	no
5244	ND2	Leber	G → A	G → S	yes	no
5460	ND2	AD	G → A	A → T	no	no
		AD	G → T	A → S	no	no
11084	ND4	MELAS	A → G	T → A	no	?
11778	ND4	Leber	G → A	R → H	yes	yes
13708	ND5	Leber	G → A	A → T	yes	no
13730	ND5	Leber	G → A	G → E	yes	yes
14459	ND6	LDYT	G → A	A → V	no	?
14484	ND6	Leber	T → C	M → V	no	yes

NIDDM: non-insulino-dependent diabetes, AD: Alzheimer's disease, PD: Parkinson's disease. MELAS: mitochondrial encephalopathy with lactic acidosis and stroke-like episodes, LDYT: Leber's disease and dystonia.
Updated data and information on mutations may be found at web sites: Mitomap (http://infinity.gen.emory.edu/) and OMIM (http:///www3.ncbi.nlm.nih.gov/omim/)

substrate (Hofhaus et al. 1996). Another group reported that the *nd4-11778* mutation is associated with a 50% decrease in the catalytic rate constant of NADH-decylubiquinone reductase activity in mitochondrial membranes isolated from a lymphoblastoid cell line carrying the *nd4-11778* mutations (Majander et al. 1996). Mitochondrial respiration with 2-oxoglutarate and L-malate was only decreased by 23% and the efficiency of oxidative phosphorylation was unaffected in their mutant cell line. It must be noted however, that this cell line also carried both the secondary *ndl-4216* and *nd5-13708* mutations, which have been shown to affect the rate of cell respiration in cybrid cells harbouring the mtDNA *nd4-11778* mutation. Production of transmitochondrial cell lines appears to be a promising tool with some drawbacks: obtaining a large homogeneous population of mutated mitochondria remains difficult: the transmitochondrial cell lines are relatively unstable and the nuclear background of the host cells may possibly interfere with the study of specific mtDNA mutations.

5.2.2
R. capsulatus Model

To overcome these problems and since no animal model of a mitochondrial disease caused by a respiratory chain deficiency is available to date, we have developped an enzymatic model to study complex I deficiencies in the purple photosynthetic bacterium *R. capsulatus*. Some of the human mutations affecting complex I may be transfered into the homologous bacterial enzyme and the biochemical consequences of the mutation may be studied.

Sequence comparisons between the human ND4 subunit and the *R. capsulatus* NUOM subunit have shown that the ND4 Arg340 residue is conserved in *R. capsulatus* and corresponds to the NUOM Arg368 residue. As a first step to validate our model, we mutagenized the *R. capsulatus nuo* operon in order to generate a *nuoM-1103* point mutant equivalent to the *nd4*-11778 mutation of the human mtDNA which leads to the Arg340His change. This mutant showed a reduced ability to grow in medium containing malate as a carbon source which indicated a clear impairment of its oxidative phosphorylation capacity. The NADH-respiration of porous bacterial cells was significantly decreased in the *nuoM-1103* mutant, while no significant reduction could be observed in isolated bacterial membranes. One possible explanation of this discrepancy is that the change of the arginine residue for another amino acid could alter the NADH chanelling between matricial dehydrogenases and complex I. As respiration of isolated bacterial membranes is measured in the presence of an excess amount of NADH, the effect of the *nuoM-1103* mutation may go undetected in this assay. Another possibility to consider is that during membrane preparation certain components of the respiratory complexes have been altered. This would result in a limiting step in the electron transfer pathway which would minimize the differences between mutant and normal membranes. Interestingly, it has been observed in the case of the *nd4-11778* mitochondrial mutation that proton-pump activity of the bacterial enzyme was not affected by the *nuoM-1103* mutation. All these results reproduce most of the biochemical features observed in patient mitochondrial harbouring the *nd4-11778* mutation (Majander et al. 1991; Hofhaus et al. 1996) and validate *R. capsulatus* complex I as a valuable model for investigating mutations of mtDNA associated with complex I deficiencies in human pathologies (Lunardi et al. 1998). Likewise, analysis of the pathogenetic mitochondrial mutation *ndI-3460*, which also causes Leber's disease, and using the bacterium *P. denitrificans* has shown that the mutated residue plays an important role in ubiquinone reduction by complex I (Zickermann et al. 1998).

6
Conclusions

Mitochondrial complex I plays a major role in cellular energetics. Therefore, functional alterations of human complex I result in a large spectrum of pathological symptoms depending on the severity of the alteration or on the affected tissue. Pathological modifications of complex I result in either (1) a direct effect on the cellular energy supply due to a decreased mitochondrial ATP synthesis; or (2) in metabolic consequences: glycolysis for example is diverted towards the production of lactic acid leading to cellular acidification; or (3) an increased production of reactive oxygen species which may target cellular components. A large number of mutations have been identified in the mitochondrial genes encoding the complex I subunits. One major problem is to define the precise involvement of the molecular defects in the disease etiology. New approaches which include transmitochondrial cell lines, development of transgenic animal models of pathologies and use of bacterial models to reproduce mitochondrial mutations, are necessary steps in understanding the molecular mechanisms underlying mitochondrial pathologies and in developing rationalized therapeutics. In addition to mutations of mtDNA, possible molecular defects affecting nuclear genes which encode most of the complex I subunits have to be considered. The recent identification of a mutation in a nuclear gene encoding a subunit of the mitochondrial complex II in patients suffering from Leigh's syndrome (Bourgeron et al. 1995) and in the gene encoding the 18-kDa subunit of complex I (van den Heuvel et al. 1998) clearly shows that the nuclear genes should not be overlooked. Today, our knowledge of the different nuclear genes and of their regulatory elements is still poor. There is no doubt, however, that considering the large number of complex I subunits encoded by the nuclear genome, some complex I deficiencies will be associated with mutations in these genes.

Acknowledgements. We thank Dr. A. Dupuis for helpful discussions and Dr. A. Fuchs for critical reading of the manuscript.

References

Albracht SPJ, de Jong AM (1997) Bovine-heart NADH: ubiquinone oxidoreductase is a monomer with 8 Fe-S clusters and 2 FMN groups. Biochim Biophys Acta 1140:105–134

Ali ST, Duncan AMV, Schappert K, Heng HHQ, Tsui LC, Chow W, Robinson BH (1993) Chromosomal localization of the human gene encoding the 51-kDa subunit of mitochondrial complex I (NDUFV1) to 11q13. Genomics 18:435–439

Baens M, Chaffanet M, Cassiman J-J, van den Berghe H, Marynen P (1993) Construction and evaluation of a hncDNA library of human 12p transcribed sequences derived from a somatic cell hybrid. Genomics 16:214–218

Bourgeron T, Rustin P, Chretien D, Birch-Machin M, Bourgeois M, Viegas-Péquignot E, Munnich A, Rötig A (1995) Mutation of a nuclear succinate dehydrogenase gene results in mitochondrial respiratory chain deficiency. Nat Genet 11:144–149

Brandt U (1997) Proton translocation by membrane-bound NADH:ubiquinone-oxidoreductase (complex I) through redox-gated ligand conduction. Biochim Biophys Acta 1318:79–91

Chevallet M, Dupuis A, Lunardi J, Van Belzen R, Albracht SPJ, Issartel JP (1997) The NUOI subunit of the *Rhodobacter capsulatus* respiratory complex I (equivalent to the bovine TYKY subunit) is required for proper assembly of the membraneous and peripheral arms of the enzyme. Eur J Biochem 250:451–458

Chomyn A, Mariottini P, Cleeter MWJ, Ragan CI, Matsuno-Yagi A, Hatefi Y, Doolittle RF, Attardi G (1985) Six unidentified reading frames of human mitochondrial DNA encode components of the respiratory-chain NADH dehydrogenase. Nature 314:592–597

Darrouzet E, Issartel J-P, Lunardi J, Dupuis A (1998) The 49 kDa subunit of NADH-ubiquinone oxidoreductase (complex I) is involved in the binding of piericidin and rotenone, two quinone-related inhibitors. FEBS Lett 431:34–38

De Coo R, Buddiger P, Smeets H, van Kessel AG, Morgan-Hughes J, Wehuis DO, Overhauser J, van Oost B (1995) Molecular cloning and characterization of the active human mitochondrial NADH:ubiquinone oxidoreductase 24 kDa gene (NDUFV2) and its pseudogene. Genomics 26:461–466

De Coo R, Buddiger P, Smeets H, Van Oost B (1997) Molecular cloning and characterization of the human mitochondrial NADH:oxidoreductase 10 kDa gene (NDUFV3). Genomics 45:434–437

Degli-Esposti M (1998) Inhibitors of NADH-ubiquinone reductase: an overview. Biochim Biophys Acta 1364:222–235

Dunbar DR, Shibasaki Y, Dobbie L, Andersson B, Brookes AJ (1997) In situ hybridisation mapping of genomic clones for five human respiratory chain complex I genes. Cytogenet Cell Genet 78:21–24

Duncan AMV, Chow W, Robinson BH (1992) Localization of the human 75 kDa I Fe-S protein of NADH-coenzyme Q reductase gene (NDUFS1) to 2q33→q34. Cytogenet Cell Genet 60:212–213

Dupuis A (1992) Identification of two genes of *Rhodobacter capsulatus* coding for proteins homologous to the ND1 and 23 kDa subunits of the mitochondrial complex I. FEBS Lett 30:215–218

Dupuis A, Peinnequin A, Chevallet M, Lunardi J, Darrouzet E, Pierrard B, Procaccio V, Issartel JP (1995) Identification of five *Rhodobacter capsulatus* genes encoding the equivalent of ND subunits of the mitochondrial NADH-ubiquinone oxidoreductase. Gene 167:99–104

Earley FG, Patel SD, Ragan CI, Attardi G (1987) Photolabeling of a mitochondrially-encoded subunit of NADH dehydrogenase with [^{3}H]dihydrorotenone. FEBS Lett 219:110–113

Frattini A, Faranda S, Bagnasco L, Patrosso C, Nulli P, Zucchi I, Vezzoni P (1997) Identification of a new member (ZNF183) of the Ring finger gene family in Xq24–25. Gene 192:291–298

Gu JZ, Lin X, Wells DE (1996) The human B22 subunit of the NADH:ubiquinone oxidoreductase maps to the region of chromosome 8 involved in branchio-oto-renal syndrome. Genomics 35:6–10

Guénebaut V., Vincentelli R, Mills D, Weiss H, Leonard KR (1997) Three-dimensional structure of NADH-dehydrogenase from *Neurospora crassa* by electron microscopy and conical tilt reconstruction. J Mol Biol 265:409–418

Hatefi M, Haavik AG, Griffiths DE (1962) Studies of the electron transfer system. XL. Preparation and properties of mitochondrial DPNH-coenzyme Q reductase. J Biol Chem 237:1676–1680

Hattori N, Suzuki H, Wang Y, Minoshima S, Shimizu N, Yoshino H, Kurashima R, Tanaka M, Ozawa T, Mizuno Y (1995) Structural organization and chromosomal localization of the human nuclear gene (NDUFV2) for the 24 kDa iron-sulfur subunit of complex I in mitochondrial respiratory chain. Biochem Biophys Res Commun 216:771–777

Hofhaus G, Johns D, Hurko O, Attardi G, Chomyn A (1996) Respiration and growth defects in transmitochondrial cell lines carrying the 11778 mutation associated with Leber's hereditary optic neuropathy. J Biol Chem 271:13155–13161

Hollingworth RM, Ahammadsahib KI (1995) Inhibitors of respiratory complex I: mechanisms, pesticidal actions and toxicology. Rev Pestic Toxicol 3:277–302

Holt IJ, Harding AE, Morgan-Hughes JA (1988) Deletions of muscle mitochondrial DNA in patients with mitochondrial myopathies. Nature 331:717–719

Hyslop SJ, Duncan AMV, Pitkänen S, Robinson BH (1996) Assigment of the PSST subunit gene of human mitochondrial complex I to chromosome 19p13. Genomics 37:375–380

King M, Attardi G (1989) Human cells lacking mtDNA: repopulation with exogenous mitochondria by complementation. Science 246:500–503

Lestienne P, Ponsot G (1988) Kearn-Sayres syndrome with muscle mitochondrial DNA deletion. Lancet 1:885

Majander A, Huoponen K, Savontaus ML, Nikoskelainen E, Wikström M (1991) Electron transfer properties of NADH:ubiquinone reductase in the Nd1/3460 and the Nd4/11778 mutations of the Leber hereditary optic neuroretinopathy (LHON). FEBS Lett 292:289–292

Majander A, Finel M, Savontaus MJ, Nikoskelainen E, Wikström M (1996) Catalytic activity of complex I in cell lines that possess replacement mutations in the ND genes in Leber's hereditary optic neuropathy. Eur J Biochem 239:201–207

Matsunaga T, Kudo J, Takahashi K, Dohmen K, Hayashida K, Okamura S, Ishibashi H, Niho Y (1996) Rotenone, a mitochondrial NADH dehydrogenase inhibitor, induces cell surface expression of CD13 and CD38 and apoptosis in HL-60 cells. Leuk Lymphoma 20:487–494

Moreadith RW, Cleeter MWJ, Ragan CI, Batshaw ML, Lehninger AL (1987) Congenital deficiency of two polypeptide subunits of the iron-protein fragment of mitochondrial complex I. J Clin Invest 79:487–494

Papa S, Skulachev VP (1997) Reactive oxygen species, mitochondria, apoptosis and aging. Mol Cell Biochem 174:305–319

Pata I, Tensing K, Metspalu A (1997) A human cDNA encoding the homologue of NADH:ubiquinone oxidoreductase subunit B13. Biochim Biophys Acta 1350:115–118

Procaccio V, Depetris D, Soularue P, Mattei MG, Lunardi J, Issartel JP (1997) cDNA sequence and chromosomal localization of the NDUFS8 human gene coding for the 23 kDa subunit of the mitochondrial complex I. Biochim Biophys Acta 1351:37–41

Procaccio V, de Sury R, Martinez P, Depetris D, Rabilloud T, Soularue P, Lunardi J, Issartel JP (1998) Mapping to 1q23 of the human gene (NDUFS2) encoding the 49 kDa subunit of the mitochondrial respiratory complex I and immunodetection of the mature protein in mitochondria. Mammal Genome 9:482–484

Ramsay RR, Krueger MJ, Youngster SK, Gluck MR, Casida JE, Singer TP (1991) Interaction of 1-methyl-4-phenylpyridinium ion (MPP+) and its analogs with the rotenone/piericidin binding site of NADH dehydrogenase. J Neurochem 56:1184–1190

Robinson BH (1998) Human complex I deficiency:clinical spectrum and involvement of oxygen free radicals in the pathogenicity of the defects. Biochim Biophys Acta 1364:271–286

Russell MW, duManoir S, Collins FS, Brody LC (1997) Cloning of the human NADH:ubiquinone oxidoreductase subunit B13: localization to chromosome 7q32 and identification of a pseudogene on 11p15. Mammal Genome 8:60–61

Shinenaga MK, Hagen TM, Ames BM (1994) Oxidative damage and mitochondrial decay in aging. Proc Natl Acad Sci USA 91:10771–10778

Sled' VD, Friedrich T, Leif H, Weiss H, Meinhardt SW, Fukumori Y, Calhoun MW, Gennis RB, Ohnishi T (1993) Bacterial NADH-quinone oxidoreductases: iron-sulfur clusters and related problems. J Bioenerg Biomembr 25:347–356

Spencer SR, Taylor JB, Coll IG, Xia CL, Pemble SE, Ketterer B (1992) The human mitochondrial NADH.ubiquinone oxidoreductase 51-kDA subunit maps adjacent to the glutathione S-transferase P1-1 gene on chromosome 11q13. Genomics 14:1116–1118

van den Heuvel L, Ruitenbeek W, Smeets R, Gelman-Kohan Z, Elpeleg O, Loeffen J, Trijbels F, Mariman E, de Bruijn D, Smeitink J (1998) Demonstration of a new pathogenic mutation in human complex I deficiency: a 5-bp duplication in the nuclear gene encoding the 18-kD (AQDQ) subunit. Am J Human Genet 62:262–268

Walker JE (1992) The NADH:ubiquinone oxidoreductase (complex I) of respiratory chains. Q Rev Biophys 25:253–324

Weidner U, Geier S, Ptock A, Friedrich T, Leif H, Weiss H (1993) The gene locus of the proton-translocating NADH-ubiquinone oxidoreductase in *Escherichia coli.* Organization of the 14 genes and relationship between the derived proteins and subunits of mitochondrial complex I. J Mol Biol 233:109–122

Yagi T, Yano T, Matsuno-Yagi A (1993) Characteristics of the energy-transducing NADH-quinone oxidoreductase of *Paracoccus denitrificans* as revealed by biochemical, biophysical and molecular biological approaches. J Bioenerg Biomembr 25:339–345

Yano T, Chu SS, Sled' VD, Ohnishi T, Yagi T (1997) The proton-translocation NADH-quinone oxidoreductase (NDH-1) of thermophilic bacterium *Thermus thermophilus* HB-8. J Biol Chem 272:4201–4211

Zhuchenko O, Wehnert M, Bailey J, Sun ZS, Lee CC (1996) Isolation, mapping, and genomic structure of an X-linked gene for a subunit of human mitochondrial complex I. Genomics 37:281–288

Zickermann V, Barquera B, Wikström M, Finel M (1998) Analysis of the pathogenic human mitochondrial mutation ND1/3460, and mutations of strictly conserved residues in its vicinity, using the bacterium *Paracoccus denitrificans.* Biochemistry 37:11792–11796

CHAPTER 6

Complex II or Succinate: Quinone Oxidoreductase and Pathology 6

P. Lestienne[1] and C. Desnuelle[2]

Contents

1 Introduction

Located across the Krebs cycle and the respiratory chain, complex II or succinate:quinone oxidoreductase (EC 1.3.5.1) is composed of four nuclear encoded protein subunits. It catalyses the oxidation of succinate into fumarate by reducing the covalently bound FAD (flavine adenine dinucleotide) on its major subunit, the flavoprotein, and by transferring two electrons from $FADH_2$ to the membrane, forming ubiquinone, or to dyes such as ferricyanide and phenazine methane sulfate (PMS). If the electron acceptor is artificial, the oxidation of succinate into fumarate may be catalysed by the hydrosoluble complex (succinate dehydrogenase, or SDH) devoid of its membrane-bound subunits. However, only the association with these membrane bound subunits within mitochondria allows its physiological function of electron transfer from the succinate to ubiquinone (Hederstedt and Ohnishi 1992).

[1] E 99-29 INSERM Université Bordeaux 2, 146 rue Léo Saignat, 33076 Bordeaux, France
[2] UMR 6494 CNRS Faculté de médecine de Nice, 06107 Nice Cedex 2, France

2 Enzymatic Properties

FAD accepts electrons from succinate in a two-step mechanism involving semiquinone flavine, which is stable enough to allow the reduction of iron-sulfur centers by sequential steps with one electron. The mean redox potential is modulated by substrate fixation.

Succinate, D and L malate and L chlorosuccinate are substrates of SDH, while malonate (Schultz et al. 1996) and oxaloacetate are powerful competitive inhibitors (Hatefi 1985). The affinity of the enzyme for its substrate depends on its oxidized or reduced state; for example the dissociation constant of succinate is about 10 μM with the oxidized enzyme, while it is 160 μM for the reduced enzyme (Kotlyar and Vinogradov 1984). Malonate bound to the enzyme's active center could mimic the transition state of succinate/fumarate. However, its binding is not significantly modified by the oxido-redox state of the enzyme.

The oxidation of succinate into fumarate is a trans-hydrogenation which explains why only the L stereoisomer of chlorosuccinate is a substrate. Malate is not a powerful substrate since it is oxidized into enol-oxaloacetate within the enzyme's active center, and its product induces the irreversible inhibition of enzymatic activity, which may compromise the enzymatic dosage. Bound oxaloacetate may be eliminated by reduction of the enzyme with organic ions at elevated temperatures.

Bovine heart SDH is very sensitive to thiol-reacting chemicals such as N-ethylmaleimide or para-chloromercuribenzoate. The substrate protects against their action, indicating that a cystein is located close to, or within the active center, binding the dicarboxylate substrates. The high reactivity of the thiol is in fact due to the proximity of an arginine whose conversion into leucine or into histidine strongly reduces the enzymatic activity to below 2% (Hederstedt and Ohnishi 1992). A triad His-Pro-Thr is conserved as well, as in other oxidoreductases. The mutation of histidine into serine prevents succinate oxidation, but allows 70% of the reducing activity of fumarate, indicating that the imidazol of the histidine is important for substract in the proton from succinate, and that another residue such as serine can function as proton donor for the fumarate reduction.

Several inhibitors play a role at the level of interaction between the membrane subunits and the quinones, such as thenoytrifluoroacetone, carboxanilides (active on eucaryote enzymes) and a series of 2-alkyl-4,6-dinitrophenols which are active with the *E. Coli* enzyme and on complex I and III (Yankovskaya et al. 1996). The tumor necrosis factor alpha prevents electron transfer at the level of this complex and at those of complexes I and IV as well (Jia et al. 1996), and the 3-nitropropionate (3-NPA) is an irreversible inhibitor (Shimano et al. 1995; Fink et al. 1995). The treatment of rat mesencephalic cells by this drug or by malonate induces a dose-dependent inhibition of dopamine import. Treatment by the 3-NPA induces a reduction of GABA cells, while malonate does not significantly reduce GABA incorporation (Zeevalk et al. 1995).

More recently, Kanedo et al. (1995) has shown that the beta amyloid protein located within senile plaques of Alzheimer's patients inhibits SDH activity (Kanedo et al. 1995), suggesting a tiny relationship between mitochondrial dysfunctions and Alzheimer's disease (cf. Chap. 16, this Vol.).

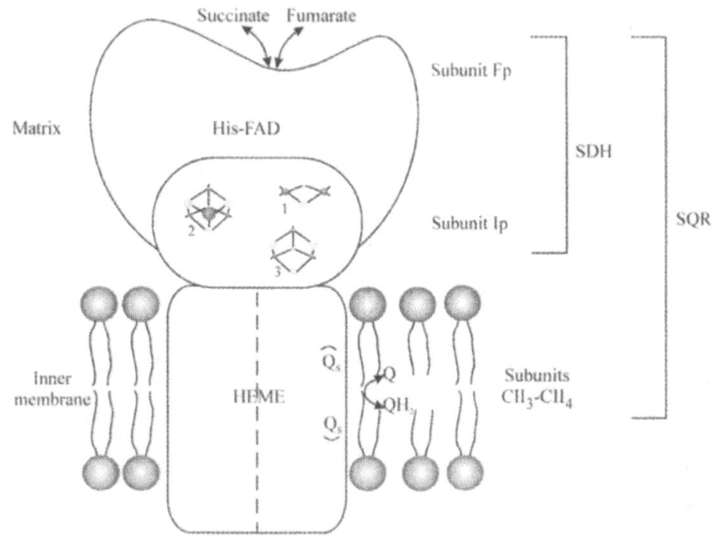

Fig. 1. Schematic representation of complex II. *SQR* and *SDH* are the succinate quinone:oxidoreductase and succinate dehydrogenase complexes respectively

3
Structure

Treatment of succinate:ubiqinone oxidoreductase with chaotropic agents such as 0.5 to 1 M $NaClO_4$ allows selective extraction of SDH, a hydrosoluble protein composed of two subunits of 27 and 70 KDa (Fig. 1). The separation of these subunits under non-denaturing conditions showed that the larger one contains a covalently bound FAD, and that the smaller peptide contains three iron-sulfur centers. This basic structure has been found in all procaryotic and eucaryotic organisms. Complex II preparations also contain two small membrane-anchored polypeptides with a type b_{560} cytochrome, distinct from the cytochrome b of complex III which is mitochondrially encoded.

3.1
The Flavoprotein Subunit Fp

It has been shown that FAD is covalently linked to Fp by a histidyl N(3)-8 α riboflavine bond (Robinson and Lemire 1996a). The FAD and its analogues may be covalently bound to the polypeptide Fp before the association with Ip and may be membrane-anchored (Robinson and Lemire 1996b); the reaction requires intra-mitochondrial import, presequence cleavage and the normal folding of the polypeptide. FAD binding is stimulated by intermediates of the Krebs cycle such as fumarate, succinate and malate. The mutation of His 90 into Ser in Fp from *Saccharomyces cerevisiae* leads to the non-covalent binding of FAD, and an assembled but non-functional complex. The disruption of the sdhA gene encoding Fp in yeast induces the loss of the Ip from mitochondria (Robinson et al. 1991).

Comparison of the sequences of yeast (Robinson and Lemire 1992), human heart (Morris et al. 1994), liver (Hirawake et al. 1994) and bovine Fp (Birch-Machin et al. 1992) showed th presence of ten conserved domains. The second domain from the N-terminal end contains histidine 56 of the mature protein in human, able to bind FAD (Morris et al. 1994). The catalytic center containing the conserved His-Pro-Thr sequence is located within the fifth conserved domain.

3.2
The Iron-Sulfur Subunit Ip

One of the first uses of polymerase chain reaction (PCR) for researching homologous genes was carried out by Gould et al. (1989) on the genes of iron-sulfur of complex II. The human gene has been characterized (Au et al. 1995), and the mature Ip subunit is composed of 252 aminoacids in humans, and is 94.1% homologous with the bovine heart Ip and 50,8% with the *E. coli* gene product sdh B (Kita et al. 1990). Three iron-sulfur centers defined by the sequences Cys XXX Cys XX Cys (11 residues) Cys and Cys XX Cys Cys XX Cys XXX Cys Pro are localized in hydrophobic parts of the protein, allowing the electron transfer from $FADH_2$ of Fp to another hydrophobic compound such as cytochrome c and/or ubiquinone. The construction of chimeric genes between human and yeast Ip may have various consequences, in particular the substitution between the first and the second iron-sulfur center prevents the formation of an active yeast complex since the Ip is degraded (Sashbini et al. 1994).

The regulation of expression of iron proteins is generally post-transcriptional due to the presence of iron-responsive elements (IRE) on the mRNA which adopts a stem and loop structure at the 5' untranslated sequence, able to bind with iron regulatory proteins (IRP). In the absence of IRP, the mRNA encoding iron proteins such as Ip is normally translated. However, binding of IRP to IRE in response to a decrease in iron levels leads to the inhibition of translation of these mRNA (Melefors 1996). Quantitative analysis of Ip from *Drosophila* cells grown under different iron concentrations showed that iron plays a critical role (Gray et al. 1996).

3.3
The Membrane Anchor Polypeptides

The two subunits, named C_{II-3} and C_{II-4} with a respective molecular weight of 11 and 9 kDa in the beef heart (Cochran et al. 1994) contain a heme b, also present in *E. coli*, which plays an important role in the structure of the complex (Nakamura et al. 1996). A few homologies have been found between the beef heart subunit C_{II-3} and *E. coli* sdhC, notably two conserved histidine residues localized at positions 34 and 90 in the beef heart (Yu et al. 1992) which may be involved in the heme ligation of cytochrome b_{560}. On the basis of sequence comparisons, a structural model of the anchor polypeptides from different origins has been proposed (Hägerhäll and Hederstedt 1996), consisting of six trans-membrane domains.

A complex II defect has been reported in a Chinese hamster mutant cell line which was complemented by a human gene located on chromosome I, but not by the cDNA of bovine Ip. The complementation was, however, possible in the presence of the bovine C_{II-3} gene. This was confirmed by showing a mutation creating a premature stop codon from the 3' end of the equivalent gene from the Chinese hamster mutant clone (Oostveen et al. 1995).

4 Evolutionary Aspects

The lack of mitochondrially encoded genes for this complex questions the evolutionary origin of the nuclear genes: whether they were present in the nuclear genome devoid of mitochondria, which hosted a protomitochondrial endosymbiont, or whether they were brought in to the early eukaryote by mitochondria-like eubacteria which have lost these gene towards the nucleus. A number of genes encoding complex II have been characterized in various species and their homologues have been found in eubacteria and arachaebacteria. For example, in *E. coli,* the sdhA and sdhB genes encode for subunits Fp and Ip, respectively, and sdhC and sdhD encode for the apocytochrome b_{556} and a hydrophobic peptide of 13 kDa, the anchor proteins (Wood et al. 1984). Furthermore, the analysis of the mitochondrial DNA sequence from two evolutionarity distant eukaryotes, *Porphyra purpura,* a multicellular alga, and that of a unicellular bacterivore *Reclinomonas americana,* demonstrated the presence of the equivalents of sdhB, sdhC, and sdhD from *E. coli* (Burger et al. 1996). Therefore, these genes were present in the ancient mitochondria and were then integrated within the nuclear genomes. This observation also reinforces the theory of a monophyletic origin of mitochondria.

5 Diseases Associated with Complex II Defects

Various, but infrequent cases of complex II defects have been reported. The clinical symptoms are very variable, suggesting Leigh's disease, encephalopathies and/or cardiomyopathies.

An enzymatic defect localized between complex II and III has been shown in a patient presenting a mitochondrial myopathy characterised by abnormal mitochondria, an elevated lactate/pyruvate ratio, but with unknown genetic lesions (Sengers et al. 1983).

Riggs et al. (1984) also reported the case of two children from the same family presenting a mitochondrial encephalopathy with lipids accumulation, and aggregated mitochondria with paracrystalline inclusions. The symptoms began after around 5 years with a lack of movement coordination, myoclonic jerks, staggering gait, and small stature at about 8 years. Analysis of respiratory chain complexes on muscle biopsies revealed a defect of succinate-cytochrome c reductase in both children.

The complete lack of an Ip subunit has been discovered in an 18-years-old girl presenting psychomotor retardation, disturbed behaviour, apathy and anorexia associated with limb-girdle weakness. Two other members of the family were affected as well, and a non-studied brother died from a dilated cardiomyopathy. The activity of the complex was reduced to 10% of normal (Desnuelle et al. 1989).

Two brothers of 19 and 25 years presenting a hypertrophic cardiomyopathy were examined. Histological studies revealed lipid accumulation and a lack of detectable SDH activity on the muscle fibers. The enzymatic activity of complex II was found to be reduced by 80% in these patients (Reichmann and Angelini 1993).

A child presenting a severe hypotony, tachypnoea, lactic acidosis, and renal tubulopathy presented a severe defect of complex II in muscle biopsy, but not in his cultivated fibroblasts (Sperl et al. 1988).

A general defect in iron-sulfur proteins, such as aconitase and Ip, associated with a 90% defect of complex II has been described in a patient presenting an exercice intolerence, muscle weakness accompanied with dyspnea and painful swelling and pigmentura. Histochemical studies revealed a severe reduction of SDH; abnormal mitochondria were present, and X-ray analysis revealed their iron-rich content (Hall et al. 1993).

A 25-years-old patient with clinical signs of Kearns-Sayre syndrome presented a complex II deficiency, through the enzyme subunits were detected upon western blot analysis (Rivner et al. 1989).

Two sisters born from consanguineous first cousins, presenting a leukodystrophy with Leigh syndrome, have been described (Bourgeois et al. 1992). One of them developed normally until 10 months, then progressively deteriorated with a lack of interest in her surroundings. At about 1 year of age, examination showed rigidity in extension in the four limbs, difficulties in following objects and apparent blindness. The blood showed hyperlactatemia. The CT scan revealed hypodensity of white matter. She stayed in a vegetative state until 19 months when she died. Her younger sister also began to regress towards 10 months. At 13 months she presented marked rigidity, and difficulties in swallowing fluids. Gas chromatography of organic acids in her urine showed an increased excretion of α ketoglutaric acid, succinic acid and creatinin. Low rates of succinate cytochrome c reductase were measured in muscle mitochondria as well as in the lymphocytes of the second patient, and in fibroblasts of the two patients.

The mRNA of SDH subunits Ip and Fp from these children were reverse transcribed, PCR amplified then sequenced. A transition C $\rightarrow$ T at nucleotide position 1684, inducing the substitution of the Arginine 544 into Tryptophan within Fp was found (Bourgeron et al. 1995). The two parents were heterozygotes for that mutation which abolishes an MspI restriction site.

The genomic analysis of the children's DNA revealed, however, a surprising result: indeed, the MspI digestion was incomplete, suggesting the presence of two couples of genes encoding the Fp subunit. This hypothesis was confirmed by hybridizing metaphase chromosomes with the cDNA of Fp: two signals were observed, one of them being localized in the distal part of the chromosome 3 (3q29), and the other being localized on chromosome 5 (5p15). In order to determine which is the transcriptionally active chromosome, the RNA and DNA were extracted from somatic hybrid human-rodent cells containing either chromosome 3 or chromosome 5. Only the reverse-transcribed mRNA from cells containing chromosome 5 have been obtained, thus revealing that the active gene is located on chromosome 5, at least in these hybrid cells.

The mutant enzyme presented a normal Km for succinate and a normal Ki for malonate. It was, however, more sensitive to inhibition by oxaloacetate. The mutation is located within the C terminal end of Fp, but is not localized in the previously conserved sequences. The analysis of complex II by immunoblot showed the disapperance of subunits Fp and Ip (Birch-Machin et al. 1996).

In Graves' hyperthyroidism, characterized by a progressive eye disorder (ophthalmopathy) with thyroid autoimmunity, the presence of auto-antibodies against the subunit Fp has been reported (Kubota et al. 1998); the authors suggest that presentation of succinate dehydrogenase is secondary to primary yet unidentified eye muscle damage.

6 Conclusion

The presence of the two couples of genes encoding SDH Fp could explain the heterogeneity of the clinical deficiencies in humans if the regulation of each of them is developmental and tissue-specific. Some defects could then be associated with cardiomyopathy and others with encephalomyopathy. The characterization of these genes, as well as studies on the regulation of their expression, will shed new light onto some causes of the dysfunctions of this nuclear gene encoded complex, one of the genes coding for Fp being present in four copies within each human cell. Interestingly, a report recently pointed to the importance of Complex II in mitochondrial protein synthesis (Escobar Galvis et al. 1998) by coupling electron transfer to protein synthesis, suggesting unexpected pleiotropic mitochondrial defects in cases of reduced activity, due to point mutations of this nuclear encoded complex.

References

Au HC, Ream-Robinson D, Bellew LA, Broomfield PL, Sagbhini M, Scheffler IE (1995) Structural organization of the gene encoding the human iron-sulfur subunit of succinate dehydrogenase. Gene 159:249–253

Birch-Machin MA, Farnswoth L, Ackrell BAC, Cochran B, Jackson S, Bindoff L, Aiken A, Diamond AG, Turnbull D (1992) The sequence of the flavoprotein subunit of bovine heart succinate dehydrogenase. J Biol Chem 267:11553–11558

Birch-Machin MA, Marsac C, Ponsot G, Parfait B, Taylor RW, Rustin P, Munnich A (1996) Biochemical investigations and immunoblot analysis of two unrelated patients with an isolated deficiency in complex II of the mitochondrial respiratory chain. Biochem Biophys Res Commun 220:57–62

Bourgeois M, Goutières F, Chrétien D, Rustin P, Munnich A, Aicardi J (1992) Deficiency in complex II of the respiratory chain, presenting as a leukodystrophy in two sisters with Leigh syndrome. Brain Dev 14:404–408

Bourgeron T, Rustin P, Chrétien D, Birch-Machin M, Bourgeois M, Viegas-Péquinot E, Munnich A, Rötig A (1995) Mutation of a nuclear succinate dehydrogenase gene results in mitochondrial respiratory chain deficiency. Nat Genet 11:144–149

Burger G, Lang F, Reith M, Gray MW (1996) Genes encoding the same three subunits of respiratory complex II are present in the mitochondrial DNA of two phylogenetically distant eukaryotes. Proc Natl Acad Sci USA 93:2328–2332

Cochran B, Capaldi RA, Ackrell BAC (1994) The cDNA sequence of beef heart C_{II-3}, a membrane-intrinsic subunit of succinate-ubiquinone oxidoreductase. Biochim Biophys Acta 1188:162–166

Desnuelle C, Birch-Machin M, Pellissier JF, Bindoff LA, Ackrell BAC, Turnbull DM (1989) Multiple defects of the respiratory chain including complex II in a family with myopathy and encephalomyopathy. Biochem Biophys Res Commun 163:695–700

Escobar Galvis ML, Allen JF, Hakanson G (1998) Protein synthesis by isolated mitochondria is dependent on the activity of respiratory complex II. Curr Genet 33:320–329

Fink SL, Dy HO, Sapolsky RM (1995) Energy and glutamate dependency of 3-nitropropionic acid neurotoxicity in culture. Exp Neurol 138:298–304

Gould SJ, Subramani S, Scheffler IE (1989) Use of the polymerase chain reaction for homology probing: isolation of partial cDNA or genomic clones encoding the iron-sulfur protein of succinate dehydrogenase from several species. Proc Natl Acad Sci USA 86:1934–1938

Gray NK, Pantopoulos K, Dandekar T, Ackrell BAC, Hentze MW (1996) Translational regulation of mammalian and *Drosophila* citric acid cycle enzymes via iron-responsive elements. Proc Natl Acad Sci USA 93:4925–4930

Hägerhäll C, Hederstedt L (1996) A structural model for the membrane-integral domain of succinate:quinone oxidoreductases. FEBS Lett 389:25–31

Hall RE, Henriksson KG, Lewis SF, Haller RG, Kennaway NG (1993) Mitochondrial myopathy with succinate dehydrogenase and aconitase deficiency. Abnormalities of several iron-sulfur proteins. J Clin Invest 92:2660–2666

Hatefi Y (1985) The mitochondrial electron transport and oxidative phosphorylation system. Annu Rev Biochem 54:1015–1069

Hederstedt L, Ohnishi T (1992) Progress in succinate:quinone oxidoreductase research. In: Ernster L (ed) Molecular mechanisms in bioenergetics. Elsevier, Amsterdam, pp 163–198

Hirawake H, Wang H, Kuramochi T, Kojima S, Kita K (1994) Human complex II (succinate-ubiquinone oxidoreductase): cDNA cloning of the flavoprotein (Fp) subunit of liver mitochondria. J Biochem 116:221–227

Jia L, Kelsey SM, Grahn MF, Jiang XR, Newland AC (1996) Increased activity and sensitivity of mitochondrial respiratory enzymes to tumor necrosis factor alpha-mediated inhibition is associated with increased cytotoxicity in drug-resistant leukemic cell lines. Blood 87:2401–2410

Kanedo I, Yamada N, Sakuharaba Y, Kamenosono M, Tutumi S (1995) Suppression of mitochondrial succinate dehydrogenase, a primary target of beta amyloid, and its derivative racemized at Ser residues. J Neurochem 65:2585–2593

Kita K, Oya H, Gennis RB, Ackrell BAC, Kasahara M (1990) Human complex II (succinate-ubiquinone oxidoreductase): cDNA cloning of iron sulfur (Ip) subunit of liver mitochondria. Biochem Biophys Res Commun 166:101–108

Kotlyar AB, Vinogradov AD (1984) Interaction of the membrane-bound succinate dehydrogenase with substrates and competitive inhibitors. Biochim Biophys Acta 784:24–34

Kubota S, Gunji K, Ackrell BAC, Cochran B, Stolarski C, Wengrwicz S, Kennerdell JS, Hiromatsu Y, Wall J (1998) The 64-kilodalton eye muscle protein is the flavoprotein subunit of mitochondrial succinate dehydrogenase: the corresponding serum antibodies are good markers of an immune-mediated damage to the eye muscle in patients with Graves' hyperthyroidism. J Clin Endocrinol Metab 83:443–447

Melefors O (1996) Translational regulation in vivo of the *Drosophila melanogaster* mRNA encoding succinate dehydrogenase iron protein via iron responsive elements. Biochem Biophys Res Commun 221:437–441

Morris AAM, Farnsworth L, Ackrell BAC, Turnbull DM, Birch-Machin MA (1994) The cDNA sequence of the flavoprotein subunit of human heart succinate dehydrogenase. Biochim Biophys Acta 1185:125–128

Nakamura K, Yamaki M, Sarada M, Nakayama S, Viat CRT, Gennis RB, Nakayashiki T, Inokuchi H, Kojima S, Kita K (1996) Two hydrophobic subunits are essential for the heme b ligation and functional assembly of complex II (succinate-ubiquinone oxidorecutase) from *Escherichia coli*. J Biol Chem 271:521–527

Oostveen FG, Au HC, Meijer P-J, Scheffler IE (1995) A Chinese hamster mutant cell line with a defect in the integral membrane protein C_{II-3} of complex II of the mitochondrial electron transport chain. J Biol Chem 270:26104–26108

Reichmann H, Angelini C (1993) Single muscle fiber analysis in 2 brothers with succinate dehydrogenase deficiency. Eur Neurol 34:95–98

Riggs JE, Schochet SS, Fakadej AV, Papadimitriou A, DiMauro S, Crosby T, Gutman L, Moxley RT III (1984) Mitochondrial encephalomyopathy with decreased succinate-cytochrome c reductase activity. Neurology 34:48–53

Rivner MH, Shamsnia M, Swift TR, Trefz J, Roesel RA, Carter AL, Yanamura W, Hommes FA (1989) Kearns-Sayre syndrome and complex II deficiency. Neurology 39:693–696

Robinson KM, Lemire BD (1992) Isolation and nucleotide sequence of the *Saccharomyces cerevisiae* gene for succinate dehydrogenase flavoprotein subunit. J Biol Chem 267:10101–10107

Robinson KM, Lemire BD (1996a) Covalent attachment of FAD to the yeast succinate dehydrogenase flavoprotein requires import into mitochondria, presequence removal, and folding. J Biol Chem 271:4055–4060

Robinson KM, Lemire BD (1996b) A requirement for matrix processing peptidase but not for mitochondrial chaperonin in the covalent attachment of FAD to the yeast succinate dehydrogenase flavoprotein. J Biol Chem 271:4061–4067

Robinson KM, Kieckebusch-Guck AV, Lemire BD (1991) Isolation and characterization of a *Saccharomyces cerevisiae* mutant disrupted for the succinate dehydrogenase flavoprotein subunit. J Biol Chem 266:2137–2150

Sashbini M, Broomfield E, Scheffler IE, Sashbini M, Broomfield E, Scheffler IE (1994) Studies on the assembly of complex II in yeast mitochondria using chimeric human/yeast genes for the iron-sulfur protein subunit. Biochemistry 33:159–165

Schultz JB, Huang PL, Matthews RT, Passov D, Fishman MC, Beal MF (1996) Striatal malonate lesions are attenuated in neuronal nitric oxide synthetase knockout mice. J Neurochem 67:430–433

Sengers RCA, Fischer JC, Trijbel JMF, Ruitenbeek W, Stadhouders AM, Laak HJT, Jaspar HHJ (1983) A mitochondrial myopathy with a defective respiratory chain and carnitine deficiency. Eur J Pediatr 140:332–327

Shimano Y, Kumazaki M, Sakurai T, Hida H, Fujimoto I, Fukuda A, Nishino H (1995) Chronically administered 3-nitropropionic acid (3-NPA), irreversible inhibitor of succinate dehydrogenase, produces selective lesions in the striatum and reduces muscle tonus. Obes Res Dec 3 (Suppl 5):779S-784S

Sperl W, Ruitenbeek W, Trijbels JMF, Sengers RCA, Stadhouders AM, Guiggenbichler JP (1988) Mitochondrial myopathy with lactic acidaemia, Franconi-De Toni Debré syndrome and a disturbed succinate:cytochrome c oxidoreductase activity. Eur J Pediatr 147:418–421

Wood D, Darlinson MG, Wilde RJ, Guest JR (1984) Nucleotide sequence encoding the flavoprotein and hydrophobic subunits of the succinate dehydrogenase of *Escherichia coli*. Biochem J 222:519–534

Yankovskaya V, Sablin SO, Ramsay RR, Singer TP, Ackrell BA, Cecchini G, Miyoshi H (1996) Inhibitor probes of the quinone binding sites of mammalian complex II and *Escherichia coli* fumarate reductase. J Biol Chem 271:21020–21024

Yu L, Wei Y-Y, Usui S, Yu C-A (1992) Cytochrome b_{560} (QPs1) of mitochondrial succinate-ubiquinone reductase. Immunochemistry, cloning, and nucleotide sequencing. J Biol Chem 234:24508–24515

Zeevalk GD, Derr-Yellin E, Nicklas WJ (1995) Relative vulnerability of dopamine and GABA neurons in mesencephalic culture to inhibition of succinate dehydrogenase by malonate and 3-nitropropionic acid and protection by NMDA receptor blockade. J Pharmacol Exp Ther 275:1124–1130

CHAPTER 7

The bc1 Complex in the Mitochondrial Respiratory Chain 7

G. BRASSEUR[1], P. BRIVET-CHEVILLOTTE[1], D. LEMESLE-MEUNIER[1], and J.-P. DI RAGO[2]

Contents

1 Introduction

In the inner mitochondrial membrane, the bc1 complex (complex III or ubiquinol cytochrome c oxidoreductase) catalyses the electron transfer from ubiquinol to cytochrome c coupled with a vectorial proton translocation from the matrix side to the positive intermembrane space (Fig. 1). The proton gradient thus created is used by he ATP synthase enzyme to drive ATP synthesis. The bc1 complex is an energy transducing component which is essential to the viability of most eukaryotic cells, which require oxidative phosphorylation to produce ATP. Genetic alterations can lead to human diseases in which the expression or functional efficiency of this complex is affected, especially in the tissues with the highest energy requirements, such as the muscles, brain and heart.

[1] Laboratoire de Bioénergétique et Ingéniérie des Protéines, Centre National de la Recherche Scientifique, 31 chemin Joseph Aiguier, 13402 Marseille Cedex 20, France
[2] Centre de Génétique Moléculaire du Centre National de la Recherche Scientifique, Laboratoire propre associé à l'Université P. et M. Curie, Avenue de la Terrasse, 91198 Gif-sur-Yvette Cedex, France

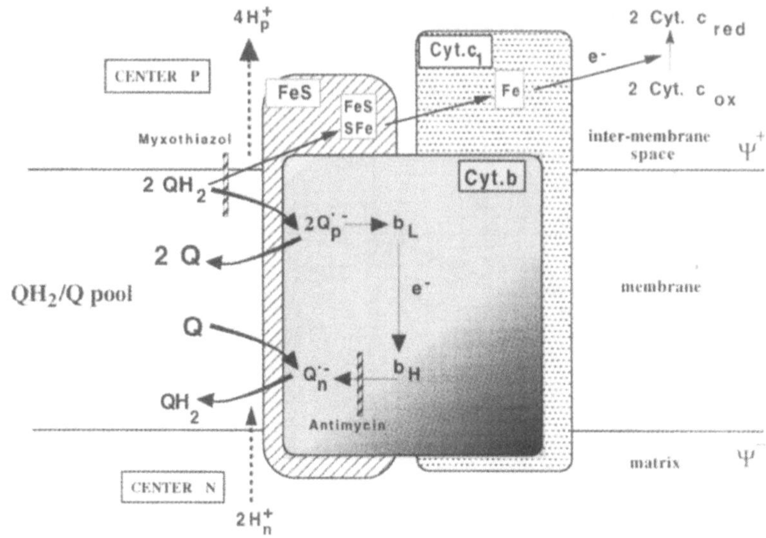

Fig. 1. Schematic structure of organization and function of the three catalytic subunits of the mitochondrial bc1 complex. The four redox groups associated with the three redox proteins are depicted, namely the two b hemes from cytochrome b in the membrane (b_L and b_H stand for low and high midpoint potential, respectively), cytochrome c_1 heme and the [2Fe–2S] cluster from Rieske FeS protein located in the positive (Ψ^+) intermembrane space. Note that seven to eight non-catalytic subnits present in the eukaryotic organisms are not shown in this figure. The ubiquinol QH_2 substrate is oxidized at *center P* and protons are released on the positive side (cf. text for more details about the bc1 complex mechanism). The first electron reduces the high potential pathway formed by FeS protein and cytochrome c1. The second electron from the semiquinone at center P reduces cytochrome hemes b_L and b_H successively, and then reduces the bound quinone into semiquinone at *center N*. Two cycles, and thus two quinol molecules, are required at *center P* to reduce one quinone into quinol at *center N*. The proton-motive force created by the bc1 complex results from oxido-reduction reactions at both *center P* and *center N*. The inhibitor myxothiazol blocks the electron transfer at *center P*, whereas antimycin blocks the b_H reoxidation at *center N*

Since the bc1 complex was first purified in the 1960s, significant progress has been made towards unterstanding the structure and function of this respiratory chain complex, in studies based on a combination of biochemical and genetic approaches. The three-dimensional structure of the mitochondrial beef heart bc1 complex was recently resolved by an X-ray diffraction method (Xia et al. 1997; Iwata et al. 1998; Zhong et al. 1998). This new information will certainly help us to understand the role of specific amino acids in the catalytic mechanism of the bc1 complex at the atomic level. Genetic studies on the eukaryotic yeast *Saccharomyces cerevisiae* should provide further information about the functional repercussions of specific structural modifications. These studies might lead to the discovery of components which may be able to mitigate the effects of mutations affecting the human bc1 complex.

2 bc1 Complex Structure

2.1 Protein Composition and the Genes Encoding These Proteins

bc1 complexes from various mammalian tissues, fungi, green algae and plants have been purified to homogeneity (Ljungdahl et al. 1987; Schägger et al. 1995; Jänsch et al. 1995). The in vitro electron transfer and proton translocation activities are similar to those observed in situ, which indicates that all the essential constituents required for the enzyme to be active are present in the purified complexes. In comparison with bacterial bc1 complexes, which are relatively simple and contain only three to four subunits (Trumpower 1990; Gray and Daldal 1995), the mitochondrial bc1 complexes in most species contain up to 11 subunits per bc1 monomer and these subunits show little variation (Table 1). All the bc1 complexes contain at least three subunits carrying the redoc centers, namely a dihemic cytochrome b, a monohemic cytochrome c1 and an iron sulfur protein named "Rieske protein" (Rieske et al. 1964) with a [2Fe–2S] cluster. All the bc1 complex proteins in eukaryotes are encoded by the nuclear genome, except for cytochrome b which is encoded by the mitochondrial DNA. Their sequences have been determined for yeast *S. cerevisiae,* potato and beef heart. In the yeast *S. cerevisiae,* from which the entire genome has been sequenced, each subunit is encoded by a single gene. In *S. tuberosum* (potato), however, isoforms have been found to exist (Braun et al. 1992; Emmermann et al. 1993; Jänsch et al. 1995), probably due to the high level of ploidy in this plant, the biological and functional significance of which is still unknown. Nuclear genetic loci have been identified only in the case of three subunits of the human bc1 complex (see Table 1).

The three redox protein sequences (cytochrome b, cytochrome c1 and FeS protein) are highly conserved among species, from yeast to mammals, as shown by the rate of amino acid identity which is about 50–60% (Table 1). The other subunit sequences are less conserved but secondary structure similarities are still found (see below).

2.2 Secondary and Tertiary Structures

Structural data have been obtained by performing two-dimensional electron microscopy (Leonard et al. 1981), chemical labelling and protease digestion (Li et al. 1981; Gonzalez-Halphen et al. 1988), and sequence comparisons (Degli Esposti et al. 1993), as well as by using genetic approaches (Slominski and Tzagoloff 1976; di Rago et al. 1989, 1990, 1993, 1995, 1997; Lemesle-Meunier et al. 1993; Graham et al. 1993; Coppée et al. 1994a,b; Brasseur et al. 1996; Saribas et al. 1998). The three dimensional structures of the beef and chicken mitochondrial bc1 complex were recently determined by X-ray diffraction at a resolution between 2.8 and 3.0 Å (Xia et al. 1997; Iwata et al. 1998; Zhang et al. 1998). The 11 subunits are visible in the most recent X-ray structure of the bovine mitochondrial complex (Iwata et al. 1998) including core I, core II, cytochrome b, cytochrome c_1, Rieske FeS protein and subunits 6–11. These atomic structures showed that the bc1 complex is a dimer in the crystal, in agreement with previous suggestions (reviewed by Brandt and Trumpower 1994). However, the

Table 1. bc1 complex subunits in *Saccharomyces cerevisiae, Bos taurus* and *Homo sapiens*

Saccharomyces cerevisiae		Bos taurus		Homo sapiens			Amino acid identity, %	
Subunit	Size	Subunit	Size	Subnit	Size	Genomic locus	B. taurus S. cerevisiae	B. baurus/ H. sapiens
Core I (P07256)[a]	457 (26)[b]	Core I (P31800)	362 (34)	Core I (P31930)	480 (34)	10.4 kb/13 exons	28.9 (256; 6.6)[c]	87.2 (298; 0.0)
Core II (P07257)	368 (16)	Core II (P23004)	453 (14)	Core II (P22695)	453 (14)	n.i.	23.4 (248; 2.0)	89.0 (453; 0.0)
Cyt b (P00163)	385	Cyt b (P00157)	379	Cyt b (P00156)	380	(mtDNA)	51.6 (374; 0.5)	78.9 (379; 0.0)
Cyt c1 (P07143)	309 (61)	Cyt c1 (P00125)	241[e]	Cyt c1 (P08574)	325 (84)	2.4 kb/7 exons	58.0 (238; 0.4)	93.4 (241; 0.0)
FeS (P08067)	215 (30)	FeS (P13272)	274 (78)	FeS (P47985)	274 (78)	19q12	53.8 (184; 0.5)	90.5 (274; 0.0)
VI (P00127)	147	VIII (P00126)	78[e]	VIII (P07919)	91	n.i.	36.5 (85; 9.4)	93.6 (78; 0.0)
VII (P00128)	127	VI (P00129)	111	VI (P14927)	111	4.5 kb/4 exons	42.0 (88; 1.1)	84.7 (111; 0.0)
VIII (P08525)	94	VII (P13271)	82	n.i.			28.0 (39; 0.0)	
X (P22289)	77 (1)	XI (P07552)	56[e]	n.i.			33.0 (21; 0.0)	
IX (P37299)	66 (1)	X (P001310)	62[e]	n.i.			43.8 (48; 0.0)	
		IX[d]	78	?				

n.i.: non identified.

[a] The name of the subunit (cytb, cytochrome b; cytc1, cytochrome c1; FeS, Rieske FeS protein) is followed in parentheses by the PIR bank accession number.

[b] The number of amino acids in the precursor protein is followed in parentheses by the number of amino acids cleaved in the mature protein.

[c] The percentage of sequence identity is followed in parentheses by the number of amino acids compared and the frequency of discontinuities in the alignments (sequences were compared with the SIM program and the BLOSUM62 matrix with default parameters, available on Internet at the following address: http://www.expasy.ch/sprot/sim-prot.html).

[d] In beef heart mitochondria, the Rieske FeS protein targeting presequence is subunit IX which is retained after cleavage as a bc1 complex subunit (Brand et al. 1993).

[e] Indicates that the sequence was determined on the basis of protein sequencing.

question as to whether the two monomers cooperate catalytically via a dimeric Q cycle, as proposed by de Vries (1986), still remains to be answered.

The membrane-spanning region of each bc1 complex monomer consists of 13 transmembrane helices. Both monomers form cavities through which the inhibitor-binding pockets can be accessed.

The core proteins I and II are both structurally similar and protrude on the negative side of the membrane. bc1 complex crystals from various organisms in both the presence and absence of inhibitors of quinone oxidation were recently obtained by Zhang et al. (1998). In some of these structures, the [2Fe–2S] cluster is closer to cytochrome c1. Zhang et al. (1998), and H. Kim et al. (1998) explained these results by a domain movement of the extrinsic part of the Rieske FeS protein in order to shuttle electrons from ubiquinol to cytochrome c1. Brasseur et al. (1997a) have suggested that the FeS protein region located after the transmembrane anchor may contribute to the mobility of the FeS protein between cytochrome b and c_1. A detailed mechanism for FeS mobility has been recently proposed by Crofts et al. (1998).

Cytochrome b consists of eight transmembrane α helices: each of the helices B and D bears two totally conserved histidines, which are the cytochrome b heme ligands (Fig. 2). Histidines 82 and 183 are the heme b_L ligands and histidines 96 and 197 are the heme b_H ligands (L stands for a low and H for a high oxido-reduction midpoint-potential, i.e. around –90 and +40 mV, respectively). These hemes are normally oriented with respect to the membrane plane (Simpkin et al. 1989), b_L being located close to the membrane surface on the positive side and heme b_H being buried more deeply in the lipid bilayer, closer to the negative side of the membrane. On the positive side of the membrane, two amphipathic helices, cd1 and cd2, form a hairpin-like structure and are part of the Q_P site (or center P; Fig. 2).

The Rieske FeS protein is located on the positive side of the membrane. The FeS protein can be dissociated from the bc1 complex and reintegrated without any drastic loss of activity (Trumpower and Edwards 1979). The Rieske protein carboxy terminal region carries the [2Fe–2S] cluster, and the amino-terminal portion consists of a transmembrane α helix, which anchors the protein to the membrane (Xia et al. 1997). A soluble fragment of the Rieske protein from beef heart mitochondria (amino acids 68–196 in the mature form) which contains the [2Fe–2S] cluster has been crystallized and its tridimensional structure determined at a resolution of 1.5 Å (Iwata et al. 1996). This fragment has an oval shape and consists mainly of β sheets. The [2Fe–2S] cluster is coordinated by two histidines (H161 and H141) and two cysteines (C139 and C158). In the cluster region, two other cysteines (C144 and C160) form a disulfide bridge which, in addition to the hydrogen bond network, stabilize the protein (Iwata et al. 1996).

In cytochrome c_1, all the sequences determined so far show a conserved CxyCH pattern, which is the site of the c type heme attachment. The two cysteines are covalently bound to the heme via thio-ether bonds with the heme vinyl groups. The iron atom is coordinated by the histidine in the CxyCH pattern (Yu et al. 1975; Nakai et al. 1990) and by a conserved methionine (M225 in yeast cytochrome c_1; Gray et al. 1992). Cytochrome c_1 has a transmembrane anchor which is located in the carboxy terminal region of the protein. The hydrophilic amino-terminal region is located in the intermembrane space and was obtained as a heme-containing fragment by action of protease without any change in its spectral or oxido-reduction characteristics (Li et al. 1981; Konishi et al. 1991).

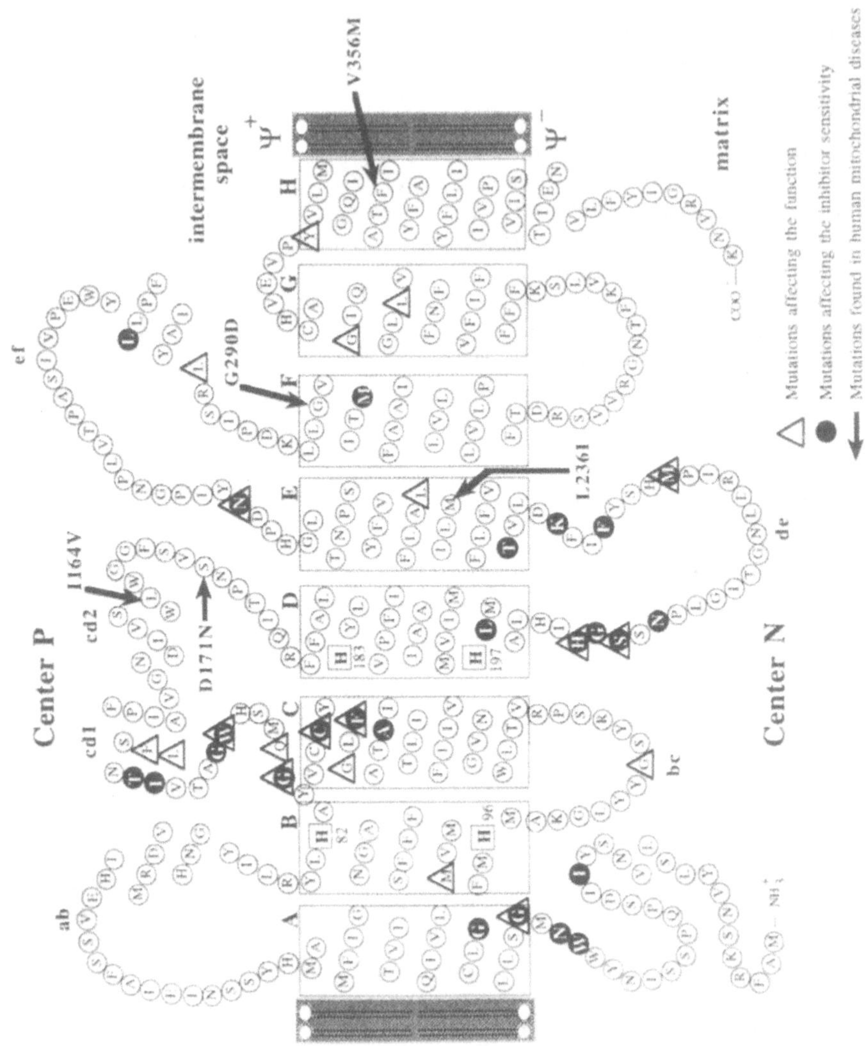

Fig. 2. Cytochrome b folding model in the mitochondrial membrane. Eight transmembrane helices (*A* to *H*) are present in mitochondrial cytochrome b (Crofts et al. 1987; Brasseur 1988) and were recently observed in the three-dimensional structure of the bc1 complex (Xia et al. 1997). The yeast cytochrome b sequence is used in this figure. Note the four totally conserved histidines (□) in helices *B* and *D*, which ligate heme b_L (His 82 and His 183) and heme b_H (His 96 and His 197). Mitochondrial mutations affecting the function are indicated by Δ, whereas inhibitor resistance or hypersensitivity mutations at either center P or center N are in a dark circle (●). Note that the bacterial cytochrome b mutations affecting the function and those involved in inhibitor resistance are not shown (for further details see: Gennis et al. 1993; Gray and Daldal 1995; Brasseur et al. 1996). Mutations D171N and V356M are suspected of being involved in Leber disease (Johns and Neufeld 1991; Brown et al. 1992). Cytochrome b mutation L236I is associated with complex III deficiency in patients with ischemic cardiomyopathy (Marin-Garcia et al. 1996). Cytochrome b mutation I164V was observed in association with a 12S rRNA gene mutation in two patients with cardiomyopathy (Ozawa et al. 1995). Cytochrome b mutation G290D was detected in a patient affected with progressive exercise intolerance linked to a bc1 complex deficiency (Dumoulin et al. 1996). Positions corresponding to the human mutated amino acids are indicated *(bold characters)* by → (L165, S172, M237, G291 and F357 in the yeast numbering)

Several other subunits are present in the bc1 complex around the catalytic core formed by cytochrome b, cytochrome c1 and Rieske FeS protein (Gonzalez-Halphen et al. 1988; Trumpower 1990). Core proteins I and II amount to about 40% of the whole mass complex and are located on the matrix side. Subunit VIII in yeast (equivalent to the beef subunit VII) is located on the same side and contains a transmembrane anchor. This bc1 complex protein was found to interact with complex II (succinate ubiquinone oxido-reductase) of the respiratory chain (Bruel et al. 1996). Yeast subunit IX and X structures have not yet been determined in the beef bc1 complex X-ray structure; based on their amino acid sequences, they may contain on α helix transmembrane anchor. Yeast subunit VI contains many acidic aminoacids and is located on the positive side of the membrane and seems to facilitate the interaction between cytochrome c1 and soluble cytochrome c (C. H. Kim et al. 1987).

Unexpectedly, purified bc1 complexes, and specifically the Core proteins I and II, from various plants have been found to be able to cleave the targeting presequence of mitochondrial proteins encoded by nuclear genes (Braun et al. 1992; Glaser et al. 1996). In yeast and mammals, this protease activity is carried out in the mitochondrial matrix by the MPP (mitochondrial processing peptidase), which consists of two soluble proteins, α-MPP and β-MPP (Yang et al. 1988). Based on sequence homologies between MPP proteins and core proteins I and II (Schulte et al. 1989), it has been suggested that this group of proteins may have a common ancestral origin and an evolutionary hypothesis has been put forward along these lines (Braun and Schmitz 1995; die Rago et al. 1997).

3 Mechanism and Inhibitors of the bc1 Complex

The Q cycle model was initially proposed by Mitchell (1976) and later supported by experimental evidence (for a review, see Brandt and Trumpower 1994). It accounts for most of the experimental data available on the mechanism of the bc1 complex. In this model, the existence of two quinone oxido-reduction sites is postulated, the one (named Q_P, center P or Q_0) is located on the positive side and the other (Q_n, center N or Q_i) on the negative side of the membrane (Fig. 1). Center P catalyses the oxidation of the two electron donor ubiquinol QH_2 (substrate) into ubiquinone (Q) on the positive side of the membrane in two strictly coupled steps, whereas at center N, on the other side of the membrane, the ubiquinone molecule is reduced into ubiquinol.

During the first half of the Q cycle (Eq. 1 below), the ubiquinol at center P is oxidized: the first electron is transferred to the [2Fe–2S] cluster of the FeS protein, yielding the unstable semiquinone intermediate which immediately reduces heme b_L of cytochrome b, and two protons are then released as the result of ubiquinol oxidation into the intermembrane space. In the high potential pathway (so named because of the high E_m of FeS and cytochrome c_1, around 300 mV), the reduced FeS protein is oxidized by cytochrome c_1, which in turn is oxidized by the soluble electron carrier cytochrome c. The unstable cytochrome c_1, which in turn is oxidized by the soluble electron carrier cytochrome c. The unstable semiquinone, formed in the first step, reduces heme b_L, yielding a quinone; in this low potential pathway (the E_m values of hemes b_L and b_H are –90 and +50 mV, respectively), heme b_L is then reoxidized by heme b_H, which reduces a quinone into a semiquinone at center N. The semiquinone $Q^{\cdot-}_n$ then has to wait for a second electron to be released by the oxidation of a second ubiquinol molecule at center P (Eq. 2). The two electrons from ubiquinol therefore diverge at center P and are transferred to two different electron acceptors, which differ by approximately 350 mV. Only one of the electrons provided by ubiquinol is transferred to cytochrome c.

$$QH_{2\,p} + cyt\ c_{ox} \rightarrow Q^{\cdot -}{}_{n} + cyt\ c_{red} + 2H^{+}p \tag{1}$$

During the second half of the Q cycle, the second quinol molecule again releases two protons into the intermembrane space, transfers one electron to the soluble cytochrome c via the FeS protein and cytochrome c_1 and recycles the second electron through the membrane via b_L and b_H hemes, reducing the intermediate semiquinone $Q^{\cdot -}{}_{n}$ formed in the first half of the cycle. The complete reduction of an ubiquinone into ubiquinol at center N leads to the uptake of two protons from the matrix (Fig. 1).

$$QH_{2\,p} + cyt\ c_{ox} + Q^{\cdot -}{}_{n} + 2H^{+}n \rightarrow Q_p + cyt\ c_{red} + QH_2\,N + 2H^{+}p \tag{2}$$

The net result of Eqs. (1) and (2) is that for each pair of electrons transferred from ubiquinol to cytochrome c, four protons are released on the positive side of the membrane and two protons are taken from the matrix (Eq. 3). The number of protons released on the positive side of the membrane for each electron transferred (the H^+/e^- ratio) is therefore two.

$$QH_2 + 2\ cyt\ c_{ox} + 2H^{+}n \rightarrow Q + 2\ cyt\ c_{red} + 4H^{+}p \tag{3}$$

The H^+/e^- ratio has been found to decrease under conditions where the electron transfer activity remains unchanged (Clejan and Beattie 1983; Degli Esposti et al. 1983; Beattie 1993). In particular, a cytochrome b point mutation located at center P has been reported to drastically decrease the efficiency of the energy transduction in the bc1 complex without affecting the electron transfer activity. This suggests that some aminoacids are critical for proton pumping and that, when they are modified, the protons may diffuse back to the negative side of the membrane, unmasking the existence of a proton wire between center P and center N (Bruel et al. 1995b). A model along these lines has been proposed to account for the decoupling of the bc1 complex induced by DCCD (dicyclohexylcarbodiimide; Brandt and Trumpower 1994; Miki et al. 1994).

The electron transfer catalysed by the bc1 complex is specifically inhibited by several inhibitors acting at either center P or center N (von Jagow and Link 1986). Myxothiazol, mucidin and stigmatellin block ubiquinol oxidation at center P. Myxothiazol binds to cytochrome b, midway between heme b_L and iron-sulfur center of Rieske protein (Xia et al. 1997). The binding of antimycin, diuron, NQNO and funiculosine to center N prevents b_H oxidation by ubiquinone (Fig. 1). The antimycin binding pocket is located in a transmembrane region of cytochrome b next to b_H (Xia et al. 1997). After binding to the bc1 complex, myxothiazol and stigmatellin shift the heme b_L absorption maximum towards the red and affect the [2Fe–2S] center characteristics. The binding of antimycin shifts the heme b_H absorption maximum towards the red and abolishes the semiquinone EPR (electron paramagnetic resonance) signal at center N.

4 Yeast is a Model System for Studying the Effects of Mutations on the bc1 Complex

The yeast *S. cerevisiae* is a suitable organism for functional mitochondrial studies. This unicellular eukaryote produces ATP, and since it is able to grow by either fermentation or respiration, any mutations occurring which affect mitochondrial respiration will not be lethal if the growing medium contains a carbon source, such as glucose. In addition, the particular mode of mitochondrial DNA transmission (only a few copies of mtDNA are transmitted to the daughter cell), and the existence of active mtDNA recombinations, facilitate the isolation and mapping of homoplasmic clones with point mutations in the various mitochondrial genes (Slonimski and Tzagoloff 1976). A large number of mutants have been isolated and found to affect the bc1 complex in this organism.

Based on mutant studies, it has been established that the subunits bearing the redox centers are essential to the catalytic activity of the bc1 complex. Several hundreds of point mutations in the cytochrome b and FeS Rieske genes have been characterized so far from bacteria to mammals, and residues involved in inhibitor binding, electron transfer, proton pumping and bc1 complex assembly have been identified (Slonimski and Tzagoloff 1976; di Rago et al. 1989, 1990, 1993, 1995, 1997; Tron and Lemesle-Meunier 1990; Lemesle-Meunier et al. 1993; Graham et al. 1993; Coppée et al. 1994a,b; Brasseur and Brivet-Chevillette 1995; Brasseur et al. 1995; Bruel et al. 1995a,b; for a review see Brasseur et al. 1996). The mutations are not randomly distributed, and seem to be concentrated in four main cytochrome b regions: the amino terminal part and the *de* loop linking transmembrane helices D and E in the case of center N and *cd1–cd2* and *ef* loops in that of center P. In addition to the mutations characterized in yeast, extensive data have been obtained on the bc1 complex from bacteria, and these have been reviewed by Gray and Daldal (1995) and Gennis et al. (1993). It is noteworthy that in cultured mouse cells, Howell et al. (1987) and Howell and Gilbert (1988) have identified some mitochondrial cytochrome b mutations involved in inhibitor resistance.

Studies on mutants have also shown the importance of the non-catalytic subunits. For example, the deletion of the gene encoding the core protein I leads to the complete absence of ubiquinol-cytochrome c reductase activity and the loss of the bc1 complex polypeptides, including cytochrome b (Tzagoloff et al. 1986, Crivellone et al. 1988). As the synthesis of these proteins is usually not affected by this deletion, the loss of the polypeptides is mainly due to degradation of the subunits by mitochondrial protease after their non-assembly in the complex. Similar effects leading to yeast respiratory deficiency have been reported to occur after mutations affecting the genes encoding core protein II (Oudshoorn et al. 1987), subunit VII (Schoppink et al. 1989), VIII (Maarse et al. 1988) or IX (Phillips et al. 1993). Less drastic effects detectable only in vitro have been observed with mutations in subunits VI (Schmitt and Trumpower 1990) and X (Brandt et al. 1994).

5
Mitochondrial Pathologies Involving Complex III Deficiency and Their Molecular Aspects

Defective complex III in the respiratory chain results in various forms of mitochondrial myopathy and other multisystemic abnormalities with remarkably heterogeneous clinical and biochemical features, mainly affecting organs with high energy requirements; the onset of the symptoms varies from one patient to another. The complex III defect is sometimes associated with other respiratory complex deficiencies. In most cases, the large amount of data available is difficult to interpret and may even be confusing, for several reasons. The nature of the genetic change (if any) responsible for deficiency, whether due to mutation(s) in the nuclear genome or in the mitochondrial one, is often difficult to establish since the pedigree analyses may not be conclusive. A more serious difficulty is raised by the heterogeneity and/or heteroplasmy of mtDNA molecules. In humans, these molecules present in thousands of copies in a single cell, readily accumulate point mutations, and other discrete changes which can be easily detected by rapid techniques like PCR. These changes, generally referred to as "polymorphic variants" of mtDNA sequences display vraious degrees of abundance and location in subpopulations of different geographic origin as well as in different tissue of the same individual. They also vary with age. In order to demonstrate that a given change in mtDNA sequence is the primary cause of observed pathologies one needs a large statistical basis on syndrome-displaying patients and syndrome-devoid controls. A mitochondrial data base (P. Golik et al., in prep.) will be of great help in this respect.

5.1
Isolated Complex III Deficiency

Many sporadic cases of complex III deficiency have been described: die Mauro et al. (1985) quoted eight cases, five of which showed a lack of reducible spectral cytochrome b; five patients had muscular symptoms and three had encephalopathy. Reichmann et al. (1986) described a case of mitochondrial myopathy due to complex III deficiency, where the reducible cytochrome b concentration was normal, while S. J. Kim et al. (1987) identified another case by immunoelectron microscopy. Darley-Usmar et al. (1983) observed a complex III deficiency, consisting of low complex activity, low reducible cytochrome b and a severe decrease in the amount of several nuclearly encoded complex III subunits including the Rieske protein, in a patient with mitochondrial myopathy. A yeast point mutation in the cytochrome b gene results in a very similar complex III phenotype (Brivet-Chevillotte and di Rago 1989). A specific complex III defect was also reported in a newborn infant with both skeletal muscle and liver mitochondria involving low cytochrome b and iron-sulfur protein levels, which led to severe lactic acidemia with a fatal outcome (Birch-Machin et al. 1989). Isolated complex III deficiency with no detectable cytochrome b in the liver was also described in a patient with facioscapulohumeral disease (Slipetz et al. 1991).

More recently, Morumans et al. (1997) described the case of six children with isolated complex III deficiency in muscle tissue, as well as in the liver in two cases; these patients showed variable symptoms (four of them had multiorgan disorders).

Mitochondrial DNA analyses revealed no deletions or point mutations in positions 3243 (corresponding to MELAS disease) or 8993 (corresponding to NARf(Leigh syndrome). The possibility could not be ruled out however, that complex III cytochrome b or nuclear subunit gene mutations may have been involved. As in this group of patients, these respiratory defects seem to often show a tissue-specific pattern of expression. This tissue specificity might be due either to the presence of different levels of mtDNA heteroplasmy in different tissues or to tissue-dependent differences in the pattern of expression of isoforms of complex III subunits encoded by nuclear genes (in which case, a mutation of expression of isoforms of complex III subunits encoded by nuclear genes (in which case, a mutation in one of these genes might be responsible for the anomaly). No isoforms of the complex III subunits have been identified in mammals so far, however, either in the liver or the heart (Vasquez-Avecedo et al. 1993) or in the kidneys (J. Berden, pers. comm.). Besides, Taylor et al. (1994) have shown that the specific respiratory complex activities vary from one organ to another, which leads to tissue-dependent differences in the control of mitochondrial electron transfer by complex III; these differences, in addition to mtDNA heteroplasmy, might contribute to the tissue specificity of complex III defects.

The primary defect leading to complex III deficiency has been identified in only a few cases. Marin-Garcia et al. (1996) detected a point mutation in the cytochrome b gene of cardiac mtDNA associated with complex III deficiency in patients suffering from cardiomyopathy. Complex III activity in the heart muscle was reduced by more than half in five out of the six patients with the L236l/15452 bp mutation, suggesting that this mutation may play a role in the pathophysiology of ischemic cardiomyopathy; this residue is located inside the membrane (Fig. 2 helix E). However, this mutation, observed by the authors in only two of 43 controls, is reported as putative polymorphism (observed in healthy people) by Wallace et al. (1995); this underlines the difficulty in establising the pathogenicity of a specific mutation.

Another cytochrome b mutation, G290D/15615 bp, linked to a complex III deficiency in progressive exercise intolerance, has been identified (Dumoulin et al. 1996). Here, the mutant mtDNA was heteroplasmic (80% mutant) in the patient's muscle but undetectable in the blood. The authors suggested that the substitution of this very highly conserved aminoacid during evolution hay have been the primary cause of the muscle disease in this patient. Immunoblot experiments showed a relative decrease in the levels of three complex III subunits: cytochrome b, iron-sulfur protein and the 9.5 kDa protein (Bouzidi et al. 1996). Glycine 290 lies at the end of the sixth transmembrane helix (Fig. 2, helix F) in center P. The replacement of this smallest amino acid might therefore result in severe repercussions in terms of the protein and complex stability. This interpretation is reinforced by studies with yeast cytochrome b mutants showing that the replacement of other glycine residues located inside the membrane (G33D, G131S, G340E) leads to complex III defects similar to those observed in patients (Lemesle-Meunier et al. 1993; Coppée et al. 1994b).

5.2
Combined Complex Deficiencies

In a number of mitochondrial myopathies, a complex III defect associated with other respiratory complex deficiencies has been described, involving complexes III and IV

(Kennaway et al. 1987; Castro-Cago et al. 1993; Bautista et al. 1995), or complexes II, III and I (Desnuelle et al. 1989). The primary molecular defects were not identified and there are several possible explanations for these phenotypes: (1) a complex III mutation which has repercussions on other complexes, as frequently observed with yeast cytochrome b point mutants, (2) mutation in a gene which plays a role in the assembly of complex III and neighbouring complexes (Wu and Tzagoloff 1989; Brasseur et al. 1997b), or (3) mutations occurring simultaneously in several complexes.

In a few cases, multiple mitochondrial point mutations have been elucidated: upon sequencing the entire mtDNA from two patients with severe cardiomyopathy, Ozawa et al. (1995) found a cytochrome b gene mutation (I16V/A15236G transition) associated with a 12S rRNA gene mutation (A827G transition), and with an additional CO1 gene mutation (I206M/A6521G transition) in one of the two patients. The effects of the mutations observed seemed to be synergetic and additive.

Mutations have been discovered in the cytochrome b gene in Leber hereditary optic neuropathy (LHON), which is a maternally inherited form of acute visual loss. While the majority of LHON cases can be attributed to two known complex I mutations (positions 3460 and 11778), other LHON patients lack these mutations; the primary cause of the disease in these patients is likely to be a mutation affecting a highly conserved amino acid present in the cytochrome b gene of all vertebrates (D171N in position 15257; loop cd2 in Fig. 2; Johns and Neufeld 1991: Brown et al. 1992); with yeast mutants, it was established that this cytochrome b region is important for center P functioning (Lemesle-Meunier et al. 1993; Bruel et al. 1995a, b; di Rago et al. 1995). This 15257 mutation is frequently associated with an ND5 gene mutation (complex I) in position 13708 – which is widely distributed among the caucasian lineage –, and with either another cytochrome b mutation (V356M in position 15812) or an ND2 mutation. These additional mutations appear to act synergistically, each increasing the probability of blindness.

Site-directed mutagenesis can be carried out on mitochondrial DNA in yeast using the biolistic gun technique (Butow et al. 1996), which consists of bombarding target cells with microscopic projectiles coated with specific mutated DNA. This method could be used on yeast to study the functional defects associated with mutations of conserved amino acids in cytochrome b, especially those assumed to be involved in human diseases, in order to tentatively improve our knowledge about their etiology.

Acknowledgements. We wish to acknowledge Professor Piotr Slonimski for helpful discussions. The English was revised by Dr. Jessica Blanc.

References

Bautista J, Munoz-Malaga A, Chinchon I, Segura D, Salazar JA, Bescansa E, Campos Y, Arenas J (1995) Adult onset mitochondrial myopathy without ophthalmoplegia. Four cases attributable to complex III and IV deficits in the respiratory chain. Neurologia 10:319–323

Beattie DS (1993) A proposed pathway of H^+ translocation through the bc_1 complex of mitochondria and chloroplasts. J Bioenerg Biomembr 25:233–244

Birch-Machin MA, Shepherd IM, Watmough NJ, Sherratt HS, Bartlett K, Darley-Usmar VM, Milligan DW, Welch RJ, Aynsley-Green A, Turnbull DM (1989) Fatal lactic acidosis in infancy with a defect of complex III of the respiratory chain. Pediatr Res 25:553–559

Bouzidi MF, Carrier H, Godinot C (1996) Antimycin resistance and ubiquinol cytochrome c reductase instability associated with a human cytochrome b mutation. Biochim Biophys Acta 1317:199209

Brandt U, Trumpower BL (1994) The protonmotive Q cycle in mitochondria and bacteria. Crit Rev Biochem Mol Biol 29:165–197

Brandt U, Yu L, Yu CA, Trumpower BL (1993) The mitochondrial targeting presequence of the Rieske ironsulfur protein is processed in a single step after insertion into the cytochrome bc_1 complex in mammals and retained as a subunit in the complex. J Biol Chem 268:8387–8390

Brandt U, Uribe S, Schägger H, Trumpower BL (1994) Isolation and characterization of QCR10, the nuclear gene encoding the 8.5-kDa subunit 10 of the *Saccharomyces cerevisiae* cytochrome bc_1 complex. J Biol Chem 269:12947–12953

Brasseur R (1988) Calculation of the three-dimensional structure of *Saccharomyces cerevisiae* cytochrome b inserted in a lipid matrix. J Biol Chem 263:12571–12575

Brasseur G, Brivet-Chevillotte P (1995) Characterization of mutations in the mitochondrial cytochrome b gene of *Saccharomyces cerevisiae* affecting the quinone reductase site (Q_N). Eur J Biochem 230:1118–1124

Brasseur G, Coppée JY, Colson AM, Brivet-Chevillotte P (1995) Structure-function relationships of the mitochondrial bc1 complex in temperature-sensitive mutants of the cytochrome b gene, impaired in the catalytic center N. J Biol Chem 270:29356–29364

Brasseur G, Saribas AS, Daidal F (1996) A compilation of mutations located in the cytochrome b subunit of the bacterial and mitochondrial cytochrome *bc*1 complex. Biochim Biophys Acta 1275:61–69

Brasseur G, Sled V, Liebl U, Ohnishi T, Daldal F (1997a) The amino-terminal portion of the Rieske iron-sulfur rotein contributes to the ubihydroquinone oxidation site catalysis of the *Rhodobacter capsulatus* bc1 complex. Biochemistry 36:11685–11696

Brasseur G, Tron P, Dujardin G, Slonimski PP, Brivet-Chevillotte P (1997b) The nuclear ABC1 gene is essential for the correct conformation and functioning of the cytochrome bc1 complex and the neighbouring complexes II and IV in the mitochondrial respiratory chain. Eur J Biochem 246:103–111

Braun HP, Schmitz UK (1995) Are the "core" proteins of the mitochondrial *bc*1 complex evolutionary relics of a processing protease? Trends Biochem Sci 20:171–174

Braun HP, Emmermann M, Kruft V, Schmitz UK (1992) The general mitochondrial processing peptidase from potato is an integral part of cytochrome *c* reductase of the respiratory chain. EMBO J 11:3219–3227

Brivet-Chevillotte P, di Rago JP (1989) Electron-transfer restoration by vitamin K3 in a complex III-deficient mutant of *S. cerevisiae* and sequence of the corresponding cytochrome b mutation. FEBS Lett 255:5–9

Brown MD, Voljavec AS, Lott MT, Torroni A, Yang CC, Wallace DC (1992) Mitochondrial DNA complex I and III mutations associated with Leber's hereditary optic neuropathy. Genetics 130:163–173

Bruel C, di Rago JP, Slonimski PP, Lemesle-Meunier D (1995a) Role of the evolutionarily conserved cyt. b tryptophan 142 in the ubiquinol oxidation catalysed by the bc1 complex in the yeast *Saccharomyces cerevisiae.* J Biol Chem 270:22321–22328

Bruel C, Maon S, Guérin M, Lemesle-Meunier D (1995b) Decoupling of the bc1 complex in *Saccharomyces cerevisiae:* point mutations affecting the cytochrome b gene bring new information about the structural aspects of the proton translocation. J Bioenerg Biomembr 27:527–539

Bruel C, Brasseur R, Trumpower BL (1996) Subunit 8 of the *Saccharomyces cerevisiae* cytochrome bc1 complex interacts with succinate-ubiquinone oxidoreductase complex. J Bioenerg Biomembr 28:59–67

Butow RA, Henke RM, Moran JV, Belcher SM, Perlman PS (1996) Transformation of *Saccharomyces cerevisiae* mitochondria using the biolistic gun. Methods Enzymol 264:265–278

Castro-Cago M, Novo-Rodriguez I, Garcia-Caballero T, Campos-Gonzalez Y, Huertas-Barbero R, Arenasmora J (1993) Miopatia mitcondrial por deficit de los complejos III y IV. Una observation familiar. An Esp Pediatr 38:82–86

Clejan L, Beattie DS (1983) Dicyclohexylcarbodiimide blocks proton ejection and affects antimycin binding but not electron transport in complex III from yeast mitochondria. J Biol Chem 258:14271–14275

Coppée JY, Brasseur G, Brivet-Chevillotte P, Colson AM (1994a) Non native intragenic reversions selcted from *Saccharomyces cerevisiae* cytochrome b-deficient mutants: structural and functional features of the catalytic center N domain. J Biol Chem 269:4221–4226

Coppée JY, Tokutake N, Marc D, di Rago JP, Miyoshi H. Colson AM (1994b) Analysis of revertants from respiratory deficient mutants within the center N of cytochrome b in *Saccharomyces cerevisiae.* FEBS Lett 339:1–6

Crivellone MD, Wu M, Tzagoloff A (1988) Assembly of the mitochondrial membrane system. Analysis of structural mutants of the yeast QH_2-cytochrome *c* reductase complex. J Biol Chem 263:14323–14333

Crofts AR, Robinson H, Andrews K, Van Doren S, Berry E (1987) Catalytic sites for reduction and oxidation of quinones. In: Papa S, Chance B, Ernster L (eds) Cytochrome systems: molecular biology and bioenergetics. Plenum Press, New York, pp 617–624

Crofts AR, Barquera B, Gennis RB, Kuras R, Guergova-Kuras M, Berry EA (1998) Mechanistic aspects of the Yo-site of the bc1 complex as revealed by mutagenesis studies and the crystallographic structure. In: Peschek GA, Loeffelhardt W, Schmetterer G (eds) The phototrophic prokaryotes. Proceedings of the 9th International Symposium on Phototrophic Prokaryotes, Viemma, Sept. 1997. Pleum Press, pp 229–239

Darley-Usmar VM, Kennaway NG, Buist NRM, Capaldi RA (1983) Deficiency in ubiquinone cytochrome c reductase in a patient with mitochondrial myopathy and lactic acidosis. Proc Natl Acad Sci USA80:5103–5106

De Vries S (1986) The pathway of electron transfer in the dimeric QH2 cyt c oxidoreductase. J Bioenerg Biomembr 18:195–224

Degli Esposti M, Meier EM, Timoneda J, Lenaz G (1983) Modification of the catalytic formation of the mitochondrial cytochrome bc1 complex by dicyclohexylcarbodiimide. Biochim Biophys Acta 725:349–360

Degli Esposti M, de Vries S, Crimi M, Ghelli A, Patarnello T, Meyer A (1993) Mitochondrial cytochrome *b:* evolution and structure of the protein. Biochim Biophys Acta 1143:243–271

Desnuelle C, Birch-Machin M, Pelissier JF, Bindoff LA, Ackrell BAC, Turnbull DM (1989) Multiple defects of the respiratory chain including complex II in a family with myopathy and encephalomyopathy. Biochem Biophys Res Commun 163:695–700

Di Mauro S, Bonilla E, Zeviani S, Makagawa M, De Vivo D (1985) Mitochondrial myopathies. Ann Neurol 17:521–538

Di Rago JP, Coppée JY, Colson AM (1989) Molecular basis for resistance to myxothiazol, mucidin (strobilurin A) and stigmatellin. Cytochrome *b* inhibitors acting at the center o of the mitochondrial ubiquinol-cytochrome *c* reductase in *Saccharomyces cerevisiae.* J Biol Chem 264:14543–14548

Di Rago JP, Netter P, Slonimski PP (1990) Intragenic suppressors reveal long distance interactions between inactivating and reactivating amino acid replacements generating three-dimensional constraints in the structure of mitochondrial cytochrome *b.* J Biol Chem 265:15750–15757

Di Rago JP, Macadre C, Lazowska J, Slonimski PP (1993) The c-terminal domain of yeast cytochrome *b* is essential for a correct assembly of the mitochondrial cytochrome *bc*1 complex. FEBS Lett 328:153–158

Di Rago JP, Hermann-Le Denmat S, Pâques F, Risler JL, Netter P, Slonimski PP (1995) Genetic analysis of the folded structure of yeast mitochondrial cytochrome *b* by selection of intragenic second-site revertants. J Mol Biol 248:804–811

Di Rago JP, Sohm F, Boccia C, Dujardin G, Trumpower BL, Slonimski PP (1997) A point mutation in the mitochondrial cytochrome *b* complex in *Saccharomyces cerevisiae.* J Biol Chem 272:4699–4704

Dumoulin R, Sagnol I, Ferlin T, Bozon D, Stepien G, Mousson B (1996) A novel gly290asp mitochondrial cytochrome b mutation linked to a complex III deficiency in progressive exercice intolerance. Mol Cell Probes 10:389–391

Emmermann M, Braun HP, Arretz M, Schmitz UK (1993) Characterization of the bifunctional cytochrome c reductase-processing peptidase complex from potato mitochondria. J Biol Chem 268:18936–18942

Gennis RB, Barquera B, Hacker B, Van Doren SR, Arnaud S, Crofts AR, Davidson E, Gray KA, Daldal F (1993) The bc_1 complexes of *Rhodobacter sphaeroides* and *Rhodobacter capsulatus.* J Bioenerg Biomembr 25:195–209

Glaser E, Sjöling S, Szigyarto C, Eriksson AC (1996) Plant mitochondrial protein import: precursor processing is catalysed by the integrated mitochondrial processing peptidase (MPP)/bc1 complex and degradation by the ATP-dependent proteinase. Biochim Biophys Acta 1275:33–37

Gonzalez-Halphen D, Lindorfer MA, Capaldi RA (1988) Subunit arrangement in beef heart complex III. Biochemistry 27:7021–7031

Graham LA, Brandt U, Sargent JS, Trumpower BL (1993) Mutational analysis of assembly and function of the iron-sulfur protein of the cytochrome *bc*1 complex in *S. cerevisiae.* J Bioenerg Biomembr 25:245–257

Gray KA, Daldal F (1995) Mutational studies of the cytochrome bc1 complex. In. Blankenship RE, Madigan MT, Bauer C (eds) Anoxygenic photosynthetic bacteria. Kluwer, Dordrecht, pp 747–774

Gray KA, Davidson E, Daldal F (1992) Mutagenesis of methionine-183 drastically affects the physicochemical properties of cytochrome c_1 complex of *Rhodobacter capsulatus.* Biochemistry 31:11864–11873

Howell N, Gilbert K (1988) Mutational analysis of the mouse mitochondrial cytochrome b gene. J Mol Biol 203:607–618

Howell N, Appel J, Cook JP, Howell B, Hausworth WW (1987) The molecular basis of inhibitor resistance in a mammalian mitochondrial cytochrome b mutant. J Biol Chem 262:2411–2414

Iwata S, Saynovits M, Link TA, Michel H (1996) Structure of a water soluble fragment of the "Rieske" ironsulfur protein of the vobine heart mitochondrial cytochrome bc1 complex determined by MAD phasing at 1.5 Å resolution. Structure 4:567–579

Iwata S, Lee JW, Okada K, Lee JK, Iwata M, Rasmussen B, Link TA, Ramaswamy S, Jap BK (1998) Complete structure of the 11-subunit bovine mitochondrial cytochome b1 complex. Science 281:64–71

Jänsch L, Kruft V, Schmitz UK, Braun HP (1995) Cytochrome *c* reductase from potato does not comprise three core proteins but contains an additional low-molecular-mass subunit. Eur J Biochem 228:878–885

Johns DR, Neufeld MJ (1991) Cytochrome b mutations in Leber hereditary optic neuropathy. Biochem Biophys Res Commun 181:1358–1364

Kennaway NG, Wagner ML, Capaldi RA, Takamiya S, Yanamura W, Ruitenberg W, Sengers RCA, Trijbels JMF (1987) Combined deficiencies of complexes III and IV of the respiratory chain, involving both nuclear and mitochondrial gene products, in skeletal muscle of a patient with lactic acidosis. J Inherit Metab Dis 10:247–251

Kim CH, Balny C, King TE (1987) Role of the hinge protein in the electron transfer between cardiac cytochrome c_1 and c. J Biol Chem 262:8103–8108

Kim H, Xia D, Yu CA, Xia JZ, Kachurin AM Zhang L, Yu L, Deisenhofer J (1998) Inhibitor binding changes domain mobility in the iron-sulfur protein of the mitochondrial bc1 complex from bovine heart. Proc Natl Acad Sci USA 95:8026–8033

Kim SJ, Lee KO, Takamiya S, Capaldi RA (1987) Mitochondrial myopathy involving ubiquinol-cytochrome c oxidoreductase (complex III) identified by immunoelectron microscopy. Biochim Biophys Acta 894:270–276

Konishi K, Van Doren SR, Kramer DM, Crofts AR, Gennis RB (1991) Preparation and characterization of the water-soluble heme-binding domain of cytochrome c_1 from the *Rhodobacter sphaeroides* bc_1 complex. J Biol Chem 266:14270–14276

Lemesie-Meunier D, Brivet-Chevillotte P, di Rago JP, Slonimski PP, Bruel C, Tron T, Forget N (1993) Cytochrome *b* deficient mutants of the ubiquinol-cytochrome *c* oxireductase in *Saccharomyces cerevisiae.* J Biol Chem 268:15626–15632

Leonard K, Wingfield P, Arad T, Weiss H (1981) Three-dimensional structure of ubiquinol:cytochrome c reductase from *Neurospora* mitochondria determined by electron microscopy of membrane crystals. J Mol Biol 149:259–274

Li Y, Leonard K, Weiss H (1981) Membrane-bound and water-soluble cytochrome c1 from *Neurospora* mitochondria. Eur J Biochem 1^16:199–205

Ljungdahl PO. Pennoyer JD, Robertson DE, Trumpower BL (1987) Purification of highly active cytochrome bc_1 complexes from phylogenetically diverse species by a single chromatographic procedure. Biochim Biophys Acta 891:227–241

Maarse AC, De Haan M, Schoppink PJ, Berden JA, Grivell LA (1988) Inactivation of the gene encoding the 11-kDa subunit VIII of the ubiquinol-cytochrome-c oxidoreductase in *Saccharomyces cerevisiae.* Eur J Biochem 172:179–184

Marin-Garcia J, Hu Y, Ananthakrishnan R, Pierpont GL, Goldenthal MJ (1996) A point mutation in the cytb gene of cardiac mtDNA associated with complex III deficiency in ischemic cardiopathy. Biochem Mol Biol Int 40:487–495

Miki T, Miki M, Orii Y (1994) Membrane potential-linked reversed electron transfer in the beef heart cytochrome bc1 complex reconstituted into potassium-loaded phospholipid vesicles. J Biol Chem 269:1827–1833

Mitchell P (1976) Possible molecular mechanisms of the protonmotive function of cytochrome systems. J Theor Biol 62:327–367

Mourmans J, Wendel U, Bentlage HA, Trijbels JM, Smeitink JA, de Coo IF, Gabreels FJ, Sengers RC, Ruitenbeek W (1997) Clinical heterogeneity in respiratory chain complex III deficiency in childhood. J Neurol Sci 149:111–117

Nakai M, Ishiwatari H, Asada A, Bogaki M, Kawai K, Tanaka Y, Matsubara H (1990) Replacement of putative axial ligands of heme iron in yeast cytochrome c_1 by site-directed mutagenesis. J Biochem (Tokyo) 108:798–803

Oudshoorn P, Van Steeg H, Swinkels BW, Schoppink P, Grivell LA (1987) Subunit II of yeast QH_2:cytochrome-*c* oxidoreductase. Eur J Biochem 163:97–103

Ozawa T, Katsumata K, Hayakawa M, Tanaka T, Itoyama S, Nunoda S, Sekiguchi M (1995) Genotype and phenotype of severe motochondrial cardiomyopathy: a recipient of heart transplantation and the genetic control. Biochem Biophys Res Commun 207:613–620

Phillips JD, Graham LA, Trumpower BL (1993) Subunit 9 of the *Saccharomyces cerevisiae* cytochrome bc_1 complex is required for insertion of EPR-detectable iron-sulfur cluster into the Tieske iron-sulfur protein. J Biol Chem 268:11727–11736

Reichmann H, Rohkamm R, Zeviani M, Servidei S, Ricker K, di Mauro S (1986) Mitochondrial myopathy due to complex III deficiency with normal reducible cytochrome b concentration. Arch Neurol 43:957–961

Rieske JS, Maclennan DH, Coleman R (1964) Isolation and properties of an iron-protein from the (reduced coenzyme Q)-cytochrome c reductase of the respiratory chain. Biochem Biophys Res Commun 15:338–344

Saribas S, Valkova M, Tokito M, Zhang Z, Berry E, Daldal F (1998) Interactions between the cytochrome b, cytochrome c1 and the Fe-S protein subunits at the ubihydroquinone ocidation (Q_o site of the bc1 complex of *Rhodobacter capsulatus.* Biochemistry 37:8105–8114

Schägger H, Brandt B, Gencic S, von Jagow G (1995) Ubiquinol cytochrome-c reductase from human and bovine mitochondria. Methods Enzymol 260:82–96

Schmitt ME, Trumpower BL (1990) Subunit 6 regulates half-of-the-sites reactivity of the dimeric cytochrome bc_1 complex in *Saccharomyces cerevisiae.* J Biol Chem 265:17005–17011

Schoppink PJ, Berden JA, Grivell LA (1989) Inactivation of the gene encoding the 14-kDa subunit VII of yeast ubiquinol cytochrome c oxidoreductase and analysis of the resulting mutant. Eur J Biochem 181:475–483

Schulte U, Arretz M, Schneider H, Tropschug M, Wachter E, Neupert W, Weiss H (1989) A family of mitochondrial proteins involved in bioenergetics and biogenesis. Nature 339:147–149

Simpkin D, Palmer G, Devlin FJ, McKenna MC, Jensen GM, Stephens PJ (1989) The axial ligands of heme in cytochromes: a near-infrared magnetic circular dichroism study of yeast cytochromes c, c_1, b and spinach cytochrome f. Biochemstry 28:8033–8039

Slipetz DM, Aprille JR, Coudyer PR, Rozen R (1991) Deficiency of complex III of the mitochondrial respiratory chain in a patient with focioscapulohumeral diseas. Am J Hum Genet 48:502–510

Stonimski PP, Tzagoloff A (1976) Localization in yeast mitochondrial DNA of mutations expressing a deficiency of cytochrome oxidase and/or coenzyme QH_2-cytochrome *c* reductase. Eur J Biochem 73:276–286

Taylor RW, Birch-Machin MA, Bartlett K, Lowerson SA, Turnbull DM (1994) The control of mitochondrial oxidations by complex III in rat muscle and liver mitochondria. Implications for our understanding of mitochondrial cytopathies in man. J Biol Chem 269:3523–3528

Tron T, Lemesle-Meunier D (1990) Two substitutions at the same position in the mitochondrial cytochrome b gene of *S. cerevisiae* induce a mitochondrial myxothiazol resistance and impair the respiratory growth of the mutated strains albeit maintaining a good electron transfer activity. Curr Genet 18:413–419

Trumpower BL (1990) Cytochrome *bc*1 complexes of microorganisms. Microbiol Rev 54:101–129

Trumpower BL, Edwards CA (1979) Purification of a reconstitutively active iron-sulfor protein (oxidation factor) from succinate-cytochrome c reductase complex of bovine heart mitochondria. J Biol Chem 254:8697–8706

Tzagoloff A, Wu M, Crivellone M (1986) Assembly of the mitochondrial membrane system. Characterization of COR1, the structural gene for the 44-kilodalton core protein of yeast coenzyme QH_2-cytochrome *c* reductase. J Biol Chem 261:17163–17169

Vasquez-Avecedo M, Antaramian A, Corona N, Gonzalez-Halphen D (1993) Subunit structures of purified beef mitochondrial cytochrome bc1 complex from liver and heart. J Bioenerg Biomembr 25:401–410

Von Jagow G, Link T (1986) Use of specific inhibitors on the mitochondrial *bc*1 complex. Methods Enzymol 126:253–271

Wallace DC, Lott MT, Brown MD, Kuoponen K, Torroni A (1995) Report of the committee on human mitochondrial DNA. In: Cuticchia AJ (ed) Human gene mapping 1995: a compendium. Johns Hopkins University Press, Baltimore, pp 910–954

Wu M, Tzagoloff A (1989) Identification and characterization of a new gene (CBP3) required for the expression of yeast coenzyme QH2-cytochrome c reductas. J Biol Chem 264:11122–11130

Xia D, Yu CA, Kim H, Xia JZ, Kachurin AM, Zhang L, Yu L, Deisenhofer J (1997) Crystal structure of the bc_1 complex from bovine heart mitochondria. Science 277:60–66

Yang M, Jensen RE, Yaffe MP, Opligger W, Schatz G (1988) Import of proteins into yeast mitochondria: the purified matrix processing protease contains two subunits which are encoded by the nuclear MAS1 and MAS2 genes. EMBO J 7:3857–3862

Yu L, Chiang YL, Yu CA, King TE (1975) A trypsin-resistant heme peptide from cardiac cytochrome c_1. Biochim Biophys Acta 379:33–42

Zhang Z, Huang L, Shulmeister VM, Chi YI, Kim KK, Hung LW, Crofts AR, Berry EA, Kim SH (1998) Electron transfer by domain movement in cytochrome bc1. Nature 392:677–684

CHAPTER 8

Cytochrome c Oxidase and Mitochondrial Pathology 8

A. Poyau and C. Godinot[1]

Contents

1 Introduction

On the basis of studies undertaken by MacMunn at the end of the last century, in 1924, Warburg introduced the term "Atmungsferment" to define the enzyme activity linked to cellular respiration. Simultaneously, Keilin, using spectral analysis, put forward a series of cell pigments that he named cytochromes. He had discovered the respiratory chain. In 1938, he showed that the bands associated with cytochrome a_3 were equivalent to Warburg's "Atmungsferment" (review: Margoliash 1988). Since then, the main difficulty in understanding the mechanisms generating energy in the presence of oxygen was due to the insertion of the respiratory chain complexes, including cytochrome c oxidase (COX), inside the mitochondrial membrane. In 1961, Mitchell pro-

[1] Due to space limitation, the authors apologize for not being able to cite all the original articles where the discoveries were described. More often than not, references are limited to reviews where the reader can find the citations of the original work.
Centre de Génétique Moléculaire et Cellulaire, UMR 5534 CNRS-UCB Lyon I, 43 Boulevard du 11 novembre 1918, 69622 Villeurbanne, France

posed the chemiosmotic theory according to which the membrane is an essential component which permits the establishment of a charge separation during electron transfer. This creates a proton electrochemical potential difference between the two faces of the inner membrane. The energy of this gradient can then be used by mitochondrial ATPase-ATP synthase, embedded in the same membrane, to synthesize ATP from ADP and phosphate. In 1977, Wikström's group studies suggested that the reduction of oxygen to water, catalyzed by COX, is accompanied by proton transfer across the membrane. These hypotheses have now been demonstrated (review: Capaldi 1990). COX, ferrocytochrome c : oxygen oxidoreductase, also called complex IV (EC 1.9.3.1) is embedded in the mitochondrial inner membrane and catalyzes the final oxidation in the respiratory chain. This enzyme accepts four electrons from cytochrome c and transfers them to molecular oxygen which is then reduced to water, consuming four protons. Simultaneously to this electron transfer, up to four protons are translocated from the mitochondrial matrix to the intermembrane space, creating the proton electrochemical gradient. The overall reaction can be written as:

$$4\ Fe^{++}\ cyt\ c + 8H^{+}_{inside} + O_2 \rightarrow 4\ Fe^{+++}\ cyt\ c + 4H^{+}_{outside} + 2\ H_2O$$

Recently, COX structure has been resolved at 2.8 Å resolution from the soil bacterium *Paracoccus denitrificans* 4-subunit enzyme (Iwata et al. 1995) and from the beef heart 13-subunit enzyme (Tsukihara et al. 1995, 1996). These two enzymes exhibit a striking structural homology at the level of the redox centers. However, the possible electron and proton pathways still need to be specified. This knowledge as well as that related to the regulation of the expression of the genes encoding COX subunits and to the mechanisms of the biogenesis and assembly of the complex should serve in the future to help understand the causes of human pathologies involving this enzyme. In this review, the main features of COX structure and function and of the genes encoding the human COX subunits will be summarized. Finally, the characteristics of the diseases concerning COX deficiencies will be described.

2 Cytochrome c Oxidase Structure and Function

2.1 Subunit Composition

COX composition varies from 3 or 4 subunits in bacteria up to 13 in mammals. In mammals, the larger subunits I, II and III are encoded by mitochondrial DNA (mtDNA) and the ten others by the nuclear genome (Merle and Kadenbach 1982). Subunits I, II and III, homologues to bacterial ones, represent the enzyme catalytic core where the metallic centers, the binding site for oxygen and cytochrome c, as well as proton and electron pathways, are located. The subunits of nuclear origin, named subunits IV, Va, Vb, VIa, VIb, VIc, VIIa, VIIb, VIIc and VIII (Fig. 1) are likely to be involved in complex biogenesis and in enzyme regulation.

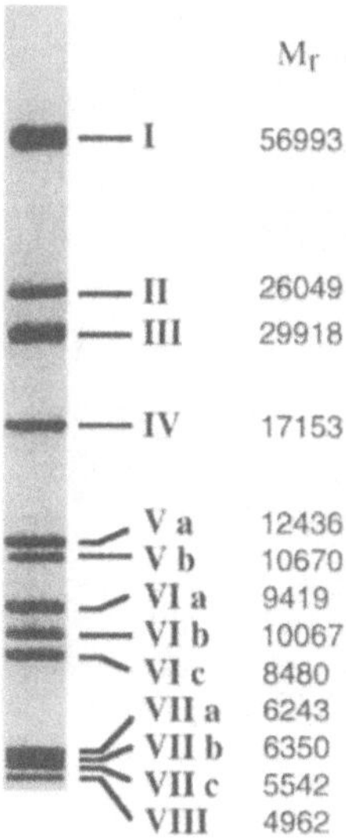

Fig. 1. Beef cytochrome c oxidase subunits separated by SDS gel electrophoresis in the presence of urea and stained with Coomassie blue. Subunit identification as described by Merle and Kadenbach (1982). *Mr* indicates the molecular ratio of each subunit in kDa (Capaldi et al. 1995). The indicated Mr of subunits VIa, VIIa and VIII correspond to those of the L isoforms

2.2 Tridimensional Structure

COX contains three redox centers: the homobinuclear copper A center (Cu_A), the heme a and the heterobinuclear Cu_B –heme a_3 center (heme a_3 - Cu_B). Iron is the ligand bound to the hemes. According to the X-ray crystallographic structure, subunit II carries the two Cu_A atoms, of mixed valence 1.5, which are at a distance of 2.6 Å *(P. denitrificans)* or of 2.7 Å (beef heart). They are stabilized by binding to Cys^{196} and Cys^{200}, to His^{161} and His^{204}, to Met ^{207} and to Glu^{198} (Speno et al. 1995, Tsukihara et al. 1995, 1996), (the numbering corresponds to the human or bovine proteins). This subunit II is inserted in the mitochondrial membrane by two transmembrane α helices capped with a ten-strand β barrel structure which maintains the Cu_A at 7 Å above the inner membrane surface, on the cytosol side. Subunit I contains 12 transmembrane α helices organized into three groups of four helices arranged according to a threefold axis of symmetry and forming a pore in the center of each group (Iwata et al. 1995; Tsukihara et al. 1996). One of these pores contains heme a, another the heme a_3 - Cu_B center and the third one is empty. The heme a_3 is linked by His^{376} (helix X) while the Cu_B is bound to His^{240} (helix VI), His ^{290} (helix VII). The ligands for heme a are His^{61} (helix II) and His^{378} (helix X; Iwata et al. 1995). The heme a and heme a_3 - Cu_B centers are located 13 Å from the membrane surface, near the cytochrome c binding site. The subunit III contains seven transmembrane helices and is bound to subunit I on the opposite side of subunit II (Fig. 2).

X-ray crystallography of the beef heart enzyme also reveals the structure of the ten subunits of nuclear origin and the presence of Mg and Zn. Most nuclear subunits include a transmembrane α helix interacting with the others or with those of the mitochondrially encoded subunits. Only subunits Va and Vb on the matrix side and VIb on the intermembrane space are exclusively located outside the membrane (Fig. 2). The subunit Vb bound to subunit I and III on the matrix side contains a zinc

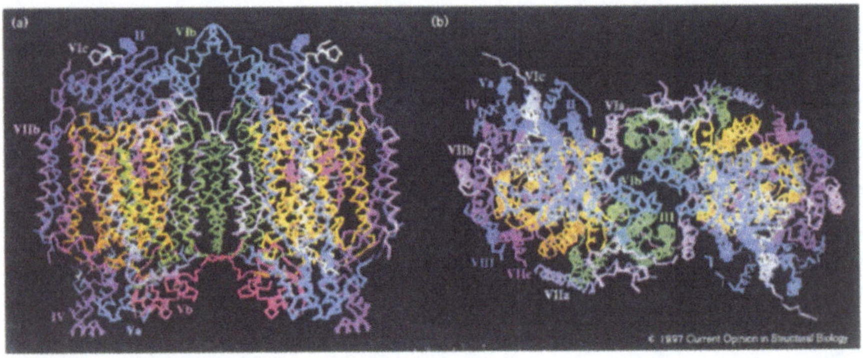

Fig. 2. X-ray crystallographic structure of beef heart cytochrome c oxidase: Cα backbone trace of COX dimer (Yoshikawa 1997). **a** Side view against the transmembrane helices. **b** Top view from the cytosolic side. Subunits I *(yellow)*, II *(dark blue)*, III *(green)*, IV *(purple)*, Va *(lightblue)*, Vb *(pink)*, VIa *(mauve)*, VIb *(turquoise)*, VIc *(gray)*, VIIa *(lavender)*, VIIb *(palepurple)*, VIIc *(dark pink)* and VIII *(indigo)*

atom linked to 4 Cys and to a zinc finger motif. This Zn site of unknown function is not present in the bacterial enzyme. The Mg is located at the level of the membrane surface on the cytoplasmic side and is coordinated with Glu^{298} of subunit II and with His^{368} and Asp^{369} of subunit I (Tsukihara et al. 1995; Figs. 2 and 3). The beef heart enzyme is a dimer whose two monomers are linked by the subunits VIa and VIb.

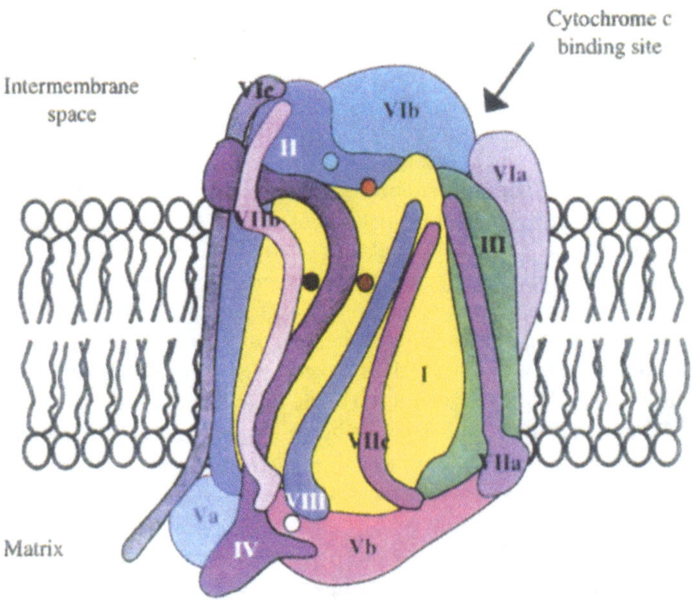

Fig. 3. Graphic representation of a beef heart cytochrome c oxidase monomer (adapted from Tsukihara et al. 1995, 1996). The various subunits are drawn with the same colors as in Fig. 2. The *turquoise, red, brown, black* and *white circles* stand for CuA, Mg, heme a_3 – Cu_B center, heme a and Zn, respectively

2.3
Electron and Proton Pathways

The monomeric form of the enzyme is as active as the dimer with respect to its electron transfer and oxygen reduction activity. On the contrary, proton pumping only occurs in the dimeric form. The two extra-membrane domains located on the cytosolic side are intertwined around two axes to form a concave surface of about 25 Å in diameter which corresponds to the reduced cytochrome c binding site. This site consists of amino acids from subunits II, VIa and VIb. In addition, several acidic residues from subunits I, II, III and VIb would interact with cytochrome c. The residues Glu^{109} and Asp^{139} of subunit II, which are conserved even in bacteria, are likely to be essential to cytochrome c binding (review Capaldi 1990; Tsukihara et al. 1996). By studying *P. denitrificans* mutants, Witt et al. (1998) have shown that the subunit II Trp^{121}, which is surrounded by a hydrophobic pocket, is the electron entry site to oxidase. The electrons issued from reduced cytochrome c are first transferred to the Cu_A center, then to the heme a center and finally to the heme a_3 – Cu_B center where the oxygen reduction site is located. A network of hydrogen bonds constitutes the most probable pathway for transfering electrons from Cu_A and heme a. In contrast, the proton pumping mechanism, as well as the various channels for water, oxygen protons, remain hypothetical (Tsukihara et al. 1996). From electrostatic calculations on *P. denitrificans* COX structure, Kannt et al. (1998) deduced that the coupling of electron transfer and proton translocation occurred at the level of the heme a_3 – Cu_B center. A hydroxide ion bound to Cu_B becomes protonated to form water upon transfer of the first electron to the heme a_3 – Cu_B center. Further reduction of this binuclear center was due to proton uptake by interaction of a cluster of 18 amino acids located around this center with the subunit II Glu^{78}. His^{325} modeled in an alternative orientation away from Cu_B should then become protonated. The complete understanding of the proton pumping pathways will probably need to unravel several structures of different enzyme conformations and will be helped by site-directed mutagenesis of amino acids located on presumed electron or proton pathways. For example, in *R. sphaeroides*, Adelroth et al. (1998) have changed Lys^{362} into Met and Thr^{359} into Ala, both located in subunit I on one of these pathways. This decreased the COX turnover and impaired proton release in the absence of dioxygen, while the electron transfer between heme a and a_3 was not modified. These mutated amino acids were involved in proton uptake during reduction of the oxidized heme a_3 – Cu_B center. In contrast, proton uptake during oxidation of the fully reduced enzyme takes place along a different pathway, through subunit II Glu^{286}.

2.4
Cytochrome c Oxidase Regulation and Biogenesis

Cytochrome oxidase activity is regulated by the availability of reduced cytochrome c, by the oxygen concentration and by the electrochemical proton gradient across the mitochondrial membrane. In addition, it depends on the presence of nuclear subunit isoforms put forward by Merle and Kadenbach (1982). The tissue specificity and evolution during development of these isoforms suggest that they can have a regulatory function. Isoforms H (heart) or L (liver) have been identified for subunits VIa, VIIa and VIII in the beef heart enzyme but only for subunits VIa and VIIa in humans (Kadenbach et al. 1995). Some isoforms might be unknown in man. The H isoform is found in heart and skeletal muscle while the L isoform is ubiquitous. A regulatory function has been characterized for the beef VIaH. This subunit can bind ATP on its N-terminal end located on the matrix side while the liver isoform VIaL cannot. ATP binding to the heart enzyme uncouples transmembrane proton flux from electron transfer, but this effect was reversed by ADP. ATP could therefore have a function in thermogenesis at rest in humans (reviews: Kadenbach et al. 1995; Taanman 1997). The tridimensional structure of beef heart COX confirms the existence of a putative nucleotide binding site between Arg^{14} and Arg^{17} of the VIaH isoform and the Trp^{275} of the helix VII of the subunit I belonging to the other monomer of the complex. This interaction would also contribute to dimer stability under physiological conditions (Tsukihara et al. 1996). The ATP/ADP ratio can also regulate COX activity through subunits IV and VIa (Anthony et al. 1993; Arnold et al. 1998): ATP is an allosteric inhibitor of COX upon binding to subunit IV. Subunit Va is the target of 3,5-diiodothyronine whose binding abolishes the allosteric inhibition of COX activity induced by ATP binding (Arnold et al. 1998).

In addition, studies performed in yeast show that at least eight nuclear proteins, such as the products of the PET genes (Poyton and McEwen 1996), COX 10 and the OXA I protein (Bonnefoy et al. 1994), which does not belong to the enzyme complex, are essential for COX assembly. It is known that at least COX 10 and OXA I are phylogenetically conserved from bacteria to man (review: Grossman and Lomax 1997). OXA I would mediate the export of the N- and C-termini of mitochondrially encoded COX subunit II from the mitochondrial matrix to the intermembrane space (Hell et al. 1997). The identification of other human proteins homologous to the yeast proteins and specifically involved in COX assembly is presently under investigation.

3 Localization, Structure and Expression of Cytochrome c Oxidase Genes

3.1 Localization of Cytochrome c Oxidase Genes

In humans, the three catalytic subunits of mitochondrial origin are encoded by the genes COI, COII and COIII, located on the heavy strand of mtDNA in positions 5904-7444, 7586-8262 and 9207-9990, respectively (Anderson et al. 1981). The incomplete stop codon of subunit III is formed during polyadenylation.

All cDNAs encoding human COX subunits of nuclear origin have been sequenced, but only the human genes encoding subunit Vb, VIaH and VIIaH have been entirely sequenced and characterized (Grossman and Lomax 1997; Bachman et al. 1997; Wolz et al. 1997). The chromosomal location of some human COX subunit genes has also been described. These data are summarized in Table 1. Most subunits of nuclear origin are synthesized as precursors and contain a cleavable signal presequence necessary for their translocation across the outer and inner mitochondrial membranes and for their maturation (Pfanner et al. 1994). However, in humans, two subunits (VIb and VIc) and one isoform (VIaL) do not possess this presequence. There-

Table 1. Chromosome location and structure of nuclear genes encoding human cytochrome c oxidase subunits. (According to Shoffner and Wallace review, 1995)

Subunit	cDNA	Mature protein size	Presequence	Gene location chromosome	Main features
IV	668 bp	147 aa	22 aa	16 q22–q24	One processed pseudogene in 14 q21–qter[a]
Va	647 bp	109 aa	41 aa		No tissue specific isoform
Vb	494 bp	98 aa	31 aa	2 cen–q13	Five exons Seven pseudogenes on chromosomes 4, 6, 7, 11, 12, 13 and 22 No tissue specific isoform
VIa	VIaL : 543 bp	VIaL : 86 aa	VIaL : -		62% homology between isoforms
	VIaH : 371 bp	VIaH : 86 aa	VIaH : 11 aa	VIaH : 16p[b]	VIaH : 3 exons and 2 introns[b]
VIb	439 bp	86 aa	-	19 q13.1	Last exon sequenced Four pseudogenes (two processed) on chromosomes 7, 15, 22 and 8 or 17 No tissue specific isoform
VIc	421 bp	75 aa	-		
VIIa	VIIaL : 408 bp	VIIaL : 60 aa	VIIaL : 23 aa	VIIaL : 4 q31–qter or 6[c]	63% homology between isoforms
	VIIaH : 341 bp	VIIaH : 58 aa	VIIaH : 21 aa	VIIaH : 19 q13.1	VIIaH : four exons[d]
VIIb	468 bp[e]	56 aa	24 aa		Processed pseudogene?
VIIc	334 bp	47 aa	16 aa		
VIII	472 bp	44 aa	25 aa	11 q12–13	

[a] Ewart et al. (1990).
[b] Bachman et al. (1997).
[c] Merante et al. (1997).
[d] Wolz et al. (1997).
[e] Sadlock et al. (1993).
aa: amino acids; bp: base pair.

fore, they must be imported into mitochondria with the help of a specific receptor which likely recognizes a domain of the mature protein (review: Shoffner and Wallace 1995).

3.2 Regulation of Isoform Expression

The H isoforms seem to be regulated at the level of transcription because muscle-specific promoter elements (Myo-D) have been identified in the genes of the beef isoforms VIaH and VIIaH. On the contrary, the regulation of the L isoform expression should be post-transcriptional, since the COLBP protein (cytochrome c oxidase L-form transcript binding protein) interacts with its mRNAs and could not be detected in beef heart or skeletal muscle. This COLBP protein has now been recognized as being identical to glutamate dehydrogenase (Preiss et al. 1995). During development, there is an interconversion among some isoforms, modifying COX properties. In rats, during transition from the fetus to the adult state, the VIaH and VIIIH subunits increase while the fetal isoforms analogous to those found in the adult liver, decrease (Schägger et al. 1995). Factors induced during development must therefore regulate the expression of these genes. Adrenal steroid hormones increase fetal kidney COX mRNAs but have no effect on those of liver and heart (Prieur et al. 1998).

3.3 Transcription Factor Binding Sites in COX Nuclear Gene Promoters

Elements acting in *cis* or in *trans* regulate the expression of many nuclear genes involved in oxidative phosphorylation and in the mammalian COX complex in particular. The promoters of the subunit IV gene in beef, rat and mouse have neither TATA box or CAAT box but they contain several binding sites for the regulatory proteins Sp1 and NRF-2 (nuclear respiratory factor 2). The same profile has also been found in the beef COX VIIa L promoter and in the rat, mouse and human COX Vb promoter. In addition, a functional binding site for NRF-1 has also been identified in the promoters of different COX subunits for several species, including humans (review: Scarpulla 1997). The consensus sequence for NRF-1 binding ([T/C]GCGCA [T/C]GCGC[A/G]) is also present in human gene promoters of complex III or complex V subunits as well as in those of RNAse MRP (maturation of RNA priming) which interferes in mtDNA replication. Similarly, the NRF-1 and NRF-2 binding sites are present in the promoter of mtTFA (mitochondrial transcription factor A) which binds to the mtDNA D-loop to stimulate its transcription (review: Scarpulla 1997). Finally, binding sites for the regulatory proteins Mt3 (enhancer) or Mt4 (silencer) present in the human mtDNA D-loop have been identified in a very distant upstream region of the human subunit Vb gene (Bachman et al. 1996). Therefore, the transcription factors must act in a coordinated way on the expression of mtDNA and COX nuclear genes.

4 Pathologies Associated with COX Deficiencies

The first known COX deficiency was described in 1972 for the Menkes disease. It was a tricopoliodystrophy due to a recessive copper metabolism linked to the chromosome X. The clinical phenotype was due to deficiencies of all copper containing proteins and therefore of COX.

The other diseases associated with a deficient COX described since 1977 have been reviewed by Di Mauro et al. (1990) and Shoffner and Wallace (1995). Some induce a muscular myopathy that can be benign or severe, the others are essentially associated with neurologic symptoms of which the most common is Leigh's syndrome (see below).

Acute myopathies appearing shortly after birth, associated with a general weakness, respiratory distress and lactic acidosis can be fatal before the age of 1 year (De Toni-Fanconi syndrome). The syndrome transmission seems to be of the autosomal recessive type. The COX activity in these patient's muscles is generally not higher than 10% of the normal value and the deficit is occasionally found in the kidney, the liver and the heart.

Other cases of benign infant myopathy begin with the same clinical symptoms and partial COX deficiency in muscle (COX-negative fibers). These children need respiratory help and gavage. However, generally between the ages of 6 months and 1 year, they spontaneously recover and are generally completely normal around the age of 2–3 years. The absence of any case of maternal transmittance for these types of myopathies suggests a nuclear origin. The existence of specific heart or liver isoforms appearing during development has been suggested to explain the evolution of the disease. The transformation of an embryonic abnormal L form into a normal H adult form would permit spontaneous recovery. However, this hypothesis has not yet been demonstrated in spite of intensive screening with molecular probes (Taanman 1997).

Leigh's syndrome is a progressive neurodegenerative disorder characterized by a subacute necrotizing encephalomyopathy. It can be observed in sporadic cases or can be of maternal or Mendelian inheritance. The syndrome may have different origins (COX, pyruvate dehydrogenase or ATPase deficiency) but COX deficiency is the most frequent cause (review: Shoffner and Wallace 1995). Up to now, all studied cases have suggested a nuclear involvement in COX-deficient Leigh's syndrome. This has been proven by fusion of human cells devoid of mtDNA with cytoplasts prepared from the fibroblasts of such patients. The obtained cybrids had a normal COX activity, which shows that their mtDNA could not induce the defect and was not responsible for the COX deficiency. The level of the COX subunit transcripts was also normal, while COX activity was decreased (Di Mauro et al. 1990), as well as the immunologically-titrated amount of several subunits, suggesting a defect in the regulation of translation, in the assembly or in the stability of the complex. Furthermore, for one patient, the sequence of all cDNAs coding for COX subunits did not contain any pathological mutation (Adams et al. 1997). Very recently, Zhu et al. (1998) have shown that SURF-1, encoding a factor involved in the biogenesis of COX, is mutated in Leigh syndrome.

Other diseases implying COX have been described. mtDNA mutations have been observed for example in genes encoding COX subunits I or III in patients suffering from the Leber's disease. The mutation G to A in position 7444 changes the stop codon into a Lys codon in COX subunit I or the mutations G to A in position 9438 and 9804 change Gly^{78} into Ser and Ala ^{200} into Thr, respectively, in COX subunit III. Leber's disease is characterized by vision loss most often occurring in young male adults. The COX mutations are associated with other mutations affecting mtDNA genes encoding NADH dehydrogenase subunits: ND1 (position 3460), ND4 (position 11778) or ND6 (14484). These complex I mutations are considered to be the primary cause of the disease but the COX mutations could be aggravating factors. Indeed, although the amino acids changed by these COX subunit mutations are very well conserved at the phylogenic scale, the first two mutations at least are homoplasmic and have been observed in normal subjects. This indicates that, when present alone, these mutations do not induce the disease. In addition, no detectable COX deficiency could be discovered in these patients. It is possible that these mutations modify the enzyme conformation or stability (review: Wallace et al. 1995).

Patients suffering from an acquired sideroblastic anemia with mitochondrial iron overload presented heteroplasmic T to C mutations in position 6742 of COX subunit I, changing a methionine to a threonine, and in position 6721, changing an isoleucine to a threonine. These well-conserved amino acids are located in the same transmembrane helix of subunit I. The authors suggest that COX might be important in iron reduction and transport through the mitochondrial membrane (Gatterman et al. 1997).

Another T to C mutation in position 9957 changes the very well-conserved Phe^{251} into Leu in COX subunit III of a child presenting a fatal progressive encephalomyopathy. This heteroplasmic mutation could be the cause of the disease. However, the consequences of this mutation on the COX structure and function have not been yet identified (Manfredi et al. 1995).

A decrease in COX activity is relatively frequent in patients presenting mutations inducing mitochondrial protein synthesis deficiency. For example, it is observed when tRNAs encoded by the mtDNA are mutated: the heteroplasmic A to G mutation in position 8344 of the $tRNA^{Lys}$ inducing the MERRF disease (myoclonic epilepsy with ragged red fibers). A decrease in the functional $tRNA^{Lys}$ level especially influences COX activity because mitochondrial COX subunits contain a higher number of Lys. A deficiency in protein synthesis of mitochondrial proteins also exists in patients presenting an mtDNA depletion. In all these cases, the biochemical deficiency is generally not limited to COX but also concerns other respiratory chain complexes.

A small deletion (15 bp) has also been described in the gene encoding the COX subunit III in a patient with a myopathy and recurrent myoglobinuria. The patient muscle also contained ragged red fibers, which are generally mainly observed for patients presenting mtDNA deletions or tRNA mutations (Keightley et al. 1996).

Finally, the decrease in COX activity in human muscle observed during aging could be related to an accumulation of various deletions or mutations of the mtDNA. However, Hayashi et al. (1994) have shown that the fusion of COX-deficient fibroblasts from aged patients with HeLa cells devoid of mtDNA restored COX activity. COX deficiency in aging would therefore rather be of nuclear origin. Another explanation could be that a reduction of COX activity would enhance H_2O_2 generation, which would, in turn, oxidize membrane lipids and mtDNA (which has less efficient repair mechanisms than the nuclear DNA) and would induce protein carbonyl modifications (Cortopassi et al. 1992). Such autocatalytic processes would accelerate the energy crisis and would further degrade COX.

5
Conclusion

In conclusion, although knowledge of COX structure has made important progress during recent years, there is still a lot to learn. The function of the subunits of nuclear origin, the regulation of nuclear gene expression and the specific cause of COX deficiencies of nuclear origin should be clarified in the years to come. Given the complexity of COX biogenesis unraveled in the groundbreaking studies made in yeast, it is likely that future progress could be made by detailed analysis of the many steps involved in COX biogenesis to further identify molecular defects in human COX deficiency.

Acknowledgments. The authors thank the Centre National de la Recherche Scientifique, the Association Française contre les Myopathies (AFM) and the Région Rhône-Alpes for financial support. AP is the recipient of a grant from the AFM.

References

Adams PL, Lightowlers RN, Turnbull DM (1997) Molecular analysis of cytochrome c oxidase deficiency in Leigh's syndrome. Ann Neurol 41:268–270

Adelroth P, Gennis KB, Brzezinski P (1998) Role of the pathway through K(I-362) in proton transfer in cytochrome c oxidase from R. sphaeroides. Biochemistry 27:2470–2476

Anderson S, Bankier AT, Barrell BG, de Bruijn MHL, Coulson AR, Drouin J, Eperon IC, Nierlich DP, Roe BA, Sanger F, Schreier PH, Smith AJH, Staden R, Young IG (1981) Sequence and organization of the human mitochondrial genome. Nature 290:457–465

Anthony G, Reimann A, Kadenbach B (1993) Tissue specific regulation of bovine heart cytochrome c oxidase activity by ADP via interaction with subunit VIa. Proc Natl Acad Sci USA 90:1652–1656

Arnold S, Goglia F, Kadenbach B (1998) 3,5-Diiodothyronine binds to subunit Va of cytochrome c oxidase and abolishes the allosteric inhibition of respiration by ATP. Eur J Biochem 252:325–330

Bachman NJ, Yang TL, Dasen JS, Ernst RE, Lomax MI (1996) Phylogenetic footprinting of the human cytochrome c oxidase subunit Vb promoter. Arch Biochem Biophys 333:152–162

Bachman NJ, Riggs PK, Siddiqui N, Makris GJ, Womack JE, Lomax MI (1997) Structure of the human gene (COX6A2) for the heart/muscle isoform of cytochrome c oxidase subunit VIa and its chromosomal location in humans, mice and cattle. Genomics 42:146–151

Bonnefoy N, Chalvet F, Hamel P, Slonimski PP, Dujardin G (1994) OXA1, a Saccharomyces cerevisiae nuclear gene whose sequence is conserved from prokaryotes to eukaryotes controls cytochrome oxidase biogenesis. J Mol Biol 239:201–212

Capaldi RA (1990) Structure and function of cytochrome c oxidase. Annu Rev Biochem 59:569–596

Capaldi RA, Marusich MF, Taanman JW (1995) Mammalian cytochrome c oxidase: characterization of enzyme and immunological detection of subunits in tissue extracts and whole cells. Methods Enzymol 260:117–132

Cortopassi GA, Shibata D, Soon NW, Arnheim N (1992) A pattern of accumulation of a somatic deletion of mitochondrial DNA in aging human tissues. Proc Natl Acad Sci USA89:7370–7374

Di Mauro S, Lombes A, Nakase H, Mita S, Fabrizi GM, Tritschler HJ, Bonilla E, Miranda AF, De Vivo DC, Schon EA (1990) Cytochrome c oxidase deficiency. Pediatr Res 28:526–541

Ewart G, Lightowlers R, Zhang YZ, Balan VJ, Kennaway N, Capaldi RA (1990) Tissue specificity and defects in human cytochrome c oxidase. Biochim Biophys Acta 1018:223–224

Gattermann N, Retzlaff S, Wang YL, Hofhaus G, Heinisch J, Aul C, Schneider W (1997) Heteroplasmic point mutations of mitochondrial DNA affecting subunit I of cytochrome c oxidase in two patients with acquired idiopathic sideroblastic anemia. Blood 90:4961–4972

Grossman LI, Lomax MI (1997) Nuclear genes for cytochrome oxidase. Biochim Biophys Acta 1352:174–192

Hayashi JI, Ohta S, Kagawa Y, Kondo H, Kaneda H, Yonekawa H, Takai D, Miyabayashi S (1994) Nuclear but not mitochondrial genome involvement in human age-related mitochondrial dysfunction. J Biol Chem 269:6878–6883

Hell K, Herrmann J, Pratje E, Neupert W, Stuart RA (1997) Oxap1 mediates the export of the N- and C-termini of pCOXII from the mitochondrial matrix to the intermembrane space. FEBS Lett 418:367–370

Iwata S, Ostermeier C, Ludwig B, Michel H (1995) Structure at 2.8 Å resolution of cytochrome c oxidase from *Paracoccus denitrificans*. Nature 376:660–669

Kadenbach B, Barth J, Akgün R, Freund R, Linder D, Possekel S (1995) Regulation of mitochondrial energy generation in health and disease. Biochim Biophys Acta 1271:103–109

Kannt A, Roy C, Lancaster D, Michel H (1998) The coupling of electron transfer and proton translocation: electrostatic calculations on *Paracoccus denitrificans* cytochrome c oxidase. Biophys J 74:708–721

Keightley JA, Hoffbuhr KC, Burton MD, Salas VM, Johnston WSW, Penn AMW, Buist NRM, Kennaway NG (1996) A microdeletion in cytochrome c oxidase (COX) subunit III associated with COX deficiency and recurrent myoglobinuria. Nat Genet 12:410–416

Manfredi G, Schon EA, Moraes CT, Bonilla E, Berry GT, Sladky JT, Di Mauro S (1995) A new mutation associated with MELAS is located in a mitochondrial DNA polypeptide-coding gene. Neuromuscul Disord 5:391–398

Margoliash E (1988) A prepared mind, infinite pains and genius. In: King TE, Mason HS, Morrison M (eds) Oxidases and related redox systems. Prog Clin Biol Res 274. Alan R Liss, New York, pp 79–84

Merante F, Duncan AM, Mitchell G, Duff C, Rommens J, Robinson BH (1997) Chromosomal localization of the human liver form cytochrome c oxidase subunit VIIa gene. Genome 40:318–324

Merle P, Kadenbach B (1982) Kinetic and structural differences between cytochrome oxidases from beef liver and heart. Eur J Biochem 125:239–244

Mitchell P (1961) Coupling of phosphorylation to electron and hydrogen transfer by a chemiosmotic type of mechanism. Nature 191:144–148

Pfanner N, Craig EA, Meijer M (1994) The protein import machinery of the mitochondrial inner membrane. Trends Biochem Sci 19:368–372

Poyton RO, McEwen JE (1996) Crosstalk between nuclear and mitochondrial genomes. Annu Rev Biochem 65:563–607

Preiss T, Sang AE, Chranowska-Lightowlers ZM, Lightowlers RN (1995) The mRNA-binding protein COLBP is glutamate dehydrogenase. FEBS Lett 367:291–296

Prieur B, Bismuth J, Delaval E (1998) Effects of adrenal steroid hormones on mitochondrial maturation during the late fetal period. Eur J Biochem 252:194–199

Sadlock JE, Lightowlers RN, Capaldi RA, Schon EA (1993) Isolation of a cDNA specifying subunit VIIb of human cytochrome c oxidase. Biochim Biophys Acta 1172:223–225

Scarpulla RC (1997) Nuclear control of respiratory chain expression in mammalian cells. J Bionerg Biomembr 29:109–119

Schägger H, Noack H, Halangk W, Brandt U, Von Jagow G (1995) Cytochrome c-oxidase in developing rat heart: enzymic properties and amino-terminal sequences suggest identity of the fetal heart and the adult liver isoform. Eur J Biochem 230:235–241

Shoffner JM, Wallace DC (1995) Oxidative phosphorylation disease. In: Scriver CTR, Beaudet AL, Sly WS, Valle D (eds) The metabolic and molecular basis of inherited disease, 7th edn. MacGraw-Hill, New York, pp 1535–1609

Speno H, Taheri MR, Sieburth D, Martin CT (1995) Identification of essential amino acids within the proposed CuA site in subunit II of cytochrome c oxidase. J Biol Chem 270:25363–25369

Taanman JW (1997) Human cytochrome c oxidase: structure function and deficiency. J Bioenerg Biomembr 29:151–163

Tsukihara T, Aoyama H, Yamashita E, Tomizaki T, Yamaguchi H, Shinzawa-Itoh K, Nakashima R, Yaono R, Yoshikawa S (1995) Structures of metal sites of oxidized bovine heart cytochrome c oxidase at 2.8 Å. Science 269:1069–1074

Tsukihara T, Aoyama H, Yamashita E, Tomizaki T, Yamaguchi H, Shinzawa-Itoh K, Nakashima R, Yaono R, Yoshikawa S (1996) The whole structure of the 13-subunit oxidized cytochrome c oxidase at 2.8 Å. Science 272:1136–1144

Wallace DC, Lott MT, Brown MD, Huoponen K, Torroni A (1995) Report of the committee on human mitochondrial DNA. in: Cuticchia AJ (ed) Human gene mapping 1995: a compen-

dium. Johns Hopkins University Press, Baltimore, pp 910–954 (or http://www.gen.emory.edu/mito-map.html)
Witt H, Malatesta F, Nicoletti F, Brunori M, Ludwig B (1998) Tryptophan 121 of subunit II is the electron entry site to cytochrome c oxidase in *Paracoccus denitrificans*. Involvement of a hydrophobic patch in the docking reaction. J Biol Chem 273:5132–5136
Wolz W, Kress W, Mueller CR (1997) Genomic sequence and organization of the human gene for cytochrome c oxidase subunit (COX7A1) VIIa-M. Genomics 45:438–442
Yoshikawa S (1997) Beef heart cytochrome c oxidase. Curr Opin Struct Biol 7:574–579
Zhu Z, Yao J, Johns T, Fu K, De Bie I, MacMillan C, Cuthbert AP, Newbold RF, Wang JC, Chevrette M, Brown GK, Brown RM and Shoubridge EA (1998) SURF-1, encoding a factor involved in the biogenesis of cytochrome c oxidase, is mutated in Leigh syndrome. Nat Genet 20:337–343

CHAPTER 9

ATPase-ATP Synthase and Mitochondrial Pathology 9

K. Buchet and C. Godinot[1]

Contents

1 Introduction

ATP, the main energy source of biological systems is mainly synthesized during oxidative phosphorylations: electron transfer along the respiratory chain located in the mitochondrial inner membrane is coupled to ATP synthesis from ADP and inorganic phosphate. The enzyme responsible for this synthesis is the ATPase-ATP synthase, F0–F1 or complex V (EC 3.6.1.34), identified in 1960 by Racker et al. To explain the coupling between electron transfer and ATP synthesis, Mitchell proposed the chemiosmotic theory in 1961, according to which an electrochemical proton gradient, established during electron transfer across the inner mitochondrial membrane represents the proton-motive force used by F0–F1 to synthesize ATP (review, Mitchell 1979). Over the last 20 years, the F0–F1 structure has been partly resolved by sequencing its subunits or their genes and by analysis of the 3-D structure of F1 (Abrahams et al. 1994;

[1] Due to space limitation, the authors apologize for not being able to cite all the articles where the original work was described. More often than not, the references are limited to reviews where the reader can find the original citations.
Pathologie mitochondriale, Centre de Génétique Moléculaire et Cellulaire Université Claude Bernard-Lyon I, 43, Boulevard du onze novembre 1918, 69622 Villeurbanne Cedex, France

review, Pedersen et al. 1994, 1996). Many biochemical studies made on the native enzyme, or after chemical or genetic modification of the enzyme, have served as a basis to elucidate the mechanisms explaining how proton transfer across F0 could induce conformational changes inside F1 that would induce ATP synthesis (reviews, Boyer 1997; Junge et al. 1997). The final understanding of the molecular mechanism of this coupling will probably await the resolution of the F0 structure. The knowledge of this structure as well as that of the mechanisms that govern the regulation of the complex biogenesis and of the expression of the genes which encode the various subunits might also be essential in identifying the origin of diseases associated with F0–F1 deficiency. This knowledge might also serve as a basis for discovering new drugs to cure the patients suffering from these diseases.

2 Structure of the F0–F1 Complex

2.1 Subunit Composition

In *E. coli*, F0–F1 contains 8 subunits, in mammals such as cattle, it is made of 16 subunits (review, Walker 1995). Two of them, subunits a and A6L are encoded by mtDNA. In all species, the catalytic F1 part, exposed as a "lolly-pop" on the inner face of the mitochondrial inner membrane, contains five different subunits which are associated according to an $\alpha_3\ \beta_3\ \gamma\ \delta\ \varepsilon$ stoichiometry. In addition, in the mammalian enzyme, a natural protein inhibitor IF1 can decrease the rate of ATP hydrolysis or synthesis upon binding to one of the β subunits (Fig. 1).

E. coli F0 only contains subunits a, b, and c. Subunits d, e, f, g, OSCP (oligomycin sensitivity conferring protein), F6 and A6L are additional components of the mammalian F0 sector and of the stalk which links F0 to F1. Subunits a and c (DCCD-binding protein) are mainly hydrophobic transmembrane proteins. The b subunit contains a 30-amino acid N-terminal tail constituting two transmembrane α helices and a 130 amino acid part extending outside the inner membrane and belonging to the stalk connecting F1 to F0. Subunit d, F6 and OSCP are bound to this extramembrane part of b and to F1. Subunit A6L is embedded in the membrane by its 25–30 hydrophobic N-terminal amino acids and binds to subunit d on the F1 side. The three transmembrane subunits e, f, g recently identified in beef heart enzyme (Walker and Collinson 1994) have N- and C-terminal ends oriented towards matrix and intermembrane space, respectively (Belogrudov et al. 1996).

The stoichiometry of the *E. coli* F0F1 complex is well characterized. Its eight subunits are assembled according to a stoichiometry $\alpha_3\ \beta_3\ \gamma\ \delta\ \varepsilon$, a, b_2, c_{10-12} (review, Fillingame 1996). In contrast, the mammalian enzyme stoichiometry containing eight other subunits (OSCP, d, A6L, F6, IF1, e, f, g) is still being discussed (Belogrudov et al. 1996; reviews, Pedersen 1994; Boyer 1997). According to sequence homologies, the mitochondrial OSCP would be equivalent to the *E. coli* δ subunit, while the mitochondrial δ subunit would correspond to the *E. coli* ε subunit. All known mammalian F0–F1 subunits have more than 85% homology. In contrast, the presequences, which are cleaved after the import of subunits of nuclear origin into the mitochondria, are not as well conserved between species.

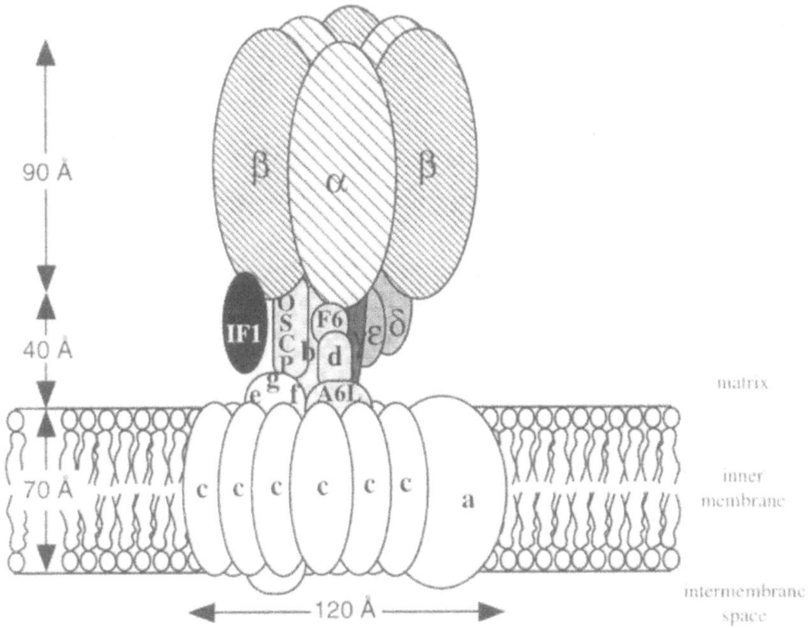

Fig. 1. Artist's view of the hypothetical overall ATPase-ATP synthase complex structure. (Adapted from Pedersen 1994; Walker 1995; Belogrudov et al. 1996; Junge et al. 1997)

3.2 Tridimensional Structure

Particles of 90 Å oriented to the mitochondrial matrix, and containing F1 were identified as early as 1962. They are linked to the inner membrane by a 40–50 Å long stalk. This structure has been confirmed by electron microscopy for the purified F0–F1 complex (review, Boyer 1997).

F1 crystals have been obtained in the presence of ligands from rat liver and beef heart enzymes. Their structures have been solved at 3.6 Å (review, Pedersen 1994, 1996) and 2.8 Å (Abrahams et al. 1994) resolution, respectively. The 3 α and β subunits alternate like the segments of an orange around the central γ subunit. The 2.8 Å structure (Fig. 2) has been obtained in the presence of azide and of the ATP analog, AMP-PNP, which both inhibit the enzyme. The N-terminal ends of subunits α and β exhibit a β-barrel structure crowning the top of F1, arranged as if it could serve as a guide for the rotating γ subunit. Nucleotide binding sites are located in the middle of F1 at the interfaces between the α and β subunits. Each α subunit and one β subunit (β_{TP}) contain bound AMP-PNP, while another β subunit (β_{DP}) contains bound ADP and the last one (β_E) is empty. The F1 molecule is asymmetrical and each α–β pair shows a different conformation, in agreement with the data obtained using monoclonal antibodies (Moradi-Améli et al. 1989). This asymmetry is induced by the presence of the other F1 subunits γ, δ and ε (Capaldi et al. 1996). The part of the γ subunit located in the center of the α–β pairs is made of an α-helix, with a C-terminal end close to the top of F1 and a N-terminal end close to F0. The δ and ε structures and that of a part

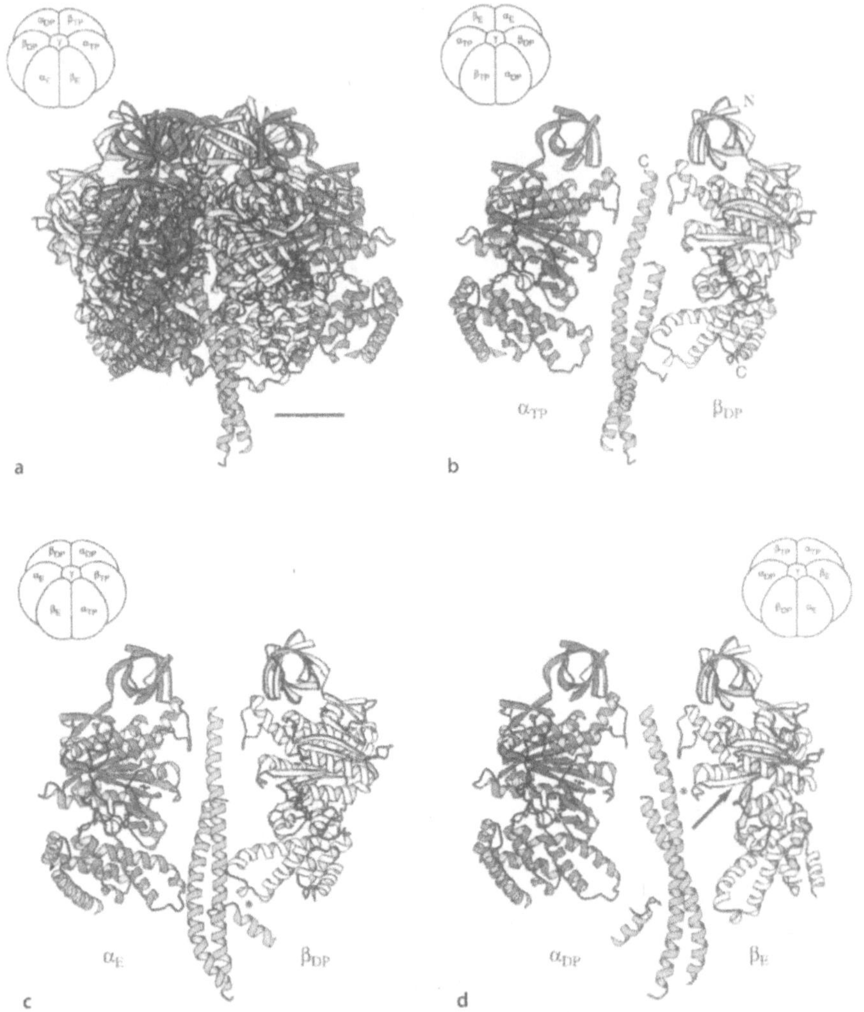

Fig. 2. Beef heart F1 ATPase structure at 2.8 Å resolution (reproduced from Walker 1995). α, β and γ subunits are shown. Adenine nucleotides are in ball and stick presentation. AMP-PNP is bound to the 3 α-subunits, and to the β-subunit defined as β_{TP} (see text). Subunit β_{DP} has bound ADP and subunit β_E has no associated nucleotide. Therefore, the structure probably represents the ADP-inhibited state of the enzyme. **a** A view of the entire F1 particle in which subunits α_E and β_E point towards the viewer. The *bar* is 20 Å long. **b** Subunits α_{TP}, γ and β_{DP}. From a similar viewpoint to **a** but rotated 180° about the axis of pseudo-symmetry. One ADP.Mg but no phosphate is bound to subunit β_{DP}. **c** Subunits α_E, γ and β_{TP}. From a similar viewpoint to **a** but rotated −60°. The *asterisk* denotes the loop containing the DELSEED sequence with which IF1 is thought to interact. **d** Subunits α_{DP}, γ and β_E from a similar viewpoint to **a**, but rotated 60°. The *arrow* indicates disruption of the β-sheet in the nucleotide-binding domain of β_E, which prevents nucleotide binding. For more details, see Abrahams et al. (1994)

of the γ subunit not ordered in the beef heart F1 crystals have not been solved. The AMP-PNP bound in the β_{TP} subunit is located in a hydrophobic pocket containing a basic residue and an ATP phosphate binding consensus sequence, GXXXXGKT(S). This structure plays an essential role in the ATP synthesis mechanism (Abrahams et al. 1994).

The structure of the *E. coli* F0–F1 ε and c subunits has been solved by ^{13}C- and ^{15}N-NMR spectroscopy. The ε N-terminal β-sheet structure interacts with F1 while its C-terminal end made of 2 α helices arranged in a hairpin structure is linked to the F0 c subunit (Wilkens et al. 1995). This c subunit contains two hydrophobic transmembrane helices with the N- and C-terminal ends oriented towards the intermembrane space. One essential Asp (Glu in some species) is located in the middle of one helix. The various c subunits are arranged in a ring inside the inner membrane (review, Fillingame 1996).

3 Function

ATP synthesis or hydrolysis does not involve phosphorylated intermediates. According to many biochemical and genetic studies, and in agreement with the 2.8 Å F1 structure, the mechanism of ATP synthesis or hydrolysis would involve an alternative change in nucleotide binding on the 3 α–β pairs (review, Boyer 1997). In the synthesis reaction, β_{DP} is ready to bind Pi in order to form bound ATP; β_{TP} has made ATP and is waiting for energy to release it, while β_E has released ATP and waits to bind a new ADP molecule. The energy would come from F0 through the γ subunit that would rotate inside the 3 α–β couples. During catalysis, each α–β couple would successively take one of the 3 conformations induced by the rotation of the γ subunit central α helix in relation to the α and β subunits surrounding it. This rotation of the γ subunit inside the 3 α–β couples was directly observed by Joji et al. (1997): these authors have bound the $\alpha_3\beta_3\gamma$ *E. coli* complex to a plate covered with nickel after adding a polyhistidine tail to the N-terminal end of each β subunit which is located on the top of F1. On the opposite end of F1, they have added a cystein to the extremity of the γ subunit and bound biotin to this cystein. This biotin could then retain a streptavidin molecule linked to a fluorescent actin filament. Following the fluorescence with an epifluorescence microscope during ATP hydrolysis, they were able to directly observe a continuous rotation of the actin filament. The rotation always occurred in the same direction, the three catalytic sites participating in ATP hydrolysis alternatively.

The F1 structure is therefore compatible with a mechanism implying an alternative conversion of nucleotide binding sites. The energy accumulated during electron transfer would be transferred through F0 to F1, not to phosphorylate ADP but to change the conformation of the α–β couples and thus either secure nucleotide binding or induce their release. Kinetic studies have indeed shown that the phosphorylation step does not require much energy (review, Boyer 1997).

The ATPase-ATP synthase activity is inhibited by several compounds acting either at the level of F1 or F0. The F1-specific inhibitors such as aurovertin, efrapeptin or the natural protein inhibitor interfere either with the nucleotide binding to F1 or to the conformational changes essential to the catalytic mechanism. Aurovertin binds to β_{TP} and β_E. In the β_{TP} site, it interacts with the Glu^{399} of the neighboring α subunit, while the aurovertin bound to β_E is too far away from α to establish this connection. The interface between α and β is too narrow in the case of β_{DP} to allow aurovertin to penetrate (Van Raaij et al. 1996). Efrapeptin is bound in a unique site in the central F1 cavity and interacts with γ. It would inhibit the F1-ATPase by preventing the conversion of β_E into one of the β conformations able to bind adenine nucleotides (Abrahams et al. 1996).

Uncouplers such as 2–4, dinitrophenol or carbonylcyanide p-trifluoromethoxyphenylhydrazone (FCCP) dissipate the transmembrane proton gradient, thus eliminating the energy source. The binding to F0 of ATP synthesis inhibitors such as oligomycin, organotins or carbodiimides prevents proton translocation through F0. Dicyclohexylcarbodiimide (DCCD) binds to a Glu (Asp in *E. coli*) located in the middle of one of the hydrophobic transmembrane α-helices of the c subunit. The binding of only one DCCD molecule is sufficient to block proton transfer. This suggests that during ATP synthesis, protons move through all the c subunits before being transferred to F1. This rotation could be symmetrical to that of the F1 γ subunit. Junge et al. (1997) and Kaim et al. (1998) propose that the c subunit carboxyl groups facing the lipid core are protonated and neutral, whereas the c subunit carboxyl group facing the a subunit can be deprotonated and charged. In addition, the transmembrane a subunit has two access channels for protons from both sides of the membrane. If one of the c subunits loses a proton, it will interact with a positively charged essential arginine residue of the a subunit. During ATP hydrolysis, ATP-driven rotation of the c subunit ring would bring one of the loaded c subunits into contact with the release channel present on the a subunit. The ion will be captured by the a subunit via a negative charge and transported outside the membrane. Simultaneously, the c subunit ring would turn one step. The rotation will proceed clockwise or counterclockwise depending on the direction of proton or ion gradient between the matrix and intermembrane space, i.e. depending on whether ATP is hydrolyzed or synthesized.

4 Localization and Structure of F0F1 Human Nuclear Genes

The genes and cDNAs of four human ATPase-ATP synthase nuclear subunits (α, β, γ and c) and the cDNAs of four others (δ, OSCP, b and F6) have been characterized (Table 1). These subunits contain an N-terminal signal sequence which varies between 22 and 66 amino acids and serves to direct proteins of nuclear origin to mitochondria. It is cleaved inside the mitochondria to give the mature subunit. Besides the functional genes, there are many processed pseudogenes, indicating very old ancestral genes. The α and β subunits are encoded by only one gene. For OSCP, b, and F6, only one cDNA has been identified. Two mRNAs of 0.7 and 1 kb which encode the same δ subunit, and are co-expressed in HeLa cells, are formed by alternative splicing of an insertion of 296 bp at 3' of the stop codon (review, Shoffner and Wallace 1995). A unique gene also encodes the two tissue-specific γ isoforms. They come from an alternative splicing of the exon 9 (Matsuda et al. 1993). The H isoform, in which the exon 9 is spliced is predominant in cardiac and skeletal muscle, while the L isoform, which contains an additional Asp at the C-terminal end, is predominant in the other tissues. The c subunit is encoded by three different functional genes, P1, P2 and P3 (Higuti et al. 1993; Yan et al. 1994), providing three isoforms with quite different signal sequences but identical

Table 1. Structure, chromosome assignment and known characteristics of nuclear genes encoding human ATPase-ATP synthase subunits. The nucleotide sequences of the various motifs are the following: TATA box (TATAAT), CAAT box (CCAAT), Sp1 (GGGCGG), NRF-2 ((A/C)GGAA(G/C)), NRF-1 ((T/C)GCGCA(T/C)GCGC(A/G)), CS1 (CAGAGGAA), CS2 (TTCAGA(T/A)T), CS3 (GGCTGCGG), Mt1 (TATTCAGGT), Mt3 (ATCTGGC), Mt4 (TGGTGTA(T/G)AG), AP1 (TGA(G/C)T(C/A)A), AP2 (CCCC(A/C)N(C/G)), AP3 ($TGTGG(A/T)_3$), GCF ($(C/G)CG(C/G)_3C$), F-ACT1 (TGGCGA), E-box (CANNTG), OXBOX (GGCTCTAAAGAGG) and REBOX (AAGAGGGC)

Sub-unit[a]	cDNA	Mature protein	Prese-quence	Gene	Gene size	Gene structure	Promoter structure	Chromo-some
F1								
α [a,b,c]	cDNA 1883 bp	510 aa	43 aa	– 1 gene – 2 processed pseudogenes	14 kb	12 exons 11 introns	– Several transcription initiation sites – No TATA or CAAT box – 2 Sp1, 2 NRF2, 1 CS1, CS2, CS3 and several AP2, GCF, F-ACT1, E-box, Mt1, Mt3, Mt4 – 1 OXBOX-like motif but no REBOX – Several Alu sequences	chr 9 and 18
β [a,d,e,f]	cDNA 1807 bp	480 aa	49 aa	– 1 gene – Several incomplete sequences	10 kb	10 exons 9 introns	– 2 Transcription initiation sites – No TATA box but 4 CAAT box – 2 Sp1, NRF2, 1 Mt1, Mt3, Mt4, CS1, CS2, CS3 – 1 Motif OXBOX-REBOX – 1 Negative control domain: 15 Alu sequences	– Gene: 12p13-qter – Sequences: chr 2 and 17
γ [c, g]	Incomplete cDNA: 1095 bp	2 Tissue-specific isoforms heart and liver 272 or 273 aa	25 aa	– 1 gene –1 processed pseudogene	23 kb	10 exons (9th exon spliced in heart) 9 introns	– 1 Transcription initiation site – No TATA box, several CAAT boxes – 2 AP1, 1 NRF1, Mt3, CS1, CS2, CS3 several Sp1, AP2, AP3, F-ACT1 – Several Alu sequences	chr 10 and 14
δ [a]	2 cDNA 0.7 or 1 kb	146 aa	22 aa	–	–	–	–	–

Table 1. (Continue)

Sub-unit[a]	cDNA	Mature protein	prese-quence	Gene	Gene size	Gene structure	Promoter structure	Chromo-some
F0								
c [h, i, j]	3 cDNA P1: 553 bp P2: 599 bp P3: 845 bp	P1, P2 and P3: 75 aa (different presequences)	P1 : 61 aa P2 : 66 aa P3 : 66 aa	– 3 Functional genes P1, P2, P3 – 14 Processed pseudogenes	P1 : 9.5 kb P2 : 15 kb	P1 and P2: 5 exons 4 introns	– 2 (P1) or 1 (P2) transcription initation sites – 2 TATA and 4 CAAT boxes for P1, none for P2 – 2 Motifs Sp1 for P1 and 3 for P2 – 10 Sequences Alu for P1 and 14 for P2	P1 : chr 17 P2 : chr 12 P3 : chr 2
OSCP [k]	cDNA 750 bp	190 aa	23 aa	–	–	–	–	Gene: 21q22.1–22.2
b [a]	cDNA 1134 bp	214 aa	42 aa	–	–	–	–	–
F6 [a]	cDNA 466 bp	76 aa	32 aa	–	–	–	–	–

[a] [a] Shoffner and Wallace (1995); [b] Akiyama et al. (1994), [c] Jabs et al. (1994); [d] Haragushi et al. (1994), [e] Villena et al. (1994); [f] Nelson et al. (1995); [g] Matsuda et al. (1993), [h] Higuti et al. (1993); [i] Yan et al. (1994); [j] Dyer and Walker (1993); [k] Chen et al. (1995).

mature forms. In rats and hamsters at least, the P2 gene is expressed constitutively, while the P1 gene responds in a tissue-specific manner to cold acclimatization, thyroid hormone or ontogenic development as a means of modulating the relative ATP synthase content (Andersson et al. 1997).

Many motifs which are more or less conserved and implied in transcription regulation are present in the promoters that have been analyzed: α, β, γ and P1 or P2 of the c subunit (Table 1). The TATA box does not exist in the α, β, γ or in the P2 genes. The CAAT box is not always present, while the Sp1 motif on which very ubiquitous factors can bind has been found in the five genes studied. It probably allows a basal expression level of mRNAs. For α, β and P1, there are several transcription initiation sites that would permit a tissue-specific regulation of transcript expression. Akiyama et al. (1994) have identified several transcription factors binding sites on the α subunit gene promoter (see Table 1). In addition, the binding of the factor ATPF1 and of the upstream regulatory factor 2 to an E-box sequence (CACGTG) and of the factor YY1 to another cis-acting element of the α subunit gene promoter stimulate the expression of the α-subunit gene (Breen et al. 1996; Breen and Jordan 1997). The level of expression of the mRNAs of the β gene that was considered for a long time to be a housekeeping gene, is in fact modulated by oxygen level, by the cell cycle phase, during muscle cell differentiation and under the influence of thyroid hormones (Haragushi et al. 1994; Izquierdo et al. 1995; review, Nelson et al. 1995; Scarpulla 1997). The motifs CS1, CS2 and CS3, which are specific to the α, β and γ promoters suggest a coordinated regulation of F1 subunits (Akiyama et al. 1994). The expression of the β and γ subunits are under control of functional NRF-1 or NRF-2 (nuclear respiratory factors 1 and 2), like many other respiratory chain proteins (review, Scarpulla 1997). In addition, motifs of *cis*-acting promoter elements Mt1, Mt3 or Mt4 are found in some nuclear genes and regulate, together with NRF-1 or NRF-2, not only the expression of either mitochondrial F1 subunits but also that of the mitochondrial transcription factor mtTFA (Akiyama et al. 1994; Villena et al. 1994; Matsuda et al. 1993). An OXBOX-REBOX motif is also present, both in the mtDNA D-loop, in the β subunit promoter, and partially in the α subunit promoter (Akiyama et al. 1994; Haragushi et al. 1994). Thus, when the energy demand is increased, these *cis*-activating elements could induce a coordinated over-expression of mitochondrial genes and of nuclear genes which encode proteins involved in oxidative phosphorylation in collaboration with ubiquitous transcription factors binding to Sp1 or CAAT in the nucleus and to the mitochondrial transcription factor mtTFA (reviews, Shoffner and Wallace 1995; Scarpulla 1997).

5 Expression of F0 Genes Encoded by the Mitochondrial Genome

Subunits A6L and a of the human ATPase-ATP synthase are encoded by two overlapping genes on mtDNA, in positions 8366-8572 (ATPase 8) and 8257-9207 (ATPase 6). These genes, like the other human mitochondrial genes, have no introns and almost no intergenic regions. The mRNAs have neither cap nor 5' untranslated sequences since the AUG codon is at the level of the transcription initiation site. The sequence of the two ATPase genes encodes proteins of 68 and 226 amino acids for A6L and a, respectively. The transcription and translation of these genes necessitate several steps (review, Wallace et al. 1995). The mtDNA of the H strand is first transcribed in a unidirectional way into a single polycistron made of the entire sequence, starting from a promoter P_H located in the D-loop. Then various endonucleases cut the precursor during its elongation on both sides of the tRNAs present in the genome. This frees the tRNAs, the ribosomal RNAs and most of the mRNAs, which are then polyadenylated.

However, in the case of the ATPase 8 and 6 genes, the absence of tRNAs between the two genes and the next gene located at 3', which encodes the cytochrome oxidase subunit III (CO III), leads to a pre-transcript corresponding to (ATPase 8 - 6 - CO III) after the cleavage of $tRNA_{Lys}$ and $tRNA_{Gly}$. Since a transcript common to the three genes has never been detected, this suggests an early cleavage of the transcript, probably at the level of a secondary structure similar to that of the tRNAs. However, the early and essential role of the tRNAs in the maturation of primary transcripts has been questioned since the discovery of intermediary RNAs containing tRNAs. Indeed, there are three different polyadenylated transcripts for the human A6L and a subunits: the RNA 14 (ATPase 8 - ATPase 6), the RNA 20 ($tRNA_{Lys}$ - ATPase 8 - ATPase 6) and the RNA 21 (3' end of ATPase 8 - ATPase 6; review, Wallace et al. 1995). The last transcript would permit an independent translation for the ATPase 6 gene. A tissue-specific regulation has also been proposed as being dependent on the percentage of the three transcripts differently co-expressed in various tissues. There are two initiation codons (AUG) in two different phases for the ATPase 8 and 6 and a stop codon at the end of the ATPase 8 (UAG), 44 bp after the AUG of the ATPase 6 gene. For the last subunit, the addition of As at the transcript end creates the stop codon UAA. The regulation of the translation of the two subunits is unknown in humans, although several nuclear factors playing a role in the post transcriptional regulation of the mitochondrial transcripts for the ATPases 6 and 8 have been identified in *S. cerevisiae* (Camougrand et al. 1995). However, a strong increase in the level of the mitochondrial transcripts encoding the ATPase 6 and 8 and the nuclear transcript encoding the β subunit has been disclosed during neonatal rat liver development (Ostronoff et al. 1995). Therefore, a coordinated regulation of the expression of the two genomes is very likely during mitochondrial biogenesis in mammals.

6 Pathologies Associated with ATPase-ATP Synthase Deficiency

6.1 Pathologies of Known Mitochondrial Origin

MtDNA deletions, including ATPase genes, have been detected in various syndromes. However, in such cases, the ATPase deficiency is not specific because the other complexes, and particularly cytochrome oxidase, are also affected. On the other hand five mutations in the ATPase 6 gene, which change amino acids conserved in various species, and induce a specific ATPase deficiency, have been described (Table 2).

The T to C or G mutation in position 8993 in the ATPase 6 gene is associated either with NARP (neurogenic ataxia retinitis pigmentosa) or to Leigh's syndrome. It changes the highly evolutionarily conserved Leu^{156} into either a proline or an arginine. This modifies the a subunit conformation, blocks the proton channel and decreases the complex stability. The main clinical manifestations of the NARP syndrome range from

Table 2. Human pathologies involving complex V

Origin	Gene	Mutation	Pathology[a]	Complex V deficiency
Mitochondrial	ATPase 6	8993 T → G	NARP, Leigh [a, b]	Yes
Mitochondrial	ATPase 6	8993 T → C	Leigh [a, b]	Yes
Mitochondrial	ATPase 6	9176 T → C	FBSN [b]	No
Mitochondrial	ATPase 6	8851 T → C	FBSN [c]	-
Mitochondrial	ATPase 6	9101 T → C	Leber [d]	Decrease in OXPHOS efficiency
Unknown	-	-	Luft [a]	Mitochondrial uncoupling, IF1 deficiency
Nuclear	CLN3 Chromosome 16	Partial deletion or mutations	Batten [e–f]	Subunit c accumulates in lysosomes
Unknown	-	-	Alzheimer [g]	Decrease in complex V amounts
Unknown	-	-	Cardiomyopathy [h]	Ethanol-induced complex V inhibition

[a] NARP: neurogenic ataxia retinitis pigmentosa; FBSN: familial bilateral striatal necrosis; OXPHOS: oxidative phosphorylation. [a] Shoffner and Wallace (1995); [b] Wallace et al. (1995); [c] De Merleir et al. (1995); [d] Lamminen et al. (1995); [e] Johnson et al. (1995); [f] International Batten Disease Consortium (1995); [g] Schägger and Ohm (1995); [h] Das and Harris (1993).

mild retinitis pigmentosa and muscular weakness to severe mental retardation, macular degeneration, olivocerebellar atrophy and lethal Leigh's syndrome. Leigh's syndrome (subacute necrotizing encephalomyelopathy) is a very heterogeneous clinical entity, implying central nervous system lesions and various oxidative phosphorylation deficiencies (review, Shoffner and Wallace 1995; Wallace et al. 1995). In some cases, a Mendelian inheritance has been proposed while the cases exhibiting a maternal transmission are associated with ATPase6 T_{8993} to G or C mutation. The segregation of these mutations are very rapid in Leigh's families and the clinical signs appear very early in life.

Recently, two new mutations in the ATPase 6 gene which are almost homoplasmic in the patients have been associated with neurologic disorders (FBSN or familial bilateral striatal necrosis). The symptoms are similar to those observed in Leigh's syndrome. The first one changes a T into a C in position 9176 of the mtDNA, which transforms the well-conserved a subunit C-terminal Leu into a Pro (review, Wallace et al. 1995). The second one mutating the T_{8851} into a C modifies the also well-conserved Trp^{109} into an Arg (De Merleir et al. 1995). However, the rate of ATP hydrolysis or ATP synthesis appeared to be normal in the fibroblasts of at least the first of these two patients.

A transition of T to C in position 9101 of the ATPase 6 gene replacing Ile^{102} by a Thr has been detected in a patient suffering from papillary microangiopathy, which is characteristic of Leber disease (Lamminen et al. 1995). The patient's lymphoblasts exhibit a reduced capacity for oxidative phosphorylation.

6.2 Other Pathologies Involving the ATPase-ATP Synthase

Several other diseases have been associated with anomalies of the ATPase-ATP synthase, although no pathological mutation has yet been discovered (Table 2).

Luft's disease was the first mitochondrial pathology ever described, and is characterized by heavy sweating, increased respiration rate, tachycardia and generalized myopathy which begins in infancy. The patient's skeletal muscle presents numerous ragged red fibers (RRF) and paracrystallin inclusions. The oxygen consumption of the muscle mitochondria is uncoupled from ATP synthesis. This could be due to mutations in F0 subunits or in the protein inhibitor IF1, since the fibroblasts of one of the patients were deficient in IF1 (review, Shoffner and Wallace 1995).

The Batten disease or NCL (neuronal ceroid lipofuscinosis) is a hereditary neurodegenerative disease, characterized by an accumulation of the F0 c subunit in the lysosomes of various tissues. The coding sequence of the three nuclear genes of this protein did not contain any mutation (Johnson et al. 1995). The gene modified in Batten disease (CLN3) has been assigned to chromosome 16. It encodes a protein of 438 amino acids of unknown function. In 81% of the patients, it is partially deleted and it is mutated in other patients (International Batten Disease Consortium 1995). Tanner et al. (1997) suggest that the deficiency of this gene would impair the normal proteolysis pathway of the c subunit.

Many mitochondrial deficiencies, implying several oxidative phosphorylation complexes, have been associated with Alzheimer disease. However, until now, although modifications of the mitochondrial ATPase-ATP synthase genes have not been specifically involved, the expression of some of its subunits seems to be reduced in Alzheimer patient brain hypocampus (Schägger and Ohm 1995).

In some cardiomyopathies a deregulation of the ATPase-ATP synthase would be induced by alcohol consumption (Das and Harris 1993).

7 Conclusion

In conclusion, many F0F1 deficiencies are related to various pathologies. More knowledge of the complex in normal subjects or patients is needed to understand the causes of such diseases and may, one day, lead to cures.

Acknowledgments. The authors thank the Centre National de la Recherche Scientifique (CNRS), the French Ministry of Education and Scientific Research (MERS), the

Association Française contre les Myopathies (AFM) and the Région Rhône-Alpes for financial support. KB was the recipient of a grant from the MERS.

References

Abrahams JP, Leslie AGW, Lutter R, Walker JE (1994) Structure at 2.8 Å resolution of F1-ATPase from bovine heart mitochondria. Nature 370:621–628

Abrahams JP, Buchanan SK, Van Raaij MJ, Fearnley IM, Leslie AGW, Walker JE (1996) The structure of bovine F1-ATPase complexed with the peptide antibiotic efrapeptin. Proc Natl Acad Sci USA93:9420–9424

Akiyama S, Endo H, Inohara N, Ohta S, Kagawa Y (1994) Gene structure and cell type-specific expression of the human ATP synthase α subunit. Biochim Biophys Acta 1219:129–140

Andersson U, Houstek J, Cannon B (1997) ATP synthase subunit c expression: physiological regulation of the P1 and P2 genes. Biochem J 323:379–385

Belogrudov GI, Tomich JM, Hatefi Y (1996) Membrane topography and near-neighbor relationships of the mitochondrial ATP synthase subunits e, f, and g. J Biol Chem 271:20340–20345

Boyer PD (1997) The ATP synthase – splendid molecular machine. Annu Rev Biochem 66:717–749

Breen GA, Jordan EM (1997) Regulation of the nuclear gene that encodes the α-subunit of the mitochondrial ATP synthase complex. Activation by upstream stimulatory factor 2. J Biol Chem 272:10538–10542

Breen GA, Vander Zee CA, Jordan EM (1996) Nuclear factor YY1 activates the mammalian F0F1 ATP synthase α-subunit gene. Gene Expr 5:181–191

Camougrand N, Pélissier P, Velours G, Guérin M (1995) NCA2, a second nuclear gene required for the control of mitochondrial synthesis of subunits 6 and 8 of ATP synthase in *Saccharomyces cerevisiae.* J Mol Biol 247:588–596

Capaldi RA, Aggeler R, Wilkens S, Grüber G (1996) Structural changes in the γ and ε subunits of the *Escherichia coli* F1F0-type ATPase during energy coupling. J Bionerg Biomembr 28:397–401

Chen H, Morris MA, Rosier C, Blouin J-L, Antonarakis SE (1995) Cloning of the cDNA for the human ATP synthase OSCP subunit by exon trapping and mapping to chromosome 21q22.1-q22.2. Genomics 28:470–476

Das AM, Harris DA (1993) Regulation of the mitochondrial ATP synthase is defective in rat heart during alcohol-induced cardiomyopathy. Biochim Biophys Acta 1181:295–299

De Meirleir L, Seneca S, Lissens W, Schoentjes E, Desprechins B (1995) Bilateral striatal necrosis with a novel point mutation in the mitochondrial ATPase 6 gene. Pediatr Neurol 13:242–246

Dyer MR, Walker JE (1993) Sequences of members of the human gene family for the c subunit of mitochondrial ATP synthase. Biochem J 293:51–64

Fillingame RH (1996) Membrane sectors of F- and V-type H+-transporting ATPases. Curr Opin Struct Biol 6:491–498

Haragushi Y, Chung AB, Neill S, Wallace DC (1994) OXBOX and REBOX overlapping promoter elements of the mitochondrial F0F1-ATP synthase β subunit gene. J Biol Chem 269:9330–9334

Higuti T, Kawamura Y, Kuroiwa K, Miyazaki S, Tsujita H (1993) Molecular cloning and sequence of two cDNAs for human subunit c of H^+-ATP synthase in mitochondria. Biochim Biophys Acta 1173:87–90

International Batten Disease Consortium (1995) The isolation of a novel gene underlying Batten disease (CLN3). Cell 82:949–957

Izquierdo JM, Jiménez E, Cuezva JM (1995) Hypothyroidism affects the expression of the β-F1-ATPase gene and limits mitochondrial proliferation in rat liver at all stages of development. Eur J Biochem 232:344–350

Jabs EW, Thomas PJ, Bernstein M, Coss C, Feirreira GC, Pedersen PL (1994) Chromosomal localization of genes required for the terminal steps of oxidative metabolism: α and γ subunits of ATP synthase and the pohosphate carrier. Hum Genet 93:600–602

Johnson DW, Speier S, Qian WH, Lane S, Cook A, Suzuki K, Daniel P, Boustany RM (1995) Role of subunit-9 of mitochondrial ATP synthase in Batten disease. Am J Med Genet 57:350–360
Junge W, Lill H, Engelbrecht S (1997) ATP synthase: an electrochemical transducer with rotatory mechanics. TIBS 22:420–423
Kaim G, Matthey U, Dimroth P (1998) Mode of interaction of the single a subunit with the multimeric c subunits during the translocation of the coupling ions by F0F1 ATPases. EMBO J 17:688–695
Lamminen T, Majander A, Juvonen V, Wikstrom M, Aula P, Nikoskelainen E, Savontaus ML (1995) A mitochondrial mutation at nt 9101 in the ATP synthase gene associated with deficient oxidative phosphorylation in a family with Leber hereditary optic neuroretinopathy. Am J Hum Genet 56:1238–1240
Matsuda C, Endo H, Ohta S, Kagawa Y (1993) Gene structure of human mitochondrial ATP synthase γ subunit. J Biol Chem 268:24950–24958
Mitchell P (1979) Keilin's respiratory chain concept and its chemiosmotic consequences. Science 206:1148–1159
Moradi-Améli M, Julliard JH, Godinot C (1989) Inhibition of mitochondrial F1-ATPase activity by an anti-α subunit monoclonal antibody which modifies interactions between catalytic and regulatory sites. J Biol Chem 264:1361–1367
Nelson BD, Luciakova K, Li R, Betina S (1995) The role of thyroid hormone and promoter diversity in the regulation of nuclear encoded mitochondrial proteins. Biochim Biophys Acta 1271:85–91
Noji H, Yasuda R, Yoshida M, Kinosita K Jr (1997) Direct observation of the rotation of F1-ATPase. Nature 386:299–302
Ostronoff LK, Izquierdo JM, Cuezva JM (1995) Mt-mRNA stability regulates the expression of the mitochondrial genome during liver development. Biochem Biophys Res Commun 217:1094–1098
Pedersen PL (1994) The machine that makes ATP. Curr Biol 4:1138–1141
Pedersen PL (1996) Frontiers in ATP synthase research: understanding the relationship between subunit movements and ATP synthesis. J Bioenerg Biomembr 28:389–395
Scarpulla RC (1997) Nuclear control of respiratory chain expression in mammalian cells. J Bioenerg Biomembr 29:109–119
Schägger H, Ohm TG (1995) Human diseases with defects in oxidative phosphorylation 2. F1–F0 ATP-synthase defects in Alzheimer disease revealed by blue native polyacrylamide gel electrophoresis. Eur J Biochem 227:916–921
Shoffner JM, Wallace DC (1995) Oxidative phosphorylation diseases. In: Scriver CR, Beaudet AL, Sly WS, Valle D (eds) The metabolic and molecular basis of inherited disease, 7th edn. MacGraw-Hill, New York, pp 1535–1609
Tanner A, Shen BH, Dice JF (1997) Turnover of F1F0ATP synthase subunit 9 and other proteolipids in normal and Batten disease fibroblasts. Biochim Biophys Acta 1361:251–262
Van Raaij MJ, Abrahams JP, Leslie AGW, Walker JE (1996) The structure of bovine F1-ATPase complexed with the antibiotic inhibitor aurovertin B. Proc Natl Acad Sci USA93:6913–6917
Villena JA, Martin I, Vinas O, Cormand B, Iglesias R, Mampel T, Giralt M, Villarroya F (1994) ETS transcription factors regulate the expression of the gene for the human mitochondrial ATP synthase β-subunit. J Biol Chem 269:32649–32654
Walker JE (1995) Determination of the structures of respiratory enzyme complexes from mammalian mitochondria. Biochim Biophys Acta 1271:221–227
Walker JE, Collinson IR (1994) The role of the stalk in the coupling mechanism of F1F0-ATPases. FEBS Lett 346:39–43
Wallace DC, Lott MT, Brown MD, Huoponen K, Torroni A (1995) Report of the committee on human mitochondrial DNA. In: Cuticchia AJ (ed) Human gene mapping: a compendium. Johns Hopkins University Press, Baltimore, pp 910–954 (or http://www.gen.emory.edu/mitomap.html)
Wilkens S, Dahlquist FW, McIntosh LP, Donaldson LW, Capaldi RA (1995) Structural features of the ε subunit of the *Escherichia coli* ATP synthase determined by NMR spectroscopy. Nat Struct Biol 2:961–967
Yan WL, Lerner TJ, Haines JL, Gusella JF (1994) Sequence analysis and mapping of a novel human mitochondrial ATP synthase subunit 9 cDNA (ATP5G3). Genomics 24:375–377

CHAPTER 10

Physiology and Pathophysiology of the Mitochondrial ADP/ATP Carrier 10

C. FIORE[1], V. TREZEGUET[2], C. SCHWIMMER[2], P. ROUX[1], F. NOEL[1], A. C. DIANOUX[1], G. J.-M. LAUQUIN[2], G. BRANDOLIN[1], and P. V. VIGNAIS[1]

Contents

1 Introduction

In eukaryotic cells, ATP generated by mitochondrial oxidative phosphorylation is exported to the cytoplasm in exchange for ADP by a protein located in the inner mitochondrial membrane, the ADP/ATP carrier. The transport capacity of the ADP/ATP carrier is very high: in an adult human body at rest, between 50 and 60 kg of ADP and ATP are transported across the mitochondrial membranes each day. Their amount increases markedly under working conditions, which means that the capacity of the carrier can be adjusted to metabolic conditions and is large enough to cope with a

[1] Laboratoire de Biochimie et Biophysique des Systèmes Intégrés, UMR 314 CNRS, Département de Biologie Moléculaire et Structurale, CEA-Grenoble, 17 rue des Martyrs, 38054 Grenoble Cedex 9, France
[2] Laboratoire de Physiologie Moléculaire et Cellulaire, UPR 9026 CNRS, Institut de Biochimie et Génétique Cellulaires, 1 rue Camille Saint-Saëns, 33077 Bordeaux Cedex, France

high energy demand. In the late 1970s, the ADP/ATP carrier was purified and reconstituted into liposomes. Its sequence revealed the presence of three repeated homologous domains. At that time, preliminary experiments aimed at deciphering its functioning and structure benefited from the use of two specific inhibitors, atractyloside and bongkrekic acid. In the 1980s, through chemical approaches, including chemical modifications, controlled proteolysis and immunochemical detection of extramembraneous loops, the structural organisation of the ADP/ATP carrier protein began to be unravelled with much focus on the nucleotide and inhibitor binding sites. The early studies had led to the characterisation of two conformers of the carrier, trapped by atractyloside and bongkrekic acid, respectively. More recently, experiments carried out with yeast mitochondria and using genetic and molecular biology, were able to pinpoint strategic amino acid residues responsible for conformational changes inherent in ADP/ATP transport. These structural aspects have been extensively reviewed (Brandolin et al. 1993, 1996; Fiore et al. 1998). The present interest of pathologists in mitochondrial myopathies, together with the considerable progress in human genetics has opened a new line of research aimed at understanding the tissue disorders brought about by dysfunctioning of mitochondrial ADP/ATP transport. In this chapter we shall summarise the general features of the ADP/ATP carrier relative to its functioning, and review recent data relative to its pathophysiology.

2 Functional Aspects

The role of the ADP/ATP carrier in oxidative phosphorylation has been hotly debated in the past. Whereas some basic findings, such as the electrogenic nature of the carrier and the sequential mechanism of the transport are well established facts, a number of physiological aspects were not unravelled until recently. These aspects are considered below.

2.1 The Role of the ADP/ATP Carrier in the Control of Cell Supply in ATP by Oxidative Phosphorylation

In the cell, the rate of mitochondrial respiration coupled to ADP phosphorylation has to meet the demands for ATP at all times. This is brought about by a feedback mechanism called respiratory control in which variations in the cellular ADP concentration are integrated by mitochondria in terms of variations in the rate of ATP production, to maintain the concentration of ATP at a constant level (Lardy and Wellman 1952; Chance and Williams 1956). The underlying mechanism of respiratory control has been the subject of numerous studies in the past. The results were successively interpreted by assuming that the respiratory control depends on the phosphate potential [ATP]/[ADP] · [Pi], then on the ATP/ADP ratio, and finally on the concentration of ADP alone (for review see Vignais et al. 1985). Recently, using ^{31}P NMR spectroscopy, Jeneson et al. (1996) analysed the variation of phosphorylated metabolites in the human forearm flexor muscle subjected to stimulation. They found that the kinetic function for ADP stimulation of mitochondrial oxidative phosphorylation flux was at least second order, and they suggested that the very high sensitivity of the ADP/ATP carrier to cytosolic ADP is due to the translocation step of cytosolic ADP. Whereas this conclusion may be valid under specific conditions, its generalisation must be qualified in the light of the control theory formulated by Kacser and Burns (1973), according to which, in a multistep pathway, the activities of all the enzymes have some influence on the overall flux through the pathway. This is the case of the ATP cycle in the cell, in which the transmembrane exchange of extramitochondrial ADP for intramitochondrial ATP represents only one step among several, which include the entry of Pi and oxidizable substrates in the matrix space, the synthesis of ATP coupled to substrate oxidation and the consumption of the newly synthesised ATP coupled to endergonic reactions in the extramitochondrial space. From data in the literature, it appears that the control strength exerted by the ADP/ATP carrier depends not only on the metabolic conditions, but also on the nature of the tissue. For example, in liver, the ADP/ATP carrier exerts a partial control over the rate of oxidative phosphorylation (Groen et al. 1982), unlike in the heart, where control is predominantly vested in the membrane-bound ATP synthase and the respiratory chain (see Doussière et al. 1984). Another complicating factor resides in the fact that enzymes in a given pathway may be expressed at different rates during perinatal maturation. For example, maturation from foetal to adult myocardium in the sheep is accompanied by a 2.5-fold increase in the level of ADP/ATP carrier protein with no detectable change in the level of F1-ATPase (Portman et al. 1997).

2.2
Functional Coupling of the ADP/ATP Carrier to Mitochondrial Microenvironment

The first evidence for the microcompartmentation of the ADP/ATP carrier stems from the observation that extramitochondrial ADP that enters the matrix space is more readily phosphorylated into ATP than intramitochondrial ADP, as if the ADP/ATP carrier and the ATP synthase are functionally linked (Vignais et al. 1975). This peculiarity is not, however, the result of a strict physical compartmentation, since the entire pool of adenine nucleotides is exchangeable over time. More likely, equilibration between intra- and extramitochondrial nucleotides is slow, due to slow diffusion in the viscous matrix, and this leads to a kinetic compartmentation (Vignais et al. 1975). Slow diffusion also explains the biphasic kinetics of ADP transport in mitochondria revealed by a rapid filtration method (Brandolin et al. 1990). In heart mitochondria, the channelling of the exported ATP to the mitochondrial creatine kinase in exchange for ADP produced from ATP after creatine phosphorylation may be also viewed as a kinetic compartmentation (Saks et al. 1994).

In the last decade, detailed examination of the physical properties of the inner mitochondrial membrane has unveiled unexpected features relative to the dependence of membrane permeability on the functioning of the ADP/ATP carrier. When mitochondria from mammalian cells are treated with millimolar concentrations of Ca^{2+}, the inner membrane becomes permeable to small molecules (≤1500 Da); this permeability is modulated by thiol-reactive agents, cyclophilin D ligands, ligands of the adenine nucleotide carrier, caspases and Bcl-2-like proteins (Susin et al. 1998; Marzo et al. 1998a; Bernardi et al. 1998). This change of permeability was ascribed to a protein complex forming a megachannel between the cytosol and the mitochondrial matrix: the permeability transition pore (PTP) or mitochondrial transition pore (MPT). This pore has been suggested to be constituted by the ADP/ATP carrier, porin, hexokinase, mitochondrial creatine kinase and cyclophilin (Marzo et al. 1998a). It has also been postulated that the ADP/ATP carrier might behave as the PTP on the basis that purified and reconstituted ADP/ATP carrier is able to form a Ca^{2+}-dependent channel that is induced by atractyloside and blocked by ADP and bongkrekic acid (Brustovetsky and Klingenberg 1996; Rück et al. 1998). The regulatory action of cyclosporin A on MTP has been demonstrated recently to be mediated by the binding of cyclophilin D to the ADP/ATP carrier (Halestrap et al. 1998). Very recently, it has been proposed that action of the death factor Bax on mitochondria, which is counteracted by Bcl-2, involves a direct interaction between Bax and the ADP/ATP carrier (Marzo et al. 1998b).

2.3
Regulatory Aspects of ADP/ATP Transport

Among the naturally occuring nucleotides, only ADP and ATP are transported by the ADP/ATP carrier. The transported nucleotides are the free ATP and ADP species rather than the Mg^{2+}-complexes. The exquisite specificity for ADP and ATP underlines the strict recognition of not only the adenine moiety, but also the diphosphate and triphosphate residues of the nucleotides, by appropriate amino acid sequences or specific arrangements of peptide segments in the carrier molecule. Under conditions of oxidative phosphorylation, the ATP delivered by ATP synthase into the matrix space is exchanged, preferentially to matricial ADP, against cytosolic ADP. The exchange stoichiometry corresponds to one ADP entering the matrix space for one ATP released outside the mitochondrion, thus maintaining the mitochondrial adenine nucleotide pool at a constant level.

Detailed kinetic studies of the initial rates of adenine nucleotide exchange led to the conclusion that the mechanism of transport is sequential, i.e. that it implies the formation of a ternary complex in which the carrier binds two nucleotides (Duyckaerts et al. 1980). Assuming that the carrier unit is a tetramer (two dimers) (Block and Vignais 1984), one dimer will take one ADP from the outside and engulf and transport it across the membrane. At the same time, one ATP, firmly bound to the inward face of the other dimer, is waiting to be transported so that the ADP which is crossing the first dimer is released into the matrix space of the mitochondrion.

When mitochondria are incubated aerobically with oxidizable substrates, Pi and ADP, the measured phosphate potential, [ATP]/[ADP] · [Pi], is higher in the extramitochondrial compartment than in the matrix space after equilibrium is reached (Heldt et al. 1972; Vignais et al. 1973). The resulting difference in phosphate potential between the matrix space and the extramitochondrial compartment is explained by the preferential exchange of ADP_{ex} for ATP_{in}. At neutral pH, the number of negative charges in ADP and ATP is 3 and 4, respectively. The $ADP^{3-}{}_{ex}/ATP^{4-}{}_{in}$ exchange is intrinsically not charge-compensated by proton movement, and therefore is electrogenic. Under physiological conditions, such an asymmetric exchange is not thermodynamically feasible, and a superimposed mechanism has to be invoked. An extrinsic mechanism used to this end is the creation of a potential in the inner mitochondrial membrane, positive outside, at the expense of the redox energy generated by the respiratory chain in respiring mitochondria. Up to one third of the redox energy of the respiratory chain can be used to compensate the difference of charge which results from the electrogenic exchange $ADP^{3-}{}_{ex}/ATP^{4-}{}_{in}$ (Duszynski et al. 1981). An important consequence of the dependence of ADP/ATP exchange on membrane potential is directionality, so that ATP must be exported from mitochondria in exchange for ADP, which is imported. Another mechanism might be partly responsible for the asymmetric ADP/ATP exchange. Its intrinsic nature is suggested by experiments on rapid exchange kinetics (Brandolin et al. 1990). In well-coupled mitochondria, the rate of ADP uptake is much higher than that of ATP uptake, whereas in uncoupled mitochondria, the difference in the rates is decreased but not completely abolished. This suggests that, besides the extrinsic control of ADP/ATP transport by membrane potential, an intrinsic mechanism regulates the direction in which ADP and ATP are translocated. The mechanism might be inherent in the uneven distribution of charged amino acid residues on each face of the carrier molecule, a larger number of positive charges being present on the matrix face of the carrier.

Another factor of complexity in the regulation of ADP/ATP transport resides in the fact that cytosolic ADP has to pass through the outer mitochondrial membrane to reach the carrier, as does the ATP discharged by the carrier in order to reach the cytosolic space. In the 1960s, it was believed that the outer mitochondrial membrane behaved as a sieve for small sized metabolites. This oversimplified view has been reassessed in the past decade. In fact, the outer mitochondrial membrane contains a number of channels, including the voltage-dependent anion carrier (VDAC) which behaves as a porin with specificity for anionic molecules in the outer membrane (for review see Benz 1994). From a study with a porin-deficient yeast mutant (Michejda et al. 1990), it was concluded that, whereas the rates of respiration of mutant cells and wild type cells are coupled to phosphorylation to the same degree, the mutant mitochondria require three to four times more ADP than the wild-type mitochondria for the transition from state 4 to state 3. This result suggests that VDAC somehow limits the diffusion of externally added ADP towards the ADP/ATP carrier. Recently, deletion studies were carried out on the two isoforms of the yeast VDAC (Lee et al. 1998). The permeability of the outer membrane was assessed by the ability of an intermembrane-space dehydrogenase to oxidise exogenously added NADH. Mitochondria lacking VDAC1 or both isoforms showed a two fold increase in the rate of NADH oxidation when the outer membrane was deliberately damaged, whereas wild-type mitochondria showed only a 10% increase. It was calculated that, in the absence of VDAC1, the outer membrane permeability to NADH is reduced 20-fold, indicating a key role of VDAC1 in outer membrane permeability.

2.4
Isoforms and Tissue Specificity of the ADP/ATP Carrier

The ADP/ATP carrier protein, encoded by nuclear genes and synthesised on cytosolic polyribosomes, is recognised by import receptors located on the outer mitochondrial membrane and then directed to the inner mitochondrial membrane (Koehler et al. 1998; Sirrenberg et al. 1998). The mitochondrial import machinery was recently reviewed (Pfanner and Meijer 1997). Here, we shall address only the genic expression of different isoforms of the ADP/ATP carrier in mammals.

2.4.1
Genomic Structure

Multiple isoforms of the ADP/ATP carrier, encoded by specific genes, have been characterised in mitochondria originating from a variety of cells (for reviews see Brandolin et al. 1993; Fiore et al. 1998). For example, three carrier isoforms have been characterised in humans and the yeast *S. cerevisiae,* and two isoforms in rat, mouse, rice and maize.

The nomenclature of the human ADP/ATP carrier genes is rather confusing, and for clarification, Table 1 specifies the names of genes and their chromosomal location. Two genes, *ANC2* and *ANC3,* are present on chromosome X whereas *ANC1* is located on the long arm of chromosome 4 (Haraguchi et al. 1993). In addition to these three functional genes, several pseudogenes were identified (at least 7 for *ANC*3). *ANC2* is located on the short arm of chromosome X, in the pseudo-autosomal region, about 1.3 Mb from the telomeric end. This gene escapes X-inactivation, which is not the case for *ANC3.* As a result, *ANC3* is present in only one transcribable copy in males and females whereas *ANC2* is present in two copies. The three human genes show a similar genomic structure, with four exons and three introns located at similar positions (Fig. 1), and spanning a 5.9 kb region. A hypomethylated CpG island has been located within the *ANC2* promoter, which also contains 13 binding sites for the transcription factor Sp1 (Slim et al. 1993; Schiebel et al. 1993). A polymorphic marker formed by CA-repeats and located at about 4 kb upstream of *ANC1* allowed detection of six different alleles.

Table 1. Nomenclature of expressed human adenine-nucleotide carrier genes

	Nomenclature		Chromosome location
T1	ANT1	ANC1	4q35-qter
T2	ANT3	ANC2	Xp22.3
T3	ANT2	ANC3	Xq13 → Xq25–q26

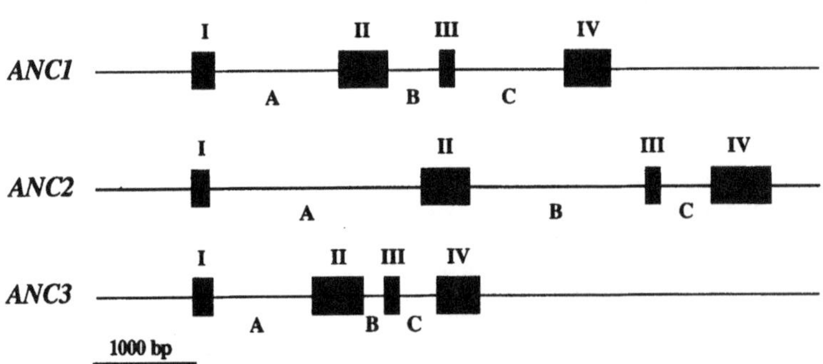

Fig. 1. Sequence comparison of the human *ANC* genes. *Full boxes* represent exons, which are numbered in Roman numerals; introns are indicated by *capital letters.* The size of exons I and IV are approximate. (Adapted from Cozens et al. 1989; Ku et al. 1990)

From detailed studies of the promoter regions, it has been proposed that differences observed between the different ADP/ATP carrier genes may explain the tissue-specific expression of the carrier (see Fiore et al. 1998). The *ANC1* promoter contains characteristic eukaryotic TATA and CCAAT boxes located immediately upstream of the transcription initiation site and a typical SV40 transcriptional enhancer 1400 bp upstream. In contrast, *ANC2* does not contain these two boxes but only 13 potential Sp1 binding sites spread over 1250 bp in the 5'-flanking region, which is characteristic of housekeeping genes. *ANC3* shares some features with *ANC1* and *ANC2*, i.e. it is devoid of the CCAAT box, but contains a TATA box and 5 Sp1 binding sites located in a 250 bp region overlapping the initiation site of transcription.

Singularly, an OXBOX enhancer has been found in *ANC1* promoter, a sequence which specifically binds transcription factors only present in muscle cells (Li et al. 1990). A REBOX silencer has also been described in *ANC1* promoter, but it binds ubiquitous transcription factors (Chung et al. 1992). As a result, these two sequences might explain the muscle-specific expression of the *ANC1* gene.

2.4.2
Transcriptional Expression and Tissue Specificity

Immunochemical studies have suggested a tissue specificity for the mammalian ADP/ATP carrier expression (Schultheiss and Bolte 1985). Comparison of Ancp sequences from different tissues in different animal species indicates that the tissue-specific isoforms are much more conserved between species than different isoforms in the same species, with a mean value of 90 versus 80% similarity, respectively. In human, ox and mouse, the *ANC1* isoform is predominantly expressed in skeletal and cardiac muscles, whereas *ANC3* is expressed at a low level, if at all, in brain, liver, kidney, heart and skeletal muscle. In contrast, *ANC2* is expressed in all tissues and in cultured fibroblasts, but in variable amounts depending on the respiratory activity of the tissue (Stepien et al. 1992; Dörner et al. 1997a). *ANC1* is expressed at low levels in proliferating myoblasts during muscle development. Its expression is markedly enhanced after completion of the proliferation phase, i.e. during induction of differentiation and subsequent fusion of differentiated myoblasts to form myotubes. At the onset of differentiation, *ANC3* is expressed at the same level as *ANC1* whereas *ANC2* is expressed half as much. Expression of these two isoforms progressively decreases during differentiation (Stepien et al. 1992). It has been argued that *ANC1* is not expressed in myoblasts (Lunardi et al. 1992).

In the cultured human cell lines HeLa, HL60, Hep3B and 143B, *ANC2* and *ANC3* are expressed in the absence of *ANC1*. Variations in the expression level of *ANC2* and *ANC3* were observed during cell growth or under specific physiological conditions (Lunardi and Attardi 1991). For example, in cultured HeLa cells, the expression level of *ANC3* is decreased to less than 50% of its original level when the cell growth reaches the stationary phase, whereas the *ANC2* expression remains approximately constant. Treatment of HL60 cells by agents inducing the arrest of cell proliferation together with differentiation, such as retinoic acid or phorbol esters, leads to a large decrease in expression of both isoforms. As a general rule, *ANC3* is highly expressed under conditions allowing cell proliferation. This is the case after injection of thyroid hormone in rats (Luciakova and Nelson 1992) or addition of serum or growth factors (PDGF, EGF) to resting NIH 3T3 cells (Battini et al. 1987).

3
Pathophysiology of the ADP/ATP Carrier

Because of the key role played by the ADP/ATP carrier in the cell economy, it is evident that dysfunctioning of this transport system should be the cause of serious metabolic aberrations. In this chapter we summarise some recent data on this topic and we propose combined fluorescent and immunologic assays to screen muscle biopsies of patients with mitochondrial myopathies caused by a deficiency in ADP/ATP transport.

3.1
Implication of the ADP/ATP Carrier in Pathologies

As mitochondria play a central role in cellular energy production, abnormalities in the mitochondrial energy generating system due to mutations in the mitochondrial or nuclear genome result in pathologies (Brown and Wallace 1994; Schon et al. 1997). These disorders are reflected by a variety of clinical symptoms ranging from pure myopathy with lactic acidosis to severe multisystem disease with central nervous system involvement (Munnich et al. 1996). The age of onset of symptoms ranges from birth to adulthood. Biochemical investigations in patients suffering from a mitochondriocytopathy are commonly performed in skeletal muscle tissue to localise the enzymatic origin of the impairment. They consist in measuring the substrate oxidation rates through the different complexes of the respiratory chain, oxygen consumption and/or ATP production by intact mitochondria. In most cases, the origin of the disease is revealed by a defect of one of the enzyme complexes of the respiratory chain. Nevertheless, in some patients with reduced substrate oxidation in muscle mitochondria, the abnormality cannot be ascribed to a defect of enzymes involved in substrate oxidation and coupled phosphorylation. In this case, a plausible hypothesis is the dysfunctioning of one of the transport systems located in the inner mitochondrial membrane, including the ADP/ATP and Pi carriers, or in the outer membrane such as the voltage dependent-anion channel (VDAC) or porin (Huizing et al. 1998).

Up to now, no deficit of the Pi carrier has been found in patients suffering from mitochondrial myopathies, but one case of a VDAC deficit was recently described (Huizing et al. 1996). In this case, investigation of the muscle mitochondria of the patient showed impaired substrate oxidation and reduced production of phosphorylated compounds. Immunochemical analysis revealed a severe decrease in VDAC expression in skeletal muscle. However, the deficiency was not detected in cultured skin fibroblasts from this patient, suggesting that the defect for this carrier was largely tissue-specific.

One case of human myopathy caused by a deficit of the ADP/ATP carrier has been reported (Bakker et al. 1993a). The patient was affected with severe lactic acidosis from the age of 3.5 years and presented shortness of breath and rapid fatigue. Since enzyme activities of the different complexes of the respiratory chain were increased, a therapeutic trial with B vitamins and carnitine was initiated. Two years later, with the strongly enhanced activities of several mitochondrial enzymes, the uncoupled mitochondrial respiration rate was found to be significantly higher than in controls. Whereas the respiration of the patient mitochondria was less stimulated by ADP than that of control mitochondria, it appeared that the defect was not due to an impairment of complex V. The deficiency in the ADP/ATP carrier was shown by immunodetection, which revealed an amount of carrier protein four fold lower than in controls. The Anc defect was restricted to striated muscles; it could not be detected in heart, liver, fibroblasts and lymphocytes. The symptoms were partially corrected by treatment with vitamin E, which is a scavenger of free radicals (Bakker et al. 1993b, c). A possible explanation is that the ADP/ATP carrier was modified by oxygen metabolites overproduced by the hyperactive respiratory chain; alternatively, the carrier inhibition may have been the result of peroxidation of mitochondrial phospholipids.

The level of expression of *ANC1* (also called *ANT1*) in muscle of patients suffering from mitochondrial myopathies has been analysed (Heddi et al. 1993). There was a three-fold increase in the *ANC1* expression level in patients affected with MERRF (myoclonic epilepsy associated with ragged-red fibers) and a 1.75-fold increase in the case of MELAS (myopathy, encephalopathy, lactic acidosis and stroke-like episodes). This contrasts with a 30% decrease of Anc1p in muscle from patients suffering from KSS (Kearns-Sayre syndrome) myopathy. It is noteworthy that SV40 transformation of human fibroblasts led to a concomitant increase in the expression of the two ADP/ATP carrier isoforms *ANC1* and *ANC3* (also called *ANT2*), but not *ANC2* (also called *ANT3*; see Table 1), and genes encoding mitochondrial proteins involved in oxidative phosphorylation, such as the β subunit of ATP synthase (Torroni et al. 1990). The same behaviour was found in three other transformed cell lines: HeLa cervical carcinoma, EBV-L lymphoblasts and HT180 fibrosarcoma (Torroni et al. 1990). Analysis of kidney carcinoma (Heddi et al. 1996) revealed a four fold decrease in *ANC3* expression, contrasting with a 4- to 30-fold increase in kidney and salivary gland oncocytoma. The *ANC3* expression in transformed cells seems to be due to the presence of a GRBOX sequence in *ANC3* promoter (Giraud et al. 1998). It has been argued that in transformed cells, as well as in rho° cells, the Anc3 protein would function by importing ATP into mitochondria (Giraud et al. 1998) and would be involved in the establishment of the membrane potential in the rho° cells (Buchet and Godinot 1998).

Schultheiss et al. (1996) and Dörner et al. (1997b) have shown the presence of autoantibodies against the ADP/ATP carrier protein in sera of patients suffering from myocarditis and dilated cardiomyopathy (DCM). These autoantibodies are specific for the heart isoform of the ADP/ATP carrier. In DCM patients with an insufficient energy supply in heart muscle, the ADP/ATP transport capacity in heart mitochondria was decreased. This decrease was accompanied by a marked elevation in total ADP/ATP carrier protein content caused by an increase in the Anc1p isoform, as determined by immunoblotting analysis. Simultaneously, the amount of *ANC2* transcripts was decreased, but the *ANC3* expression was unchanged. The specificity of these defects was illustrated by the fact that patients with ischemic or valvular heart disease showed no alteration in ADP/ATP carrier function or expression.

The ability of ADP/ATP carrier autoantibodies to inhibit the nucleotide exchange activity of heart mitochondria was demonstrated in vitro using guinea pigs immunised with the purified ADP/ATP carrier protein or mice affected with myocarditis following infection with Coxsackie B3 enterovirus (Schulze et al. 1989). In both cases, the ADP/ATP transport of heart mitochondria was dramatically decreased, and, in agreement with this defect, a nucleotide analysis showed a higher ATP/ADP ratio in isolated heart mitochondria and a lower ATP/ADP ratio in cytosol from cardiac cells compared with controls. In addition, the amount of ADP/ATP carrier protein was increased, due to an increase in *ANC1* expression, similar to that found for DCM patients (Schultheiss et al. 1996; Dörner et al. 1997b). These effects were explained by direct inhibition of ADP/ATP carrier by autoantibodies after cytosolic internalisation of immunoglobulin (Schulze et al. 1989).

Recently, a "knockout" mouse deficient in the heart/muscle isoform of the ADP/ATP carrier (Anc1p) was generated (Graham et al. 1997). Histological studies revealed ragged-red fibers in skeletal muscle and a cardiac hypertrophy, with a dramatic proliferation of mitochondria in both tissues. Mitochondria isolated from skeletal muscle exhibited a severe defect in coupled respiration. In mutant adults, a lactic acidosis and a severe intolerance to exercise were observed. Clearly, Anc1p mutant mice display the histological, metabolic and physiologic characteristics of mitochondrial myopathy and cardiomyopathy. Surprisingly, the absence of Anc1p, which is normally highly expressed in heart, skeletal muscle and brain, was not compensated by an increase in the other ADP/ATP carrier isoform (Anc2p). Mutant mice are fertile and exhibit normal growth characteristics when compared with wild-type mice up to at least eight months of age. They probably survive because of the low but sufficient level of Anc2p expression in heart and brain.

3.2 Methodological Aspects for the Characterisation of ADP/ATP Carrier Defects

As discussed above, some pathologies are likely to be associated with the defect of mitochondrial carriers, and specifically with the ADP/ATP carrier which is directly involved in ATP production. Because of the small size of biopsies, the functional capacities of mitochondrial carriers are difficult to assess by kinetic measurements. This is why investigators have chosen to determine the presence of mitochondrial carrier proteins, including the ADP/ATP carrier, in muscle samples by immunodetection or reverse transcription-polymerase chain reaction. However, these methods do not give access to the functional parameters of the transport systems. In the case of the ADP/ATP carrier, binding of specific ligands, namely the substrates ADP and ATP, and the inhibitors ATR and BA, is considered to reflect the carrier's ability to transport. We therefore set up a fluorometric test to quantify the amount of ADP/ATP carrier in muscle biopsies, based on the interaction of the carrier with specific ligands.

Because ATR and some of its derivatives exhibit an exquisite specificity and affinity towards the ADP/ATP carrier, they were chosen for determining the carrier amount within mitochondria, either in isolated or in permeabilised cells. Several ATR derivatives were used for binding studies, radiolabelled ATR or fluorescent derivatives of ATR such as naphthoyl-ATR (N-ATR) and methylanthranyloyl-ATR (Mant-ATR; Boulay et al. 1983; Roux et al. 1996; Fig. 2A). The fluorescence approach is much more sensitive than the radioactive assay since it requires approximately 100 times less biological material. The principle of the fluorometric titration consists in measurement of fluorescence changes corresponding to the specific CATR-induced release of N-ATR or Mant-ATR bound to the carrier. Their amplitudes are proportional to the amount of added CATR until a saturation effect is reached. Measurement of the fluorescence changes elicited by successive additions of subsaturating amounts of CATR in the same cuvette allows titration of the CATR binding sites and therefore determination of the ADP/ATP carrier concentration (Fig. 2B).

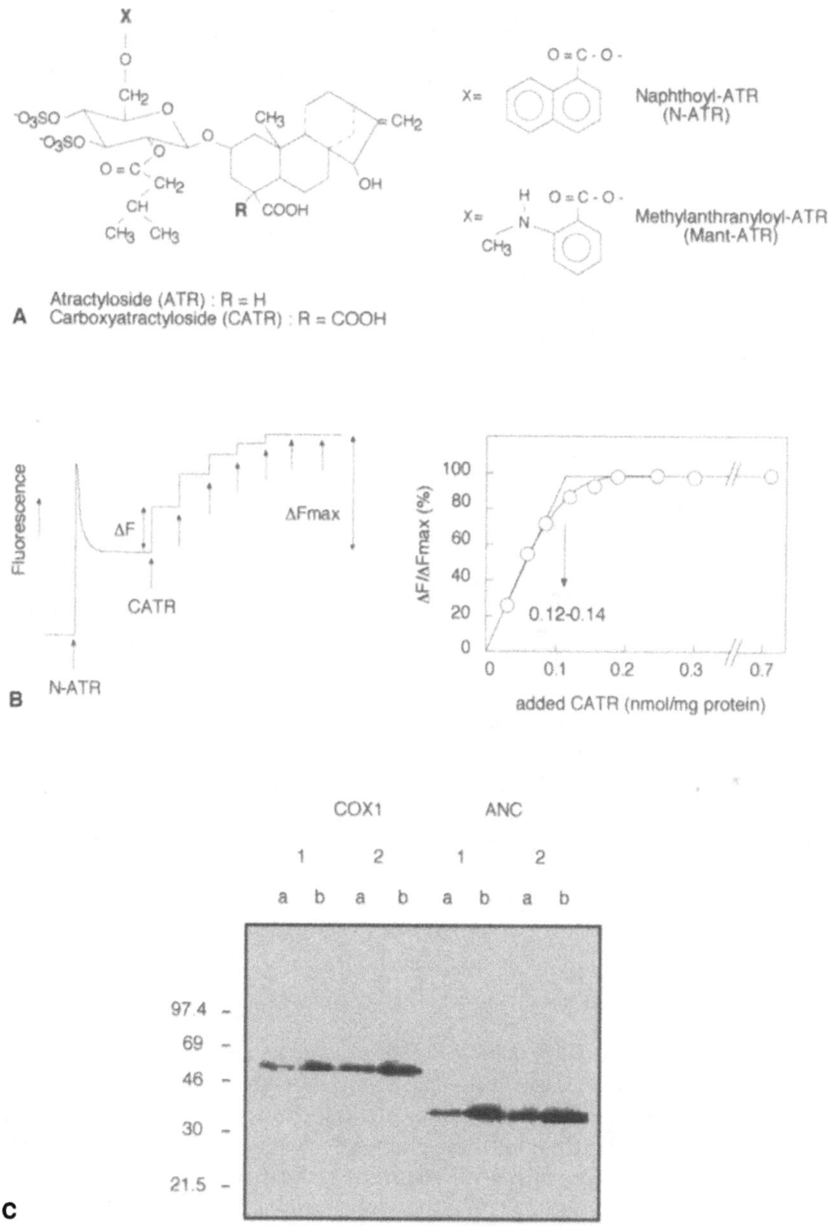

Fig. 2. Fluorometric and immunochemical assays for the characterisation of ADP/ATP carrier defects. **A** Structure of fluorescent derivatives of atractyloside. **B** Fluorometric titration of the ADP/ATP carrier in rabbit muscle homogenate. The principle of the method and a titration curve are shown on the *left* and on the *right*, respectively. **C** Immunochemical detection of the ADP/ATP carrier in muscle sample issued from fluorometric assay. Subunit I of cytochrome oxidase (COX1) is used as a reference marker for the quantification of ANC. The western blot analysis shows detection of 25 ng (*a*) and 50 ng (*b*) of estimated Ancp, in two different samples (*1* and *2*). Molecular weight markers are indicated

Titration of the ADP/ATP carrier by the fluorometric technique on rabbit muscle homogenates, without preliminary isolation of mitochondria, but in the presence of digitonin to permeabilise cells, gave values of 0.12–0.14 nmol of CATR binding sites/mg of protein, similar to those calculated from binding experiments carried out with radiolabelled ATR. To ensure that the totality of the CATR binding sites are titrated, the amount of ADP/ATP carrier in muscle tissue homogenates and in mitochondria isolated from these homogenates were determined together with the content of heme aa_3 taken as a mitochondrial reference. For both types of biological material, the ratio of the number of CATR binding sites (nmol/mg protein) to the heme aaa_3 content (nmol/mg protein) was close to 2, indicating that the totality of the CATR binding sites are titrated (Roux et al. 1996).

Up to the present time, some 70 samples of human skeletal muscle from healthy subjects have been analysed by the fluorometric test (Fiore, unpubl. results). Citrate synthase was used as a mitochondrial reference enzyme and assayed in the muscle homogenate by the DTNB method (Moriyama and Srere 1971). The ADP/ATP carrier was quantified by measuring the number of CATR binding sites in the membrane fraction of the homogenate. The ratio of the citrate synthase activity (expressed in nmol/min) and the number of CATR binding sites (nmol) calculated for the whole muscle sample was fairly constant, amounting to 0.31 ± 0.08. If a patient is found with a deficit in the number of CATR binding sites, this has to be correlated with a possible decrease of the ADP/ATP carrier amount. For this purpose, we set up a sensitive immunochemical assay to directly analyse the muscle samples issued from the fluorometric titration, without isolation of mitochondria which can lead to a possible loss of material. In this case, subunit I of cytochrome oxidase was found to be a reliable and convenient marker reference because it is a nuclear-encoded protein located in the inner mitochondrial membrane, like the ADP/ATP carrier (Fig. 2C).

It is clear that fluorescent titration, combined with immunodetection, is a promising approach for screening ADP/ATP carrier defects in as yet uncharacterised myopathies.

4 Conclusion

To summarise, in spite of major kinetic, biochemical and genetic advances in the past decade, several important issues of mitochondrial ADP/ATP transport remain unsolved. For example, we do not know how binding and translocation of ADP and ATP are controlled by interaction of these nucleotides with specific amino acid residues in the carrier. Knowledge of the 3-D structure of the carrier is obviously needed to propose a realistic model of ADP/ATP transport. In the last 5 years, a few cases of dysfunction of the ADP/ATP carrier in humans have been reported. With these recent findings, a new chapter in the story of mitochondrial myopathies is emerging.

Acknowledgements. The work presented in this chapter was developed in the Grenoble and Bordeaux laboratories and was supported by grants of the Association Française contre les Myopathies, the Commissariat à l'Energie Atomique, the Région Rhône-Alpes, the Région Aquitaine, the University Joseph Fourier of Grenoble and the University Victor Segalen of Bordeaux.

References

Bakker HD, Scholte HR, Van den Bogert C, Ruitenbeek W, Jeneson JA, Wanders RJ, Abeling NG, Dorland B, Sengers RC, Van Gennip AH (1993a) Deficiency of the adenine nucleotide translocator in muscle of a patient with myopathy and lactic acidosis: a new mitochondrial defect. Pediatr Res 33:412–417

Bakker HD, Scholte HR, Van den Bogert C, Jeneson JA, Ruitenbeek W, Wanders RJ, Abeling NG, Van Gennip AH (1993b) Adenine nucleotide translocator deficiency in muscle: potential therapeutic value of vitamin E. J Inherit Metab Dis 16:548–552

Bakker HD, Scholte HR, Jeneson JA (1993c) Vitamin E in a mitochondrial myopathy with proliferating mitochondria. Lancet 342:175–176

Battini R, Ferrari S, Kaczmarek L, Calabretta B, Chen ST, Baserga R (1987) Molecular cloning of a cDNA for a human ADP/ATP carrier which is growth-regulated. J Biol Chem 262:4355–4359

Benz R (1994) Permeation of hydrophilic solutes through mitochondrial outer membrane: review on mitochondrial porins. Biochim Biophys Acta 1197:167–196

Bernardi P, Basso E, Colonna R, Costantini P, Di Lisa F, Eriksson O, Fontaine E, Forte M, Ichas F, Massari S, Nicolli A, Petronilli V, Scorrano L (1998) Perspectives on the mitochondrial permeability transition. Biochim Biophys Acta 1365:200–206

Block MR, Vignais PV (1984) Substrate-site interactions in the membrane-bound adenine-nucleotide carrier as disclosed by ADP and ATP analogs. Biochim Biophys Acta 767:369–376

Boulay F, Brandolin G, Lauquin GJ-M, Vignais PV (1983) Synthesis and properties of fluorescent derivatives of atractyloside as potential probes of the mitochondrial ADP/ATP carrier protein. Anal Biochem 128:323–330

Brandolin G, Marty I, Vignais PV (1990) Kinetics of nucleotide transport in rat heart mitochondria studied by a rapid filtration technique. Biochemistry 29:9720–9727

Brandolin G, Le Saux A, Trézéguet V, Lauquin GJ-M, Vignais PV (1993) Chemical, immunological, enzymatic, and genetic approaches to studying the arrangement of the peptide chain of the ADP/ATP carrier in the mitochondrial membrane. J Bioenerg Biomembr 25:459–472

Brandolin G, Le Saux A, Roux P, Trézéguet V, Fiore C, Schwimmer C, Dianoux AC, Lauquin GJ-M, Vignais PV (1996) The ADP/ATP transport system: a paradigm of metabolic transport in mitochondria. In: Claphan D, Erlich BE (eds) Organellar ion channels and transports. Rockefeller University Press, New York, pp 173–185

Brown MD, Wallace DC (1994) Molecular basis of mitochondrial DNA disease. J Bioenerg Biomembr 26:273–289

Brustovetsky N, Klingenberg M (1996) Mitochondrial ADP/ATP carrier can be reversibly converted into a large channel by Ca^{2+}. Biochemistry 35:8483–8488

Buchet F, Godinot C (1998) Functional F1-ATPase essential in maintaining growth and membrane potential of human mitochondrial DNA-depleted ϱ° cells. J Biol Cell 273:22983–22989

Chance B, Williams GR (1956) The respiratory Chain and oxidative phosphorylation. Adv Enzymol 17:65–134

Chung AB, Stepien G, Haraguchi Y, Li K, Wallace DC (1992) Transcriptional control of nuclear genes for the mitochondrial muscle ADP/ATP translocator and the ATP synthase beta subunit. Multiple factors interact with the OXBOX/REBOX promoter sequences. J Biol Chem 267:21154–21161

Cozens AL, Runswick MJ, Walker JE (1989) DNA sequence of two expressed nuclear genes for human mitochondrial ADP/ATP translocase. J Mol Biol 206:261–280

Dörner A, Pauschinger M, Badorff A, Noutsias M, Giessen S, Schulze K, Bilger J, Rauch U, Schultheiss HP (1997a) Tissue-specific transcription pattern of the adenine nucleotide translocase isoforms in humans. FEBS Lett 414:258–262

Dörner A, Schulze K, Rauch U, Schultheiss HP (1997b) Adenine nucleotide translocator in dilated cardiomyopathy: pathophysiological alterations in expression and function. Mol Cell Biochem 174:261–269

Doussière J, Ligeti E, Brandolin G, Vignais PV (1984) Control of oxidative phosphorylation in rat heart mitochondria. The role of the adenine nucleotide carrier. Biochim Biophys Acta 766:492–500

Duszynski J, Bogucka K, Letko G, Küster U, Kunz W, Wojtczak L (1981) Relationship between the energy cost of ATP transport and ATP synthesis in mitochondria. Biochim Biophys Acta 637-217-223

Duyckaerts C, Sluse-Goffart CM, Fux JP, Sluse FE, Liebecq C (1980) Kinetic mechanism of the exchanges catalyzed by the adenine-nucleotide carrier. Eur J Biochem 106:1–6

Fiore C, Trézéguet V, Le Saux A, Roux P, Schwimmer C, Dianoux AC, Noël F, Lauquin GJ-M, Brandolin G, Vignais PV (1998) The mitochondrial ADP/ATP carrier: structural, physiological and pathological aspects. Biochimie 80:137–150

Giraud S, Bonod-Bidaud C, Wesolowski-Louvel M, Stepien G (1998) Expression of human ANT2 gene in highly proliferative cells: GRBOX, a new transcriptional element, is involved in the regulation of glycolytic ATP import into mitochondria. J Mol Biol 281:409–418

Graham BH, Waymine KG, Cottrell B, Trounce IA, MacGregor GR, Wallace DC (1997) A mouse model for mitochondrial myopathy and cardiomyopathy resulting from a deficiency in the heart/muscle isoform of the adenine nucleotide translocator. Nat Genet 16:226–234

Groen AK, Wanders RJA, Westerhoff HV, Van der Meer R, Tager JM (1982) Quantification of the contribution of various steps to the control of mitochondrial respiration. J Biol Chem 257:2754

Halestrap AP, Kerr PM, Javadov S, Woodfield KY (1998) Elucidating the molecular mechanism of the permeability transition pore and its role in reperfusion injury of the heart. Biochim Biophys Acta 1366:79–94

Haraguchi Y, Chung AB, Torroni A, Stepien G, Shoffner JM, Wasmuth JJ, Costigan DA, Polak M, Altherr MR, Winokur ST (1993) Genetic mapping of human heart-skeletal muscle adenine nucleotide translocator and its relationship to the facioscapulohumeral muscular dystrophy locus. Genomics 16:479–485

Heddi A, Lestienne P, Wallace DC, Stepien G (1993) Mitochondrial DNA expression in mitochondrial myopathies and coordinated expression of nuclear genes involved in ATP production. J Biol Chem 268:12156–12163

Heddi A, Faure-Vigny H, Wallace DC, Stepien G (1996) Coordinate expression of nuclear and mitochondrial genes involved in energy production in carcinoma and oncocytoma. Biochim Biophys Acta 1316:203–209

Heldt HW, Klingenberg M, Milovanco M (1972) Differences between the ATP/ADP ratio in the mitochondrial matrix and the extramitochondrial space. Eur J Biochem 30:434–440

Huizing M, Ruitenbeek W, Thinnes FP, DePinto V, Wendel U, Trijbels JMF, Smit LME, Ter Laak HJ, Van den Heuvel LP (1996) Deficiency of the voltage-dependent anion channel (VDAC): a novel cause of mitochondriocytopathy. Pediatr Res 39:760–765

Huizing M, Ruitenbeek W, Van den Heuvel LP, Dolce V, Iacobazzi V, Smeitink JAM, Palmieri F, Trijbels JMF (1998) Human mitochondrial transmembrane metabolite carriers: tissue distribution and its implication for mitochondrial disorders. J Bioenerg Biomembr 30:277–284

Jeneson JAL, Wiseman RW, Westerhoff HV, Kushmerick MJ (1996) The signal transduction function of oxidative phosphorylation is at least second order in ADP. J Biol Chem 271:27995–27998

Kacser H, Burns JA (1973) The control flux. In: Davies DD (ed) Rate control of biological processes. Cambridge University Press, London, pp 65–104

Koehler CM, Jarosch E, Tokatlidis K, Schmid K, Schweyen RJ, Schatz G (1998) Import of mitochondrial carriers mediated by essential proteins of the intermembrane space. Science 279:369–373

Ku DH, Kagan J, Chen ST, Chang CD, Baserga R, Wurzel J (1990) The human fibroblast adenine nucleotide translocator gene. Molecular cloning and sequence. J Biol Chem 265:16060–16063

Lardy HA, Wellman H (1952) Oxidative phosphorylation. Role of inorganic phosphate and acceptor systems in control of metabolite rates. J Biol Chem 195:215–224

Lee AC, Xu X, Blachly-Dyson E, Forte M, Colombini M (1998) The role of yeast VDAC genes on the permeability of the mitochondrial outer membrane. J Membr Biol 161:173–181

Li K, Hodge JA, Wallace DC (1990) OXBOX, a positive transcriptional element of the heart-skeletal muscle ADP/ATP translocator gene. J Biol Chem 265:20585–20588

Luciakova K, Nelson BD (1992) Transcript levels for nuclear-encoded mammalian mitochondrial respiratory-chain components are regulated by thyroid hormone in an uncoordinated fashion. Eur J Biochem 207:247–251

Lunardi J, Attardi G (1991) Differential regulation of expression of the multiple ADP/ATP translocase genes in human cells. J Biol Chem 266:16534–16540
Lunardi J, Hurko O, Engel WK, Attardi G (1992) The multiple ADP/ATP translocase genes are differentially expressed during human muscle development. J Biol Chem 267:15267–15270
Marzo I, Brenner C, Zamzami N, Susin SA, Beutner G, Brdiczka D, Rémy R, Xie ZH, Reed JC, Kroemer G (1998a) The permeability transition pore complex: a target for apoptosis regulation by caspases and Bcl-2-related proteins. J Exp Med 187:1261–1271
Marzo I, Brenner C, Zamzami N, Jürgensmeier JM, Susin SA, Vieira HLA, Prévost MC, Xie Z, Matsuyama S, Reed JC, Kroemer G (1998b) Bax and adenine nucleotide carrier cooperate in the mitochondrial control of apoptosis. Science 281:2027–2031
Michejda J, Guo XJ, Lauquin GJ-M (1990) The respiration of cells and mitochondria of porin deficient yeast mutants is coupled. Biochem Biophys Res Commun 171:354–361
Moriyama T, Srere PA (1971) Purification of rat heart and rat liver citrate synthases. J Biol Chem 246:3217–3223
Munnich A, Rotig A, Chretien D, Saudubray JM, Cormier V, Rustin P (1996) Clinical presentation and laboratory investigations in respiratory chain deficiency. Eur J Pediatr 155:262–274
Pfanner N, Meijer M (1997) Mitochondrial biogenesis: the Tom and Tim machine. Curr Biol 7:R100–R103
Portman MA, Xiao Y, Song Y, Ning XH (1997) Expression of adenine nucleotide translocator parallels maturation of respiratory control in heart in vivo. Am J Physiol 273:H1977–H1983
Roux P, Le Saux A, Fiore C, Schwimmer C, Dianoux AC, Trézéguet V, Vignais PV, Lauquin GJ-M, Brandolin G (1996) Fluorometric titration of the mitochondrial ADP/ATP carrier protein in muscle homogenate with atractyloside derivatives. Anal Biochem 234:31–37
Rück A, Dolder M, Wallimann T, Brdiczka D (1998) Reconstituted adenine nucleotide translocase forms a channel for small molecules comparable to the mitochondrial permeability transition pore. FEBS Lett 426:97–101
Saks VA, Khuchua ZA, Vasilyeva EV, Belikova OY, Kuznetsov AV (1994) Metabolic compartmentation and substrate channelling in muscle cells. Role of coupled creatine kinases in vivo regulation of cellular respiration – a synthesis. Mol Cell Biochem 133/134:155–192
Schiebel K, Weiss B, Wohrle D, Rappold G (1993) A human pseudoautosomal gene, ADP/ATP translocase, escapes X-inactivation whereas a homologue on Xq is subject to X-inactivation. Nat Genet 3:82–87
Schon EA, Bonilla E, DiMauro S (1997) Mitochondrial DNA mutations and pathogenesis. J Bioenerg Biomembr 29:131–149
Schultheiss HP, Bolte HD (1985) Immunological analysis of autoantibodies against the adenine nucleotide translocator in dilated cardiomyopathy. J Mol Cell Cardiol 17:603–617
Schultheiss HP, Schulze K, Dörner A (1996) Significance of the adenine nucleotide translocator in the pathogenesis of viral heart disease. Mol Cell Biochem 163/164:319–327
Schulze K, Becker BF, Schultheiss HP (1989) Antibodies to the ADP/ATP carrier – an autoantigen in myocarditis and dilated cardiomyopathy – penetrate into myocardial cells and disturb energy metabolism in vivo. Circ Res 64:179–192
Sirrenberg C, Endres M, Folsch H, Stuart RA, Neupert W, Brunner M (1998) Carrier protein import into mitochondria mediated by the intermembrane proteins Tim10/Mrs11 and Tim12/Mrs5. Nature 391:912
Slim R, Levilliers J, Ludecke HJ, Claussen U, Nguyen VC, Gough NM, Horsthemke B, Petit C (1993) A human pseudoautosomal gene encodes the Ant3 ADP/ATP translocase and escapes X-inactivation. Genomics 16:26–33
Stepien G, Torroni A, Chung AB, Hodge JA, Wallace DC (1992) Differential expression of adenine nucleotide translocator isoforms in mammalian tissues and during muscle cell differentiation. J Biol Chem 267:14592–14597
Susin SA, Zamzami N, Kroemer G (1998) Mitochondria as regulators of apoptosis: doubt no more. Biochim Biophys Acta 1366:11–165
Torroni A, Stepien G, Hodge JA, Wallace DC (1990) Neoplastic transformation is associated with coordinate induction of nuclear and cytoplasmic oxidative phosphorylation genes. J Biol Chem 265:20589–20593

Vignais PV, Vignais PM, Lauquin GJ-M, Morel F (1973) Binding of adenine nucleotide and agonist ligand to the mitochondrial ADP/ATP carrier. Biochimie: 55:763–778
Vignais PV, Vignais PM, Doussière J (1975) Functional relationship between the ADP/ATP carrier and the F1-ATPase in mitochondria. Biochim Biophys Acta 376:219–230
Vignais PV, Block MR, Boulay F, Brandolin G, Lauquin GJ-M (1985) Molecular aspects of structure function relationships in mitochondrial adenine nucleotide carrier. In: Benga G (ed) Structure and properties of cell membranes. CRC Press, Boca Raton, pp 139–179

CHAPTER 11

The Normal and Pathological Structure, Function and Expression of Mitochondrial Creatine Kinase

11

E. Clottes[1,2], O. Marcillat[1], M. J. Vacheron[1], C. Leydier[1], and C. Vial[1]

Contents

1 Introduction

Creatine kinase (CK) isoenzymes are expressed in tissues with important and rapid energy requirements. These enzymes catalyse the reversible phosphorylation of creatine by ATP.

$$\text{Creatine (Cr)} + \text{ATPMg}^{2-} \leftrightarrow \text{Phosphocreatine}^{2-} \text{(PCr)} + \text{ADPMg}^{-} + \text{H}^{+}$$

During the 1920s and 1930s, an initial concept of the role of the creatine kinase system in muscle energetics emerged from three observations: the respiration of a muscle homogenate is stimulated by creatine (Belitzer and Tsibakova 1939); muscle cells contain a phosphagen (later identified as phosphocreatine) which is consumed during muscle contraction (Eggleton and Eggleton 1927; Fiske and Subbarow 1927); and an enzyme, creatine kinase, catalyses the transphosphorylation reaction between PCr and ADP (Lohman 1934). Phosphocreatine was thus considered as being an energy reservoir.

Using electrophoretic techniques it was later shown that CK exists as a family of isoenzymes: one type is found in muscle cells (MM), another in brain cells (BB) and a third one, with an intermediate electrophoretic migration (MB), is found in cardiac

[1] Biomembranes et Enzymes Associés, UPRESA 5013 CNRS, Université Claude Bernard Lyon I, 43, Boulevard du 11 novembre 1918, 69622 Villeurbanne Cedex, France
[2] Present adress: IUT Clermont 1, Dept. Génie biologique, Ensemble Universitaire des Cézeaux, B.P. 86-63172 Aubière Cedex, France

and smooth muscles (Bürger et al. 1964). These three cytoplasmic isoforms are homo- or hetero-dimers composed of either M or B-type subunits (Dawson et al. 1965). In vertebrates, about 15 primary structures have been reported and these subunits have a high degree of sequence identity, either between species or between types of subunits: 89–99% between M subunits, 88–98% between B subunits and 77–82% between M and B subunits (Babbitt et al. 1986; Mühlebach et al. 1994). They all consist of 380 amino acids with a molecular mass of about 43 kDa. In man, subunits M and B are encoded for by two nuclear genes located on chromosomes 19 and 14 respectively (McBride et al. 1982; Nigro et al. 1987; Stallings et al. 1988).

Brain and muscle cells contain a fourth enzyme which, in contrast to the above mentioned forms, has a cathodic migration and is associated with the outer face of the mitochondrial inner membrane (Jacobs et al. 1964). The presence of a specific mitochondrial isoenzyme and numerous physiological and biochemical data led to a new proposal where phosphocreatine is considered as being a high energy compound shuttling between mitochondria (where it is synthesized), and energy-consuming sites in the cytoplasm.

A scheme of the phosphocreatine shuttle is presented in Fig. 1. A part of the cytosolic CK activity is associated with subcellular structures: myofibrils (Bessman et al. 1980; Saks et al. 1984; Wallimann et al. 1984; Schafer and Perriard 1988; Ventura-Clapier et al. 1994), sarcoplasmic reticulum (Rossi et al. 1990; Korge and Campbell 1994), plasma or nuclear membranes (Grosse et al. 1980; Saks et al. 1984; Hardin et al. 1992; Manos and Bryan 1993; Friedman and Roberts 1994) which use phosphocreatine to phosphorylate ADP into ATP in the vicinity of the ATPase active sites. It has been repeatedly demonstrated that certain functions (muscle contraction, ion transport) coupled with ATP hydrolysis are more efficient when ATP is produced through an associated CK catalyzed reaction rather than by any other ATP regenerating system. More detailed information on the development of the shuttle concept and on the role of the CK reaction in the control of respiration and oxidative phosphorylations can be

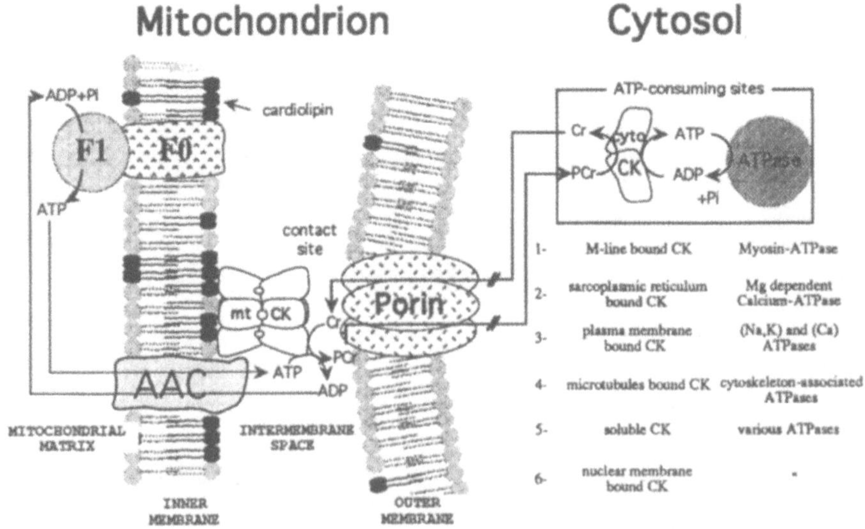

Fig. 1. The phosphocreatine shuttle

found in the reviews (Saks et al. 1978; Bessman and Geiger 1981; Bessman and Carpenter 1985; Wallimann et al. 1992).

2 Expression and Localization

The existence of two different mitochondrial isoforms, sarcomeric mt-CK and ubiquitous mt-CK, has been demonstrated. They are encoded by two nuclear genes (Haas and Strauss 1990) which, in man, are located on chromosomes 5 and 15 respectively (McBride et al. 1982; Klein et al. 1991). The expression of the two genes has been studied in several species at both the level of mRNA (Trask and Billadello 1990; Fontanet et al. 1991; Payne et al. 1991, 1993; Ch'ng and Ibrahim 1994; Payne and Strauss 1994a, b) and proteins (Hall and DeLuca 1975; Norwood et al. 1983; Hoerter et al 1991; Wegmann etal. 1991). This expression is strictly tissue-specific and appears to be regulated in a concerted fashion during development. Indeed, large amounts of BB-CK and ubiquitous mt-CK are found in adult brain, retina and to a lesser extent in kidney, intestine or uterus smooth-muscle and lung, MM-CK and sarcoplasmic mt-CK being detected only in skeletal or cardiac muscle. The concerted expression of these pairs of genes has been well described in myocardium during pressure overload-induced hypertrophy (Fontanet et al. 1991) and in the placenta and uterus during pregnancy (Payne et al. 1993).

Studies on the developmental changes of CK isoenzyme expression gave similar results, whatever the species. During foetal life, the BB isoenzyme is the dominant form, while in cardiac or skeletal muscle the expression of the M subunit begins shortly before birth, the exact time depending on the species and the type of muscular cell, then increases to become the dominant form in mature cells. Mitochondrial subunits are not expressed during early foetal life and they accumulate at later stages than their cytosolic counterparts. It is interesting to stress that in cardiac muscle, where the maturation of the phosphocreatine shuttle has been well studied, sarcomeric mt-CK synthesis is synchronous with the incorporation of MM-CK into the M-line of myofibrils (Hoerter et al. 1991). This event is also coordinated with the complete development of the contractile properties of the myocardium. The expression of mitochondrial isoenzymes is also correlated with the transition of a glycolytic metabolism to an oxidative one (Payne and Strauss 1994a).

The relative proportion of mt-CK as compared with the other isoenzymes, varies according to the tissue and the species (and depends also, perhaps, on the methods used for its estimation). The mean range is 2–15% of the total activity in most tissues, the highest percentages being most often found in the myocardium (Saks et al. 1974; Bittl et al. 1987; Sylven et al. 1991). This value is increased by chronic stimulation in the animal (Schmitt and Pette 1985) or by training in man (Apple and Rogers 1986).

Although it has been claimed that the molar ratio of mt-CK over the ADP/ATP carrier (AAC) was 1, this value is probably overestimated (Kuznetsov and Saks 1986). The AAC is supposed to represent 10% of the mitochondrial proteins whereas mt-CK probably does not amount to more than 1% of the total (Vial et al. 1986; Schlegel et al. 1988; Quemeneur et al. 1990). Its specific activity is, however, sufficient to allow it the use of all the ATP produced by oxidative phosphorylation and exported by the ADP/ATP carrier to phosphorylate creatine (Jacobus and Lehninger 1973; Saks 1980).

3 Mitochondrial CK Structure

In contrast to the cytoplasmic isoforms, mitochondrial isoenzymes have an amino-terminal presequence which targets the protein to the intermembrane space (Haas and Strauss 1990; Payne et al. 1991). This target peptide, which is 20–40 amino acids long, is cleaved after the import of the enzyme. The similarity in the sequences of the subunits M or B is 88–99% between species, cf amounts to 82–84% between ubiquitous and sarcoplasmic subunits and to 60–65% between cytosolic and mitochondrial forms (Mühlebach et al. 1994). In all the CK isoenzymes known so far, six areas of very high similarity are separated by seven less conserved regions. The highly conserved regions are likely to be involved in common functions of the CK isoenzymes, such as binding of substrates, catalytic activity or dimer formation (Mühlebach et al. 1994). As a matter of fact, it has been demonstrated, using chemical modification, cross-linking or site-directed mutagenesis experiments, that some of the amino acid residues of these regions were located in the active site: the accessible Cys 278 (Buechter et al. 1992), Lys 19 (Mahowald 1969), Arg 288 (Lin et al. 1994), His 291 (Chen et al. 1996), Asp 335 (James et al. 1990) and the peptides 219-237 and 275-287 (Olcott et al. 1994). The seven less conserved regions could be responsible for isoform-specific properties, such as binding to subcellular structures or octamer formation.

Using partial denaturation and monoclonal antibodies, it has been demonstrated that the CK monomer comprised two distinct domains (Morris and Cartwright 1990). However, this monomer has a very compact structure in its native state since it is not readily cleaved by proteases such as trypsin or chymotrypsin, in spite of the presence of 55 basic residues and 25 aromatic residues. Only three proteolytic enzymes (proteinase K, subtilisine and thermolysine), cleave the various isoforms in their native state at the level of a large, exposed loop (Wyss et al. 1993; Leydier et al. 1997). Upon fixation of the CK substrates, this flexible loop would move and become inaccessible to the active site of these proteases (Fritz-Wolf et al. 1996).

The recent publication of data on the three-dimensional structure of the chicken mitochondrial isoenzyme revealed useful information about the structures of the monomer, the dimer and the octamer which extend previous experimental data (Fritz-Wolf et al. 1996). The monomer consists of a small domain (residues 1–100) containing mainly six helices, and a larger domain (residues 120–380) which contains an eight-stranded antiparallel β-sheet flanked by seven α-helices. The active site is located in the cleft between the two domains and comprises the six highly conserved regions which form a compact core. The contact zone between monomers, which leads to the formation of the dimer, include isoform-specific residues and also conserved ones (Ile 52-Asn 58, Lys 191, Arg 204-Trp 206). Site-directed mutagenesis of Trp 206 (or Trp 210 in the MM isoenzyme) decreases the stability of the dimer, suggesting an important contribution of this residue to the monomer-monomer interaction (Gross et al. 1994; Perraut et al. 1998).

The existence of a high molecular weight-form of mt-CK has been a matter of debate for a long time, but it is now acknowledged that mt-CK exists under two interconvertible oligomeric forms, a dimer and an octamer (Marcillat et al. 1987; Schlegel et al. 1988; Schnyder et al. 1988). Trp 264, a mitochondrial isoenzyme specific residue, is located in a hydrophobic cluster which plays a role in the dimer-dimer interactions leading to the formation of the octamer (Gross et al. 1994). An N-terminal heptapeptide also contributes to these interactions, since mutations in this area shifts the equilibrium towards the dimeric form (Kaldis et al. 1994).

The M_r value of the dimer is about 86 000, its Stokes radius is 3.8 nm and it has a sedimentation coefficient of 5.6 S; the corresponding values for the octamer are 340 000, 6 nm and 12 S respectively (see Table I of Wyss et al. 1992). The reported isoelectric points of mt-CK range between pH 8 and 9.5, i.e. at values distinctly higher than those of the cytoplasmic isoenzymes. The apparent pI of the octameric form is usually higher than that of the dimer (see Table II of Wyss et al. 1992): these differences allow for easy separation of the various isoenzymes by electrophoretic or ion-exchange techniques.

Electron microscopy coupled to image analysis reveals that the mitochondrial octamer has a cube-like structure with a side-length of about 10 nm (9.3 × 9.3 × 8.6 nm). A 2 nm central channel extends through the cube (Schnyder et al. 1991). The N-terminal end of the monomers protrudes into this cavity, whereas the C-terminal parts are located at the top and bottom faces of the octamer. As a consequence, several positively charged residues are found on these faces which can interact with negatively charged phospholipids; these residues are five lysines and one arginine located in the C-terminal region and one histidine and one lysine of the N-terminal part of the monomer (Fritz-Wolf et al. 1996). The presence of these positively charged regions on each face of the cube could explain the ability of mt-CK to mediate the interaction of inner and outer mitochondrial membranes (Rojo et al. 1991a, b).

4
Interaction with Membranes

The enzyme is bound to the outer side of the inner membrane (Jacobus and Lehninger 1973; Brooks and Suelter 1987). Its concentration is especially high at the level of contact sites between the inner and outer membranes (Schlegel et al. 1988; Adams et al. 1989; Biermans et al. 1990; Kottke et al. 1991). At the contact sites, a functional interaction of mt-CK with AAC and porin has been proposed (Brdiczka 1991). After rupture of the outer membrane, mt-CK can be released by high salt concentration, ADP or ATP, or negatively charged organomercurials (Jacobus and Lehninger 1973; Hall et al. 1979; Font et al. 1981; Lipskaya et al. 1981; Vial et al. 1986; Brooks and Suelter 1987). The octamer is the only form associated with the mitochondrial membrane, and it has been shown that exclusively octameric mt-CK can be re-bound to mitochondrial membranes (Lipskaya et al. 1981; Marcillat et al. 1987; Quemeneur et al. 1988). After solubilization, the octamer (CKm_2) can be converted to four active dimers (CKm_1) by incubation at alkaline pH or in the presence of its substrates, as shown in Fig. 2. This conversion is a reversible phenomenon (Lipskaya et al. 1986; Lipskaya and Rybina 1987; Marcillat et al. 1987; Schlegel et al. 1988, 1990). However, there are no experimental data on a possible regulatory mechanism based on this interconversion.

Mt-CK is able to interact with monolayers constituted of inner mitochondrial membrane phospholipids (Rojo et al. 1991a; Schnyder et al. 1994). This interaction, the evidence for which is an increase in surface pressure upon incorporation of mt-CK, is stronger with the octamer than with the dimer; this shows the higher affinity of the octamer for the phospholipids.

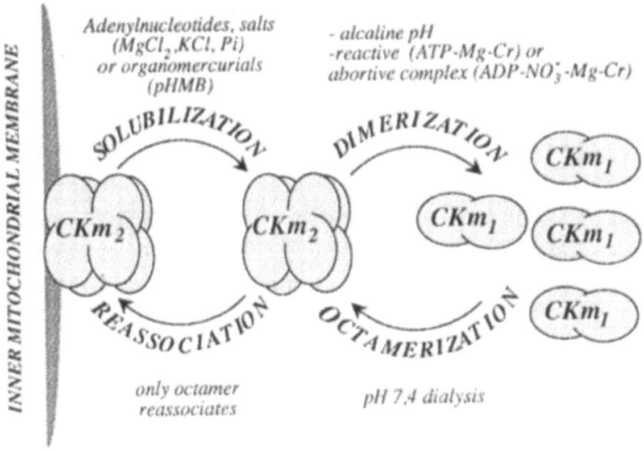

Fig. 2. Solubilization and interconversion of the two oligomeric forms of mitochondrial creatine kinase

Experiments have also been performed with monolayers made of outer membrane phospholipids and with mixtures containing phosphatidylcholine and various amounts of anionic phospholipids, phosphatidylserine, phosphatidylinositol and cardiolipin (Rojo et al. 1991a). The interaction of mt-CK with monolayers or liposomes was dependent on the amount of anionic phospholipid and was strongest with cardiolipin, which is the major phospholipid of the inner membrane (Rojo et al. 1991b; Vacheron et al. 1997). Mt-CK was able to create an intermembrane contact between a monolayer of phospholipids of the outer membrane and unilamellar vesicles of the phospholipids of the outer membrane and unilamellar vesicles of the phospholipids of the inner membrane (Nicolay et al. 1990). The ability of mt-CK to mediate intermembrane adhesion was specific to the octameric form. Other cationic proteins of the intermembrane space, or polylysine, which strongly interact with acid phospholipid vesicles, as well as cytosolic isoenzymes of CK, failed to induce contact formation (Rojo et al. 1991b). In mitochondria, the formation of the contact areas could be induced by the fusion of inner and outer membranes at defined sites, thereby creating a microcompartmentation. The fusion would imply a phase transition of the phospholipids from a bilayer configuration to a non-bilayer configuration. Indeed, cardiolipin is known to adopt a hexagonal phase in the presence of Ca^{2+} (Tilcock 1986). The effect of calcium concentration on the number of contact sites has recently been investigated and the maximum activity of mt-CK was reached at the calcium concentration giving the highest number of contact sites (Bakker et al. 1994).

Cardiolipin, the specific anionic phospholipid of the inner mitochondrial membrane, could be the membrane receptor of mt-CK and the interaction between the protein and cardiolipin would thus be of an electrostatic nature. Lysine and arginine residues located at the top and the bottom faces of the cube of mt-CK may be involved in the binding of mt-CK octamers to the cardiolipin of the inner mitochondrial membrane and to negatively charged phospholipids of the outer membrane. However, the binding of mt-CK to membranes could also be mediated by other membrane proteins, such as the adenine nucleotide translocase of the inner mitochondrial membrane, which has been assumed to interact with mt-CK (Wallimann et al. 1992), and/or porin which, in vitro, is able to form a complex with mt-CK (Brdiczka et al. 1994).

5
Genes

Human genes coding for the two mitochondrial CK isoforms have been cloned. The sarcomeric gene spans 37 kb with 11 exons (the size of which varies from 77 to 241 bp) separated by 0.5 to 11 kb introns. The structure of the 5' part of this gene resembles those described for other sarcomeric contractile proteins with a first untranslated exon at some distance from the second one; in the sarcomeric mt-CK gene, the two first exons are not translated and translation and transcription start points are 17 kb apart. As reported by Klein et al. (1991), this shared organization could be important for their coordinate expression. The sarcomeric mt-CK transcript has a long 5'-untranslated sequence (345 bp), including a GC-rich region which could possess a structure similar to that found in ornithine decarboxylase mRNA 5' end and which could play a role in the regulation of translation. The 5' flanking region contains one TATA box, three potential CCAAT boxes and several thyroid hormone response elements, but their functionality has not yet been documented. It also contains several cAMP response elements and it has been shown that cAMP dramatically increases gene transcription in C_2C_{12} cells (Payne and Strauss 1994a).

There are also several possible binding sites for the muscle specific transcription factors MYOD/MEF-1 and MEF-2 but these sites do not seem to be responsible for the sarcomer-specific expression of the gene (Klein et al. 1991; Payne and Strauss 1994b). The occurrence of two GFII motifs has also been reported; these motifs are potential sites for a regulation of the expression of nuclear genes coding for mitochondrial proteins (Dorsman et al. 1988). No sequence which could explain the coordinated expression of M- and sarcoplasmic mt-CK isoforms has yet been characterized (Klein et al. 1991; Payne and Strauss 1994a, b).

The ubiquitous gene is smaller (5.5 kb) and has a house-keeping gene-like structure: no TATA or CCAAT boxes, a GC-rich region upstream of the transcription start site and potential binding sites for the transactivation factors Sp1, AP2 (Haas et al. 1989). This gene also contains glucocorticoid and oestrogen response elements, which may be functional since transcription is rapidly induced by β-oestradiol in vitro (Payne et al. 1993). The murine gene (Steeghs et al. 1995a) possesses a similar structure and sequence analysis revealed Mt_1, Mt_3, Mt_4 sequences analogous to that described by Suzuki et al. (1990).

6
Variations of mt-CK Activity in Pathology

In animal models of muscular dystrophy, a decrease in mt-CK activity is always observed (Mahler 1979; Bennett et al. 1985), although studies on humans are rare (Smeitink et al. 1994; Stadhouders et al. 1994; Bouzidi et al. 1996). Deficiency of mt-CK has never been observed in the skeletal muscle of patients with disturbances in the complexes responsible for mitochondrial energy production. On the contrary, an increase in the mt-CK activity has been sometimes described in mitochondrial myopathies associated with the presence of ragged red fibers. Electron microscopic inspection of these muscle samples frequently reveals the presence of mitochondrial crystalline inclusions which, in some cases, could be labelled by antibodies directed against mt-CK (Stadhouders et al. 1994). Similar crystalline mitochondrial inclusions are also induced in rat muscle by ischemia (Hanzlikova and Schiaffino 1977) or by treatment with β-guanidinopropionate (βGPA), a structural analog of creatine (Ohira et al. 1988). It should be noted that long-term AZT therapy used for the treatment of AIDS patients also induces a myopathy with crystal-like inclusions (Lewis and Dalakas 1995). Cardiomyocytes grown in the presence of βGPA or in the absence of creatine also develop such crystalline inclusions (Eppenberger-Eberhardt et al. 1991). However, in nearly all the cases of cardiomyopathies or in experimental models, the mitochondrial CK activity was diminished (Ingwall et al. 1990; Veksler et al. 1991).

The appearance of these mitochondrial inclusions is like purified mt-CK crystals obtained under various experimental conditions. Stadhouders et al. (1994) proposed that the presence of mt-CK in the mitochondrial inclusions may result from the overexpression and accumulation of this protein as a response to a cytosolic low-energy stress condition.

Recently, Wieringa's group has obtained mice carrying a null mutation of the gene coding for ubiquitous mt-CK (Steeghs et al. 1995a, b). These mice are viable, they show no obvious abnormalities and are completely fertile. This result was unexpected in view of the prominent role of the CK system in energy transfer reactions. The importance of this function was further supported by a recent report on the neurological disorders experienced by a patient who has an inborn error in the metabolism of creatine (Stöckler et al. 1994). Since the problems were relieved by the administration of creatine, they were assumed to be due to a dysfunction of the phosphocreatine shuttle. On the other hand, this result needs to be compared with the data obtained in transgenic mice bearing a null mutation of the M isoform of CK gene: no overt dysfunction was observed in these animals but their muscular contractile kinetics were slowed down and their energy supply pattern was modified (Ventura-Clapier et al. 1995).

The occurrence of mt-CK in serum (macro CK type 2) has been detected in various pathologies and it is important to note that its presence may disturb the experimental protocols used to routinely estimate the MB-CK isoenzyme. Its presence is well documented in patients with several types of malignant tumors localized in the liver (which normally does not express CK, Kanemitsu et al. 1982, 1983, 1984), digestive tract, lung (Wu et al. 1983; Kanemitsu et al. 1984), breast (Wu et al. 1983; Stein et al. 1985) and prostate (Ström and Bendz 1986) and in all these cases, the prognosis is usually poor. The identity of the circulating form, compared with that found in mitochondria from the tumor, has been demonstrated in several cases (Kanemitsu et al. 1982; Okano et al. 1987). Serum mt-CK could thus be a valuable disease marker (Lee et al. 1994). A high level of mt-CK in serum has also been observed in Reye's syndrome (James and Harrison 1979). It is occasionally detected as a consequence of myocardial damage occurring after surgery, or myocarditis (Wu et al. 1983), or ischemia (Stein et al. 1985; Sylven et al. 1985).

In recent years, a number of results have shown that creatine or creatine phosphate analogs were able to inhibit the growth of tumor cells (Baggetto et al. 1992; Martin et al. 1994). The reason for this effect is currently not very clear but the increasing interest of several pharmaceutical firms in the evaluation of creatine analogs as anticancer drugs shows that the specific inactivation of creatine kinase isoenzymes could be a promising approach in therapeutics, in parallel with conventional treatments.

References

Adams V, Bosch W, Schlegel J, Wallimann T, Brdiczka D (1989) Further characterization of contact sites from mitochondria of different tissues: topology of peripheral kinases. Biochim Biophys Acta 981:213–225

Apple FS, Rogers MA (1986) Mitochondrial creatine kinase activity alterations in skeletal muscle during long-distance running. J Appl Physiol 61:482–485

Babbitt PC, Kenyon GL, Kuntz ID, Cohen FE, Baxter JD, Benfield PA, Buskin JD, Gilbert WA, Hauschka SD, Hossle JP, Ordahl CH, Pearson M, Perriard JC, Pickering LA, Putney SD, West BL, Ziven RA (1986) Comparisons of creatine kinase primary structures. J Protein Chem 5:1–14

Baggetto LG, Clottes E, Vial C (1992) Low mitochondrial proton leak due to high membrane cholesterol content and cytosolic creatine kinase as 2 features of the deviant bioenergetics of Ehrlich and AS30-D tumor cells. Cancer Res 52:4935–4941

Bakker A, Bernaert I, Debie M, Ravingerova T, Ziegelhoffer A, Vanbelle H, Jacob W (1994) The effect of calcium on mitochondrial contact sites. A study on isolated rat hearts. Biochim Biophys Acta 1224:583–588

Belitzer VA, Tsibakova ET (1939) On the mechanism of phosphorylation coupled with respiration. Biokhimiya 4:516–534

Bennet VD, Hall N, DeLuca M, Suelter CH (1985) Decreased mitochondrial creatine kinase activity in dystrophic chicken breast muscle alters creatine-linked resiratory coupling. Arch Biochem Biophys 240:380–391

Bessman SP, Carpenter CL (1985) The creatine-creatine phosphate energy shuttle. Annu Rev Biochem 54:831–862

Bessman S, Geiger PI (1981) Transport of energy in muscle: the phosphorylcreatine shuttle. Science 211:448–452

Bessman S, Yang WCT, Geiger PJ, Erickson-Viitanen S (1980) Intimate coupling of creatine phosphokinase and myofibrillar adenosinetriphosphatase. Biochem Biophys Res Commun 96:1414–1420

Biermans W, Bakker A, Jacob W (1990) Contact sites between inner and outer mitochondrial membrane: a dynamic microcompartment for creatine kinase activity. Biochim Biophys Acta 1018:225–228

Bittl JA, DeLayre J, Ingwall JS (1987) Rate equation for creatine kinase predicts the in vivo reaction velocity: 31P NMR surface coil studies in brain, heart, and skeletal muscle of the living rat. Biochemistry 26:6083–6090

Bouzidi MF, Enjolras N, Carrier H, Vial C, Lopez-Mediavilla C, Burt-Pichat B, Couthon F, Godinot C (1996) Variations of muscle mitochondrial creatine kinase activity in mitochondrial diseases. Biochim Biophys Acta 1316:61–70

Brdiczka D (1991) Contact sites between mitochondrial envelope membranes. Structure and function in energy-transfer and protein-transfer. Biochim Biophys Acta 1071:291–312

Brdiczka D, Kaldis P, Wallimann T (1994) In vitro complex formation between the octamer of mitochondrial creatine kinase and porin. J Biol Chem 269:27640–27644

Brooks SPJ, Suelter CH (1987) Association of chicken mitochondrial creatine kinase with the inner mitochondrial membrane. Arch Biochem Biophys 253:122–132

Buechter DD, Medzihradszky KF, Burlingame AL, Kenyon GL (1992) The active site of creatine kinase – affinity labelling of cysteine 282 with N-(2,3-epoxypropyl)-N-amidinoglycine. J Biol Chem 267:2173–2178

Bürger A, Richterich P, Aebi H (1964) Die Heterogenität der Kreatin-Kinase. Biochem Z 339:305–314

Ch'ng JLC, Ibrahim B (1994) Transcriptional and post-transcriptional mechanisms modulate creatine kinase expression during differentiation of osteoblastic cells. J Biol Chem 269:2336–2341

Chen LH, Borders CL, Vasquez JR, Kenyon GL (1996) Rabbit muscle creatine kinase: consequences of the mutagenesis of conserved histidine residues. Biochemistry 35:7895–7902

Dawson DM, Eppenberger HM, Kaplan N (1965) Creatine kinase: evidence for a dimeric structure. Biochem Biophys Res Commun 21:346–349

Dorsman JC, Van Heeswijk WC, Grivell LA (1988) Identification of two factors which bind to the upstream sequences of a number of nuclear genes coding for mitochondrial proteins and to genetic elements important for cell division in yeast cells. Nucleic Acids Res 16:7287–7301
Eggleton P, Eggleton GP (1927) The physiological significance of "phosphagen". J Physiol 63:155–161
Eppenberger-Eberhardt M, Riesinger I, Messerli M, Schwarb P, Muller M, Eppenberger HM, Wallimann T (1991) Adult rat cardiomyocytes cultured in creatine-deficient medium display large mitochondria with paracrystalline inclusions, enriched for creatine kinase. J Cell Biol 113:289–302
Fiske CH, Subbarow Y (1927) The nature of the "inorganic phosphate" in voluntary muscle. Science 65:401–403
Font B, Vial C, Goldschmidt D, Eichenberger D, Gautheron DC (1991) Heart mitochondrial creatine kinase solubilization effects of mitochondrial swelling and SH group reagents. Arch Biochem Biophys 212:195–203
Fontanet HL, Trask RV, Haas RC, Strauss AW, Abendschein DR, Billadello JJ (1991) Regulation of expression of M, B, and mitochondrial creatine kinase messenger-RNAs in the left ventricle after pressure overload in rats. Circ Res 68:1007–1012
Friedman DL, Roberts R (1994) Compartmentation of brain-type creatine kinase and ubiquitous mitochondrial creatine kinase in neurons – evidence for a creatine phosphate energy shuttle in adult rat brain. J Comp Neurol 343:500–511
Fritz-Wolf K, Schnyder T, Wallimann T, Kabsch W (1996) Structure of mitochondrial creatine kinase. Nature 381:341–345
Gross M, Furter-Graves EM, Wallimann T, Eppenberger HM, Furter R (1994) The tryptophan residues of mitochondrial creatine kinase: roles of Trp-223, Trp-206, and Trp-264 in active-site and quaternary structure formation. Protein Sci 3:1058–1068
Grosse R, Spitzer E, Kupriyanov VV, Saks VA, Repke KRH (1980) Coordinate interplay between (Na^+ + K^+)-ATPase and creatine phosphokinase optimizes (Na^+/K^+)-antiport across the membrane of vesicles formed from the plasma membrane of cardiac muscle cell. Biochim Biophys Acta 603:142–156
Haas RC, Strauss AW (1990) Separate nuclear genes encode sarcomere-specific and ubiquitous human mitochondrial creatine kinase isoenzymes. J Biol Chem 265:6921–6927
Haas RC, Korenfeld C, Zhang Z, Perryman B, Roman D, Strauss AW (1989) Isolation and characterization of the gene and cDNA encoding human mitochondrial creatine kinase. J Biol Chem 264:2980–2987
Hall N, DeLuca M (1975) Developmental changes in creatine phosphokinase isoenzymes in neonatal mouse hearts. Biochem Biophys Res Commun 66:988–994
Hall N, Addis P, DeLuca M (1979)Mitochondrial creatine kinase. Physical and kinetic properties of the purified enzyme from beef heart. Biochemistry 18:1745–1751
Hanzlikova V, Schiaffino S (1977) Mitochondrial changes in ischemic skeletal muscle. J Ultrastruct Res 60:121–133
Hardin CD, Raeymaekers L, Paul RJ (1992) Comparison of endogenous and exogenous sources of ATP in fueling Ca^{2+} uptake in smooth muscle plasma membrane vesicles. J Gen Physiol 99:21–40
Hoerter JA, Kuznetsov A, Ventura-Clapier R (1991) Functional development of the creatine kinase system in perinatal rabbit heart. Circ Res 69:665–676
Ingwall JS, Atkinson DE, Clarke K, Fetters JK (1990) Energetics correlates of cardiac failure: changes in the creatine kinase system in the failing myocardium. Eur Heart J 1:108–115
Jacobs H, Heldt HW, Klingenberg M (1964) High activity of creatine kinase in mitochondria from muscle and brain and evidence for a separate mitochondrial isoenzyme of creatine kinase. Biochem Biophys Res Commun 16:516–521
Jacobus WE, Lehninger AL (1973) Creatine kinase of rat heart mitochondria. J Biol Chem 248:4803–4810
James GP, Harrison RL (1979)Creatine kinase isoenzymes of mitochondrial origin in human serum. Clin Chem 25:943–947
James P, Wyss M, Lutsenko S, Wallimann T, Carafoli E (1990) ATP binding site of mitochondrial creatine kinase affinity labelling of Asp-335 with CIRATP. FEBS Lett 273:139–143

Kaldis P, Furter R, Wallimann T (1994) The N-terminal heptapeptide of mitochondrial creatine kinase is important for octamerization. Biochemistry 33:952–959
Kanemitsu F, Kawanishi I, Mizushima J (1982) The origin of a cathode-migrating creatine kinase found in serum from a cancer patient. Clin Chim Acta 122:377–383
Kanemitsu F, Kawanishi I, Mizushima J, Okigaki T (1984) Mitochondrial creatine kinase as a tumor-associated marker. Clin Chim Acta 138:175–183
Kanemitsu F, Kawanishi I, Mizushima JA (1983) New creatine kinase found in mitochondrial extracts from malignant liver tissue. Clin Chim Acta 128:233–240
Klein SC, Haas RC, Perryman MB, Billadello JJ, Strauss AW (1991) Regulatory element analysis and structural characterization of the human sarcomeric mitochondrial creatine kinase gene. J Biol Chem 266:18058–18065
Korge P, Campbell KB (1994) Local ATP regeneration is important for sarcoplasmic reticulum Ca2+ pump function. Am J Physiol 267:C357–C366
Kottke M, Adams V, Wallimann T, Nalam VK, Brdiczka D (1991) Location and regulation of octameric mitochondrial creatine kinase in the contact sites. Biochim Biophys Acta 1061:215–225
Kuznetsov AV, Saks VA (1986) Affinity modification of creatine kinase and ATP-ADP translocase in heart mitochondria: determination of their molar stoichiometry. Biochem Biophys Res Commun 134:359–366
Lee KN, Csako G, Bernhardt P, Elin RJ (1994) Relevance of macro creatine kinase type 1 and type 2 isoenzymes to laboratory and clinical data. Clin Chem 40:1278–1283
Lewis W, Dalakas MC (1995) Mitochondrial toxicity of antiviral drugs. Nat Med 1:417–422
Leydier C, Andersen J, Couthon F, Forest E, Denoroy L, Vial C, Clottes E (1997) Proteinase K processing of rabbit muscle creatine kinase. J Protein Chem 16:67–74
Lin LJ, Perryman MB, Friedman D, Roberts R, Ma TS (1994) Determination of the catalytic site of creatine kinase by site-directed mutagenesis. Biochim Biophys Acta 1206:97–104
Lipskaya TYU, Rybina IV (1987) Properties of mitochondrial creatine kinase from skeletal muscle. Biokhimiya 52:594–603
Lipskaya T, Temple VD, Belousova LV, Molokova EV, Rybina IV (1981) Investigation of the interaction of mitochondrial creatine kinase with the membranes of the mitochondria. Biochemistry USSR 45:877–886
Lipskaya TYU, Kedishvili NY, Kalenova ME (1986) Conditions of interconversion of oligomeric forms of heart mitochondrial creatine kinase. Biokhimiya 50:1339–1348
Lohman K (1934) Über die enzymatische Aufspaltung der Kreatinphosphorsäure. Biochem Z 271:264–277
Mahler M (1979)Progressive loss of mitochondrial creatine phosphokinase activity in muscular dystrophy. Biochem Biophys Res Commun 88:895–906
Mahowald TA (1969) Identification of an ε-NH2 group of lysine and a SH group of cysteine near the reactive cysteine residue in rabbit muscle creatine kinase. Fed Proc 28:601
Manos P, Bryan GK (1993) Cellular and subcellular compartmentation of creatine kinase in brain. Dev Neurol 15:271–279
Marcillat O, Goldschmidt D, Eichenberger D, Vial C (1987) Only one of the two interconvertible forms of mitochondrial creatine kinase binds to heart mitoplasts. Biochim Biophys Acta 890:233–241
Martin KJ, Chen SF, Clark GM, Degen D, Wajima M, Vonhoff DD, Kaddurah-Daouk R (1994) Evaluation of creatine analogues as a new class of anticancer agents using freshly explanted human tumor cells. J Natl Cancer Inst 86:608–613
McBride OW, Battey J, Hollis GF, Swan DC, Siebenlost U, Leder P (1982) Localization of human variable and constant region immunoglobulin heavy chain genes on subtelomeric band q32 of chromosome 14. Nucleic Acids Res 10:8155–8170
Morris GE, Cartwright AJ (1990) Monoclonal antibody studies suggest a catalytic site at the interface between domains in creatine kinase. Biochim Biophys Acta 1039:318–322
Mühlebach SM, Gross M, Wirz T, Wallimann T, Perriard JC, Wyss M (1994) Sequence homology and structure prodictions of the creatine kinase isoenzymes. Mol Cell Biochem 133:245–262
Nicolay K, Rojo M, Wallimann T, Demel R, Hovius R (1990) The role of contact sites between inner and outer mitochondrial membrane in energy transfer. Biochim Biophys Acta 1018:229–233

Nigro JM, Schweinfest CW, Rajkovic A, Pavlovic J, Jamal S, Dottin RP, Hart JT, Kamarck ME, Rae PMM, Carty MD, Martin Deleon P (1987) cDNA cloning and mapping of the human creatine kinase M gene to 19q13. Am J Hum Genet 40:115–125

Norwood WI, Ingwall JS, Norwood CR, Fossel ET (1983) Developmental changes of creatine kinase metabolism in rat brain. Am J Physiol 244:C205–C210

Ohira Y, Kanzaki M, Chen CS (1988) Intramitochondrial inclusions caused by depletion of creatine kinase in rat skeletal muscles. JN J Physiol 28:159–166

Okano K, Yamamoto K, Ohba Y, Matsumura K, Miyaji T (1987) Source of elevated serum mitochondrial creatine kinase activity in patients with malignancy. Clin Chim Acta 169:159–164

Olcott MC, Bradley ML, Haley BE (1994) Photoaffinity labeling of creatine kinase with 2-azido- and 8-azidoadenosine triphosphate: identification of two peptides from the ATP-binding domain. Biochemistry 33:11935–11941

Payne RM, Strauss AW (1994a) Expression of the mitochondrial creatine kinase genes. Mol Cell Biochem 133:235–243

Payne RM, Strauss AW (1994b) Developmental expression of sarcomeric and ubiquitous mitochondrial creatine kinase is tissue-specific. Biochim Biophys Acta 1219:33–38

Payne RM, Haas RC, Strauss AW (1991) Structural characterization and tissue-specific expression of the messenger RNAs encoding isoenzymes from two rat mitochondrial creatine kinase genes. Biochim Biophys Acta 1089:352–361

Payne RM, Friedman DL, Grant JW, Perryman MB, Strauss AW (1993) Creatine kinase isoenzymes are highly regulated during pregnancy in rat uterus and placenta. Am J Physiol 265:E624–E635

Perraut C, Clottes E, Leydier C, Vial C, Marcillat O (1998) Role of quaternary structure in the stability of muscle creatine kinase monomer: tryptophan 210 is important for dimer cohesion. Proteins Struct Funct Genet 32:43–51

Quemeneur E, Eichenberger D, Goldschmidt D, Vial C, Beauregard G, Potier M (1988) The radiation inactivation method provides evidence that membrane-bound mitochondrial creatine kinase is an oligomer. Biochem Biophys Res Commun 153:1310–1314

Quemeneur E, Eichenberger D, Vial C (1990) Immunological determination of the oligomeric form of mitochondrial creatine kinase in situ. FEBS Lett 262:275–278

Rojo M, Hovius R, Demel R, Wallimann T, Eppenberger HM, Nicolay K (1991a) Interaction of mitochondrial creatine kinase with model membranes – a monolayer study. FEBS Lett 281:123–129

Rojo M, Hovius R, Demel RA, Nicolay K, Wallimann T (1991b) Mitochondrial creatine kinase mediates contact formation between mitochondrial membranes. J Biol Chem 266:20290–20295

Rossi AM, Eppenberger HM, Volpe P, Cotrufo R, Wallimann T (1990) Muscle-type MM creatine kinase is specifically bound to sarcoplasmic reiculum and can support Ca^{2+} uptake and regulate local ATP/ADP ratios. J Biol Chem 265:5258–5266

Saks VA (1980) Creatine kinase isoenzymes and the control of cardiac contraction. In: Jacobus WE, Ingwall JS (eds) Heart creatine kinase. Williams and Wilkins, Baltimore, pp 109–124

Saks VA, Chernousova GB, Voronkov L, Smirnov VN, Chazov EI (1974) Study of energy transport mechanism in myocardial cells. Circ Res 34:138–148

Saks VA, Rosenshtraukh LV, Smirnov VN, Chazov EI (1978) Role of creatine phosphokinase in cellular function and metabolism. Can J Physiol Pharmacol 56:691–706

Saks VA, Ventura-Clapier R, Khuchua ZA, Preobrazhensky AN, Emelin IV (1984) Creatine kinase in regulation of heart function and metabolism. Further evidence for compartmentation of adenine nucleotides in cardiac myofibrillar and sarcolemmal coupled ATPase-creatine kinase. Biochim Biophys Acta 803:254–264

Schafer BW, Perriard JC (1988) Intracellular targeting of isoproteins in muscle cytoarchitecture. J Cell Biol 106:1161–1170

Schlegel J, Zurbriggen B, Wegmann G, Wyss M, Eppenberger HM, Wallimann T (1988) Native mitochondrial creatine kinase forms octameric structures. Isolation of two interconvertible mitochondrial creatine kinase forms, dimeric and octameric mitochondrial creatine kinase: characterization, localization, and structure-function relationships. J Biol Chem 263: 16942–16953

Schlegel J, Wyss M, Eppenberger HM, Wallimann T (1990) Functional studies with the octameric and dimeric form of mitochondrial creatine kinase. J Biol Chem 265:9221–9227
Schmitt T, Pette D (1985) Increased mitochondrial creatine kinase in chronically stimulated fast-twitch rabbit muscle. FEBS Lett 188:341–344
Schnyder T, Engel A, Lustig A, Wallimann T (1988) Native mitochondrial creatine kinase forms octameric structures. Characterization of dimers and octamers by ultracentrifugation, direct mass measurements by scanning transmission electron microscopy, and image analysis of single mitochondrial creatine kinase octamers. J Biol Chem 263:16954-16962
Schnyder T, Gross H, Winkler H, Eppenberger HM, Wallimann T (1991) Structure of the mitochondrial creatine kinase octamer – high-resolution shadowing and image averaging of single molecules and formation of linear filaments under specific staining conditions. J Cell Biol 112:95–101
Schnyder T, Cyrklaff M, Wallimann T, Fuchs K (1994) Crystallization of mitochondrial creatine kinase on negatively charged lipid layers. J Struct Biol 112:136–147
Smeitink J, Ruitenbeek W, Sengers R, Wevers R, Vonlith T, Trijbels F (1994) Mitochondiral creatine kinase activity in patients with disturbed energy generation in muscle mitochondria. J Inherit Metab Dis 17:67–73
Stadhouders AM, Jap PHK, Winkler HP, Eppenberger HM, Wallimann T (1994) Mitochondiral creatine kinase: a major constituent of pathological inclusions seen in mitochondrial myopathies. Proc Natl Acad Sci USA 91:5089–5093
Stallings RL, Olson E, Strauss AW, Thompson LH (1988) Human creatine kinase genes on chromosomes 15 and 19, and proximity of the gene for the muscle form to the genes for apolipoprotein C2 and excision repair. Am J Hum Genet 43:144–151
Steeghs K, Peters W, Bruckwilder M, Croes H, Vanalewijk D, Wieringa B (1995a) Mouse ubiquitous mitochondrial creatine kinase: gene organization and consequences from inactivation in mouse embryonic stem cells. DNA Cell Biol 14:539–553
Steeghs K, Oerlemans F, Wieringa B (1995b) Mice deficient in ubiquitous mitochondrial creatine kinase are viable and fertile. Biochim Biophys Acta 1230:130–138
Stein W, Bohner J, Renn W, Maulbetsch R (1985) Macro creatine kinase type 2: results of a prospective study in hospitalized patients. Clin Chem 31:1959–1964
Stöckler S, Holzbach U, Hanefeld F, Marquardt I, Helms G, Requart M, Hänicke W, Frahm J (1994) Creatine deficiency in the brain: a new treatable inborn error of metabolism. Pediatr Res 36:409–413
Ström S, Bendz R (1986) Creatine kinase and lactate dehydrogenase isoenzymes in stage D prostatic carcinoma. Clin Chim Acta 159:219–228
Suzuki H, Hosokawa Y, Toda H, Nishikimi M, Ozawa T (1990) Common protein-binding sites in the 5' flanking regions of human genes for cytochrome c1 and ubiquinone binding proteins. J Biol Chem 265:8159–8163
Sylven C, Kallner A, Henze A, Larsen F, Liska J, Mogensen L (1985) Release patterns of CK-MB and mitochondrial CK following myocardial ischaemia. Clin Chim Act 151:111–119
Sylven C, Lin L, Kallner A, Sotonyi P, Somogyi E, Jansson E (1991) Dynamics of creatine kinase shuttle enzymes in the human heart. Eur J Clin Invest 21:350–354
Tilcock CPS (1986) Lipid polymorphism. Chem Phys Lipids 40:109–125
Trask RV, Billadello JJ (1990) Tissue-specific distribution and developmental regulation of M and B creatine kinase mRNAs. Biochim Biophys Acta 1049:182–188
Vacheron MJ, Clottes E, Chautard C, Vial C (1997) Mitochondrial creatine kinase interaction with phospholipid vesicles. Arch Biochem Biophys 344:316–324
Veksler VI, Murat I, Ventura-Clapier R (1991) Creatine kinase and mechanical and mitochondrial functions in hereditary and diabetic cardiomyopathies. Can J Physiol Pharmacol 69:852–858
Ventura-Clapier R, Veksler V, Hoerter JA (1994) Myofibrillar creatine kinase and cardiac contraction. Mol Cell Biochem 133:125–144
Ventura-Clapier R, Kuznetsov AV, D'Albis A, Vandeursen J, Wieringa B, Veksler VI (1995) Muscle creatine kinase-deficient mice. I. Alterations in myofibrillar function. J Biol Chem 270:19914–19920

Vial C, Marcillat O, Goldschmidt D, Font B, Eichenberger D (1986) Interaction of creatine kinase with phosphorylating rabbit heart mitochondria and mitoplasts. Arch Biochem Biophys 251:558–566

Wallimann T, Schlöser T, Eppenberger HM (1984) Function of M-line-bound creatine kinase as intramyofibrillar ATP regenerator at the receiving end of the phosphorylcreatine shuttle in muscle. J Biol Chem 259:5238–5246

Wallimann T, Wyss M, Brdiczka D, Nicolay K, Eppenberger HM (1992) Intracellular compartmentation, structure and function of creatine kinase isoenzymes in tissues with high and fluctuating energy demands – the phosphocreatine circuit for cellular energy homeostasis. Biochem J 281:21–40

Wegmann G, Huber R, Zanolla E, Eppenberger HM, Wallimann T (1991) Differential expression and localization of brain-type and mitochondrial creatine kinase isoenzymes during development of the chicken retina – Mi-Ck as a marker for differentiation of photoreceptor cells. Differentiation 46:77–87

Wu AHB, Herson WC, Bowers GN (1983) Macro creatine kinase type 1 and 2: clinical significance in neonates and children as compared with adults. Clin Chem 29:201–204

Wyss M, Smeitink J, Wevers RA, Wallimann T (1992) Mitochondrial creatine kinase – a key enzyme of aerobic energy metabolism. Biochim Biophys Acta 1102:119–166

Wyss M, James P, Schlegel J, Wallimann T (1993) Limited proteolysis of creatine kinase – implications for 3-dimensional structure and for conformational substates. Biochemistry 32:10727–10735

CHAPTER 12

Pyruvate Dehydrogenase Deficiencies

12

C. MARSAC, D. FRANÇOIS, F. FOUQUE, and C. BENELLI

Contents

Summary

The mammalian pyruvate dehydrogenase complex (PDHC) catalyses the oxidative and irreversible decarboxylation of pyruvate, giving acetyl-CoA, Co_2, and NADH + H^+. It is located in the mitochondrial inner membrane-matrix space. It plays an essential role in aerobic energy metabolism by controlling the supply of acetyl-CoA arising from carbohydrates and amino acids which are oxidised or converted to fat.

PDHC deficiencies primarily affect the nervous system and are associated with congenital lactic acidosis. The clinical presentation is highly heterogeneous, varying from mild episodic ataxia to severe, rapidly fatal lactic acidosis. The most frequent forms include developmental delay, mental retardation, lactic acidosis and cerebral abnormalities leading to death in early infancy. The most frequent PDHC deficiencies are due to a defect of the E1α subunit. Other defects are very rare and do not differ clinically from the PDH-E1α defects. Fifty-two different mutations have been found in both sexes in the E1α-gene located on the X-chromosome. In girls, due to the variable inactivation of this chromosome in different tissues, the biochemical expression of the deficiency may be difficult to interpret. A lot of mutations are sporadic and most are not found in the mother's somatic cells. We report here an overview of all mutations described in the literature up until now.

Inserm U75 et U30, Hôpital et Faculté Necker, 149 rue de Sèvres, 75015 Paris, France

1 Introduction

1.1 Structure

PDHC is a key enzyme of mitochondrial energy metabolism. Its major function is to produce energy in the brain which relies exclusively upon glucose metabolism in the first weeks of fetal life. It is one of the mitochondrial 2-oxoacid dehydrogenase complexes with 2-oxo-glutarate dehydrogenase and branched-chain 2-oxoacid dehydrogenase complexes. All these complexes contain multiple copies of the same three catalytic subunits: E1α2β2 (a 2-oxoacid decarboxylase), E2 (an dihydrolipoamide acetyltransferase), and E3 (dihydrolipoamide dehydrogenase, common to the three complexes; Kerr et al. 1996; Figs. 1 and 2). Moreover, another protein called PDH-X protein serves as a binding component between E3 and E2 (De Marcucci and Lindsay 1985). E1, E2, E3 and PDH-X are encoded by different genes (Table 1).

1.2 Regulation

There are two ways in which the activity of PDHC may be rapidly and reversibly controlled: end-product inhibition and enzyme interconversion by covalent modification through dephosphorylation/phosphorylation monitoring the activation/inactivation of the E1 component (Reed 1981; Fig. 3). PDH-phosphatase requires Mg^{2+} and Ca^{2+} for its activation while PDH kinase (ATP dependent) is strongly inhibited by pyruvate or dichloroacetate (a structural analog of pyruvate) and activated by an increase of NADH/NAD^+, acetyl-CoA/CoA and ATP/ADP ratios. Maintaining glucose homeo-

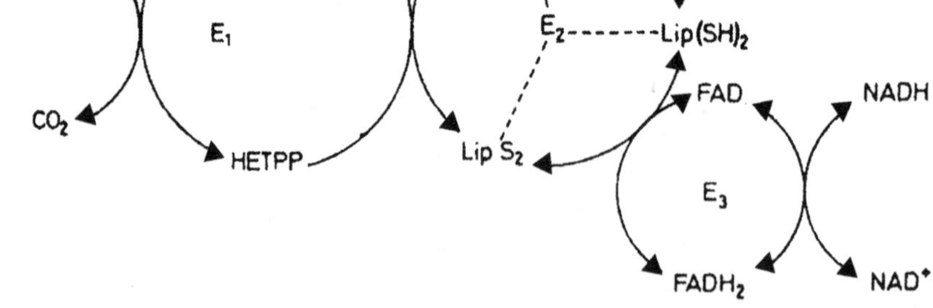

Fig. 1. Sequence of reactions catalyzed by PDHC complex in human. *E1* Pyruvate decarboxylase; *TPP* thiamine pyrophosphate; *HETPP* α-hydroxyethyl pyrophosphate; *E2* lipoate acetyl transferase with bound lipoic acid in oxidized (lip S_2), acetylated (Ac S lip SH) and reduced [lip-$(SH)_2$] forms; *E3* dihydrolipoate dehydrogenase

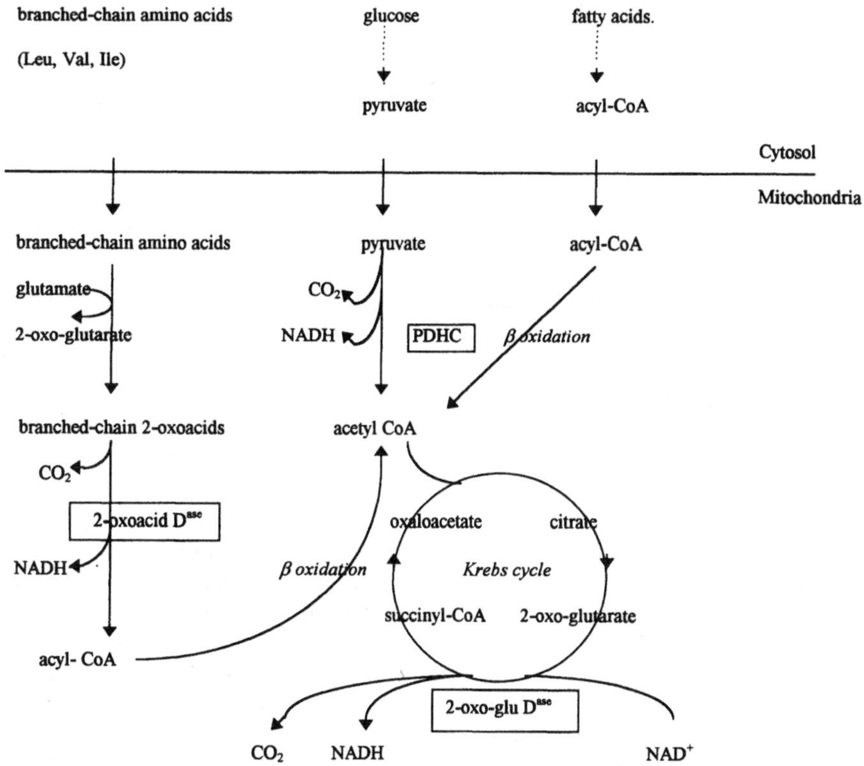

Fig. 2. Place of the α 2-oxo-acid dehydrogenase in the energy metabolism. *PDHC* pyruvate dehydrogenase complex. *2-oxo-acid D^ase^* branched amino acids 2-oxo acid dehydrogenase. *2-oxo-glu D^ase^* 2-oxo-glutarate dehydrogenase

Table 1. Reaction, description, and chromosome localization of the pyruvate dehydrogenase complex subunits: Pyruvate + TPP + CoA + NAD^+ → acetyl-CoA + CO_2 + NADH + H^+

Subunits	E1α2β2 tetramer MW E1α = 41 000 ME E1β = 36 000 30 copies	E2 monomer MW = 52 000 apparent MW = 74 000 60 copies	E3 dimer MW = 55 000 6 copies	PDH-X protein MW = 50 000
Function	TPP-decarboxylase	Transacetylase	Dehydrogenase	Binds E3 to E2
Chromosome localization	X-E1α[a] (p22.1–22.2) 4-E1α[a] (q22–23) 3-E1β[a] (p13–q23)	11 (q21–23)	7 (q31.1–q32)	11 (p12–13)[b,c]

[a] Patel and Harris (1995).
[b] Aral et al. (1997).
[c] Ling et al. (1998).

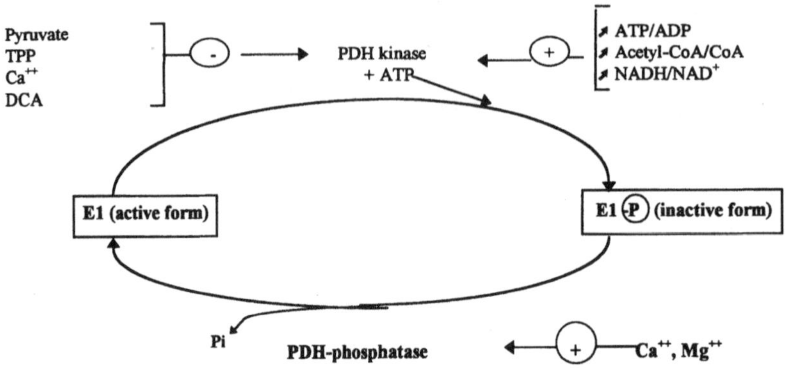

Fig. 3. PDH regulation by an inactivating PDH-kinase (plus ATP) and by an activating PDH-phosphatase (plus Ca^{++} and Mg^{++}). *DCA* dichloroacetate, *TPP* thiamine pyrophosphate. Phosphorylation/dephosphorylation depend on $NADH/NAD^+$, acetyl-CoA/CoA, and ATP/ADP intramitochondrial ratios

stasis is of special importance during starvation, when the supply of glucose is limited. The down-regulation by increased phosphorylation, together with end-product inhibition as observed in starvation, and its reversal after resumption of feeding, illustrates how this control operates. Obviously, there are marked differences between different tissues depending on their respective capabilities in switching from carbohydrate to lipid utilisation and vice versa. The adaptability in the choice of fuel is especially developed in heart muscle and kidney. By contrast, only 20% in liver, and 80% of the PDHC in brain are in the active form after a diet rich in carbohydrates and these proportions are only slightly changed on starvation and refeeding. Unequivocal effects of insulin have been demonstrated in adipose tissue. Increase of both inactive and active forms have been shown in rat adipocytes (Da Silva et al. 1993). There is a good correlation between biosynthesis of the subunit proteins and synthesis of their corresponding mRNAs. All tissues, except brain, are able to metabolize glucides, lipids and amino acids as energy substrates. Brain depends mainly on glucose as a fuel.

2 Pyruvate Dehydrogenase Deficiencies

2.1 Clinical Presentation

Presentations of PDHC deficiencies are highly variable (Robinson et al. 1980; Brown et al. 1994) and can be distinguished from deficiencies in respiratory chain complexes or defects in pyruvate carboxylase or in one enzyme of the Krebs cycle. All are causes of primary lactic acidosis in children. In general, but not in all cases, there is a good correlation between the severity of the clinical features and the abnormal levels of lactate in the blood. Symptoms are worsened with a diet rich in carbohydrates and improved with a diet rich in lipids and poor in glucose. The most severe forms are found in hypotonic children with encephalopathy, who die during their first year. The most frequent forms are psychomotor retardation, which is compatible with greater survival. Most boys present with Leigh syndrome whereas cerebral atrophy and ventricular dilatation are more frequently observed in girls. A third category of patients includes mildly affected boys presenting with chronic or episodic ataxia. In fact, clinical features vary with sex, and the nature of the mutation found in the E1α PDH gene differs between boys and girls (Dahl 1995).

Chronic alcoholism in pregnant women may be the cause of secondary PDHC deficiencies due to lack of thiamine pyrophosphate (coenzyme of PDHC-E1). Some symptoms are similar to those described in primary defects (facial dysmorphism with flat nasal bridge, agenesis of the corpus callosum).

2.2 Biological and Structural Abnormalities

In pyruvate dehydrogenase deficiencies, pyruvate oxidation is only partial in the brain and increased lactate levels are found in both cerebrospinal fluid (CSF) and blood, with a normal lactate/pyruvate ratio. Cerebral structural abnormalities begin very early during the fetal life (ventricular dilatation, microcephaly, putamen necrosis, corpus callosum agenesis). Muscles are morphologically normal.

2.3 Biochemical Defects

Five components of the PDH complex have been described causing a deficiency of PDHC activity involved in human pathologies: 4 defects of PDH-phosphatase activity, less than 10 cases with PDH-E3 defects, 1 case with a PDH-E2 defect, 6 cases with protein-X defects, and about 200 cases with PDHC-E1 defects.

Defects in PDH-phosphatase are very rare and do not differ clinically from the PDHC defects (Sorbi and Blass 1982; Ito et al. 1992b; Liu et al. 1993).

PDH-E3 defects differ from other defects by increased levels of 2-oxo-glutarate and branched-chain amino acids in the blood. In the urine, some organic acids can be present as α-hydroxybutyrate and α-hydroxyisovalerate. The diagnosis is made by the decrease of the PDH-E3 activity in cultured patient skin fibroblasts or different tissues.

Deficiencies in PDH-E2 and PDH-X are rare and do not clinically differ from those with PDH-E1 deficiencies. The case with PDH-E2 deficiency is associated with a congenital lactic acidosis and a residual E2 transacetylase activity corresponding to 32% of control activities with normal PDH-E1 and PDH-E3 activities (Robinson et al. 1990). For six patients (five unrelated) with PDH-X defects, diagnoses were performed by Western blot analysis, showing a complete absence of protein X and normal levels of other PDH subunits. In all these patients, residual PDHC activities were always very low (about 10%).

About 200 cases with PDHC-E1α defects have been described. The first observation of PDHC-E1α was reported by Blass et al. (1970). Diagnosis of PDHC is performed by the radioactive $^{14}CO_2$ measurement from 1-^{14}C radioactive pyruvate with TPP, NAD^+ and CoA-SH as coenzymes. The residual activities found in heterozygote girls can reflect the relative proportion of inactivation of the normal X-chromosome: cells with normal activated allele can have a replicating advantage over cells with a normal and inactivated allele and decreased PDHC-E1 activity.

In primary PDHC-E1 deficiency, thiamine treatment is often given as a supplement to diet, but improves the clinical features in only very few patients (Scholte et al. 1992; Bonne et al. 1993; Murakami et al. 1995; Pastoris et al. 1996). Administration of dichloroacetate can sometimes decrease the elevated levels of lactate in the blood.

2.4 Genetic Defects

Two mutations (K3337E and P453L) have been identified in the E3 gene (located in chromosome 7) in a patient with no PDH-E3 activity and 10–30% PDHC residual activity (DeVivo et al. 1979).

Genetic studies of six patients (five unrelated) with PDH-X defects have recently become possible (Harris et al. 1997; Aral et al. 1997; Ling et al. 1998). All were affected by different mutations (Aral et al. 1997; Ling et al. 1998).

Table 2. Mutations in exons 1 to 9 in E1 α gene

Exon	Sex	Mutation (aa)	References
Exon 1	2 M + mother	R10P	Takakubo et al. (1995a)
Exon 3	M	H44R	Naito et al. (1994a)
Exon 3	3 M	R72C	Lissens et al. (1996); Chun et al. (1995) Marsac et al. (1997)
Exon 3	M	R88S	Bonne et al. (1993)
Exon 3	M	R88C	(personal unpublished data)
Exon 3	F	G89S	Matsuda et al. (1995)
Exon 4	F	H113D	Lissens et al. (1996)
Exon 4	2 M + mother	R127W	Fuji et al. (1994)
Exon 5	F	G162R	Lissens et al. (1996)
Exon 5	2 F	V167M	Chun et al. (1993)
Exon 6	2 M Familial cases	V171del	Dahl et al. (1992a) De Meirleir et al. (1994)
Exon 6	F	G174A + exon 6 del	Chun et al. (1995)
Exon 6	F	A175P	Takakubo et al. (1993a)
Exon 6	M	A199T	Chun et al. (1993)
Exon 7	2 M	F205L	Dahl et al. (1992b); Chun et al. (1995)
Exon 7	M	M210V	Tripatara et al. (1996)
Exon 7	M	P217L	Hemalatha et al. (1995)
Exon 7	M	T231A	Chun et al. (1993)
Exon 7	M	Y243N	Matthews et al. (1994)
Exon 7	M	Y243S	(personal unpublished data)
Exon 8	M	D258A	Matthews et al. (1993b)
Exon 8	7 M + 4/7	R263G	Kerr et al. (1988); Wexler et al. (1992); Chun et al. (1993, 1995); Lissens et al. (1996)
Exon 8	2 M + mothers	R263Q	Awata et al. (1994)
Exon 9	F	M282L	Matthews et al. (1994)
Exon 9	F	R288fs	Matthews et al. (1994)
Exon 9	F	G291R	Matsuda et al. (1995)
Exon 9	F	H292L	Chun et al. (1993)

aa: amino acid; bp: base pair; del: deletion; fs: frameshift

Table 3. Mutations in exon 10 in E1 α gene

Sex	Mutation (aa)	References
F	S300fs	Matthews et al. (1993a)
2F	302C	De Meirleir et al. (1993); Lissens et al. (1996); Fujii et al. (1996) Dahl et al. (1992)
F	R302fs	Hansen et al. (1993)
M	E306ins	De Meirleir et al. (1992)
F	I307ins	Hansen et al. (1994)
M	R311del	Tripatara et al. (1996)
F	R311del	Chun et al. (1995)
6F	S312fs	Lissens et al. (1995); Fujii et al. (1996) Dahl et al. (1990); Chun et al. (1993); Chun et al. (1995)
F	K313del	Hansen et al. (1991)
M	D315N	Matthews et al. (1994)
F	P316L	Takakubo et al. (1995b)
F	P316fs	Dahl et al. (1992b)
M	D322fs	Lissens et al. (1996)
M	N326ins	Fujii et al. (1996)
F	N328fs	Chun et al. (1995)

aa: amino acid; bp: base pair; del: deletion; fs: frameshift; ins: insertion.

So far, more than 70 mutations have been described in the E1α gene, which is located on the X-chromosome and is made of 11 exons. Some of them have been published in the literature, using either the first amino acid of the presequence or the first amino acid of the sequence. Here, we name all the mutations beginning from the first amino acid of the sequence according to Takakubo et al. (1995a) and to Lissens et al. (1996). Tables 2, 3, 4, and Fig. 4 recapitulate reviews of all the mutations described

Table 4. Mutations in exon 11 in E1 α gene

Sex	Mutation (aa)	References
F	Lys344 ins	Ito et al. (1995)
F	E358fs	Chun et al. (1991)
F	P360fs	Chun et al. (1995)
M	G365ins	Fujii et al. (1996)
F	S371fs	Lissens et al. (1996)
F	E376fs	Chun et al. (1993)
5M and 1 F	R378H	Hansen et al. (1991); Chun et al. (1995); Matthews et al. (1994); Wexler et al. (1997)
F	R378C	Fujii et al. (1996)
M	R378C	(personal unpublished data)
F	Q382fs	Endo et al. (1991); Ito et al. (1992a); Takakubo et al. (1993b)
F	W383del	Marsac et al. (1997)
3M	K387fs	Tripatara et al. (1996); Fujii et al. (1996) Hansen et al. (1991)
4M	S388ins	Naito et al. (1994b); Chun et al. (1995); Fujii et al. (1996)
M	V389fs	Endo et al. (1989)

aa: amino acid; bp: base pair; del: deletion; fs: frameshift; ins: insertion.

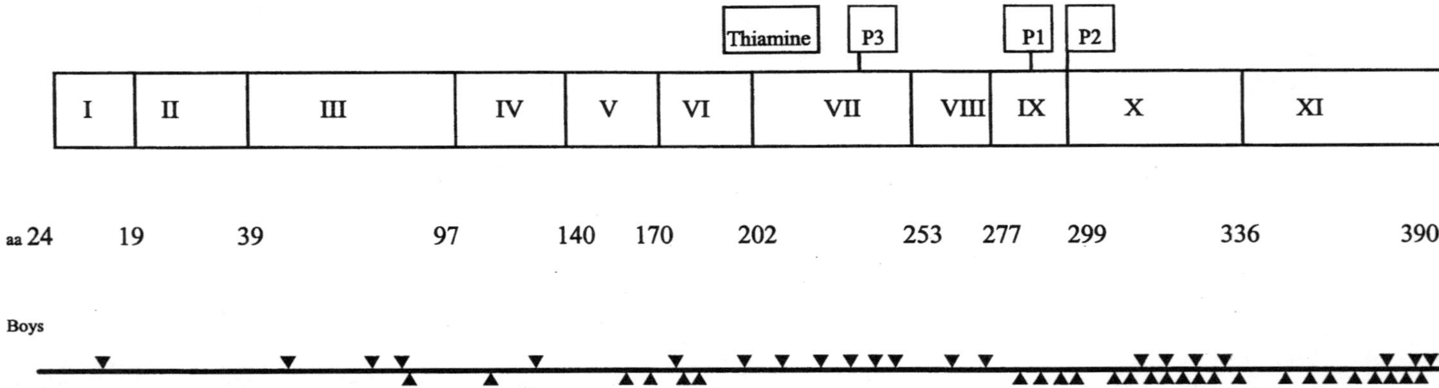

Fig. 4. Showing pathogenic mutations (deletions, insertions, point mutations) in the PDH-E1α gene in unrelated patients. The 11 exons are numbered in Roman digits. P1, P2, and P3 indicate the 3 phosphorylation sites of the enzyme. Under the exons are shown the positions of the amino acids (aa) at the end of the corresponding codons. The mutations found in boys are separated by a bold line from those found in girls

in this gene (Kerr et al. 1988; Endo et al. 1989, 1991; Dahl et al. 1990; 1992a, b; Chun et al. 1991, 1993, 1995; Hansen et al. 1991, 1993, 1994; De Meirleir et al. 1992, 1993, 1994; Ito et al. 1992, 1995; Wexler et al. 1992a, 1997; Bonne et al. 1993; Brown and Brown 1993; Matthews et al. 1993a, b; 1994; Takakubo et al. 1993a, b, 1995; Dahl and Brown 1994; Awata et al. 1994; Fujii et al. 1994; Naito et al. 1994a, b; Lissens et al. 1995, 1996; Matsuda et al. 1995; Hemalatha et al. 1995; Tripatara et al. 1996; Marsac et al. 1997). We correlated the different mutations with sex (Fig. 4). Forty-two mutations have been described once, and six at least three times (R72C, R236G, R302C, 312f, Q382fs, S388fs). An equal number of boys and girls are affected by PDH-E1α defects because of the variable normal X-inactivation pattern in heterozygous female patients. In girls, the deficiency can be highly variable between tissues (Lissens et al. 1996). If the normal allele is inactivated in cerebral cells, clinical expression will be severe (microcephaly). In boys, total deficiency is incompatible with life and only partial deficiencies affect, with a variable expression, the cerebral metabolism. Mutations found in boys differ from those found in females except one mutation (R378) found in both sexes. Most deletions and insertions are described in exons 10 and 11 in girls (23 girls out of 32) except the total deletion of exon 8 found in two boys and one girl (De Meirleir et al. 1994; Chun et al. 1995). Most point mutations have been identified in exons 1 to 9, particularly in boys (25 boys out of 36). Mutation R263G in exon 8 and mutation R72C in exon 3 are described in eight boys (seven unrelated) and four boys, respectively, all presenting with similar clinical features (late form of Leigh syndrome, ataxis). In only 10% of cases, transmission by the mother was found (R10P, R127P, V171del, R263G, R302G). Mutant mothers were slightly or not clinically affected. A great number of mutation are de novo mutations. Possible prenatal diagnosis can be proposed in hereditary cases and in cases with mutations in PDH-X protein.

The great number of mutations in the E1α gene gives the opportunity to better understand the functional role of some sites of the PDH complex. E1α and E1β proteins are linked together to form a functional E1 enzyme. They are unstable (degradation of E1α and E1β are visible on electrophoretic profiles of Western blots) when a mutation is present in the binding zone (Endo et al. 1991; Chun et al. 1991, 1995; Dahl et al. 1992a; Takakubo et al. 1993b, 1995b; Fujii et al. 1994; Matthews et al. 1994; Naito et al. 1994a; Lissens et al. 1995). Some mutations in the thiamine site (amino acid 199 in exon 6) and others which modify the function of the site without being in this site, also give degradation of E1α and E1β subunits (Chun et al. 1993, 1995).

3 Conclusion

A defect in the pyruvate dehydrogenase complex is one of the major causes of abnormalities of mitochondrial energy metabolism in the child, giving a variable, but constant increase of lactate levels in the blood and/or CSF. The defect can be expressed very early in the fetal brain. A review of 55 different mutations in 72 unrelated patients are presented here. Most of them are unique. Identification of a great number of mutations can permit better understanding of the relationship between structure and function of this complex, which is a key enzyme of energy metabolism in children, and particularly in the brain.

References

Aral B, Benelli C, Ait-Ghezala G, Amessou M, Fouque F, Maunoury C, Créau N, Kamoun P, Marsac C (1997) Mutations in PDX1, the human lipoyl-containing component X of the pyruvate dehydrogenase complex gene on chromosome 11p1, in congenital lactic acidosis. Am J Hum Genet 61:1318–1326

Awata H, Endo F, Tanoue A, Kitano A, Matsuda I (1994) Characterization of a point mutation in the pyruvate dehydrogenase E1 alpha gene from two boys with primary lactic acidaemia. J Inherit Metab Dis 17:189–195

Blass JP, Avigan J, Uhlendorf BW (1970) A defect of pyruvate decarboxylase in a child with an intermitent movement disorder. J Clin Invest 49:423–432

Bonne G, Benelli C, DeMeirleir L, Lissens W, Chaussain M, Diry M, Clot JP, Ponsot G, Geoffroy V, Leroux JP, Marsac C (1993) E1 pyruvate dehydrogenase deficiency in a child with motor neuropathy. Pediatr Res 33:284–288

Brown GK, Otero LJ, LeGris M, Brown RM (1994) Pyruvate dehydrogenase deficiency. J Med Genet 31:875–879

Brown RM, Brown GK (1993) X chromosome inactivation and the diagnosis of X-linked disease in females. J Med Genet 30:177–184

Chun K, MacKay N, Petrova-Benedict R, Robinson BH (1991) Pyruvate dehydrogenase deficiency due to a 20-bp deletion in exon 11 of the pyruvate dehydrogenase (PDH) E1α gene. Am J Hum Genet 49:414–420

Chun K, MacKay N, Petrova-Benedict R, Robinson BH (1993) Mutatins in the X-linked E1α subunit of pyruvate dehydrogenase leading to deficiency of the pyruvate dehydrogenase complex. Hum Mol Genet 4:449–454

Chun K, MacKay N, Petrova-Benedict R, Federico A, Fois A, Cole DEC, Robertson E, Robinson BH (1995) Mutations in the X-linked E1α subunit of pyruvate dehydrogenase: exon skipping, insertion of duplicate sequence, and missense mutations leading to the deficiency of the pyruvate dehydrogenase complex. Am J Hum Genet 56:558–569

Da Silva LA, De Marcucci OL, Kuhnle ZR (1993) Dietary polyunsaturated fats suppress the high-sucrose-induced increase of rat liver pyruvate dehydrogenase levels. Biochem Biophys Physiol 103A: 407–411

Dahl HHM (1995) Pyruvate dehydrogenase E1α deficiency: males and females differ yet again. Am J Hum Genet 56:553–557

Dahl HHM, Brown GK (1994) Pyruvate dehydrogenase deficiency in a male caused by a point mutation (F205L) in the E1α subunit. Hum Mutat 3:152–155

Dahl HHM, Maragos C, Brown RM, Hansen LL, Brown GK (1990) Pyruvate dehydrogenase deficiency caused by deletion of a 7-bp repeat sequence in the E1α gene. Am J Hum Genet 47:286–293

Dahl HHM, Brown GK, Brown RM, Hansen LL, Ker DS, Wexler ID, Patel MS, De Meirleir L, Lissens W, Chunk K, MacKay N, Robinson BH (1992a) Mutations and polymorphisms in the pyruvate dehydrogenase E1α gene. Hum Mutat 1:97–102

Dahl HHM, Hansen LL, Brown RM, Danks DM, Rogers JG, Brown GK (1992b) X-linked pyruvate dehydrogenase E1α subunit deficiency in heterozygous females: variable manifestation of the same mutation. J Inherit Metab Dis 15:835–847

De Marcucci O, Lindsay JG (1985) Component X. An immunologically distinct polypeptide associated with mammalian pyruvate dehydrogenase multi-enzyme complex. Eur J Biochem 149:641–648

De Meirleir L, Lissens W, Vamos E, Liebaers I (1992) Pyruvate dehydrogenase (PDH) deficiency caused by a 21-base pair insertion mutation in the E1α subunit. Hum Genet 88:649–652

De Meirleir L, Lissens W, Denis R, Wayenberg JL, Michotte A, Brucher JM, Vamos E, Gerlo E, Liebaers I (1993) Aberrant splicing of exon 6 in the pyruvate dehydrogenase-E1α mRNA linked to a silent mutation in a large family with Leigh's encephalomyelopathy. Pediatr Res 36:707–712

De Vivo DC, Haymond MW, Obert KA, Nelson JS, Pagliara AS (1979) Defective activations of the pyruvate dehydrogenase complex in subacute necrotizing encephalomyelopathy (Leigh disease). Ann Neurol 6:483–494

Endo H, Hasegawa K, Narisawa K, Tada K, Kagawa Y, Ohta S (1989) Defective gene in lactic acidosis: abnormal pyruvate dehydrogenase E1α-subunit caused by a frameshift. Am J Hum Genet 44:358–364

Endo H, Miyabayashi S, Tada K, Narisawa K (1991) A four-nucleotide insertion at the E1α gene in a patient with pyruvate dehydrogenase deficiency. J Inherit Metab Dis 14:793–799

Fujii T, Van Coster RN, Old SE, Medori R, Winter S, Gubits RM, Matthews PM, Brown RM, Brown GK, Dahl HHM, De Vivo DC (1994) Pyruvate dehydrogenase deficiency: molecular basis for intrafamilial heterogeneity. Ann Neurol 36:83–89

Fujii T, Alvarez G, Sheu KF, Kranz-Eble P, De Vivo D (1996) Pyruvate dehydrogenase deficiency: The relation of the E1α mutation to the E1β subunit deficiency. Pediat Neurol 14:328–334

Geoffroy V, Fouque F, Benelli C, Poggi F, Lissens W, Marsac C, Sanderson SJ, Lindsay JG (1996) Defect in the X-lipoyl-containing component of the pyruvate dehydrogenase complex in a patient with a neonatal lactic acidemia. Pediatrics 97:267–272

Hansen LL, Brown GK, Kirby DM, Dahl HHM (1991) Characterization of the mutations in three patients with pyruvate dehydrogenase E1α deficiency. J Inherit Metab Dis 14:140–151

Hansen LL, Brown GK, Brown RM, Dahl HHM (1993) Pyruvate dehydrogenase deficiency caused by a 5 base pair duplication in the E1α subunit. Hum Mol Genet 6:805–807

Hansen LL, Horn N, Dahl HHM, Kruse TA (1994) Pyruvate dehydrogenase deficiency caused by a 33 base pair duplication in the PDH E1α subunit. Hum Mol Genet 6:1021–1022

Harris RA, Bowker-Kinley MM, Wu P, Jeng J, Popov M (1997) Dihydrolipoamide dehydrogenase-binding protein of the human pyruvate dehydrogenase complex. DNA-derived amino acid sequence, expression, and reconstitution of the pyruvate dehydrogenase complex. J Biol Chem 272:19746–19751

Hemalatha SG, Kerr DS, Wexler ID, Lusk MM, Kaung M, Du Y, Kolli M, Schelper L, Patel MS (1995) Pyruvate dehydrogenase complex-deficiency due to a (P188L) within the thiamine pyrophosphate binding loop of the E1α subunit. Hum Mol Genet 4:315–318

Ito M, Huq AHMM, Naito E, Saijo T, Takeda E, Kuroda Y (1992a) Mutation of E1α gene in a female patient with pyruvate dehydrogenase deficiency due to rapid degradation of E1 protein. J Inherit Metab Dis 15:848–856

Ito M, Kobashi H, Naito E, Saijo T, Kakeda E, Huq AHMM, Kuroda Y (1992b) Decrease of pyruvate dehydrogenase phosphatase activity in patients with congenital lactic acidemia. Clin Chim Acta 209:1–7

Ito M, Naito E, Yokota I. Takeda E, Matsuda J, Hirose M, Sejima H, Aiba H, Hojo H, Kuroda Y (1995) Molecular genetic analysis of a female patient with pyruvate dehydrogenase deficiency: detection of a new mutation and differential expression of mutant gene product in cultured cells. J Inherit Metab Dis 18:547–557

Kerr DS, Berry SA, Lusk MM, Ho L, Patel MS (1988) A deficiency of both subunits of pyruvate dehydrogenase which is not expressed in fibroblasts. Pediatr Res 24:95–100

Ker DS, Wexler ID, Tripatara A, Patel MS (1996) Human defects of the pyruvate dehydrogenase complex. In: Patel MS, Roche TE, Harris RA (eds) Keto acid dehydrogenase complexes. Birkhäuser, Basel, pp 249–269

Ling M, McEachern G, Seyda A, MacKay M, Scherer SW, Bratinova S, Beatty B, Giovannucci-Uzielli ML, Robinson BH (1998) Detection of a homozygous four base pair deletion in the protein X gene in a case of pyruvate dehydrogenase complex deficiency. Hum Mol Genet 7:501–505

Lissens W, Desguerre I, Benelli C, Marsac C, Fouque F, Haenggeli C, Ponsot G, Seneca S, Liebaers I, De Meirleir L (1995) Pyruvate dehydrogenase deficiency in a female due to a 4 base pair deletion in exon 10 of the E1α gene. Hum Mol Genet 4:307–308

Lissens W, De Meirleir L, Seneca S, Benelli C, Marsac C, Poll-The BT, Briones P, Ruitenbeek W, Van Diggelen O, Chaigne D, Ramaekers V, Liebaers I (1996) Mutation analysis of the pyruvate dehydrogenase E1α gene in eight patients with a pyruvate dehydrogenase complex deficiency. Hum Mutat 7:46–51

Liu TC, Kim H, Arizmendi C, Kitano A, Patel MS (1993) Identification of two missense mutations in a dihydrolipoamide dehydrogenase deficiency patient. Proc Natl Acad Sci USA 90:5186–5190

Marsac C, Stansbie D, Bonne G, Cousin J, Jehenson P, Benelli C, Leroux JP, Lindsay G (1993) Defect in the lipoyl-bearing protein X subunit of the pyruvate dehydrogenase complex in two patients with encephalomyelopathy. J Pediatr 123:915–920

Marsac C, Benelli C, Desguerre I, Diry M, Fouque F, De Meirleir L, Ponsot G, Seneca S, Poggi F, Saudubray JM, Zabot MT, Fontan D, Lissens W (1997) Biochemical and genetic studies of four patients with pyruvate dehydrogenase E1α deficiency. Hum Genet 99:785–792

Matsuda J, Ito M, Naito E, Yokota I, Kuroda Y (1995) DNA diagnosis of pyruvate dehydrogenase deficiency in female patients with congenital lactic acidaemia. J Inherit Metab Dis 18:534–546

Matthews PM, Brown RM, Otero LJ, Marchington DR, Leonard JV, Brown GK (1993a) Neurodevelopmental abnormalities and lactic acidosis in a girl with a 20-bp deletion in the X-linked pyruvate dehydrogenase E1α subunit gene. Neurology 43:2025–2030

Matthews PM, Marchington DR, Squier M, land J, Brown RM, Brown GK (1993b) Molecular genetic characterization of an X-linked form of Leigh's syndrome. Ann Neurol 33:652–655

Matthews PM, Brown RM, Otero LJ, Marchington DR, LeGris M, Howes R, Meadows LS, Shevell M, Scriver CR, Brown GK (1994) Pyruvate dehydrogenase deficiency: clinical presentation and molecular genetic characterization of five new patients. Brain 117:435–443

Murakami N, Iso A, Naoto E, Kuroda Y, Nonaka I (1995) Thiamine responsive congenital lactic acidemia and type I muscle fiber atrophy. Brain Dev 17:78

Naito E, Ito M, Takeda E, Yokota I, Yoshijima S, Kuroda Y (1994a) Molecular analysis of abnormal pyruvate dehydrogenase in a patient with thiamine-responsive congenital lactic acidemia. Pediatr Res 36:340–346

Naito E, Ito M, Yokota I, Matsuda J, Yara A, Kuroda Y (1994b) Pyruvate dehydrogenase deficiency caused by a four-nucleotide insertion in the E1α subunit gene. Hum Mol Genet 3:1193–1194

Pastoris O, Savasta S, Foppa P, Catapano M, Dossena M (1996) Pyruvate dehydrogenase deficiency in a child responsive to thiamine treatment. Acta Paediatr 85:625–628

Patel MS, Harris RA (1995) Mammalian α-2-oxo acid dehydrogenase complexes: gene regulation and genetic defects. FASEB J 9:1164–1172

Reed LJ (1981) Regulation of mammalian pyruvate dehydrogenase complex by phosphorylation-dephosphorylation cycle. Curr Top Cell Regul 18:95–106

Robinson BH (1995) Lactic acidemia (disorders of pyruvate carboxylase, pyruvate dehydrogenase). In: Scriver CR, Beaudet AL, Sly WS, Valle D (eds) Metabolic and molecular basis of inherited disease, 7th edn. McGraw-Hill, New York, pp 1479–1499

Robinson BH, Taylor J, Sherwood W (1980) The genetic heterogeneity of lactic acidosis occurrence of recognisable inborn errors of metabolism in a paediatric population with lactic acidosis. Pediatr Res 14:956–962

Robinson BH, MacKay N, Petrova-Benedict R, Ozalp I, Coskun T, Stacpoole PW (1990) Defects in the E2 lipoyl transacetylase and X-lipoyl containing component of the pyruvate dehydrogenase complex in patients with lactic acidemia. J Clin Invest 85:1821–1824

Scholte HR, Busch FM, Luyt-Houwen IEM (1992) Vitamin-responsive pyruvate deficiency in a young girl with external ophthalmoplegia, myopathy and lactic acidosis. J Inherit Metab Dis 15:331–334

Sorbi S, Blass JP (1982) Abnormal activation of pyruvate dehydrogenase in Leigh disease fibroblasts. Neurology 32:555–558

Takakubo F, Thorburn D, Dahl HHM (1993a) A novel mutation and a polymorphism in the X chromosome located pyruvate dehydrogenase E1α gene (PDHA1). Hum Mol Genet 2:1961–1962

Takakubo F, Cartwright P, Hoogenraad N, Thorburn DR, Collins F, Lithgow T, Dahl HHM (1995a) An amino acid substitution in the pyruvate dehydrogenase E1α gene, affecting mitochondrial import of the precursor protein. Am J Hum Genet 57:772–780

Takakubo F, Thorburn DR, Brown RM, Brown GK, Dahl HHM (1995b) A novel mutation (P316L) in a female with pyruvate dehydrogenase E1α deficiency. Hum Mutat 6:274–275

Tripatara A, Ker DS, Lusk MM, Kolli M, Tan J, Patel MS (1996) Three new mutations of the pyruvate dehydrogenase alpha subunit: a point mutation (M181V), 3 bp deletion (-R282), and 16 bp insertion/frameshift (K358VS → TVDQS). Hum Mutat 8:180–182

Wexler ID, Hemalatha SG, Liu TC, Berry SA, Kerr DS, Patel MS (1992) A mutation in the E1α subunit of pyruvate dehydrogenase associated with variable expression of pyruvate dehydrogenase complex deficiency. Pediatr Res 32:169–174

Wexler ID, Hemalatha SG, McConnell RD, Buist NR, Dahl HHM, Berry SA, Cederbaum SD, Patel MS, Ken TS (1997) Outcome of pyruvate dehydrogenase treated with ketogenic diet. Neurology 49:1655–1661

CHAPTER 13

Electrochemical Gradient and Mitochondrial DNA in Living Cells

13

J. Coppey, C. Durieux, and M. Coppey-Moisan

Contents

1 Introduction

For imaging cells with fluorescent probes, an extremely low excitation light level can be used if the microscopy system is equiped with an intensification device. The low light level allows, when observing living cells, preservation of the native state with regard to lysosome compartmentation, electrical inner membrane potential, structure and intrinsic reactivity of chromatinized DNA, cytoplasmic rRNA and mitochondrial

Laboratoire de Biochimie des Acides Nucléiques, Section de Recherche, Institut Curie, 26 rue d'Ulm, 75248 Paris Cedex 5, France (mt)DNA. Following contact with ethidium, cycling cells examined under this condition had fluorescent mitochondria. The fluorescence came from ethidium intercalated in mtDNA, since cycling $\varrho°$ (devoid of mtDNA) cells harbour mitochondria, revealed by potential sensitive fluorescent probes, which did not fluoresce following ethidium incubation. As B-DNA where ethidium intercalation gives rise to enhanced fluorescence, it seems that mtDNA is under B-conformation in cycling and living cells. Moreover, mtDNA seems to stay in B-conformation, since the mitochondrial fluorescence maintains a high intensity for several days.

When abolishing the electrochemical gradient across the inner membrane by inhibitors of electron transport or by ionophores, the ethidium fluorescence disappeared from mitochondria within 2 min. As the enhanced fluorescence is due to the fact that the chromophores of intercalated ethidium are physically separated from environmental quenchers (water), the steep extinction of this fluorescence shows that the chromophores have been rendered accessible to local quenchers. Such a process can result from a global structural change in mtDNA. During the change, the fluorescence from a minor groove probe (DAPI) in interaction with the same mtDNA molecules did not decrease, indicating that mtDNA kept doublestrandedness troughout. Upon removal of the inhibitors/ionophores, the electrochemical gradient spontaneously recovered. In parallel, the ethidium fluorescence reappeared in mitochondria.

Hence, in living, cycling cells, mtDNA seems (1) to be in a stable state as a B-form, (2) to undergo a structural change when the electrochemical gradient across the inner membrane is cancelled, and (3) to take back the starting B-form when the cells recover an electrochemical gradient. In mitochondria of unperturbed respiring human skin fibroblasts labelled with ethidium and with DAPI, both mtDNA states seem to coexist in a given mitochondrion. Moreover, there appears to occur a spontaneous switch back and forth from one state, the B-form, to the other state. This happens over 1 min of observation.

1.1 Low Excitation Light Level Allows Preservation of the Living State in Fluorescence Microscopy

Technical advances, such as optical apparatus, light intensifiers (microchannel plates) and digital imaging (Coppey-Moisan et al. 1994), endow fluorescence microscopy with a good temporal resolution (40 images per s, that of video) together with an excellent spatial resolution. A single copy 2 kb probe gives a bright spot, which is nicely individualized on each sister chromatid (Viegas-Pequignot et al. 1989). On the other hand, the extremely low level of light excitation used (down to 10 000 less than that for conventional fluorescence microscopy) allows the native state to be preserved during the observation in terms of:

1. Compartmentation. For instance, when the lyozonomal membrane blows out due to a local photodynamical reaction triggered by even modest excitation light level, the functional organization of chromatinized DNA is destroyed (Delic et al. 1991). Subsequently, the nuclear DNA leaks out of the nuclear area and the cell will die (Delic et al. 1992).

2. Energetics of mitochondria. The electrical potential through the inner membrane is maintained. It can be evaluated by the accumulation in the mitochondrial matrix of lipophilic cations, rhodamine 123 (rd 123) (Coppey-Moisan et al. 1996), or tetramethyl rhodamine ester (TMRM). TMRM is a dye which rapidly and reversibly equilibrates across membranes in a voltage-independent manner (Ehrenberg et al. 1988; Wadia et al. 1998) which allows an estimate of the potential value. Measuring the amount of accumulated TMRM, and applying the Nernst equation gives values in fair agreement with data reported by others (Durieux et al., submitted). Moreover, as the low light level used minimizes the extent of photoinduced crosslinks between the lipophilic cations and protein constituents of the inner membrane, the cations freely move across the membrane, making possible to monitor variations of the potential in living cells.
   For ethidium, the intracellular distribution of the fluorescence which features all types of cycling cells examined so far in the absence of major perturbation, is depicted in Fig. 1: with a low excitation light level (attenuated by a factor of 100), mtDNA in the cytoplasm and in DNA nucleoli are fluorescent, whilst chromatinized DNA does not exhibit detectable fluorescence (Fig. 1B); with an excitation light level 100 times greater, there is no more mitochondrial fluorescence, the necleoli are rendered more fluorescent and the fluorescence appears in chromatinized DNA (Fig. 1A).
   Thus, an extremely low excitation light level is critical for observing the fluorescence of ethidium in mitochondria. In parallel, DNA was visualized by DAPI (4', 6-diamino-2-phenylindole), a fluorescent probe (Wilson et al. 1990), which freely enters into living cells (Delic et al. 1992). The enhanced fluorescence of DAPI means the presence of DNA, and indicates its doublestrandedness. The unchanged patterns of nuclear DAPI fluorescence after either excitation light level show that the architecture of chromatin DNA as well as its doublestrandedness are not modified.
3. Dynamic reactivity of DNAs in their native state. The parameters described in 1. and 2. constitute critical determinants for cells to maintaining the behaviour of their DNAs in the living state, in mitochondria and in nucleus (Delic et al. 1991, 1992; Coppey-Moisan et al., in prep.).

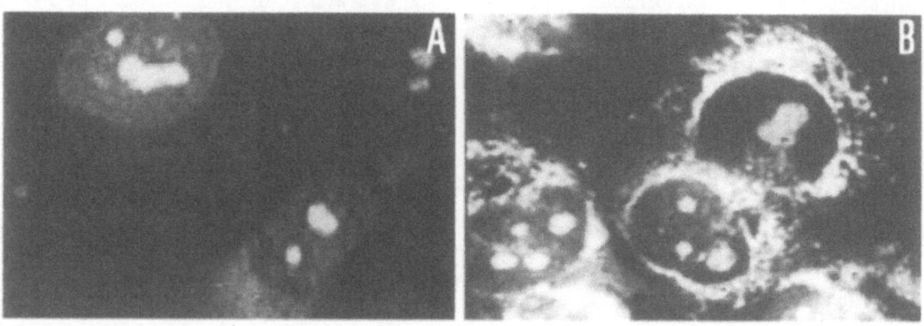

Fig. 1. Effect of the light excitation intensity on the distribution of ethidium fluorescence in a transformed cell line. Epithelial cells from a human pancreatic tumor incubated with ethidium were observed: **A** with a light excitation fluence close to that used in conventional microscopy (no OD filter); **B** with a light fluence attenuated by a factor of 100 (OD filter of 2). Note in **A** the absence of mitochondrial fluorescence, a nucleolar and nuclear fluorescence, in **B** mitochondria fluorescence (corresponding to ethidium intercalation in mtDNA), an intense nucleolar fluorescence and no fluorescence in nuclear DNA

We followed, using these conditions, the variations of condensation, structure and conformations of DNA in its microenvironments either unperturbed or perturbed by inhibitors ionophores. This was carried out by fluorescence intensity measurements. To probe the conformations of DNA, we used ethidium, a drug which intercalates into DNA; to probe the condensation and structure of DNA, we used DAPI, a drug which interacts with the minor groove.

1.2
Interactions Between Ethidium and DNA

1.2.1
In Vitro Studies

The physico-chemical properties of DNA allows it to take many conformations in vitro (A, B, C, Z, S) or compaction state according to the environment:

1. Ions (Pohl and Jovin 1972; Pecka and Nordheim 1982), water activity (Eickbush and Moudrianakis 1978; Preisler et al. 1995).
2. Protein interactions (Härd and Kearns 1990; Winzeler and Small 1991),
3. Stretching (Cluzel et al. 1996; Smith et al. 1996).

Ethidium, with its properties of intercalation and of fluorescence has constituted a reliable tool for monitoring conformational changes in DNA, torsional dynamics (Genest and Wahl 1978) and superhelix vaiations (Naimushin et al. 1994), for analysing the B to Z transition (Gilbert et al. 1991), and for studying the interaction of DNA with the histone octamer (McMurray and van Holde 1991). Ethidium freely intercalates into DNA under B-conformation because under this conformation the intrinsic elasticity of the double helix permits the local deformation required for intercalation to take place (see Coppey-Moisan et al. 1996). The fluorescence of ethidium molecules intercalated in B-DNA is enhanced because their chromophores are physically separated from local quenchers such as water (Olmsted and Kearns 1977). Despite the fact that intercalation has been examined from numerous points of view, recent investigations still reveal new facets of the underlying process: control of its kinetics by the reaction itself (Meyer-Almes and Porschke 1993), and involvement of large-scale dynamics when the intercalation occurs over long DNA molecules (Monaco and Gardiner 1993).

1.2.2
In Living Cells

The conformations of DNA are as yet unknown in living cells. By taking advantage of the fact that ethidium intercalation into B-DNA gives rise to enhanced fluorescence, we could, by imaging mitochondria, get an insight into the conformation of mitochondrial (mt)DNA in living cells. We chose to study DNA in this organelle because: (1) in cycling cells, ethidium stably intercalates into mtDNA and gives a strong localized fluorescence; DAPI also interacts with mtDNA and also gives enhanced fluorescence, (2) the microenvironnement of mtDNA is permanently regulated by ion movements through the inner membrane: this gives rise to proton gradients which can be modulated by respiration inhibitors/ionophores, and (3) mtDNA is in close contact with the inner membrane, which suggests the existence of functional links, as yet unknown, between the two partners.

2 Materials

The cell lines chosen were: (1) a line of human skin fibroblast cells (a normal 30-years-old woman), subcultured for less than 20 doubling times in minimum essential medium, supplemented with 2 mM glutamine and 10% fetal calf serum (FCS, Bio-Whittaker), and (2) two lines of transformed cells, Vero monkey kideny and a line derived from a human pancreatic tumor, were subcultured in Dulbecco modified medium supplemented with 2 mM glutamine and 10% FCS.

All culture media did not contain antibiotics and antifungicidal drugs, two conditions needed for the cells to exhibit normally shaped mitochondria and optimal inner membrane potentials. All lines were routinely checked for the absence of mycoplasma. They were followed in the growth phase.

Described elsewhere are the conditions for labelling DNA with fluorophores and mitochondria with inner membrane potential sensitive lipophilic cations, for perturbing the electrochemical gradient and the particularities of the microscopy system at low light level fluorescence (Coppey-Moisan et al. 1996 and refs. therein).

2.1 mtDNA Undergoes a Structural Change When the Electrochemical Gradient is Cancelled

A structural change occurred in almost all mtDNA molecules examined when the transmembrane electrochemical gradient was abolished by inhibitors of electron transport, sodium azide or antimycin A, or by ionophores, carbonylcyanid p- trifluoromethoxyphenylhydrazone (FCCP) (H^+ ionophore), valinomycin (K^+ ionophore) or nigericin (ionophore exchanging K^+ for H^+). Each compound was applied at a weak concentration and for short time periods. This ensured the occurrence, following rinsing out of each compound, of spontaneous recovery of the electrochemical gradient. The structural change in mtDNA was revealed by the variations of fluorescence intensity of intercalated ethidium. The potential in the inner membrane was monitored in parallel by the fluorescence intensity of rd123 accumulated in the mitochondrial matrix (Johnson et al. 1981).

When the electrochemical gradient was abolished, the mitochondrial ethidium fluorescence rapidly fell to complete extinction in less than 2 min in all mitochondria (several hundred mitochondria were followed). When azide or ionophores were rinsed out from the cultures, the mitochondrial fluorescence returned back in parallel with the potential, or more generally with the electrochemical gradient (Coppey-Moisan et al. 1996); Fig. 2 illustrates a set of data with azide. Following abolition of the membrane potential for a relatively long period, rd123 diffused out of mitochondria, then subsequently out of cells where the drug was diluted away. In contrast, ethidium

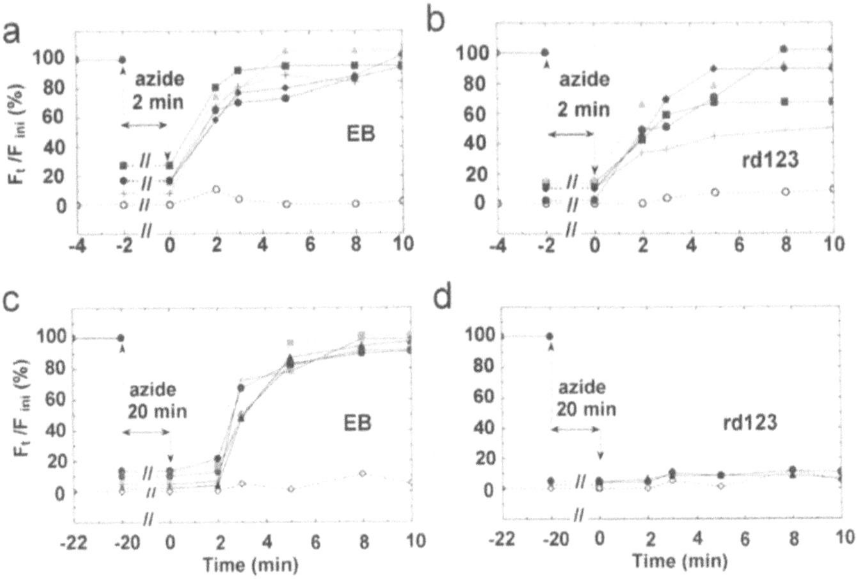

Fig. 2. The fluorescence of ethidium and of rd123 recover differently in mitochondria after rinsing azide out of the cultures. Following 2 or 20 min of azide contact, the time variations of the fluorescence of ethidium and of rd123 were followed in several mitochondria. The *full symbols* correspond to the fluorescence intensity over the mitochondrial matrix (mean values of five mitochondria), normalized per pixel. F_{ini} is the fluorescence intensity just after contact with azide; F_t is the fluorescence intensity at several time points after a contact of 2 or of 20 min with azide at 10 mM. The *empty symbols* correspond to the fluorescence intensity normalized per pixel over cytoplasmic regions close to the mitochondria observed

stayed in all cases close to the mtDNA, probably in interaction with it. In practice, azide was applied for 2 or for 20 min. Following a 20-min contact, the mitochondrial ethidium fluorescence completely recovered after a lag period, without need to reload the cultures with ethidium. The kinetics of recovery was slightly slowed down compared with after a 2-min contact (Fig. 2, left). In contrast, after a 2-min contact, rd123 accumulated back into miochondria with no need to reload the cultures, but after 20 min contact, there was no reaccumulation of rd123 in mitochondria (Fig. 2, right). However, following the long-lasting treatment and a reloading with rd123, a recoverey of the membrane potential could be seen. We deduce that rd123 has leaked out of the cells after the long abolition of the membrane potential, whilst ethidium stayed close to the mtDNA, perhaps still in interaction with it.

To further examine the kinetic parameters of this effect, we used TMRM, which allows, at equilibrium, an estimate of the membrane potential. The variations of ethidium and TMRM fluorescences were recorded during and following FCCP application for 15 min at 100 nM. Figure 4 shows that the recovery of membrane potential preceded that of ethidium fluorescence in mtDNA. This work is detailed in another paper (Durieux et al., submitted).

During the change, mtDNA remained doublestranded, since DAPI in interaction with the same molecules showed unmodified fluorescence patterns. Figure 3 illustrates this point in fibroblast cells in which ethidium and DAPI mitochondrial fluor-

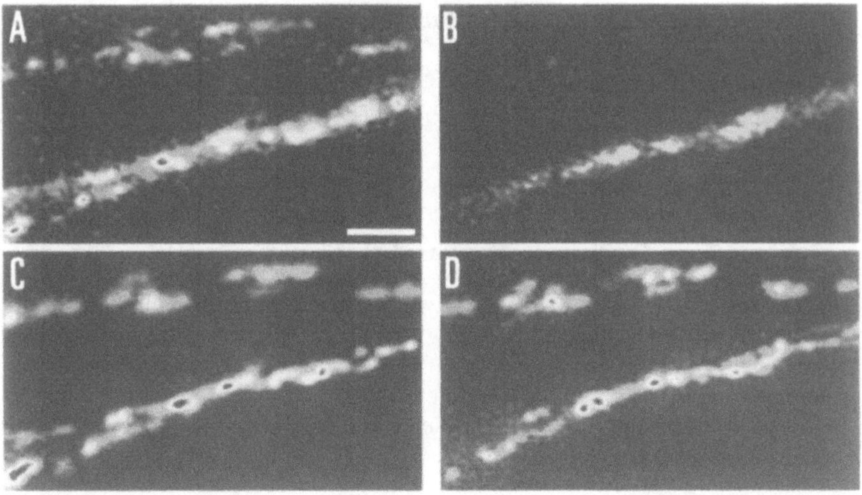

Fig. 3. Differential effects of FCCP on the fluorescences of DAPI and ethidium in the same mitochondrion. Fibroblast cell cultures were incubated with DAPI at 10 µg/ml for 2 hours, then with ethidium bromide at 0.4 µg/ml for 30 min. The molecules remaining in the supernatant were rinsed out. Shown are the fluorescence images in the same mitochondria of **A** ethidium, **C** DAPI; following contact of 100 nM FCCP for 15 min, the ethidium fluorescence disappears (**B**), that of DAPI stays unchanged (**D**)

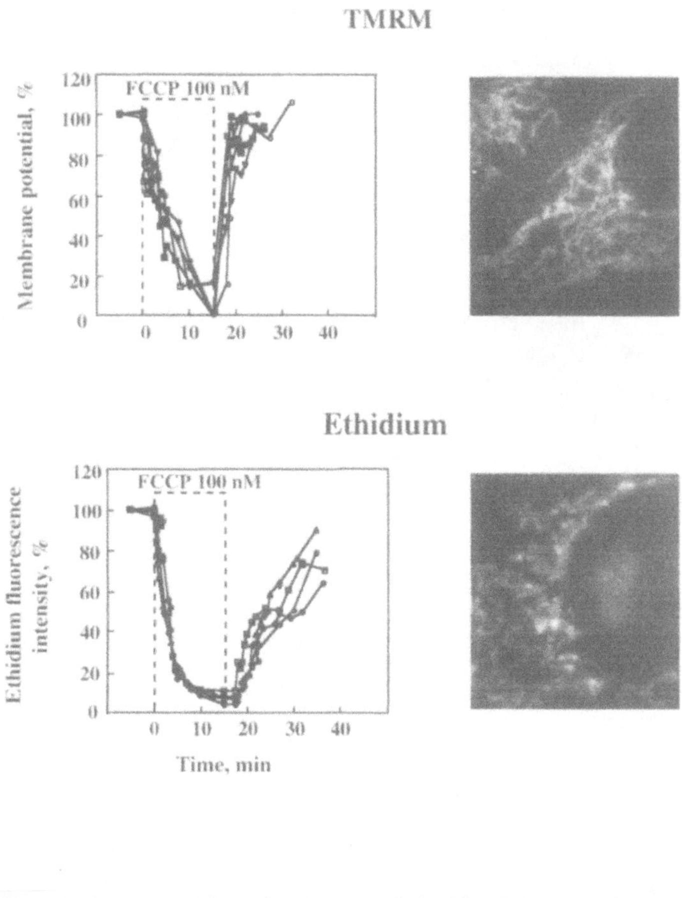

Fig. 4. Parallel modulations of TMRM fluorescence (membrane potential) and of ethidium fluorescence (mtDNA structural change) in four S2 transformed cells left in contact for 15 min with FCCP. The numerical values are shown in the *left* panels. Typical fluorescence patterns in one cell are shown in the *right* panels

escences show comparable intensities (Fig. 3A, C); after FCCP treatment, the ethidium fluorescence vanishes (Fig. 3B), while that of DAPI remains unchanged (Fig. 3D). We deduce that the fluorescence extinction of ethidium reveals a structural change of mtDNA. We will examine this possibility by measuring the lifetimes of ethidium fluorescence in mitochondria.

2.2
The Structural Change Spontaneously Occurs in mtDNA of Respiring Fibroblasts

In the mitochondria of respiring fibroblast cell, the distribution of ethidium fluorescence exhibits discontinuous patterns (Fig. 5A) which fluctuate over a minute (cf. Fig. 5A and C). In contrast, the distribution of DAPI fluorescence in the same mitochondrion (this can be obtained because mitochondria are labelled with ethidium and DAPI) exhibits a fluorescence pattern which is more continuous and more stable over a minute (cf. Fig. 5B and D).

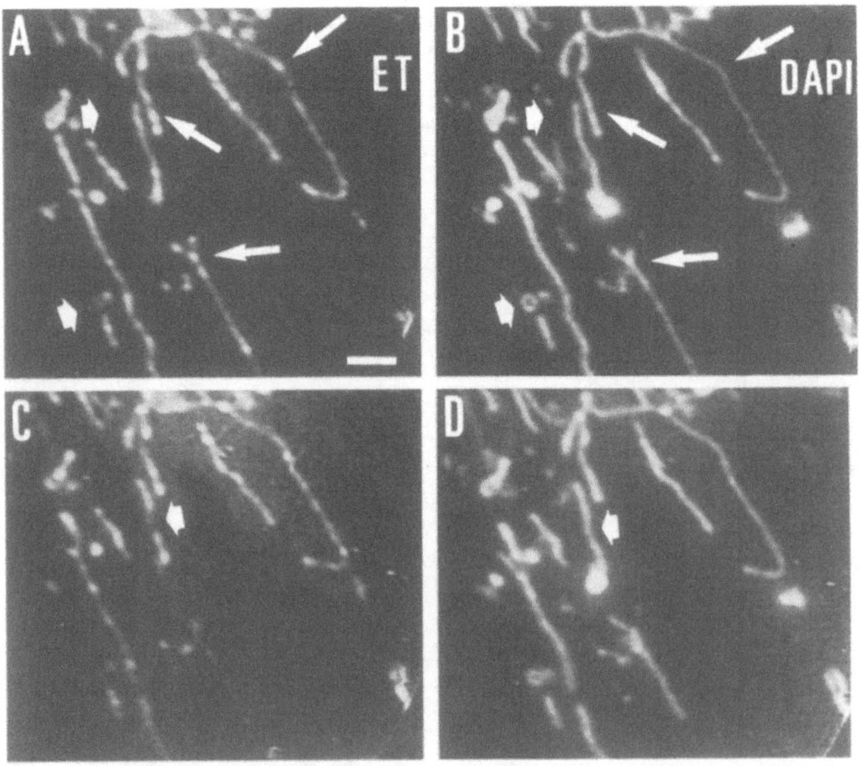

Fig. 5. Coexistence of two states of mtDNA in mitochondria in a respiring fibroblast cell. The fluorescence images of ethidium **A** and of DAPI **B** were successively acquired (see fig. 3) at time interval of 20 s. **C** and **D** are the same area as in **A** and **B** but photographed 1 min later. *Thin arrows* point to mitochondrial regions along which the ethidium fluorescence is typically heterogeneous whilst that of DAPI exhibits much more homogeneous patterns. *Thick arrows* point to mitochondrial regions harbouring mtDNA (revealed by DAPI labelling) which do not exhibit ethidium labelling

3
The Change in mtDNA is Linked to an Energetic Perturbation in Mitochondria

After short contacts of cultured cells with ethidium, the fluorescence of molecules intercalated in mtDNA keeps its intensity unchanged for several days, indicating that, in functioning mitochondria, mtDNA is stable under B-conformation. The abrupt extinction of mitochondrial fluorescence followed by its reappearance whenever the electrochemical gradient ($\Delta\Psi$ and/or ΔpH) falls and subsequently recovers, reveals a global dynamic change in mtDNA. The extinction cannot be due to either a direct effect of the inhibitors/ionophores or a displacement of the equilibrium ethidium-DNA resulting, for instance, from leakage of the drug out of mitochondria (Coppey-Moisan et al. 1996).

Also ruling out this last possibility are the behaviours of ethidium and of rd123, which differ from each other. After a long application of FCCP or azide, ethidium recovers its mitochondrial fluorescence without having to reload the cultures, whilst the membrane potential also recovers. But to reveal it, cultures have to be reloaded with rd123. Since both ethidium and rd123 are cationic lipophilic molecules with comparable steric hindrances, this observation adds further support to the possibility that abolishing the electrochemical gradient triggers a structural change in mtDNA.

3.1
Abolishing the Electrochemical Gradient Modifies the Mitochondrial Matrix

We have discussed the possibility that a global structural change in mtDNA could expose the chromophores of intercalated molecules to quenchers of its fluorescence (Coppey-Moisan et al. 1996). Whatever the molecular nature of this change, its occurrence is likely to be linked to modifications of the microenvironment of mtDNA. Indeed, when the electrochemical gradient is abolished, the mitochondrial matrix dehydrates intensely and undergoes shrinkage (Hackenbrock 1968). Moreover, it is known that the mitochondrial matrix, poorly hydrated in basal conditions (Srere 1981), contains proteins at a concentration high enough to constitute a modulator of the viscosity in this matrix (Scalletar et al. 1991).

3.2
The Change Can Involve Interactions Between mtDNA and the Inner Membrane

The change in mtDNA could involve such interactions. Electron microscopy observations strongly suggest that mtDNA is in contact with the inner membrane (Albring et al. 1977). On the other hand, observations made in vitro show that interactions between DNA and lipid bilayers may change the fluorescence intensity of preintercalated ethidium. This fluorescence vanishes when DNA is put in contact with cationic liposomes; it reappears when the positive charges of the liposomes are neutralized (Gershon et al. 1993). The fluorescence extinction needs physical contacts to be established between DNA and the charged lipid bilayer (Xu and Szoka 1996). For the extinction to occur, the bilayer has to possess some rigidity. Moreover, the extinction process seems to be due to a transconformation of DNA form B- towards the C-form (Akao et al. 1996).

3.3
Effects of the Change in mtDNA

When transconforming from B towards C, the double helix becomes narrower because of supertwisting and stretching (Ivanov et al. 1973; Arnott and Selsing 1975). There is no obvious steric hindrance for such a topological change to occur if DNA contains intercalated ethidium. We can thus envision that if DNA takes, let us say, the C conformation ethidium molecules remain intercalated but that their fluorescence will be quenched, because they are exposed to local quenchers. The conformational change, topologically minimal (stretching/supertwisting) could be driven (1) by a differing fluidity of the inner membrane linked (because of the phophorylation extent) to the density of positive charges, a parameter indeed critical at least in vitro on model membranes since it can drive the extinction of fluorescence of intercalated ethidium (Akao et al. 1996), (2) by the water activity in mitochondrial matrix. Indeed, structuring water governs the conformation of DNA (Berman 1991, 1994), a process directly shown to provoke its transconformation (Preisler et al. 1995). The movements of water being free across the inner mitochondrial membrane, water tends to re-equilibrate the osmotic pressure when the electrochemical gradient is abolished (Nichols and Fergusson 1992). As the bulk of water is made of structuring water in the mitochondrial matrix (references in Mantré 1996), the energetic balance of hydration could trigger a water movement involving electron carriers, the surface of the inner membrane and mtDNA. Proteins of the respiratory chain could be donors or acceptors of water, as happens for the COX protein (Kornblatt and Hui Bon Hoa 1990).

However, the precise structural change responsible for extinguishing the fluorescence of preintercalated ethidium in mtDNA remains as yet unknown. Additional data (lifetime and anisotropy decay measurements on ethidium intercalated in mtDNA) are required to learn about the molecular basis of such a change.

3.4
Perspectives

Mitochondrial $\Delta\psi$ decrease constitutes a critical event involved in programmed cell death (PCD) (for review see Mignotte and Vayssière 1998; Green and Reed 1998). On the other hand, imaging of mitochondria has shown tremendous advances in vitro and in living cells (Zhang et al. 1998; Camilleri-Bröet et al. 1998; Hüser et al. 1998). Moreover, the possibility of following variations of the membrane potential, as we have shown, constitutes a promising challenge in neuronal cells.

Our aim is therefore to characterize further the correlation between $\Delta\psi$ and mitochondrial bioenergetic variations with mtDNA changes and the implication of these parameters in apoptosis in neurodegenerative deseases. Indeed, PCD appears to constitute one key event in triggering cellular events associated with major neurodegenerative diseases such as Alzheimers, in a way which remains to be explored (Barinaga 1998).

Acknowledgments. We are deeply indebted to A. C. Brunet for her skilled technical assistance. This work was supported by a contract with Association de Recherche sur le Cancer and by an European Union contract n° Bio4CT972177.

References

Akao T, Fukumoto T, Ihara H, Ito A (1996) Conformational change in DNA induced by cationic bilayer membranes. FEBS Lett 391:215–218
Albring M, Griffith J, Attardi G (1977) Association of a protein structure of probable membrane derivation with HeLa cell mitochondrial DNA near its origin of replication. Proc Natl Acad Sci USA74:1348–1352
Arnott S, Selsing E (1975) The conformation of C-DNA. J Mol Biol 98:265–269
Barinaga M (1998) Is apoptosis key in Alzheimer's disease? Science 281:1303–1304
Berman HM (1991) Hydration of DNA. Curr Opin Struct Biol 1:423–427
Berman HM (1994) Hydration of DNA: take 2. Curr Opin Struct Biol 4:345–350
Camilleri-Bröet S, Vanderwerff H, Caldwell E, Hockenberg D (1998) Distinct alterations in mitochondrial mass and function characterize different models of apoptosis. Exp Cell Res 239:277–292
Cluzel P, Lebrun A, Heller C, Lavery R, Viovy J-L, Chatenay D, Caron F (1996) DNA: an extensible molecule. Science 271:792–794
Coppey-Moisan M, Delic J, Magdelénat H, Coppey J (1994) Principle of digital imaging microscopy. Methods Mol Cell Biol 33:359–393
Coppey-Moisan M, Brunet AC, Morais R, Coppey J (1996) Dynamical change of mitochondrial DNA induced in the living cell by perturbing the electrochemical gradient. Biophys J 71:2139–2338
Cuniberti C, Guenza M (1990) Environment-induced changes in DNA conformation as probed by ethidium bromide fluorescence. Biophys Chem 38:11–22
Delic J, Coppey J, Magdelénat H, Coppey-Moisan M (1991) Impossibility of acridine orange intercalation in nuclear DNA of the living cell. Exp Cell Res 194:147–153
Delic J, Coppey J, Ben Saada M, Magdelénat H, Coppey-Moisan M (1992) Probing the nuclear DNA in living cells with fluorescent intercalating dyes. J Cell Pharmacol 3:126–131
Ehrenberg B, Montana V, Wei MD, Wuskell JP, Loew LM (1988) Membrane potential can be determined in individual cells from the Nernstian distribution of cationic dyes. Biophys J 53:785–794
Eickbush TH, Moudrianakis EN (1978) The compaction of DNA helices into either continuous supercoils or folded-fiber rods and toroids. Cell 13:295–306
Fisher RP, Lisowsky T, Parisi MA, Clayton DA (1992) DNA wrapping and bending by a mitochondrial high mobility group-like transcriptional activator protein. J Biol Chem 267:3358–3367
Genest D, Wahl P (1978) Fluorescence anisotropy decay due to rational Brownian mition of ethidium intercalated in double stranded DNA. Biochim Biophys Acta 521:502–509
Gershon H, Ghirlando R, Guttman SB, Minsky A (1993) Mode of formation and structural features of DNA-cationic liposome complexes used for transfection. Biochemistry 32:7143–7151
Green DR, Reed JC (1998) Mitochondria and apoptosis. Science 281:1309–1312
Hackenbrock CR (1968) Ultrastrucutral bases for metabolically linked mechanical activity in mitochondria. J Cell Biol 37:345–369
Härd T, Kearns DR (1990) Reduced DNA flexibility in complexes with a type II DNA binding protein. Biochemistry 29:959–965
Hüser J, Rechenmacher CE, Blatter LA (1998) Imaging the permeability pore transition in single mitochondria. Biophys J 74:2129–2137
Ivanov VI, Minchenkova LE, Schyolkina AK, Poletayev AI (1973) Different conformations of double-stranded nucleic acid in solution as revealed by circular dichroism. Biopolymers 12:89–110
Johnson LV, Walsh ML, Bockus BJ, Chen LB (1981) Monitoring of relative mitochondrial membrane potential in living cells by fluorescence microscopy. J Cell Biol 88:526–535
Kornblatt JA, Hui Bon Hoa G (1990) A nontraditional role for water in the cytochrome c oxidase reaction. Biochemistry 29:9370–9376
Mantré P (1996) L'eau dans la cellule. Masson, Paris
McMurray CT, van Holde KE (1991) Binding of ethidium to the nucleosome core particle. I. Binding and dissociation reactions. Biochemistry 30:5631–5643

Meyer-Almes FJ, Porschke D (1993) Mechanism of intercalation into the DNA double helix by ethidium. Biochemistry 32:4246–4253

Mignotte B, Vayssière JL (1998) Mitochondria and apoptosis. Eur J Biochem 252:1–15

Monaco RR, Gardiner WC (1993) T-jump fluorescence relaxation study of the binding of ethidium cation to natural DNA. Biochem Biophys Res Commun 196:975–983

Naimushin AN, Clendenning JB, Kim U-S, Song L, Fujimoto BS, Stewart DW, Schurr MJ (1994) Effect of ethidium binding and superhelix density on the apparent supercoiling free energy and torsion constant of pBR322 DNA. Biophys Chem 52:219–226

Nicholls DG, Ferguson SJ (1992) Bioenergetics 2. Academic Press, New York

Olmsted J, Kearns DR (1977) Mechanism of ethidium bromide fluorescence enhancement on binding to nucleic acids. Biochemistry 16:3647–3654

Peck LJ, Nordheim A, Rich A, Wang JC (1982) Flipping of cloned d(pCpG)n.d(pGpC)n DNA sequences from right to left handed helical structure by salt, Co(III) or negative supercoiling. Proc Natl Acad Sci USA79:4560–4564

Pohl FM, Jovin TM (1972) Salt-induced co-operative conformational change of a synthetic DNA: equilibrium and kinetic studies with poly (dG-dC). J Mol Biol 67:375–396

Preisler RS, Chen HH, Colombo MF, Choe Y, Short BJ, Rau DC (1995) The B form to Z form transition of poly(dG-m6dC) Is sensitive to neutral solutes through an osmotic stress. Biochemistry 34:14400–14407

Scalletar BA, Abney JR, Hackenbrock CR (1991) Dynamics, structure and function are coupled in the mitochondrial matrix. Proc Natl Acad Sci USA88:8057–8061

Selvin PR, Cook DN, Pon NG, Bauer WR, Klein MP, Hearst JE (1992) Torsional rigidity of positively and negatively supercoiled DNA. Science 255:82–85

Smith SB, Cui Y, Bustamante C (1996) Overstretching B-DNA: the elastic response of individual double-stranded and single-stranded DNA molecules. Science 271:795–799

Srere PA (1981) Protein crystals as a model for mitochondrial matrix proteins. TIBS 6:4–7

Viegas-Péquignot E, Dutrillaux B, Magdelénat H, Coppey-Moisan M (1989) Mapping of single-copy DNA sequences on human chromosomes by in situ hybridization with biotinylated probes: enhancement of detection sensitivity by intensified-fluorescence digital-imaging microscopy. Proc Natl Acad Sci USA 86:582–586

Wadia JS, Chalmers-Redman ME, Ju WJH, Carlile GW, Philips JL, Fraser AD, Tatton WG (1998) Mitochondrial membrane potential and nuclear changes in apoptosis caused by serum and nerve growth factor withdrawal: time course and modivication by (-)deprenyl. J Neurosci 18:932–947

Wilson WD, Tanious FA, Barton HJ, Jones RL, Fox K, Wydra RL, Strekowski L (1990) DNA sequence dependent binding modes of 4',6-diamidino-2-phenylindole (DAPI). Biochemistry 29:8452–8461

Winzeler EA, Small EW (1991) Fluorescence anisotropy decay of ethidium bound to nucleosome core particles. Biochemistry 30:5304–5313

Zhang C, Sriratana A, Minamikawa T, Nagley P (1998) Photosensitisation properties of mitochondrially localised green fluorescent protein. Biochem Biophys Res Commun 242:390–395

CHAPTER 14

A Heteroplasmic Strain of *D. Subobscura*. An Animal Model of Mitochondrial Genome Rearrangement 14

S. Alziari, N. Petit, E. Lefai, F. Beziat, P. Lecher, S. Touraille, R. Debise, and F. Morel

Contents

1 Introduction

Although its coding ability is very limited, the mitochondrial genome bears genes responsible for products that are essential to cell biochemistry and energy balance. These genomes can undergo point or more extensive alterations (deletions or duplications) that have been correlated with serious pathologies such as MERRF, MELAS (Suomalainen 1977) and syndromes such as those of Kearns-Sayre (Lestienne and Ponsot 1988; Zeviani et al. 1988) and Pearson (Rötig et al. 1989).

In the more extensive modifications, the altered genomes are generally present in the heteroplasmic state (Holt et al. 1988; Lestienne and Ponsot 1988), and the more of them there are, the worse the symptoms.

Equipe Génome Mitochondrial, UMR 6547, Université Blaise-Pascal-Clermont II, 63177 Aubière Cédex, France

In most of the cases described, these deletions arise sporadically. The role of the nuclear genome in these modifications is not yet fully known. Zeviani et al. (1989) have described cases of multiple deletions in mitochondrial DNA where the transmission was autosomal and dominant. Recently, Suomalainen et al. (1995) identified a locus situated on the long arm of chromosome 10 in a Finnish family and Kaukonen et al. (1996) a locus situated on the short arm of chromosome 3 in three Italian families. These loci were linked genetically to multiple deletions. No gene has yet been identified in these regions. This approach by family study is important and highly informative. However, it is cumbersome to implement, and limited in scope.

A mutant strain of the filamentous fungus *Posospora anserina* was one of the first models studied. Belcour et al. (1991) and Dequard-Chablat and Sellem (1994) have shown that its mitochondrial genomes can undergo extensive rearrangements associated with a premature death syndrome (see Chap. 15).

The mutant strain of the fly *Drosophila subobscura* that we are studying displays a stable extensive rearragement of part of its mitochondrial genomes; most (80%) have lost more than 30% of their coding zone by deletion. This mutation is transmitted identically and at the same level, to progeny. Also, this strain displays no obvious phenotypes.

This animal model should allow a fine analysis of several fundamental points such as the biochemical and molecular consequences of this deletion, the respective roles of the two genomes, mitochondrial and nuclear, in the maintenance of this mutation and its transmission to progeny, and finally the characterization of nuclear genes which may be involved in the mechanisms of rearrangement.

2
The Animal Model

2.1
Presentation of the Mutant

The mutant strain of *D. subobscura* was isolated from the wild from a single female (Volz-Lingenhöhl et al. 1992). It possesses two populations of mitochondrial genomes; a minor one (approximately 20%) of the same size as that found in the wild strain (15.9 kb) and a major one (approximately 80%), which has undergone a deletion of 5 kb, i.e., more than 30% of its coding zone (Fig. 1).

Ten genes are affected by the mutation:

1. Five genes coding for subunits of the respiratory complex NADH-ubiquinone oxidoreductase (ND1, ND4, ND4L, ND5, ND6).
2. One gene coding for the single subunit of mitochondrial origin in complex III, ubiquinone-cytochrome c oxidoreductase (Cyt b).
3. Four genes coding for tRNA (Pro, His, Thr, Ser), their sole representation in this genome.

Genes ND1 and ND5 are fused by the deletion. The two segments are not in the "reading frame", however, and the frame shift gives rise to "stop" codons at the first nucleotides in the remaining part of gene ND5. Short repeated sequences of six nucleotides

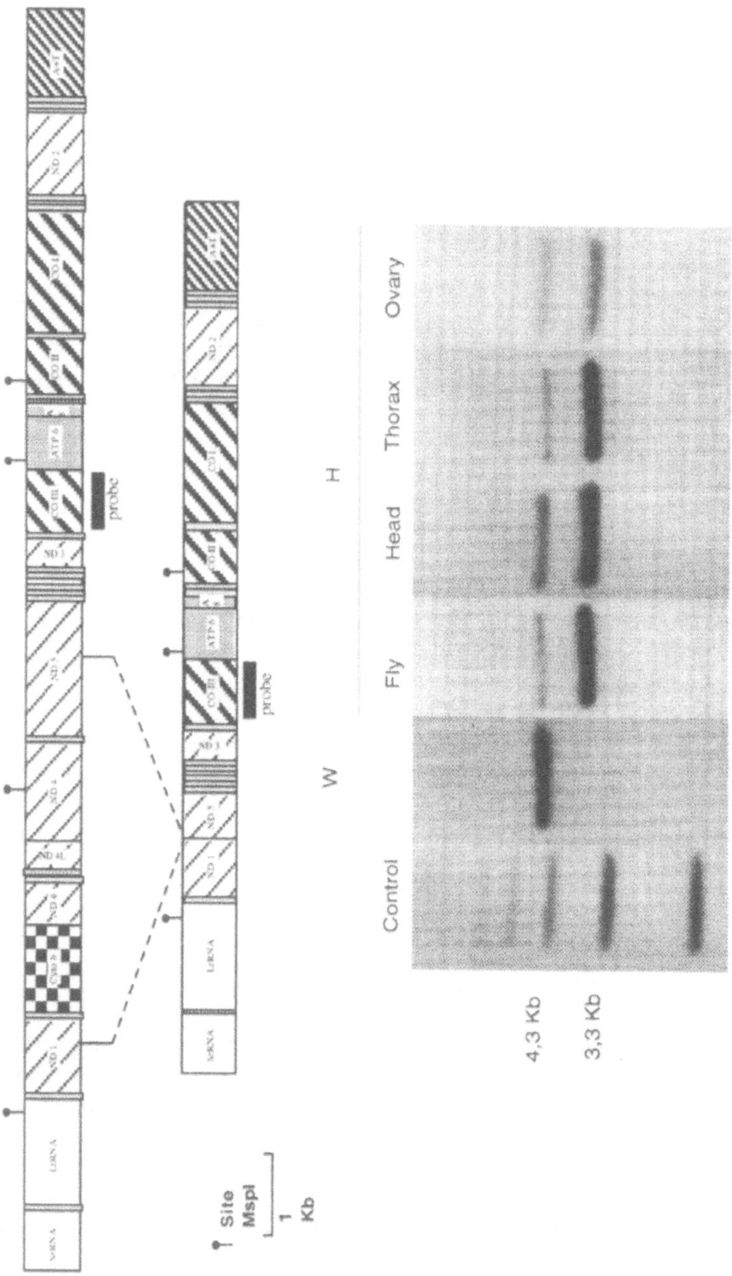

Fig. 1. Study of heteroplasmy by hybridization with COIII probe. The *upper line* represents the linearized organization of the wild-type mtDNA of *D. melanogaster*, and the deleted genome shown below causes a fusion of genes ND1–ND5. The total DNA extracted from flies of the wild strain, *W*, the mutant strain *H* or the different fractions are cut with Msp1, analysed on gel and hybridized with the COIII probe. The band at 4.3 kb is characteristic of the intact mtDNA, the band at 3.3 kb of the deleted mtDNA. The ratio of hybridization signals indicates the percentage of deleted mtDNA in the mutant strain; flies (79 ± 5%), heads (74 ± 5%), thorax (79 ± 6%), ovaries (63 ± 5%)

(perfectly repeated) or 11 nucleotides (imperfectly repeated) were found in the regions bordering the deletion (Volz-Lingenhöhl et al. 1992).

The results of cleavage of the mitochondrial genomes using various restriction enzymes, which cleave the inact or deleted genomes only once, or by analysis by electrophoresis of native DNA extracted from wild or mutant strains and hybridized with specific probes (Petit et al. 1998) ruled out duplicaiton of intact and deleted genomes. The two types of molecule are physically independent.

This strain has therefore undergone a massive loss of genes coding for subunits of enzyme complexes that play an essential role in cell energy balance, or genes necessary for the expression of this genome (tRNA). Mitochondria containing only deleted genomes will have a nonfunctional respiratory chain, and will not support protein synthesis. Such mitochondria must therefore display marked modifications.

Nevertheless, this mutant does not appear to be impaired in reproduction (number of eggs, hatch rate), duration of growth stages or flying ability. The enzyme activities of the complexes of the respiratory chain diminish with ageing in the wild strain (Morel et al. 1995). The lifetime and time course of these activities are the same in the mutant strain. In addition, the mutation has been inherited, identically and with the same heteroplasmic ration, by over 80 generations of progeny so far.

These unusual properties mark out this mutant strain from similar cases of deletion encountered in human pathology. As a result, it exhibits three interesting features:

1. Unlike the cases described in humans, this strain has had to develop mechanisms to offset the molecular and biochemical consequences of its mutation. Clearly, knowledge of these mechanisms is important.
2. This model is apparently not pathological. However, since the mutant has survived and passes on its mutation to further generations, it offers an opportunity to study the mechanisms responsible for alterations of mitochondrial genomes (if still present) and their maintenance at a constant level.
3. It enables us to evaluate the respective roles of the mitochondrial and nuclear genomes in these phenomena and identify the genes that might be implicated.

The first investigations concerned the characterisation of the mutant strain, compared with the different wild strains of *D. subobscura*.

2.2 Heteroplasmic Distribution and Mitochondrial Genome Cell Contents (Béziat et al. 1993, 1997)

The distribution of the different types of genome in different mitochondria-enriched fractions from different tissues was studied by southern blot (Fig. 1). In the fractions enriched in somatic tissues (thorax, head), the proportion of deleted genomes was about 80% (Fig. 1). In contrast, in female or male abdomens, or in the ovaries, this proportion was significantly lower (60 to 70%). In stage 14 oocytes this heteroplasmy was 60%.

The mtDNA cell contents (per nuclear genome) of mutant and wild strains were estimated by hybridization (slot blot) of the "total" DNA of nuclear (18s) or mitochondrial (12s) probes. Comparison of hybridization signal ratios indicated that these contents were increased by 50% in all the reactions studied (including the tissues containing germ cells). Hence, compared with the wild strain, there is a limited increase in cell concentrations of mitochondrial genes. This may be a first compensation step, though it cannot totally make up for the massive gene loss.

2.3
Concentration Measurements for the Different Transcripts (Béziat et al. 1993, 1997)

Steady-state concentrations were estimated by northern blot after extraction of "total" RNA and hybridization with fragments of the genes affected or not by deletion. The hybridization signals were measured by comparison with the signals obtained with the 18s probe. The rations of the signals of mutant and wild strains for each type of transcript were then calculated. The quantificaiton of the hybridization signals by image analysis indicates that these concentrations are identical for the transcripts of genes unaffected by deletion (12s, COIII, COI): They are lowered for the transcripts of genes affected by deletion. However, these reductions are less marked than expected (approximately 70%, allowing for the increase in cell mitochondrial genomes). Also, they differ according to the transcripts (Cyt b = 35%, ND1 = 45%, ND4-ND4L = 45%, ND5 = 65%). These results were obtained in the different tissues.

The transcript/gene ratios are thus increased in the mutant strain for the genes affected by the deletion, and reduced for the genes not affected. Hence a second level of compensation is operative, probably by modifications to the half life of the transcripts.

A new transcript can be observed in all the tissues of the mutant strain. It is revealed by probes overlapping the deletion. It has the expected size for the ND1-ND5 fusion transcript. The presence of this transcript indicates that the deleted genomes are transcribed. The concentration of this new transcript, estimated by northern blot is low, however (equivalent to that of ND1); either these deleted genomes are weakly expressed, implying a mechanism of selective transcription, or the half-life of this transcript is short.

2.4
Mitochondrial Biochemistry and ATP Synthesis Capacity (Debise et al. 1993; Béziat et al. 1997)

Two respiratory complexes, NADH dehydrogenase and cytochrome C reductase are directly affected by the mutation. Also, four tRNAs are involved, which may lead to severe impairment of mitochondrial protein synthesis, and the biochemical activities of complexes not directly affected by the mutation may also be adversely affected.

The measurements carried out on isolated mitochondria of the two strains showed that cytochrome oxidase (complex IV), not affected by the deletion, displays a maximal activity equivalent to that measured in the wild strain. Those of complexes I and III were diminished by 40% and 20% respectively (Fig. 2A).

Oxygraphic measurements of the activity of the respiratory chain (Fig. 2B) show that in the presence of substrates that can be oxidized by complex I (glutamate-malate), the respiratory activity of mitochondria isolated from the mutant strain was always reduced by about 30%. On the other hand, it was identical in the two strains with substrates able to supply electrons to complex III (via ubiquinone), such as α-glycerophosphate. The residual activity of complex III in the mutant strain is therefore apparently sufficient to ensure normal operation of the respiratory chain from this complex.

Direct measurements of the capacity of isolated mitochondria for ATP synthesis were carried out by bioluminescence using the luciferin-luciferase system. The results

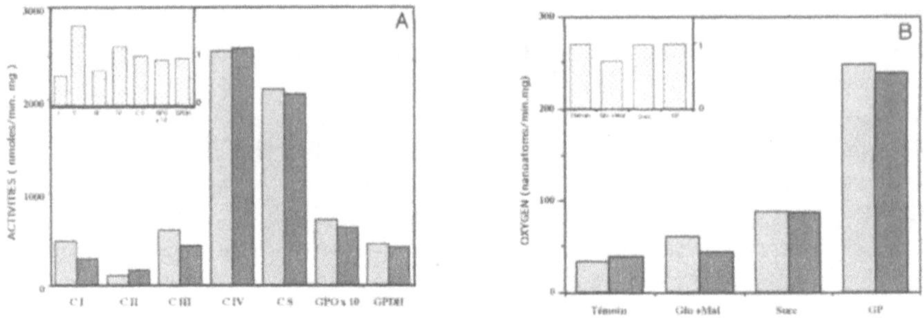

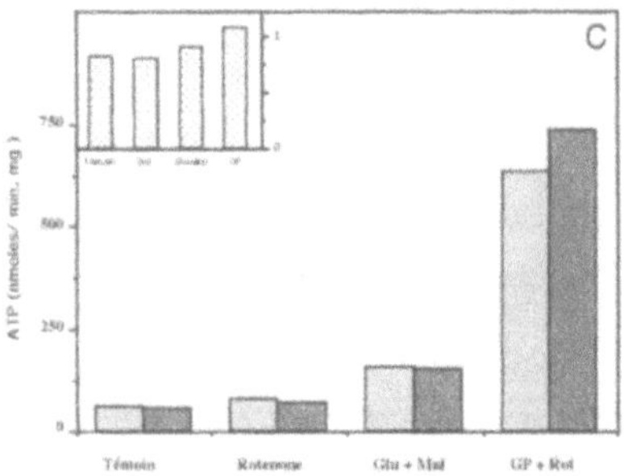

Fig. 2. **A** Enzymatic activities of complexes *I, II, III, IV,* citrate synthase *(CS),* glycerophosphate oxidase *(GPO)* and glycerophosphate deshydrogenase *(GPDH).* **B** Oxygen reduction by isolated mitochondria incubated with different substrates: *Glu + Mal* glutamate + malate; *Succ* succinate; *GP* alpha-glycerophosphate. **C** ATP synthesis b isolated mitochondria incubated with different substrates: *Glu + Mal* glutamate + malate; *GP* glycerophosphate. ( ) Wild type; ( ) mutant type. *Inserts* show the activity ratios H/W

(Fig. 2C) show that these synthesis capacities were identical in the two strains irrespective of the substrate, including those specific to complex I. Identical results were obtained with the different fractions studied. In the thorax fraction, however, with some substrates, such as proline dehydrogenase, ATP synthesis may even be stronger in mitochondria of the mutant strain.

These results indicate that there is no energy deficit in the mutant strain. The operation of the respiratory chain, even if it is diminished from complex I, thus maintains cell energy balance. These important findings may account for the apparent harmlessness of the mutation.

2.5
Ultractructural Studies and In Situ Hybridization (Lecher et al. 1994, 1996)

Electron microscopy showed that unlike the observations commonly made in human pathology, the structure of the mitochondria in different tissues of the mutant strain was undamaged. There was also no accumulation of mitochondria in the subsarcolemmal regions of the muscles.

The in situ detection of the cytochrome oxidase activity by cytochemistry shows also that there are no mitochondria that have lost this activity (no COX-mitochondria). Protein synthesis therefore takes place in all the mitochondria of the mutant strain. The two types of genome therefore coesixt in each mitochondrion.

This interpretation was confirmed by in situ hybridization experiments. The presence of transcripts of Cyt b or ND4 genes (both affected by the mutation) was detected in all the mitochondria of the intestinal epithelial cells.

Although intact and delected genomes coexist in all the mitochondria, there is nevertheless apparently no functional dominance of any one type of genome, as postulated in human studies (Shoubridge et al. 1990).

2.6
Evolution of Heteroplasmy During Growth (Petit et al. 1998)

In stage 14 oocytes, the representation of deleted genomes is significantly weaker than in the adult fly or in the somatic tissues. This parameter therefore evolves during growth. Measurements made from early embryos to the emergence of the imago have shown that this increase takes place at the larval stage; the level for type 1 larvae was 63% and type 3 larvae displayed the highest level (80%), which remained constant thereafter until the death of the fly. This increase is therefore specific to a particular growth stage.

The evolution of the mitochondrial genome contents of the cell (relative to the nuclear genomes) was positive from larvae 1 in both wild and mutant strains (in which it is always higher, as already reported for different tissues). At the larval stage, nuclear replications are very abundant. This stage therefore corresponds to a time of abundant replication of mitochondrial genomes. The percentage of deleted mitochondrial genomes increases, therefore, during strong replication of mtDNA, and then stabilizes. The mitochondrial replicative apparatus may be involved in these variations, either indirectly (preferential replication of mitochondrial genomes) or directly (implication of an element of the replicative apparatus in the production of deleted genomes).

3
Identification of the Genomes Involved

This mutant strain is thus uniquely able to compensate for the consequences of a mutation that in humans is seriously disabling at such a high level of heteroplasmy. Although it is probably not an animal model of a Kearns-Sayre type pathology, it nevertheless possesses a large number of features that make it a choice model for the study of mtDNA rearrangements:

1. It is easily raised in laboratory conditions identical to wild strains.
2. Its generation time is short (about 3 weeks).
3. The availability of numerous wild strains allows biochemical and molecular comparisons with the mutant strain.
4. The different growth stages from the embryo stage onwards are accessible for experimentation.
5. Lastly, a broad range of highly informative approaches has been developed for *Drosophila.*

Except for ribosomal and transfer RNAs, all the elements necessary for replication, expression and segregation of mitochondrial genomes are coded by the nuclear genome. This is therefore very probably implicated in maintaining the deleted mitochondrial genomes at a very high level from larval stage 3 and (if they are still active) in the mechanisms of production of delted genomes.

3.1
Back-Crosses

The role of the nuclear genome is shown by reciprocal interstrain crossbreeding: a female of one strain is crossed with a male of another strain. The females obtained are then corssed again with the males of this other strain. In a few generations, the nuclear context will be modified. The mitochondria, supplied essentially by the founding gemale, will thus be placed in different nuclear contexts.

3.1.1
Mutant Female Crossed with Wild Male

When the mitochondria of the mutant strain are placed in a wild nuclear context, the heteroplasmy, initially 80%, falls in five to seven generations to 50 to 60%, and then the level remains stable in the next generations (same after 25 generations). The "return" to a mutant nuclear context by crossbreeding with mutant males for several generations brings about a renewed increase in heteroplasmy (to be published).

Thus, the nuclear genome plays a role in maintaining the level of heteroplasmy. This role is probably complex, because the deleted molecules are not completely eliminated (stabilization at around 50%). Total elimination may need more generations (slow dilution of deleted genomes replicated as intact genomes).

3.1.2
Wild Female Crossed with Mutant Male

In this cross, the mitochondria of the mutant strain, which contain no deleted mitochondrial genomes, are placed in a mutant nuclear context. The presence of very low levels of deleted genomes is observed, but these genomes require PCR and hybridization to be identified. Their quantification and any increase in their concentrations in successive generations is difficult to establish, but they do not seem to be inherited through the male line.

Although further research is necessary in this area, these results indicate that a still active mechanism, but with a very low efficiency, may generate deleted molecules, increasing the population of these molecules slowly and gradually. However, it is also possible that the mitochondrial genomes of the mutant strain bear special sequences that favour deletion, as shown in several cases (Brockington et al. 1993; Torroni et al. 1994). Sequencing of certain regions of intact mitochondrial genomes of the mutant strain and of several wild strains is currently in progress.

3.2 Microinjections of Mitochondria into Embryos

Microinjections of cytoplasm from "donor" embryos into "receiver" embryos have been described; they mainly concern *D. melangoaster* (Santamaria 1987; De Stordeur et al. 1989; Niki et al. 1989). The cytoplasm, containing mitochondria of early embryos of a "donor" strain, is first harvested microscopically and then rapidly injected into embryos of the "receiver" strain, before the cell formation stage, in the apical region where the germ cells differentiate. The resulting "founding" females are then crossbred with males of the same strain. This delicate method, which sharply modifies the nuclear contexts of the injected mitochondria, has been developed in our laboratory. Mitochondria of the mutant strain were injected into embryos of different wild strains, and vice versa. Analysis of the genomes in the different lines obained from the founding females revealed the presence of injected mitochondria in some of them. Their maintenance in these lines and the structure of their genomes are being studied.

Mitochondria of the mutant strain were also injected into embryos of *D. melanogaster*. Here again, we detected the presence of deleted mtDNA from *D. subobscura* in the lines from founding females. The isolation of these lines is important because this enables us to use the nomerous genetic tools that exist for this species, which is the classical model for geneticists.

3.3 Classical Genetic Tools

Certain tools developed by geneticists may be applied to the mutant strain. For the species *D. subobscura*, more than 300 loci have been identified (Krimbas 1993), distributed over the six pairs of chromosomes. Of these, 175 are assigned to linking groups, and 137 are directly observable. It should therefore be possible to characterize the chromosome(s) bearing the relevant genes. However, the problem is compounded by the absence of any obvious phenotype. These tools are far fewer in number than those available for *D. melanogaster*, for which several thousand mutant strains are known. Hence, the lines of *D. melanogaster* described above containing mitochondria from mutant *D. subobscura* will help in the development of this approach, which although lenghty and complex, should prove powerful.

3.4 Identification of Potentially Implicated Nuclear Genes

The identification of genes can be more narrowly targeted, particularly for those that may be implicated in the mechanisms of gene deletion. Increased heteroplasmy at the larval stage, where a strong replication of mitochondrial genomes occurs, indicates that if the process is still active, the genes coding for the proteins of the mitochondrial replisome, or involved in post-replicative rearrangements, are good candidates:

1. Mitochondrial DNA polymerase has been characterized in *Drosophila* by L. Kaguni and coworkers (Williams et al. 1993; Williams and Kaguni 1995). Analysis of the

implication of this first gene is in progress, jointly with the team of R. Garesse at the University of Madrid [Spain]. The genes of the different strains are being cloned in this lab. Their sequences and expressions will be compared.
2. The second "candidate" gene codes for single strand mtDNA linking proteins (mtSSB). The cDNA has been isolated and sequenced in *D. melanogaster* (Stroumbakis et al. 1994). Its expression may depend on the growth stage (Schultz et al. 1998). Hence it is a good potential candidate. We are studying its possible role. Probes specific to *D. subobscura* have been constructed and gene is being cloned and sequenced.

Finally, rat mitochondrial topoisomerase has been characterised by Fairfield et al. (1985). The sequence of this gene can serve as a basis for the isolation of the equivalent gene in *Drosophila.*

To test the possible role of these genes, two types of tool can be used; cell transfection and transgenesis.

Cell lines from homogenates of embryo cells of mutant and wild strains are being established. Cell lines from mutant embryos always display the same degree of deletion involving 80% of the mtDNA. Once they are firmly established, these cells (together with wild type cells) will be transfected with cDNA of potential genes. The consequences of their overexpression on the structures of the mitochondrial genomes will then be analysed.

Transgenesis of *Drosophila* will be carrid out using the transposable element P (Paricio et al. 1994). The cDNA to be tested will thus be introduced into the strain of *D. melanogaster,* in which such a system is commonly used. This method of transgenesis would also be used on strains of *D. subobscura.*

4
Conclusion

This *Drosophila* model should first of all enable us to identify the nuclear genes likely to be involved in the rearragement and the control of the integrity of mitochondrial genomes. Equivalent genes very probably occur in humans and may be implicated in the deletions that correlate with pathologies of the Kearns-Sayre or Pearson types, or in multiple deletions. It should be possible to identify them using *Drosophila* probes. The work being conducted on this model can thus contribute to research on human pathologies that involve mitochondria.

The different biological tools developed to study this model should also enable us to identify other genes that play key roles in the maintenance of heteroplasmy, in the segregation of the two types of genome in the mitochondria, and in the increase in the mtDNA cell content. These genes are certainly implicated in the control, by the nuclear genome, of replication and the expression of the mitochondrial genome, and thereby mitochondrial biogenesis. Little is known about these control and intracellular exchange phenomena. Also, as proposed by Clayton (1992), it is likely that some pathologies can be correlated with dysfunction of one or more of these processes.

This animal model can accordingly help improve our knowledge in these areas.

Acknowledgements. This work is supported by grants from the CNRS, the Université Blaise-Pascal-Clermont II, the Association Française contre les Myopathies (AFM) and the EU Contract CHRX 940491 "Mitochondrial Biogenesis in Development and Disease".

References

Belcour L, Begel O, Picard M (1991) A site-specific deletion in mitochondrial DNA of *Podospora* is under the control of nuclear genes. Proc Natl Acad Sci USA 88:3579–3583

Béziat F, Morel F, Volz-Lingenhol A, Saint-Paul N, Alziari S (1993) Mitochondrial genome expression in a mutant strain of *D. subobscura*, an amimal model for large-scale mtDNA deletion. Nucleic Acids Res 21:387–392

Béziat F, Touraille S, Debise R, Morel F, Petit N, Lécher P, Alziari S (1997) Biochemical and molecular consequences of massive mitochondrial gene loss in different tisues of a mutant strain of *D. subobscura*. J. Biol Chem 272:22583–22590

Brockington M, Sweeney MG, Hammans SR, Morgan-Hughes JA, Harding AE (1993) A tandem duplication in the D-loop of human mitochondrial DNA is associated with deletions in mitochondrial myopathies. Nat Genet 4:67–71

Clayton DA (1992) Structure and function of the mitochondrial genome. J Inherit Metab Dis 15:439–447

De Stordeur E, Solignac M, Monnerot M, Mounolou JC (1989) The generation of transplasmic *Drosophila simulans* by cytoplasmic injection: effects of segregation and selection on the perpetuation of mitochondrial DNA heteroplasmy. Mol Gen Genet 220:127–132

Debise R, Touraille S, Durand R, Alziari S (1993) Biochemical consequences of a large deletion in the mitochondrial genome of a *Drosophila subobscura* strain. Biochem Biophys Res Commun 196:355–362

Dequard-Chablat M, Sellem CH (1994) The S12 ribosomal protein of *Podospora anserina* belongs to the S19 bacterial family and controls the mitochondrial genome integrity through cytoplasmic translation. J Biol Chem 269:14951–14956

Fairfield FR, Bauer WR, Simpson MV (1985) Studies on mitochondrial type I topoisomerase and on its function. Biochim Biophys Acta 824:45–57

Holt IJ, Harding AE, Morgan-Hugues JA (1988) Deletion of muscle mitochondrial DNA in patients with mitochondrial myopathies. Nature 331:717–719

Kaukonen JA, Amati P, Suomalainen A, Rotig A, Piscaglia MG, Salvi F, Weissenbach J, Fratta G, Comi G, Peltonen L, Zeviani M (1996) An autosomal locus predisposing to multiple deletions of mtDNA on chromosome 3p. Am J Hum Genet 58:763–769

Krimbas CB (1993) *Drosophila subobscura*. Kovac, Hamburg

Lecher P, Béziat F, Alziari S (1994) Tissular distribution of heteroplasmy and ultrastructural studies of mitochondria from a *Drosophila subobscura* mitochondrial deletion mutant. Biol Cell 80:25–33

Lecher P, Petit N, Beziat F, Alziari S (1996) Localization by ultrastructural in situ hybridization of mitochondrial transcriptsin epithelial cells of a *Drosophila subobscura* deletion mutant. Eur J Cell Biol 71:42427

Lestienne P, Ponsot G (1988) Kearns-Sayre syndrome with muscle mitochondrial DNA deletion. Lancet 1:885

Morel F, Mazet F, Touraille S, Alziari S (1995) Changes in the respiratory chain complexes activities and in the mitochondrial DNA content during ageing in *D. subobscura*. Mech Ageing Dev 84:171–181

Niki Y, Chigusa SI, Matsuura ET (1989) Complete replacement of mitochondrial DNA in *Drosophila*. Nature 341:551–552

Paricio N, Martinezsebastian MJ, Defrutos R (1994) A heterochromatic P sequence in the *D. subobscura* genome. Genetica 92:177–186

Petit N, Touraille S, Debise R, Morel F, Renoux M, Lecher P, Alziari S (1998) Developmental changes in heteroplasmy level and mitochondrial gene expression in a *Drosophila suboscura* mitochondrial deletion mutant. Curr Genet (in press)

Rötig A, Colonna M, Bonnefont JP, Blanche S, Fischer A, Saudubray JM, Munnich A (1989) Mitochondrial DNA deletion in Pearson's marrow/pancreas syndrome. Lancet 1:902–903

Santamaria P (1987) Injecting eggs. In: Roberts DB (ed) *Drosophila:* a practical approach. IRL Press, Oxford, pp 159–173

Schultz RA, Swoap ST, McDaniel LD, Zhang B, Colin Koon E, Garry DJ, Kang L, Williams RS (1998) Differential expression of mitochondrial DNA replication factors in mammalian tissues. J Biol Chem 273:3447–3451

Shoubridge EA, Karpati G, Hastings KE (1990) Deletion mutants are functionally dominant over wild-type mitochondrial genome in skeletal muscle fiber segments in mitochondrial disease. Cell 62:43–49

Stroumbakis ND, Li Z, Tolias PP (1994) RNA- and single strand DNA-binding (SSB) proteins expressed during *Drosophila melanogaster* oogenesis: a homolog of bacterial and eukaryotic mitochondrial SSBs. Gene 143:171–177

Suomalainen A (1997) Mitochondrial DNA and disease. Ann Med 29:235–246

Suomalainen A, Kaukonen J, Amati P, Timonen R, Haltia M, Weissenbach J, Zeviani M, Somer H, Peltonen L (1995) An autosomal locus predisposing to deletions of mitochondrial DNA. Nat Genet 9:146–151

Torroni A, Lott MT, Cabell MF, Chen YS, Lavergne L, Wallace DC (1994) mtDNA and the origin of Caucasians: identification of ancient Caucasian-specific haplogroups, one of which is prone to a recurrent somatic duplication in the D-loop region. Am J Hum Genet 55:760–776

Volz-Lingenhöhl A, Solignac M, Sperlich D (1992) Stable heteroplasmy for a large-scale deletion in the coding region of *Drosophila subobscura* mitochondrial DNA. Proc Natl Acad Sci USA89:11528–11532

Williams AJ, Kaguni LS (1995) Stimulation of *Drosophila* mitochondrial DNA polymerase by single strand DNA-binding proteins. J Biol Chem 270:860–865

Williams AJ, Wernette CM, Kaguni LS (1993) Processivity of mitochondrial DNA polymerase from *Drosophila* embryos – effects of reaction conditions and enzyme purity. J Biol Chem 268:24855–24862

Zeviani M, Moraes CT, DiMauro S, Nakase H, Bonilla E, Schon EA, Rowland LP (1988) Deletion of mitochondrial DNA in Kearns-Sayre syndrome. Neurology 38:1339–1346

Zeviani M, Servidei S, Gellera C, Bertini E, DiMauro S, DiDonato S (1989) An autosomal dominant disorder with multiple deletions of mitochondrial DNA starting at the D-loop region. Nature 339:309–311

CHAPTER 15

Stability of the Mitochondrial Genome of *Podospora anserina* and Its Genetic Control

15

L. Belcour[1], A. Sainsard-Chanet[1], C. Jamet-Vierny[2], and M. Picard[2]

Contents

1 Introduction

Certain human degenerative diseases are associated with rearrangements of the mitochondrial genome, the accumulation of which may occur sporadically or be control-

[1] Centre de Génétique Moléculaire, Laboratoire Associé à l'Université Pierre et Marie Curie, Centre National de la Recherche Scientifique, 91198 Gif-sur-Yvette, France
[2] Institut de Génétique et Microbiologie (UMR 8621), Université Paris-Sud, Bât. 400, 91405 Orsay Cedex, France

led by nuclear genes (see Chap. 3, this Vol.; for review, see Larsson and Clayton 1995). These diseases thus raise fundamental questions which remain the subject of much discussion: which mechanisms generate these rearrangements and how are they transmitted during development (i.e. somatic divisions)? In the face of these complex problems, it is clear that research efforts in human medicine and genetics must be complemented and supported by work using model systems which can be more easily understood.

The filamentous fungus *Podospora anserina* is one of the best-adapted systems to fulfil this role. It is easily amenable to classical and molecular genetics. It is an obligate aerobe. The whole sequence of its mitochondrial genome is known (Cummings et al. 1990). Its vegetative system (the mycelium) is relatively simple and situates it at the borderline between unicellular eukaryotes and multicellular organisms: it shares with the latter the ability to present heteroplasmic states (cohabitation of wild-type and altered mitochondrial genomes). Its sexual reproduction (which involves true fertilization) is associated with a maternal transmission of mitochondria. Finally, and above all, *P. anserina* exhibits degenerative processes leading to death (Rizet 1953a, b; Belcour et al. 1991). These processes (senescence, premature death) are associated with the accumulation of rearranged forms of the mitochondrial genome: short sequence amplifications (senDNAs) or deletions (for review, see Griffiths 1992). In all cases, these degenerative processes are controlled by genetic and environmental parameters. Analysis of the mechanisms causing these rearrangements, and study of the nuclear genes involved in their genesis and/or transmission, constitute high-priority objectives for laboratories working in the field. The results obtained with *P. anserina* should aid our understanding of diseases associated with mitochondrial genome instability in more complex eukaryotes. It is indeed probable that at least some of the genetic factors involved in this instability have been conserved during evolution, even if the organization of mitochondrial genomes has markedly diverged.

2
The Mitochondrial Chromosome of *Podospora anserina*

In *P. anserina,* as in metazoans, mitochondrial genetic information is borne on a single, circular DNA molecule. This mitochondrial chromosome has been entirely sequenced (Cummings et al. 1990). Although its organization differs greatly from that found in humans, the mitochondrial chromosome of *P. anserina* is similar in terms of its genetic content. It encodes two ribosomal RNAs, 25 transfer RNAs (instead of 22 in humans), and the same 13 inner membrane polypeptides involved in oxydative phosphorylation (see Fig. 1). However, the size of the chromosome, which varies from one strain to another (Cummings et al. 1990), is about 100000 base pairs, and is thus six times larger than the human chromosome (Anderson et al. 1981). This difference in size is due to the presence in *P. anserina* of numerous introns (up to 36 in certain strains) and the existence of intergenic regions which can be larger than 1 kb. Certain introns add to the genetic content of this genome, since they encode proteins which exhibit reverse transcriptase and/or endonuclease activities. As in other filamentous fungi, the processes involved in the replication and expression of this genome are as yet not fully understood.

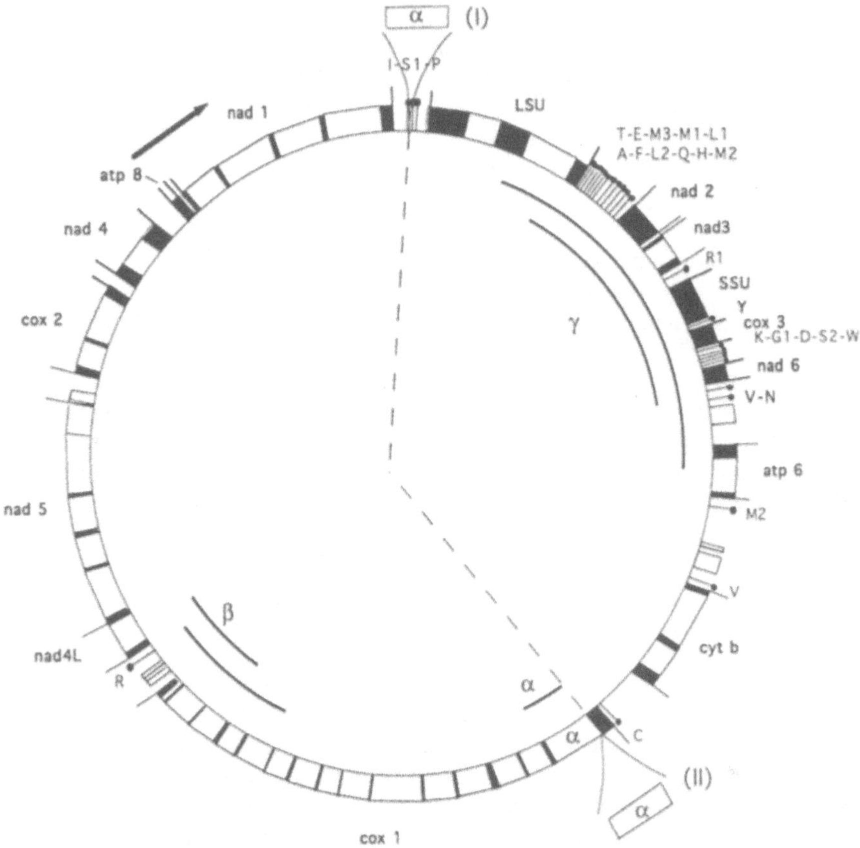

Fig. 1. Map of the mitochondrial chromosome of *Podospora anserina,* strains *s* (Cummings et al. 1990). The complete nucleotide sequence is 94 192 bp long. The exons are presented in *black*, the introns in *white*. ⚲ represents tRNAs, which are named by a *letter* corresponding to the symbol of the amino acid they are able to charge. The lines emerging from the circle delineate the genes. *LSU* and *SSU* are the genes encoding the large and the small rRNAs, respectively. The other symbols are identical to those used for the homologous human genes. The *dotted line* marks the region of the chromosome which is deleted during the premature death syndrome, α, β, γ show the regions able to give senDNAs during senescence. The arrow indicates the direction of transcription. The positions where intron α is able to transpose are indicated in (I): transposition between genes encoding $tRNA^{Ile}$ (I) and $tRNA^{Ser}$ (S1); in (II): tandem transposition

3 Senescence-Related Rearrangements of Mitochondrial DNA

The vegetative growth of this fungus, i.e. the extension of its mycelium, is always limited by a degenerative phenomenon, which was given the name "senescence" by Rizet (1953a, b).

He showed that all wild-type strains of *P. anserina* exhibit this event, although each isolate may show different kinetics and characteristics (Rizet 1953b; Marcou 1961). In *P. anserina,* senescence is associated with rearrangements of mitochondrial DNA (for review, see Dujon and Belcour 1989; Griffiths 1992). These rearrangements consists in the amplification of multimeric, circular molecules (senDNAs), of three non-overlapping DNA regions called α, β and γ (Stahl et al. 1978; Cummings et al. 1979; Jamet-Vierny et al. 1980; Belcour et al. 1981; Wright et al. 1982; Fig. 1). SenDNAα is found systematically in all senescent cultures of wild-type strains (Kück et al. 1981; Vierny et al. 1982). This results in amplification of the first intron of the *cox1* gene (called *cox1-i1* or α; Osiewacz and Esser 1984). In contrast, some nuclear mutant strains do not accumulate senDNAα but other sen DNAs during the senescence process, indicating that the accumulation of senDNAα is subject to nuclear control (Silar et al. 1997; Borghouts et al. 1997). SenDNAs β and γ appear frequently but not always in senescent cultures. Furthermore, the sequences repeated in different senDNAs β or γ do not usually have the same size or the same extremities (Koll et al. 1985; Jamet-Vierny et al. 1997a). Nevertheless, each of the two families of sequence exhibit a common region. Since senDNAs β and γ have similar properties, it is probable that the mechanisms involved in both their genesis and accumulation are similar. For this reason, of these two categories, only senDNAβ has been the subject of detailed study (Jamet-Vierny and Shechter 1994; Jamet-Vierny et al. 1997a).

3.1
Properties of Intron α and of senDNAα

Intron α is a group II intron, 2539 bp long (Osiewacz and Esser 1984; Michel and Lang 1985). Group II introns are ribozymes (catalytic RNAs) capable of self-splicing via a lariat form (for review, see Michel and Ferrat 1995). Intron α exhibits a long open reading frame (ORF) which is in the same frame as the preceding exon. The ORF corresponds to a protein related to the reverse transcriptase of retroviruses and retrotransposons (Michel and Lang 1985; Xiong and Eickbush 1990). This intronic protein displays the HNH motif characteristic of group I intron endonucleases and bacteriocines (Shub and Goodrich-Blair 1994; Gorbalenya 1994). Like other intronic proteins, the α protein is very weakly expressed. As yet, it has only been detected using specific antibodies in certain strains bearing the AS3-1 mutation which is responsible for early senescence and massive accumulation of senDNAα (Sellem et al. 1990).

3.2
Reverse Transcriptase Activity of the α Protein

Fassbender et al. (1994) demonstrated the reverse transcriptase activity of the α protein. In the yeast *Saccharomyces cerevisiae,* these authors replaced the gene encoding the reverse transcriptase of the Tyl retrotransposon with the intronic α ORF (corrected according to the universal code), and showed that the virus-like particles harvested from yeast cells transformed by this plasmid showed a reverse transcriptase activity. In addition, PCR experiments identified mitochondrial DNA molecules which were specifically deleted from their intronic sequences and thus provided indirect proof of the effective presence of reverse transcriptase activity in vivo capable of using mature RNA as a matrix in the mitochondria of *P. anserina.* It should be noted that these reverse transcription products are more abundant in mitochondria extracted from senescent cultures than in those from young cultures (Sainsard-Chanet et al. 1993). Although probable, the intronic origin of this in vivo activity has not been demonstrated and the properties of this reverse transcriptase are as yet unknown. However, it seems reasonable to suppose that this activity may play an important role in the instability of the mitochondrial genome of *P. anserina.*

3.3 Mobility of Intron α

It has been known since 1993 that certain group II introns are mobile elements capable of transposition in the mitochondrial chromosome. This mobility was initially demonstrated for intron α of *P. anserina* (Sellem et al. 1993) and intron *ai1* of *S. cerevisiae* (Mueller et al. 1993). In both cases, the mobility hypothesis is based on the study of site-specific deletions with one terminal corresponding to the 5' extremity of the intron (see paragraph below). PCR experiments showed that the intron sequence in these organisms was capable of transposing to the corresponding site at the other terminal of the deletion (location I on Fig. 1). The transposition mechanism for group II introns has not yet been eludicated. However, in vitro study of the mechanisms leading to the "homing" of these introns (insertion of the intron sequence at its corresponding exon-exon junction) has provided valuable data. The first stage of "homing" is a DNA double-strand break achieved jointly by the ribozyme and the intronic potein at the insertion site. The specificity of the break depends to a great extent on the presence of an IBS (intron binding site) sequence, corresponding to a short exon motif sited upstream of group II introns, pairing of which with a complementary intron sequence is necessary to the splicing of these introns (Zimmerly et al. 1995; Yang et al. 1996; 1998; Eskes et al. 1997; Guo et al. 1997).

It is probable that the transposition of group II introns constitutes an erroneous "homing". The ribozyme may be capable of attacking an IBS-type DNA sequence which is not located at the same site on the chromosome as the true IBS sequence. Through its transposition capability, the α sequence must therefore (like other group II introns) be considered as a mobile element, and this property probably plays an important role in the plasticity and instability of the mitochondrial genome.

3.4 Amplification of senDNAα and Senescence

While increasingly abundant data are now available concerning group II introns (for reviews, see Lambowitz and Belfort 1993; Michel and Ferrat 1995), the mechanisms ensuring amplification of the α-sequence in the form of multimeric circular molecules, and the relationships between this amplification and the senescence syndrome, are still poorly understood. DNA molecules containing the circular junction of other group II introns: *cox1-i4* of *P. anserina* (Sainsard-Chanet et al. 1994), *cox1-i1* of *S. cerevisiae* and *cytb-i1* of *S. pombe* (Schmidt et al. 1994), have been demonstrated using PCR. In contrast to the α sequence however, these molecules are present at very low levels.

The hypothesis advanced to explain the ability of group II introns to produce circular DNA molecules is based on their property of mobility. Indeed, transpositon of the intron behind its own IBS sequence at the exon-intron junction (location II on Fig. 1) probably leads to the formation of multimers of the intron sequence. Recombination between these tandem repeats may release circular DNA molecules containing the entire intron sequence, which may be in the form of multimers. Although this hypothesis provides an explanation for the genesis of circular intronic DNA molecules, it does not, at least directly, explain the amplification of senDNAα. Controlled cytoplasmic fusion experiments between a donor culture containing senDNAα and recipient culture devoid of senDNAα have demonstrated that this amplification probably did not result from autonomous replication of the α sequence (Sainsard-Chanet et al. 1994).

A hypothesis of positive autoregulation has been proposed (Belcour et al. 1994). This supposes that the transcripts containing intron multimers are inefficiently spliced so that the intron protein which is itself involved in the transposition mechanism is synthesized (from the premessenger) more abundantly. Such a phenomenon would exhibit autoregulation: the probability of transposition, and therefore of instability, would increase with the number of transposition events. Another hypothesis proposed that senDNAα amplification and the onset of senescence are triggered by the existence of an unknown factor which is probably not the α sequence itself (Silar et al. 1997).

3.5 SenDNAsβ Are Generated by Intramolecular Crossover

The DNA sequences overlapping the circularization points of 36 senDNAsβ of independent origin have been determined (Jamet-Vierny et al. 1997a). The results indicate that short, direct repeats (from 3 to 27 nucleotides) are present in all cases, either specifically at the terminals of senDNAsβ, or very close to them. A link has been described in other organisms between the presence of direct repeats and mitochondrial DNA deletions (De Zamaroczy et al. 1983; Gross et al. 1984; Mita et al. 1990), and various recombination mechanisms proposed, including intramolecular crossover or slipped mispairing (a shift during replication, due to illegitimate pairing between repeats present on one of the strands and the other repeat present on the complementary strand). However, no definite proof has been obtained in favour of a specific mechanism. Defective molecules, with a structure which is exactly complementary to that of senDNAβ, have been identified using PCR in mitochondrial DNA extracted from certain senescent cultures of *P. anserina*. These experiments demonstrated that in *P. anserina*, deleted mitochondrial DNA molecules may be generated through reciprocal exchange occurring at the level of very short repeated sequences. Futhermore, the fact that in certain cases the repetitions are imperfect makes it possible to establish that the crossover may be associated with a conversion involving several nucleotides (Jamet-Vierny et al. 1997a).

3.6
SenDNAsβ Are Able to Replicate

A simple hypothesis to explain the amplificaiton of senDNAsβ during the senescence process is to suppose that they are self replicative. Strong arguments in favour of this suggestion were provided by cytoplasmic fusion experiments (anastomosis) between a "donor" implant from a senescent culture containing a senDNAβ with known terminals, and a young "recipient" culture devoid of senDNAβ (Jamet-Vierny et al. 1997b). The mitochondrial DNA in the two cultures employed did not have an entirely identical structure, differing notably in terms of their polymorphism in the β region. Following anastomosis, molecules of senDNAβ with terminals identical to those of the "donor" senDNAβ were found in massive amounts in the recipient culture, associated with the mitochondrial DNA of the latter. Furthermore, the sequence of accumulated senDNAβ, proved to be identical, in termo of polymorphism, to that of the "donor" senDNAβ. This suggests that the accumulated senDNAβ resulted from self replication of the "donor" senDNAβ. It seems probable that the self replication ability of senDNAsβ is related to the presence of a replication origin located in their common region.

3.7
Amplification of senDNAsβ and Senescence

Only cultures already designated to become senescent (Marcou 1961) contain high levels of senDNAβ. Recent data (Jamet-Vierny et al. 1999) have indicated that (1) senDNAsβ may however be present at low levels in "young" cultures, and (2) senDNAsβ introduced by anastomosis into young cultures do not systematically replicate but in certain cases may progressively become diluted. These observations suggest that the amplification of senDNAsβ does not simply result from an intrinsic ability (due, for example, to the small size of the amplified monomer) to replicate more rapidly than the mitochondrial chromosome, but may be linked to the presence of a determinant of senesence, the nature of which is yet to be established.

SenDNAsβ are never found independently of senDNAα in wild-type senescent cultures. This suggests a cause and effect relationship between the specific properties of intron α, from which the senDNAα is derived, and the presence of senDNAsβ. This relationship is currently contradicted by a certain number of findings which suggest that (1) the genesis of senDNAsβ does not seem to be linked with activity (either reverse transcriptase or endonuclease) of the protein coded by intron α (Jamet-Vierny et al. 1997a), and (2) the amplification of senDNAsβ may occur in the absence of intron α or senDNAα (Jamet-Vierny et al. 1997b). Finally it has been shown that some mutant strains amplify mainly senDNAβ during the senescence process (Silar et al. 1997).

4
Mitochondrial Chromosome Deletions and Cellular Degenerative Processes of Mutant Strains

The cellular degeneration of certain *P. anserina* mutant strains is associated with the reproducible accumulation of defective mitochondrial genomes which differ from those present in senescent cultures of the wild-type strain.

4.1
Deletion Characteristic of the *AS1-4* Mutation

The *AS1-4* nuclear mutation, located in the gene encoding the cytosolic ribosomal protein S12, was selected for its ability to increase the accuracy of cytosolic translation (Picard-Bennoun 1976). During their cellular degeneration, *AS1-4* mutant strains systematically accumulate mitochondrial chromosomes bearing a site-specific deletion which covers approximately one third of the genome (Belcour et al. 1991; Fig. 1). However, although most abundant, the defective chromosomes always remain associated with a minority of intact chromosomes. The *AS1-4* mutation may be associated with one or other of the two alleles (*rmp* + or *rmp-*) of a naturally polymorphous gene, and in both cases, cell death is associated with the presence of the deletion. However, although the degenerative syndrome of *AS1-4 rmp-* strains occurs very soon after germination of the ascospore, and for this reason has been called "premature death", *AS1-4 rmp+* strains have a life span which may be very long, and they die in a sporadic fashion (Contamine et al. 1996).

One of the deletion endpoints corresponds exactly to the 5' end of intron α, while the other lies in an intergenic region between genes encoding $tRNA^{Ser}$ and $tRNA^{Ile}$ (S1 and I in Fig. 1). Intron α is capable of transposition at this position (Sellem et al. 1993). This transposition is considered to be the first stage in the mechanism responsible for genesis of the deletion (Fig. 1). It is probably followed by homologous recombination between the transposed and resident copies of intron α, to give rise to the defective chromosome. Recently, other deletions have been observed in certain cultures of *AS1-4 rmp-* strains (Sainsard-Chanet et al. 1998). These deletions may accumulate and replace the "classic" deletion described previously; they are then responsible for the same degenerative syndrome.

These new deletions may be placed in two categories depending on whether intron α is joined to an IBS-like site or not. It thus appears that at least two mechanisms are involved in the genesis of large-scale deletions in the mitochondrial DNA of *P. anserina*. One of them seems to be based on the transposition properties of intron α, the other one is not. A DNA double-strand break at the 3' end of the exon, upstream of inton α, has been described (Sainsard-Chanet et al. 1994). The mitochondrial and/or nuclear origin of the endonuclease responsible for this break, and its molecular consequences, are not yet fully understood. It is nevertheless tempting to suggest that this break, which is located at the junction between intron α and its upstream exon, plays a role in the genesis of deletions which accumulate in the nuclear context of *AS1-4*. Some deletions, and particularly the "classic" deletion, are present in the wild-type strain, but deleted chromosomes are always present in very small amounts and can only be detected through PCR experiments. The mechanisms responsible for the formation of these deletions are thus present in the wild-type strain, and it seems more likely that the mechanisms responsible for the accumulation of defective genomes are modified in the mutant nuclear context. This situation is reminiscent of the natural death syndrome in *N. crassa,* where a nuclear recessive mutation causes rapid rearrangements of the mtDNA (Bertrand et al. 1993) and the situation found in some human hereditary diseases associated with the accumulation of mitochondrial deletions in specific tissues (see Chap. 3).

4.2
Deletion Characteristic of the *PaMDM10-1* Mutation

Ten mutations which increase the lifespan of *AS1-4 rmp-* strains were selected. These mutations lie in six genes (Contamine and Picard 1998). One of these genes was cloned and sequenced (Jamet-Vierny et al. 1997c): it is probably the homologue of the *MDM10* gene of *S. cerevisiae*, encoding a mitochondrial outer membrane protein involved in the morphology and distribution of mitochondria (Sogo and Yaffe 1994). The gene from *P. anserina* was therefore called *PaMDM10*.

The mitochondrial DNA of senescent cultures obtained from various strains bearing the *PaMDM10-1* mutation was analyzed. This study revealed the systematic accumulation of two types of defective, circular molecules (14.8 and 20.6 kb) which generally were not amplified in senescent cultures of the wild-type strain. Nevertheless, these defective molecules were detected through PCR in young and senescent cultures of the wild-type strain. The exact structure and the breakpoints of defective molecules were determined. This showed that direct repeats were present at the ends of these two types of molecules. As in the case of senDNAsβ (see above), intramolecular crossover, occuring between two direct repeats, may be the origin of the two types of defective molecules.

5
Phenotypic and Mitochondrial Reversions of Degenerative Phenomena

In *P. anserina*, the senescent state can be cured by treatment at a low temperature (Marcou 1954) or using ethidium bromide. Mycelia which restart normal growth following culture on a medium containing ethidium bromide (10 μg/ml for several days) are "rejuvenated" (Koll et al. 1984). They have lost all traces of senDNAs and recovered the longevity characteristic of their genotype. The mechanisms through which this intercalating agent acts are not understoood. However, ethidium bromide has no effect on the premature death syndrome, whether it is applied before or after it is triggered. In contrast, two inhibitors of protein synthesis (paromomycin and kasugamycin) are capable of significantly delaying its onset (Contamine et al. 1996).

Spontaneous revertants, able to restore vegetative growth after the expression of senescence and premature death, have proved to be mitochondrial mutants. Indeed, analysis of the mitochondrial chromosome of these revertants demonstrated the constant presence of partial or total deletion of intron α, and more specifically of a region overlapping this intron and its upstream exon (Belcour and Vierny 1986; Belcour et al. 1991; Sainsard-Chanet and Begel 1990). Consequently, these revertants lack both the intron properties and also cytochrome oxidase activity. They use an alternative pathway for respiring. We have also shown that mitochondrial point mutations are capable of causing considerable modifications to the life span of the fungus, either lengthening or shortening it (Belcour and Begel 1978, 1980).

6 Nuclear Control of Longevity

The results presented above suggest that mitochondria play a key role in degenerative phenomena of *P. anserina*. However, like mitochondrial biogenesis and functions, these processes are under the control of the nuclear genome. As an example, Schatz (1995) estimated that approximately 1000 proteins encoded by the nucleus are imported into the mitochondria, while only 13 are encoded and synthesized in the mitochondrion itself. In *S. cerevisiae*, the detection and systematic study of nuclear mutations confering respiratory deficiency (*PET* genes) gave rise to the estimation that these genes numbered several hundreds (Séraphin et al. 1987; Tzagoloff and Dieckmann 1990). In particular, all the proteins which ensure the replication and expression of the mitochondrial genome, and the import of proteins into mitochondria, are encoded by the nucleus. The latter may thus substantially influence the stability of the mitochondrial chromosome and as a consequence the incidence and quality of its rearrangements. The study of mitochondrial genome instability and of the transmission of some of its "variants" (to the detriment of the wild-type genome) constitutes a considerable challenge, as this is certainly the most complex multifactorial system which may be encountered in an eukaryotic cell.

In *P. anserina*, the importance of the nuclear genome to the triggering and expression of syndromes linked to mitochondrial genome rearrangements has become progressively clearer thanks to four complementary approaches (none of which is intrinsically original, as they have all been used to study other functions in various organisms): fortuitous observations, systematic research without preconceived ideas, studies concerning the impact of general cell function, and finally, targetted research efforts.

6.1 Fortuitous Observations

Once the phenomenon of senescence had been observed, Rizet (1953b) noted that in the geographical isolate used as a reference, the syndrome appeared earlier in a *mat-* mating-type strain than in a *mat+* strain. Marcou (1961) broadened this observation to other geographical isolates, without generalizing it; indeed, in certain isolates, strains exhibit the same longevity regardless of their mating type. We now know that a naturally polymorphous gene, tightly linked to the *mat* locus (but distinct from the *mat* genes) is responsible for this difference in the timing of senescence in certain strains of *P. anserina* (Coppin et al. 1993). Furthermore, it should be noted that this difference is markedly dependent upon the culture medium (Contamine et al. 1996; Silar and Picard 1994).

Similarly, the genetic parameters which control premature death were discovered by chance. During systematic analysis of mutations affecting the accuracy of cytosolic translation, it was seen that strains bearing the *AS1-4* mutation died very early if they were of the *mat-* mating type. Analysis of the mitochondrial genome of these strains revealed that their "premature" death was associated with the accumulation of a site-specific deletion of mtDNA (see Sect. 4; Belcour et al. 1991). Multidisciplinary analysis helped to discover the roles of these two genetic parameters (the *AS1-4* mutation and the *mat-* haplotype) in triggering premature death.

The *AS1-4* mutation lies in a gene encoding a cytosolic ribosomal protein (S12) (Dequard-Chablat et al. 1986). This protein has been sufficiently conserved during evolution to ensure that the human protein (S15) can be substituted for the *Podospora* protein (Dequard-Chablat and Rötig 1997). Since the S12 protein could not be detected in mitochondria (Dequard-Chablat and Sellem 1994) it is reasonable to suppose that the *AS1-4* mutation indirectly controls (via cytosolic translation) the massive genesis and/or selection of the defective mitochondrial genome which characterizes premature death. This conclusion has encouraged the search for genes directly involved in the syndrome (see Sect. 6.4).

Different approaches have helped to clarify the role of the *mat-* halotype in triggering premature death. Culture conditions play an important role in expression of the syndrome. It has thus been possible to save certain *AS1-4 mat-* cultures and also to observe that *AS1-4 mat+* strains (until then considered to be immortal) could exhibit the syndrome, in the sense that their death was correlated with the accumulation of the defective mitochondrial genome characteristic of premature death (Contamine et al. 1996). The gene associated with the *mat* haplotype is distinct from the *mat* genes themselves and is naturally polymorphous. This gene has been called *rmp* (a French acronym for **r**égulateur de la **m**ort **p**rématurée), and its two alleles, *rmp+* and *rmp-* (linked to *mat+* and *mat-*, respectively). Genetic data strongly suggest that *rmp-* is functional, while *rmp+* carries a nonsense mutation. Natural suppression of this mutation probably gives rise to the cases of premature death observed sporadically in *AS1-4 rmp+ (mat+)* strains. In the same way, situations where *AS1-4 rmp- (mat-)* strains escape from the syndrome, may be explained by poor expression of either the *rmp-* gene or another gene required for expression of this syndrome (Contamine et al. 1996). The next step will be to clone the *rmp* gene. This cloning will also shed light on whether *rmp* controls the kinetics of expression of both premature death and senescence (Rizet 1953b; Marcou 1961) or whether two linked genes are involved. The indirect data so far available have not enabled a clear choice between these two hypotheses (Coppin et al. 1993; Contamine et al. 1996).

6.2
Systematic Research Without Preconceived Ideas

Once the role played by the *mat* haplotype in senescence expression had been discovered, certain laboratories tested for the possible effect of the few nuclear mutations available at the time. Initially, these mutations affected the morphology of the vegetative system (Marcou 1961; Esser and Keller 1976; Tudzynski and Esser 1979). Most of these mutations delayed the senescence process (often discretely but nonetheless significantly). In contrast, certain double-mutants could exhibit a spectacular increase in longevity (up to 60 times that of the reference strain). One of these genes, *grisea*, has recently been cloned. It probably encodes a transcription factor which is activated by copper (Osiewacz and Nuber 1996; Borghouts et al. 1997; Borghouts and Osiewacz 1998). Thereafter, mutations affecting the sexual development of the fungus (A. Raynal, pers. comm.) or the cytosolic translation apparatus (Picard-Bennoun 1985; Belcour et al. 1991) were tested. In the first instance, there was also a trend towards increased longevity, while in the second, the two types of effect (increase or decrease in longevity) could be observed. These attempts showed that numerous genes, without a clear relationship with mitochondria, could affect the longevity of *P. anserina* (see also Rossignol and Silar 1996). However, these opportunistic approaches only provided a biased description of the situation, the bias being due to the fact that the mutations analyzed were those available in the laboratories at the time, which naturally reflected the particular scientific interests of the research team.

More recently, a systematic study was undertaken. It consisted of estimating the number of nuclear genes which could accelerate or delay the senescence process, without any preconceived ideas concerning the function of these genes (Rossignol and Silar 1996). The experiment was based on the observation that in *P. anserina* (as in other filamentous fungi) exogenous DNA integrates into the genome in a more or less random fashion (Brygoo and Debuchy 1985; Rossignol and Picard 1991). This phenomenon is probably accentuated when the transgenic DNA does not present any significant homology with the genome of the fungus. Furthermore, it can be assumed that this integration can be mutagenic. The experiment thus consisted in transfecting protoplasts from a wild-type strain with a plasmid bearing a foreign selection gene (conferring resistance to hygromycin). Transformants were harvested and this enabled phenotypic analysis of 159 insertion points of the plasmid. Longevity was significantly modified in nine cases. In order to estimate the percentage of nuclear genes which might affect longevity, the nine revertant integrations were compared with the corrected number of integration events. This correction had to take account of two parameters: an estimation (taken from other fungi) that coding regions represent 70% of the genome, and the fact that 85% of genes are nonessential to survival and could thus have been altered during this experiment. The conclusin was that nearly 10% of genes ($9/[159 \times 0.7 \times 0.85]$), i.e. several hundreds, play a more or less direct role in *P. anserina* longevity (Rossignol and Silar 1996). This value might seem very high, but it should be borne in mind that several hundred proteins, encoded by the nuclear genome, must enter the mitochondrial compartment to ensure its correct functioning.

6.3 Impact of Cytosolic Translation

We have already considered the major role played by the translational apparatus in premature death. It was therefore reasonable to question its effect on senescence. A preliminary analysis of some forty mutations affecting translation did in fact reveal that most of them had an effect (which was sometimes spectacular) on the timing of senescence, without the nature of the mitochondrial events being analyzed (Picard-Bennoun 1985; Belcour et al. 1991). This problem was recently revisited in detail by cloning the relevant genes in their wild-type and mutant forms (in many cases, the effect was allele-specific) and by analyzing the structure of the mitochondrial DNA accumulated at the time of senescence. The data currently available concern six genes. They belong to two classes according to the effect of their mutations on longevity and the nature of mitochondrial rearrangements observed during senescence.

In the first class, we find three genes: *AS4*, encoding the elongation factor EF-1α (Silar and Picard 1994), *su1* and *su2*, each encoding one of the components of the termination factor (Gagny and Silar 1998). Mutations of these genes delay *(AS4)* or accelerate *(su1, su2)* the process of senescence, while maintaining a senDNA pattern similar to that observed in wild-type strains (see sects 3.4 and 3.7; Silar and Rossignol, pers. comm.). The second class contains the *su12*, *AS6* and *AS3* genes. The first two encode a ribosomal protein, while *AS3* probably encodes a modification enzyme of a ribosomal component (Dequard-Chablat et al. 1986). The protein encoded by the *su12* gene is the *Podospora* homologue of the *E. coli* S4 protein (Silar et al. 1997). Mutations of the *AS3* gene accelerate the senescence process, which in this case is characterized by a spectacular accumulation of senDNAα (Sainsard-Chanet et al. 1993). One mutant of the *su12* gene *(su12-1C1)* and one of the *AS6* gene *(AS6-5)* have been the subject of detailed analysis. The former could only be studied in the *mat-* context (the *su12* gene being very closely linked to the *mat* locus), while the latter was studied in the two contexts (*mat+* and *mat-*). Longevity was considerably increased in all three situations. When it exhibited senescence, the *AS6-5 mat+* strain had a mitochondrial DNA pattern similar to that of a senescent wild-type strain, showing in particular an accumulation of senDNAα. In contrast, the *su12-1C1 mat-* and *AS6-5 mat-* strains exhibited a different pattern, characterized by the accumulation of senDNAβ and no accumulation of senDNAα (Silar et al. 1997). This observation is of particular interest with respect to the mechanisms postulated for amplification of these two senDNAs (see above: amplifications associated with senescence; Belcour et al. 1994; Jamet-Vierny et al. 1997b). One can imagine that the *su12-1C1* and *AS6-5* mutations disturb cytosolic translation in quite a specific fashion, giving rise to an imbalance in the synthesis of mitochondrial proteins, some of which are more affected than others. These changes may inhibit (or delay) the accumulation of senDNAα, without affecting other rearrangements.

6.4 Targetted Research Efforts

Obviously, this approach is aimed at identifying the genes directly involved in the passive (massive genesis) or active (suppressiveness) accumulation of rearranged forms of mtDNA associated with senescence or premature death. With this objective in mind, we chose a method which has proved its value in numerous situations: the isolation of suppressors. So far, this has only been applied to the premature death syndrome, but it could be used in all cases of early senescence, or in other words in all situations where the method is not too heavy to apply.

Consequently, following mutagenesis, we searched for nuclear mutations capable of delaying the death of *AS1-4 rmp-* strains. Ten mutations were thus identified; they lie in six genes, but none of them affect the *AS1* or *rmp* genes. In four of these genes (one with two alleles, one with four alleles and the two others with a single mutation), the mutations abolished premature death; the death of *AS1-4 rmp-* strains (bearing these mutations) was no longer associated with the accumulation of the characteristic defective molecule, but with the accumulation of senDNAα (characteristic of senescence). The last two mutations identified following mutagenesis (and located in two different genes) were distinct from the others. In one case, *(AS1-4 rmp- rgs43)* the strain showed no sign of cellular degeneration, while in the other *(AS1-4 rmp- rgs27)*, when death occured, it was still accompanied by the deletion characteristic of premature death. All these mutations, screened on the basis of the partial or total suppression of premature death, modify the longevity of $AS1^+$ strains (Contamine and Picard 1998). The genes identified by the *rgs43* and *rgs27* mutations were cloned. The first encodes a protein of the mitochondrial outer membrane, TOM70 (Jamet-Vierny et al. 1997c), characterized in yeast and in the fungus *Neurospora crassa* as being one of the components of the receptor involved in importing cytosolic proteins into mitochondria (for review, see Lithgow et al. 1995). The second encodes a protein which had so far only been identified in the yeast *S. cerevisiae* (MDM10; Sogo and Yaffe 1994). This protein belongs to the mitochondrial outer membrane, and its absence leads to the formation of giant mitochondria. This characteristic is found in *P. anserina* strains bearing the *rgs27* mutation (Jamet-Viernyet al. 1997c).

Two scenarios could explain the role of mitochondrial outer membrane proteins in the accumulation of a defective genome. The first applies mainly to the MDM10 protein (*rgs 27* mutation). The hypothesis of an interaction between this protein and the cytoskeleton was proposed by Sogo and Yaffe (1994) who observed that, in yeast, mutations of the *MDM10* gene cause a change not only to mitochondrial morphology but also to mitochondrial transmission. This hypothesis can be adapted to the situation observed in *Podospora,* if we imagine that the defect in the mitochondria/cytoskeleton interactin caused by the *rgs27* mutation gives rise to the preferential transmission of certain mitochondria during growth. The second scenario concerns the TOM70 protein (*rgs 43* mutation). Its role in importing cytosolic proteins into mitochondria is well documented (for review, see Lithgow et al. 1995). The accumulation of defective mitochondrial DNA in the *AS1-4* strain is probably caused by an imbalance in the synthesis of proteins which are devoted to the mitochondria, as has been proposed for other mutations of the translation system (see above: Silar et al. 1997). This imbalance may be alleviated by a defect in the import process caused by the *rgs 43* mutation.

The two scenarios are not mutually exclusive: at present, there is no evidence that the MDM10 protein is not involved in the import process or that TOM70 does not interact with the cytoskeleton. Nevertheless, one remarkable fact concerning the *rgs27* mutation must be taken into account in any scenario: in a wild-type context for the *AS1* gene, this mutation is responsible for a cellular degeneration associated with an accumulation of particular defective mitochondrial genomes. This concerns two specific and reproducible deletions which had not previously been observed (Jamet-Vierny et al. 1997c). It is plausible that defects in interaction with the cytoskeleton and/or changes in the import process may have different consequences in a strain (*AS1*$^+$) which synthesizes its mitochondrial proteins in a "normal" fashion (compared with the *AS1-4* strain): thus it would be possible to observe the preferential transmission of other mitochondrial genomes. However, there is a long way to go before we will fully understand the intimate mechanisms involved in this selectivity, either in the *AS1-4* strain or in the *AS1*$^+$ *rgs27* strain. In the first case, the characterization of other suppressors of premature death could provide enlightenment, while in the second, the answer may be found through the search for revertants of *AS1*$^+$ *rgs27* strains (which, in a *mat-* context die very rapidly). It would be equally interesting to obtain nuclear suppressors of the "classic" senescence process as an aid to understanding how senDNAs accumulate.

7
Conclusion: Mitochondria and Cellular Degenerative Processes

Instability of the mitochondrial genome is systematically associated with two degenerative phenomena, senescence and premature death, which have been studied in the filamentous fungus *Podospora anserina*. We do not yet understand the molecular mechanisms involved in mitochondrial chromosome rearrangements, or the types of control exerted by the nuclear genome on these degenerative processes.

Similar phenomena indicated either by a definitive halt or major changes to vegetative growth are known in other fungal species (for review, see Griffiths 1992). These phenomena have been given a variety of names (vegetative death, natural death, *ragged*, *poky* and *stopper* mutants, senescence, degenerative disease), but all are associated with rearrangements of mitochondrial DNA. Amplifications in the form of multimers of certain mitochondrial chromosomal regions are always found during senescence in *P. anserina* (see above), in *ragged* mutants of *Aspergillus amstelodami* (Lazarus and Küntzel 1981) and during the degenerative process of *Ophiostoma novo-ulmi*, the pathogenic fungus responsible for elm disease (Charter et al. 1993). In *Neurospora*, recombination between natural mitochondrial plasmids (*Kalilo*-type linear plasmids encoding RNA and DNA polymerases, *Varkud*-type circular plasmids encoding reverse transcriptase) and the fungal mitochondrial chromosome gives rise to rearrangements which are incompatible with either survival or normal growth (for review, see Griffiths 1992). Finally, as in the premature death of *P. anserina* (described above), the expression of nuclear genes may be responsible for the definitive arrest of vegetative growth. In the case of the "natural death" of *Neurospora crassa*, expression of the recessive *nd* gene leads to fragmentation of the mitochondrial chromosome, which can be interpreted as the result of an increase in the level of mitochondrial recombination (Bertrand et al. 1993).

Mitochondrial involvement in the genesis of degenerative phenomena is not limited to fungi. In humans, numerous degenerative diseases, generally, but not exclusively, affecting the neuromuscular system, result from deletions of mitochondrial DNA. Most of these diseases are sporadic, although Mendelian or maternal transmission have been demonstrated in certain cases. In one particularly interesting situation, dominant nuclear genes, yet unidentified, induce a number of different mitochondrial deletions. More recently, it has been observed that mitochondrial DNA deletions, including those responsible for mitochondrial diseases, appear and "accumulate" during normal ageing in humans and other mammals (Cortopassi et al. 1992; Eimon et al. 1996; see also Chap. 16, this Vol.), as well as in the nematode *C. elegans* (Melov et al. 1995).

In view of the complexity of the cross-talk between the nucleus and the mitochondria, one can expect that other deleterious biological phenomena may result from a defect or a change in this dialogue between compartments.

A remarkable example of the interactions between the mitochondrial and nuclear genomes can be seen in higher plants in the case of male sterility with female transmission (CMS; for review, see Vedel et al. 1994). CMS is known in more than 150 species and causes systematic degeneration of the cell line which gives rise to the pollen. In all the well-documented cases (maize, petunia, tobacco, sunflower, beetroot), male sterility is due to the expression of an abnormal mitochondrial gene which is present and/or expressed only in the sterile line.

Apoptosis, or more generally programmed cell death, is a phenomenon widely present in metazoans and involves the activity of known nuclear genes which have been well conserved during evolution (Hale et al. 1996). A direct role for the mitochondria in triggering apoptosis has recently been suggested on the basis of experiments using different experimental models (Kroemer et al. 1995; Liu et al. 1996; Petit et al. 1996; Frade and Michaelidis 1997; see Chap. 17, this Vol.). Strong permeabilization of the inner mitochondrial membrane causes a drop in membrane potential and may constitute a common initiatory step in the phenomenon.

In conclusion, the clear similarity between these different degenerative phenomena raises the question of a putative evolutionary filiation between them.

Acknowledgements. We are very grateful to all our colleagues in the two laboratories who participated in experimental work and communicated the results of their unpublished studies. The work described in this chapter received the financial support of: Association pour la Recherche contre le Cancer, Association Française contre les Myopathies, Institut Natinal de la Santé et de la Recherche Médicale (contract n°. 920101), the Fondation pour la Recherche Médicale and Ministère de l'Education Nationale, de l'Enseignement Supérieur, de la Recherche (Action Concertée Coordonnée des Sciences du Vivant No. SV4 9504063).

References

Anderson S, Banquier AT, Barrell BG, de Bruijn HL, Coulson IC, Drouin IC, Eperon DP, Nierlich BA, Roe BA, Sanger PH, Schreier PH, Smith JH, Staden R, Young IJ (1981) Sequence and organization of the human mitochondrial genome. Nature 290:457–465

Belcour L, Begel O (1978) Lethal mitochondrial genotypes in *Podospora anserina:* a model for senescence. Mol Gen Genet 163:113–123

Belcour L, Begel O (1980) Life-span and senescence in *Podospora anserina:* effect of mitochondrial genes and functions. J Gen Microbiol 119:505–515

Belcour L, Vierny C (1986) Variable DNA-splicing sites of a mitochondrial intron: relationship to the senescence process in *Podospora.* EMBO J 5:609–614

Belcour L, Begel O, Mossé MO, Vierny C (1981) Mitochondrial DNA amplification in senescent cultures of *Podospora anserina:* variability between the retained, amplified sequences. Curr Genet 3:13–21

Belcour L, Begel O, Picard M (1991) A site-specific deletion in the mitochondrial DNA of *Podospora* is under the control of nuclear genes. Proc Natl Acad Sci USA 88:3579–3583

Belcour L, Sainsard-Chanet A, Sellem CH (1994) Mobile group II introns, DNA circles, reverse transcriptase and senescence. Genetica 93:225–228

Bertrand H, Wu Q, Seidel-Rogol BL (1993) Hyperactive recombination in the mitochondrial DNA of the *natural death* nuclear mutant of *Neurospora crassa.* Mol Cell Biol 13:6778–6788

Borghouts C, Osiewacz HD (1998) GRISEA, a copper-modulated transcription factor from *Podospora anserina* involved in senescence and morphogenesis, is an ortholog of MAC1 in *Saccharomyces cerevisiae.* Mol Gen Genet 260:492–502

Borghouts C, Kimpel E, Osiewacz HD (1997) Mitochondrial DNA rearrangements of *Podospora anserina* are under the control of the nuclear gene *grisea.* Proc Natl Acad Sci USA 94:10768–10773

Brygoo Y, Debuchy R (1985) Transformation by integration in *Podospora anserina.* I. Methodology and phenomenology. Mol Gen Genet 200:128–131

Charter NW, Buck KW, Brasier CM (1993) De-novo generation of mitochondrial DNA plasmids following cytoplasmic transmission of a degenerative disease in *Ophiostoma novo-ulmi.* Curr Genet 2:505–514

Contamine V, Picard M (1998) Escape from premature death due to nuclear mutations in *Podospora anserina:* repeal versus respite. Fungal Genet Biol 23:223–236
Contamine V, Lecellier G, Belcour L, Picard M (1996) Premature death in *Podospora anserina:* sporadic accumulation of the deleted mitochondrial genome, translational parameters and innocuity of the mating types. Genetics 144:541–555
Coppin E, Arnaise S, Contamine V, Picard M (1993) Deletion of mating-type sequences in *Podospora anserina* abolishes mating without affecting vegetative functions and sexual differentiation. Mol Gen Genet 241:409–414
Cortopassi GA, Shibata D, Soong NW, Arnheim N (1992) A pattern of accumulation of a somatic deletion of mitochondrial DNA in aging human tissues. Proc Natl Acad Sci USA 89:7370–7373
Cummings DJ, Belcour L, Grandchamp C (1979)Mitochondrial DNA from *Podospora anserina.* II. Properties of mutant DNA and multimeric circular DNA from senescent cultures. Mol Gen Genet 171:239–250
Cummings DJ, McNally KL, Domenico JM, Matsuura ET (1990) The complete DNA sequence of the mitochondrial genome of *Podospora anserina.* Curr Genet 17:375–402
De Zamaroczy M, Faugeron-Fonty G, Bernardi G (1983) Excision sequences in the mitochondrial genome of yeast. Gene 21:193–202
Dequard-Chablat M, Rötig A (1997) Homologous and heterologous expression of a ribosomal protein gene in *Podospora anserina* requires an intron. Mol Gen Genet 253:546–552
Dequard-Chablat M, Selem C (1994) The S12 ribosomal protein of *Podospora anserina* belongs to the S19 bacterial family and controls the mitochondrial genome integrity through cytoplasmic translation. J Biol Chem 269:14951–14956
Dequard-Chablat M, Coppin.-Raynal E, Picard-Bennoun M, Madjar JJ (1986) At least seven ribosomal proteins are involved in the control of translational accuracy in a eukaryotic organism. J Mol Biol 190:167–175
Dujon B, Belcour L (1989) Mitochondrial DNA instabilities and rearrangements in yeasts and fungi. In: Berg DE, Howe MM (eds) Mobile DNA. American Society for Microbiology, Washington, DC, pp 861–878
Eimon PM, Chung SS, Lee CM, Weindruch R, Aiken JM (1996) Age-associated mitochondrial DNA deletions in mouse skeletal muscle: comparison of different regions of the mitochondrial genome. Dev Genet 18:107–113
Eskes R, Yang T, Lambowitz AM, Perlman PS (1997) Mobility of yeast mitochondrial group II introus: engineering a new site spenfrutry and retrohoming via full reverse splicing. Cell 88:865–874
Esser K, Keller W (1976) Genes inhibiting senescence in the ascomycete *Podospora anserina.* Mol Gen Genet 144:107–110
Fassbender S, Brichl KH, Ciriacy M, Kück U (1994) Reverse transcriptase activity of an intron encoded polypeptide. EMBO J 13:2075–2083
Frade JM, Michaelidis TM (1997) Origin of eukaryotic programmed cell death: a consequence of aerobic metabolism? Bioessays 19:827–832
Gagny B, Silar P (1998) Identification of the genes encoding the cytosolic translation release factors from *Podospora anserina* and analysis of their role during the life cycle. Genetics 149:1763–1775
Gorbalenya AE (1994) Self-splicing group I and group II introns encode homologous (putative) DNA endonucleases of a new family. Protein Sci 3:1117–1120
Griffiths AJF (1992) Fungal senescence. Annu Rev Genet 26:351–372
Gross SR, Hsieh T-S, Levine PH (1984) Intramolecular recombination as a source of mitochondrial chromosome heteromorphism in *Neurospora.* Cell 38:233–239
Guo H, Zimmerly S, Perlman PS, Lambowitz AM (1997) Group II intron endonucleases use both RNA and protein subunits for recognition of specific sequences in double-stranded DNA. EMBO J 16:6835–6848
Hale AJ, Smith CA, Sutherland LC, Stoneman VEA, Longthorne VL, Culhane AC, Williams GT (1996) Apoptosis: molecular regulation of cell death. Eur J Biochem 236:1–26
Jamet-Vierny C, Shechter E (1994) Senescence specific mitochondrial DNA molecules in *Podospora anserina:* evidence for transcription and normal processing of the introns. Curr Genet 25:538–544

Jamet-Vierny C, Begel O, Belcour L (1980) Senescence in *Podospora anserina:* amplification of a mitochondrial DNA sequence. Cell 21:189–194

Jamet-Vierny C, Boulay J, Briand JF (1997a) Intramolecular crossing-overs generate deleted mitochondrial DNA molecules in *Podospora anserina.* Curr Genet 31:162–170

Jamet-Vierny C, Boulay C, Begel O, Silar P (1997b) Contribution of various classes of defective mitochondrial DNA molecules to senescence in *Podospora anserina.* Curr Genet 31:171–178

Jamet-Vierny C, Contamine V, Boulay J, Zickler D, Picard M (1997c) Mutations in genes encoding the mitochondrial outer membrane proteins Tom70 and Mdm10 of *Podospora anserina* modify the spectrum of mitochondrial DNA rearrangements associated with cellular death. Mol Cell Biol 17:6359–6366

Jamet-Vierny C, Rossignol M, Haedens V, Silar P (1999) What triggers senescence in *Podospora anserina?* Fungal Genet Biol (in press)

Koll F, Begel O, Keller AM, Vierny C, Belcour L (1984) Ethidium bromide rejuvenation of senescent cultures of *Podospora anserina:* loss of senescence-specific DNA and recovery of normal mitochondrial DNA. Curr Genet 8:127–134

Koll F, Belcour L, Vierny C (1985) A 1100-bp sequence of mitochondrial DNA is involved in senescence process in *Podospora:* study of senescent and mutant cultures. Plasmid 14:106–117

Kroemer G, Petit P, Zamzami N, Vayssiére JL, Mignotte B (1995) The biochemistry of programmed cell death. FASEB J 9:1277–1287

Kück U, Stahl U, Esser K (1981) Plasmid-like DNA is part of mitochondria DNA in *Podospora anserina.* Curr Genet 3:151–156

Lambowitz AM, Belfort M (1993) Introns as mobile genetic elements. Annu Rev Biochem 62:587–622

Larsson NG, Clayton DA (1995) Molecular genetic aspects of human mitochondrial disorders. Annu Rev Genet 29:151–178

Lazarus CM, Küntzel H (1981) Anatomy of amplified mitochondrial DNA in *ragged* mutants of *Aspergillus amstelodami:* excision points within a protein gene and a common 215 bp segment containing a possible origin of replication. Curr Genet 4:99–107

Lithgow T, Glick BS, Schatz G (1995) The protein import receptor of mitochondria. TIBS 20:98–101

Liu X, Kim CN, Yang J, Jemmerson R, Wang X (1996) Induction of apoptosis program in cell-free extracts: requirement of dATP and cytochrome c. Cell 86:147–157

Marcou D (1954) Sur le rajeunissement par le froid des souches de *Podospora anserina.* C R Acad Sci Paris 239:1153–1155

Marcou D (1961) Notion de longévité et nature cytoplasmique du déterminant de la sénescence chez quelques champignons. Ann Sci Nat Bot 12:653–764

Melov S, Lithgow GJ, Fischer DR, Tedesco PM, Johnson TE (1995) Increased frequency of deletions in the mitochondrial genome with age of *Caenorhabditis elegans.* Nucleic Acids Res 23:1419–1425

Michel F, Ferrat JL (1995) Structure and activities of group II introns. Annu Rev Biochem 64:435–461

Michel F, Lang BF (1985) Mitochondrial class II introns encode proteins related to the reverse transcriptases of retroviruses. Nature 316:641–643

Mita S, Rizzuto R, Morares CT, Shanske S, Anaudo E, Fabrizi GM, Koga Y, Di Mauro S, Schon E (1990) Recombination via flanking direct sequences is a major cause of large scale deletions of human mitochondrial DNA. Nucleic Acides Res 18:561–567

Mueller MW, Allmaier M, Eskes R, Schweyen R (1993) Transposition of group II intron aiI in yeast and invasion of mitochondrial genes at new locations. Nature 366:174–176

Osiewacz HD, Esser K (1984) The mitochondrial plasmid of *Podospora anserina:* a mobile intron of a mitochondrial gene. Curr Genet 8:299–305

Osiewacz HD, Nuber U (1996) *Grisea,* a putative copper-activated transcription factor from *Podospora anserina* involved in differentiation and senescence. Mol Gen Genet 252:115–124

Petit PX, Susin SA, Zamazami N, Mignotte B, Kroemer G (1996) Mitochondria and programmed cell death: back to the future. FEBS Lett 396:7–13

Picard-Bennoun M (1976) Genetic evidence for ribosomal antisuppressors in *Podospora anserina.* Mol Gen Genet 147:299–306

Picard-Bennoun M (1985) Introns, protein syntheses and aging. FEBS Lett 184:1–5

Rizet G (1953a) Sur l'impossibilité d'obtenir la multiplication végétative ininterrompue et illimitée de l'ascomycète *Podospora anserina*. C R Acad Sci Paris 237:838–840

Rizet G (1953b) Sur la longévité des souches de *Podospora anserina*. C R Acad Sci Paris 237:1106–1109

Rossignol J-L, Picard M (1991) *Ascobolus immersus* and *Podospora anserina:* sex, recombination, silencing, and death. In: Bennett JW, Lasure LL (eds) More gene manipulations in fungi. Academic Press, New York, pp 266–290

Rossignol M, Silar P (1996) Genes that control longevity in *Podospora anserina*. Mech Age Dev 90:183–193

Sainsard-Chanet A, Begel O (1990) Insertion of an LrDNA gene fragment and of filler DNA at a mitochondrial exon-intron junction in *Podospora*. Nucleic Acids Res 18:779–783

Sainsard-Chanet A, Begel O, Belcour L (1993) DNA deletion of mitochondrial introns is correlated with the process of senescence in *Podospora*. J Mol Biol 234:1–7

Sainsard-Chanet A, Begel O, Belcour L (1994) DNA double-strand break in vivo at the 3' extremity of exons located upstream of group II introns- senescence and circular DNA introns in *Podospora* mitochondria. J Mol Biol 242:630–643

Sainsard-Chanet A, Begel O, d'Aubenton-Carafa Y (1998) Two coexisting mechanisms account for the large-scale deletions of mitochondrial DNA in *Podospora anserina* that involve the 5' border of a group II intron. Curr Genet 34:326–335

Schatz G (1995) Mitochondria: beyond oxidative phosphorylation. Biochim Biophys Acta 1271:123–126

Schmidt WM, Schweyen RJ, Wolf K, Mueller MW (1994) Transposable group II introns in fission and budding yeast - site-specific genomic instabilities and formation of group II IVS plDNAs. J Mol Biol 243:157–166

Sellem CH, Sainsard-Chanet A, Belcour L (1990) Detection of a protein encoded by a class II mitochondrial intron of *Podospora anserina*. Mol Gen Genet 224–232–240

Sellem CH, Lecellier G, Belcour L (1993) Transposition of a group II intron. Nature 366:176–178

Séraphin B, Boulet A, Simon M, Faye G (1987) Construction of a yeast strain devoid of mitochondrial introns and its use to screen nuclear genes involved in mitochondrial splicing. Proc Natl Acad Sci USA84:6810–6814

Shub DA, Goodrich-Blair H (1994) Amino acid sequence motif of group I intron endonucleases is conserved in open reading frames of group II introns. TIBS 19:402–404

Silar P, Picard M (1994) Increased longevity of the EF-1 alpha high fidelity mutants in *Podospora anserina*. J Mol Biol 235:231–236

Silar P, Koll F, Rossignol M (1997) Cytosolic ribosomal mutations which abolish accumulation of circular intron in the mitochondria without preventing senescence of *Podospora anserina*. Genetics 145:697–705

Sogo L, Yaffe MP (1994) Regulation of mitochondrial morphology and inheritance by Mdm10p, a protein of the mitochondrial outer membrane. J Cell Biol 126:1361–1373

Stahl U, Lemke PA, Tudzynski P, Kück U, Esser K (1978) Evidence for plasmid-like DNA in a filamentous fungus, the ascomycete *Podospora anserina*. Mol Gen Genet 162:341–143

Tudzynski P, Esser K (1979)Chromosomal and extrachromosomal control of senescence in the ascomycete *Podospora anserina*. Mol Gen Genet 173:71–84

Tzagoloff A, Dieckmann CL (1990) PET genes of *Saccharomyces cerevisiae*. Microbiol Rev 54:211–225

Vedel F, Pla M, Vitart V, Guierres S, Chétrit P, De Paepe R (1994) Molecular basis of nuclear and cytoplasmic male sterility in higher plants. Plant Physiol Biochem 32:601–618

Vierny C, Keller A-M, Begel O, Belcour L (1982) A sequence of mitochondrial DNA is associated with the onset of senescence in a fungus. Nature 297:157–1569

Wright RM, Horrum MA, Cummings DJ (1982) Are mitochondrial structural genes selectively amplified during senescence in *Podospora anserina?* Cell 29:505–515

Xiong Y, Eickbush TH (1990) Origin and evolution of retroelements based upon their reverse transcriptase sequences. EMBO J 9:3353–3362

Yang J, Mohr G, Perlman PS, Lambowitz AM (1998) Group II intron mobility in yeast mitochondria; target DNA-primed reverse transcription activity of aiI and reverse splicing into DNA transposition sites in vitro. J Mol Biol 282:505–523

Yang T, Zimmerly S, Perlman PS, Lambowitz AM (1996) Efficient integration of an intron RNA in double-stranded DNA by reverse splicing. Nature 381:332–335
Zimmerly S, Guo H, Eskes R, Yang J, Perlman PS, Lambowitz AM (1995) A group II intron RNA is a catalytic component of a DNA endonuclease involved in intron mobility. Cell 83:529–538

CHAPTER 16

The Control of Ageing and Mitochondria 16

P. Lestienne and J. Veziers

Contents

1 Introduction

By their central role in energetic metabolism and in nucleotide biosynthesis, mitochondria appear to play an important role in cell degeneration and ageing, primarily due to a reduction of the production of ATP, a universal energetic compound which also controls reactions such as the phosphorylation of proteins, which plays an important function in cell regulation.

Ageing can be defined as the progressive accumulation of molecular changes during a lifetime which are associated with or responsible for the increasing number of diseases which come with age. Some general articles have been published on this subject (Jazwinski 1996; Sohal and Weindruch 1996; Smith and Pereira-Smith 1996; Finch and Tanzi 1997).

2 The Theories of Ageing

Two principal theories have been proposed to account for ageing:
The Germinal Theory. This postulates that the genes involved in development and in life span actively participate in ageing according to a determined process. This mecha-

E 99-29 INSERM, physiologie mitochondriale, Université Bordeaux 2, 146 rue Léo Saignat, 33076 Bordeaux, France nism, beginning during early life and ensuring a constant renewal of living species, is the price paid for evolution by ensuring a constantly unstable genetic diversity.

The Somatic Theory. This proposes that over time, mutations or post-translational modifications accumulate in various tissues, affecting their main function and inducing important secondary effects such as cell transformation, or the inability to achieve their program function as in some neurological disorders. This theory may account for errors of replication, transcription, translation and post-translational events. Chance can be considered a major factor, in which one's way of life and the environment play major roles.

By their central function in the cell's energetic metabolism, mitochondria play a fundamental role in these two theories, since their information content represents quantitatively (several thousands of copies per diploid cell), as well as qualitatively (the thirteen indispensable respiratory subunits) an important part of cell's DNA content and function.

In 1955, D. Harman pointed out the potential role played by free radicals in ageing (reviewed by Harman 1994). The main cause of ageing and associated diseases results from free radical production which may be modified by genetic and environmental factors. These modifications occur with ionizing radiation, or non-enzymatic and enzymatic reactions involving energetic reactions such as in photosynthesis and respiration.

3 Free Radicals and Mitochondrial DNA

The mitochondrial DNA is localized near the respiratory chain where a very active molecule, due to electron leakage, combines with oxygen to produce a very active species: superoxide anion ($O_2^{\circ-}$). In the presence of manganese superoxide dismutase (SOD_2), and of protons, the superoxide anions are converted into hydrogene peroxide (H_2O_2) which are decomposed into O_2 and H_2O by thiol specific antioxidant, thioredoxine and thioredoxine reductase, a selenoenzyme, in the presence of NADPH, or by the selenium dependent glutation peroxidase in the presence of reduced glutatione (Fig. 1). However, if these reactions are not active enough, the hydroxyl radical (OH°) may be formed, either in the presence of Fe^{2+} by the Fenton reaction, or by the Haber-Weiss reaction in the presence of H^+ leading to H_2O and O_2.

The hydroxyl radicals (OH°) are a very powerful oxidant reacting with mitochondrial quinones, fatty acids, proteins, and nucleic acids inducing strand break and mutation (Imlay and Linn 1988). As many as 100 different nucleotide modifications have been identified upon exposure to (OH°). In 95% of cases the formation of 8-oxoguanosine (8-OxoG) induces a mispairing with the complementary base (Kuchino et al. 1987) leading to a transversion (G-C $\rightarrow$ T-A; Moriya and Grollman 1993).

The measurement of 8-OxoG in rat liver DNA revealed a striking difference between nuclear and mitochondrial DNA modifications: one base out of 130 000 bases in nuclear DNA against one base out of only 8000 bases in the mitochondrial DNA (Richter et al. 1988). Other studies on base modifications, such as the formation of 7-Methylguanine adducts, revealed an increase which was three times higher in the

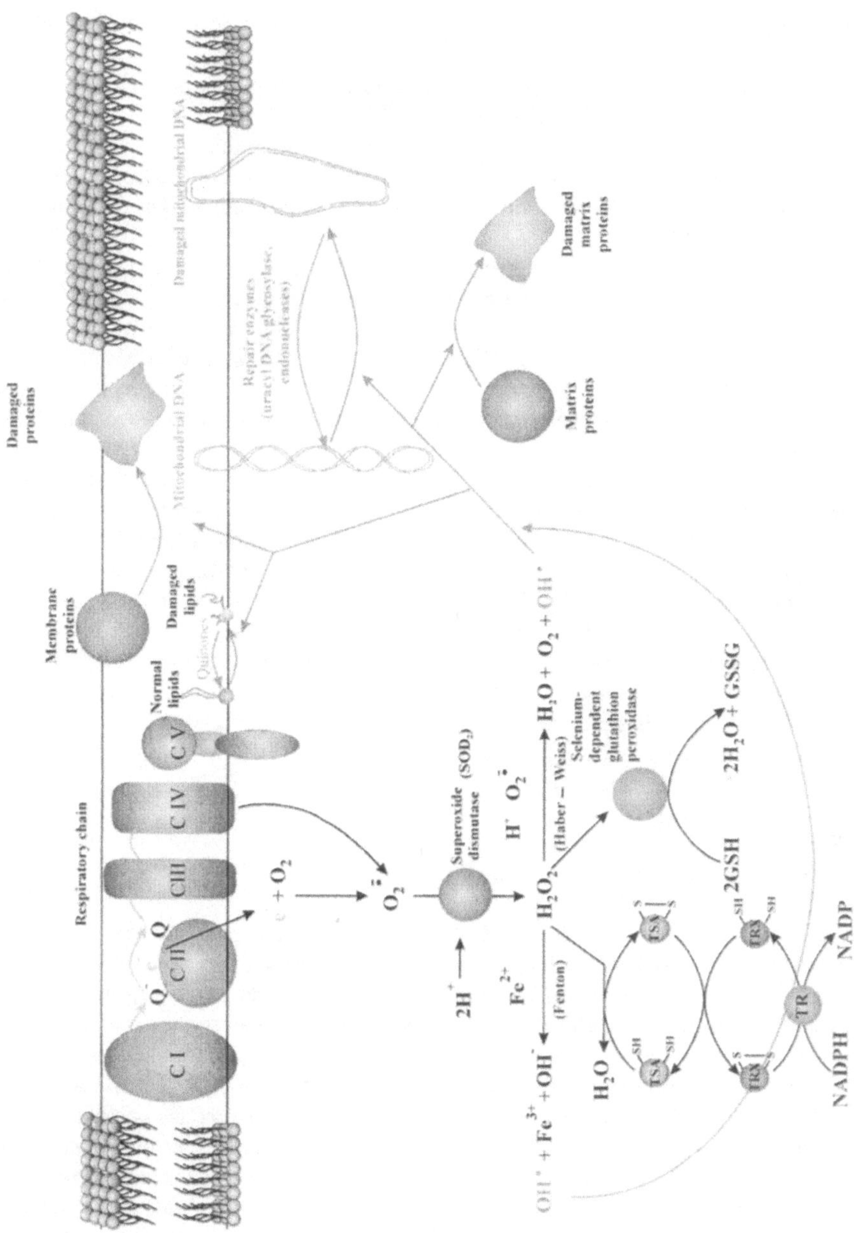

Fig. 1. Schematic representation of the mechanisms involved in superoxide anions ($O_2^{\circ-}$) formation and transformation into harmful hydroxyl anions (OH°). Detoxification reactions of hydrogen peroxide occur by two different reactions involving either the thiol-specific antioxidant (TSA), the thioredoxine (TRX) and the thioredoxine reductase (TR) in the presence of NADPH, or the selenium-dependent glutathion peroxidase in the presence of glutathione (GSH)

mitochondrial DNA, compared with the nuclear DNA (Park and Ames 1988); furthermore, there is a 2.5-fold increase between 6 and 24 months of age, demonstrating a reduced DNA reparation in mitochondria, compared with the nucleus, and an accumulation of these lesions during ageing.

Other research of DNA oxidation products, such as thymidine glycol, showed that their accumulation reflects the metabolic activity of the respiratory chain; for example, mice excrete 18 times more thymidine glycol per unit weight than humans (Adelman et al. 1988).

These differences may result from reduced reparation-recombination in mitochondria (Clayton et al. 1974; Prakash 1975), though three uracyl glycosylases have been identified (Anderson and Friedberg 1980; Gupta and Sirover 1981; Caradonna et al. 1996), as well as a UV-inducible endonuclease (Tomkinson et al. 1990). The induction of free radicals by alloxan on cell cultures showed that the damaged mtDNA is repaired within 4 h (Drigger et al. 1993). Similar conclusions were reached by Hegler et al. (1993).

The treatment of house flies with X-rays or with hyperoxy induced a three-fold modification of the mtDNA by 8-OxoG, compared with the nuclear DNA; furthermore, ageing seemed to be correlated with these modifications (Agarwal and Sohal 1994), but these modifications are lower than the results obtained on rats (16-fold, Richter et al. 1988). These differences may be related to the difference in the mutation rates between mitochondria and the nucleus, which are similar in *Drosophila* (Powell et al. 1986), but are ten times higher in mammals (Vauwter and Brown 1986). Transgenic *Drosophila* for supplementary cytoplasmic superoxide dismutase and catalase have a 30% longer lifetime, indicating that increased detoxification of free radicals does indeed increase longevity in insects (Orr and Sohal 1994).

4 Origins of Respiratory Chain Defects During Ageing

A progressive decline of the activities of the respiratory chain complexes from human skeletal muscle associated with ageing was reported by Trounce et al. (1989). In order to know whether mitochondrial defects result from nuclear or mitochondrial gene product(s) during ageing, Hayashi et al. (1994) observed cytochrome c oxidase defects as well as a reduced mitochondrial protein synthesis. The mitochondria from aged patients were introduced into HeLa cells devoid of mtDNA (Rho° cells). Mitochondrial protein synthesis was restored, as well as cytochrome c oxidase activity. Then, nuclei of HeLa cells were introduced into fibroblasts from older subjects, which restored both mitochondrial protein synthesis and cytochrome c oxidase activity. In cell cultures, a progressive loss of HeLa chromosomes as well as a decrease in cytochrome c oxidase activity were observed. These data show that, in these cases, the nuclear chromosomes from the aged donors are responsible for respiratory chain dysfunction, notably mitochondrial protein synthesis.

Another comparable experiment was later conducted, but with opposite conclusions. The introduction of mitochondria from aged donors into HeLa Rho° cells was accompanied by a significant decrease in the cell's oxygen consumption, compared with young donors (Laderman et al. 1996).

5
Mitochondrial DNA Alterations in Metazoa During Ageing

The first report of mtDNA rearrangements with ageing was described in 1988 by Pikò et al., who hybridized mtDNA from young and old rats, and observed the hybrids by electron microscopy. About 2% of hybrids derived from adults, and 11% derived from senescents with loops and knots indicating some additions or deletions of about 500 nucleotides during ageing.

While using PCR with the primers allowing the detection of the common deletion of 4977 base pairs, a low number of these deletions have been found in human brain and heart, below 0.1%, while it was absent in embryonic tissues (Cortopassi and Arnheim 1990). According to their results, the proportion of deleted molecules increases from 100 to 100 000 in a lifetime.

Another deletion of 7436 base pairs has been shown in the hearts of aged subjects (Sugiyama et al. 1991). Its amount ranged between 3 and 9% in subjects from 80 to 90 years old. A parallel accumulation of 8-oxoG was also detected (Hayakawa et al. 1992).

Different parts of the brain were analyzed, showing important regional variations in the content of deletions (Soon et al. 1992; Corral-Debrinski et al. 1992). The highest amounts were found in the putamen, the substantia nigra and caudate; these regions contain the monoamine oxidase located on the outer mitochondrial membrane which produces hydrogen peroxide during dopamine metablism, and hence is a potential source of free radicals. On the other hand, the cerebral grey matter does not accumulate the deletion of 4977 base pairs during aging. Even though the highest amount of deletion reported is very low (0.12% in the putamen at 80 years old), compared with mitochondrial myopathies (more than 50%), these results could partly account for some neurological disorders during ageing if mtDNA damage accumulates in specific cells, perturbing specific functions. In situ hybridization may shed light on this possibility.

The common deletion has been detected in aged women's ovaries (Suganuma et al. 1993), and in the lung, where it begins to increase at about the age of 50 (Fahn et al. 1996). It has also been observed in the skin, where it accumulates throughout the lifetime (Yang et al. 1994).

Ageing has also been associated with tandem duplications of the mtDNA control region containing the promoters (Lee et al. 1994), as well as a number of various sized mtDNAs which lack origin of replication (Hayakawa et al. 1996). Other particular mitochondrial DNA structures have been shown in mice heart and brain whose amount decreases with caloric restriction (Melov et al. 1997), as well as multiple deletions of 8.04 kb in humans (Baumer et al. 1994).

Multiple mtDNA deletions have been also detected in ageing in mice (Brossas et al. 1994; Tanhauser and Laipis 1995), in rat (Gadaleta et al. 1992), and even in the nematode *Caenorhabditis elegans* (Melov et al. 1994).

6
Mitochondria and Neurodegenerative Diseases

Various studies indicate that oxidative phosphorylation could play a role in Parkinson's and Alzheimer's diseases. Treatment of normal fibroblasts with the uncoupler

CCCP (carbonyl cyanide m-chloro phenyl hydrazone) induces a ten fold increase of cells reacting with the antibody specific to the Alzheimer's "paired helical filaments", as well as a 150-fold increase of cells reacting against the monoclonal antibody 50 of Alzheimer's, suggesting a relationship between mitochondrial dysfunction and Alzheimer's disease (Blass et al. 1990). Defects in complex I were reported in Parkinson's patients (Mizumo et al. 1989; Shapira et al. 1990) as well as complexes II and III (Haas et al. 1995). These three complexes contain iron-sulfur centers for their activities which could trigger hydroxyl radical production by the Fenton or by the Haber-Weiss reactions upon hydrogen peroxide production by SOD_2, and hence the inhibition of their activities. The mitochondrial DNA structure appears normal in the substantia nigra of Parkinson's patients by Southern blot analysis (Lestienne et al. 1990) and by PCR with age-matched subjects (Lestienne et al. 1991).

A large-scale study of Caucasian Parkinson's and Alzheimer's patients showed the presence of a moderately conserved variant (A $\rightarrow$ G) in the $tRNA^{Gln}$ at a position of 4336 in 5.2% (9/173) of cases, compared with 0.7% in the "control" population. In addition, an insertion of five base pairs at positions 956–965 of the 12 S rRNA, a missense variant converting a highly conserved methionine into valine in subunit 1 of complex I, and a heteroplasmic variant at position 3196 of the 16 S rRNA, were found (Shoffner et al. 1993). The frequency of the $tRNA^{Gln}$ variant was confirmed and amplified in more late-onset Alzheimer's patients (Hutchin and Cortopassi 1995).

Several heteroplasmic mutations in the mitochondrial genes encoding cytochrome c oxidase subunits I and 2 have been associated with Alzheimer's disease (Davies et al. 1997). However, further investigations revealed that they correspond to nuclear pseudo-genes (Hirano et al. 1997; Wallace et al. 1997).

The disruption of the MnSOD gene in two independent lines of mice have led to their deaths within a few weeks. Extensive mitochondrial damage was observed, with lipid accumulation in the liver, neurologic dysfunctions similar to Leigh's disease (Melov et al. 1998) as well as dilatated cardiomyopathy. In addition, reduced activities of two iron-sulfur enzymes susceptible to oxygen-reactive species, aconitase and complex II, were noted upon histological studies (Li et al. 1995). The treatment of these mice with a metalloporphyrin mimicking the SOD activity, MnTBAP (Day et al. 1995), dramatically enhances the lifetime of these mice, although the drug does not cross the blood-brain barrier. The study of the mtDNA of 21-day-old treated mice showed their apparently normal structure (Melov et al. 1998), a result coherent with previous findings in humans (Lestienne et al. 1990, 1991).

7 Prospects

Telomere Shortening (Harley et al. 1990) as well as mtDNA alterations appear to be important new genetic markers of ageing. However, both the cause and the effects of mtDNA alterations on cell physiology remain to be elucidated.

Evolution of organisms with higher metabolic activities has been accompanied by an increased production of free radicals potentially altering their mtDNA structure; possible processes to prevent these deleterious effects may be mtDNA shortening, along with more mtDNA copies. Without intramitochondrial mtDNA compartmentation, the low macroscopic amount of mtDNA modifications could be compensated in *trans* by the normal mtDNA gene product(s), a mechanism which has been found in radiation-resistant bacteria which compensate for their DNA damage by an increased chromosome copy number without specific reparation processes.

A premature form of ageing has been ascribed to the nuclear gene of nuclear DNA helicase in Werner's syndrome (Yu et al. 1996). The finding that telomerase increases the life span of cells (Bodnar et al. 1998) strongly supports the role played by nuclear gene product(s) involved in the maintenance of DNA structure as a major contribution to ageing. Considering all this, it may be hypothesized that some nuclear genes controlling mitochondrial biogenesis, and mitochondrial DNA structure, could be located sub-telomerically.

Telomere shortening could directly or indirectly inactivate these genes, leading either to respiratory chain defects (and hence an increase in superoxide anions and free radical production), or to down-regulation of genes whose products are specifically involved in mtDNA replication and/or repair (Leading to mtDNA alterations). This vicious circle could be partly prevented by known antioxidant cofactors such as manganese and selenium, as well as reactive oxygen species scavanger drugs such as MnTBAP and caloric restriction.

One report describes the association between mtDNA mutations and premature telomere shortening (Oexel and Zwirner 1997). Since telomere lengths are regulated by development and type of tissue (Prowse and Greider 1995), this hypothesis may account for previous observations of tissue-specific mtDNA alterations (Corral-Debrinski et al. 1992; Melov et al. 1997), as well as of apparent conflicting results on the contribution of each genome to human ageing (Hayashi et al. 1994; Laderman et al. 1996). Finding mitochondrial DNA genotypes associated with longevity would suggest that the mtDNA sequence does plays a role in longevity, as reported recently by Tanaka et al. (1998), showing increased frequency of a mutation in subunit 2 of NADH dehydrogenase.

References

Adelman R, Saul RL, Ames B (1988) Oxidative damage to DNA: relation to specific metabolic rate and life span. Proc Natl Acad Sci USA 85:2706–2708

Agarwal S, Sohal RS (1994) DNA oxidative damage and life expectancy in houseflies. Proc Natl Acad Sci USA 91:12332–12335

Anderson CT, Friedberg EC (1980) The presence of nuclear and mitochondrial uracyl-glycosylases in extracts of human KB cells. Nucleic Acids Res 8:875–877

Baumer A, Zhang C, Linnane AW, Nagley P (1994) Age-related human mitochondrial DNA deletions: a heterogeneous set of deletions arising at a single pair of directly repeated sequences. Am J Hum Genet 54:618–630

Blass JP, Baker AL, Ko L, Black RS (1990) Induction of Alzheimer antigens by an uncoupler of oxidative phosphorylation. Arch Neurol 47:864–869

Bodnar AG, Ouellette M, Frolkis M, Holt SE, Chiu C-P, Morin GB, Harley CB, Shay JW, Lichtsteiner S, Wright WE (1998) Extension of life-span by introduction of telomerase into normal human cells. Science 279:349–352

Brossas JY, Barreau E, Courtois Y, Tréton J (1994) Multiple deletions in mitochondrial DNA are present in senescent mouse brain. Biochem Biophys Res Commun 202:654–659

Caradonna S, Ladner R, Hansburt M, Kosciuk M, Lynch F, Muller S (1996) Affinity purification and comparative analysis of two distinct human uracil-DNA glycosylases. Exp Cell Res 222:345–359

Clayton D, Doda JN, Friedberg EC (1974) The absence of a pyridmidine dimer repair in mammalian mitochondria. Proc Natl Acad Sci USA 71:2778–2781
Corral-Debrinski M, Horton T, Lott M, Shoffner JM, Flint BM, Wallace DC (1992) Mitochondrial DNA deletions in human brain: regional variability and increase with advanced age. Nat Genet 2:324–329
Cortopassi GA, Arnheim N (1990) Detection of a specific mitochondrial DNA deletion in tissues of older humans. Nucleic Acids Res 18:6927–6933
Cortopassi GA, Shibota D, Soon NW, Arnheim N (1992) A pattern of accumulation of somatic deletions of mitochondrial DNA in aging human tissues. Proc Natl Acad Sci USA 89:7370–7373
Davis RE, Miller S, Herrntadt C, Ghosh SS, Fahy E, Shinobu LA, Galasko D, Thal LJ, Beal MF, Howell N, Parker WD (1997) Mutations in mitochondrial cytochrome c oxidase genes segregate with late-onset Alzheimer disease. Proc Natl Acad Sci USA 94:4526–4531
Day BJ, Shawen S, Liochev SI, Crapo JD (1995) A metalloporphyrin superoxide dismutase mimetic protects against paraquat-induced endothelial cell injury, in vitro. J Pharmacol Exp Ther 275:1227–1232
Drigger W, Ledoux S, Wilson GL (1993) Repair of oxidative damage within the mitochondrial DNA of RINr 38 cells. J Biol Chem 268:22042–22045
Fahn H-J, Wang L-S, Hsieh R-H, Chang S-C, Kao S-H, Huang M-H, Wei Y-H (1996) Age related 4977 bp deletion in human lung mitochondrial DNA. Am J Respir Crit Care Med 154:1141–1145
Finch CE, Tanzi RE (1997) Genetics of aging. Science 278:407–411
Gadaleta MN, Rainaldi G, Lezza AMS, Milella F, Fracasso F, Cantatore P (1992) Mitochondrial DNA copy number and mitochondrial DNA deletion in adult and senescent rat. Mutat Res 275:181–193
Gupta PK, Sirover M (1981) Stimulation of the nuclear uracyl DNA glycosylase in human fibroblasts. Cancer Res 41:3133–3136
Gutteridge JMC (1992) Ageing and free radicals. Med Lab Sci 49:313–319
Haas RH, Bchir MB, Nasirian F, Nakano K, Ward D, Pay M, Hill R, Shults CW (1995) Low platelet mitochondrial complex I and complex II and III activities in early untreated Parkinson's disease. Ann Neurol 37:714–722
Harley CB, Futcher AB, Greider CW (1990) Telomeres shorten during ageing of human fibroblasts. Nature 345:458–460
Harman D (1994) "Aging": prospects for the further increase in the functional life span. Age 17:119–146
Hayakawa M, Hattori K, Sugiyama S, Ozawa AT (1992) Age associated oxygen damage and mutations in mitochondrial DNA of human hearts. Biochem Biophys Res Commun 189:979–985
Hayakawa M, Katsumata K, Yoneda M, Tanaka M, Sugiyama S, Ozawa T (1996) Age-related extensive fragmentation of mitochondrial DNA into minicircles. Biochem Biophys Res Commun 226:369–377
Hayashi JI, Ohta S, Kagawa Y, Kondo H, Kaneda H, Yonekawa H, Takai D, Miyabayashi S (1994) Nuclear but not mitochondrial genome involvement in human age-related mitochondrial dysfunction. J Biol Chem 269:6878–6883
Hegler J, Bittner D, Boiteux S, Epe B (1993) Quantification of oxidative DNA modifications in mitochondria. Carcinogenesis 14:2309–2312
Hirano M, Shtilbans A, Mayeux R, Davidson MM, DiMauro S, Knowles JA, Schon EA (1997) Apparent mtDNA heteroplasmy in Alzheimer's disease patients and in normals due to PCR amplification of nucleus-embedded mtDNA pseudgenes. Proc Natl Acad Sci USA 94: 14894–14899
Hutchin T, Cortopassi GA (1995) Mitochondrial DNA clone associated with increased risk for Alzheimer disease. Proc Natl Acad Sci USA 92:6892–6895
Imlay J, Linn S (1988) DNA damage and oxygen radical toxicity. Science 240:1302–1309
Jazwinski SM (1996) Longevity genes and aging. Science 273:54–59
Kuchino Y, Mori F, Kasai H, Inou H, Iwai S, Miura K, Ohtsuka E, Nishimura S (1987) Misreading of DNA template containing 8-hydroxydeoxyguanosine at the modified bases and at adjacent residues. Nature 327:77–79

Laderman KA, Penny JR, Mazzucchelli F, Bresolin N, Scarlato G, Attardi G (1996) Aging-dependent functional alterations of mitochondrial DNA (mtDNA) from human fibroblasts transferred into mtDNA-less cells. J Biol Chem 271:15891–15897
Lee H-C, Pang C-Y, Hsu H-S, Wei Y-H (1994) Ageing-associated tandem duplication in the D-loop of mitochondrial DNA of human muscle. FEBS Lett 354:79–83
Lestienne P, Nelson I, Riederer P, Jillinger K, Reichmann H (1990) Normal mitochondrial genome in brain from patients with Parkinson's disease and complex I defect. J Neurochem 55:1810–1812
Lestienne P, Nelson I, Riederer P, Jillinger K, Reichmann H (1991) Mitochondrial DNA in postmortem brain from patients with Parkinson's disease. J Neurochem 56:1819
Li Y, Huang T-T, Carlson EJ, Melov S, Ursell PC, Olson JL, Noble LJ, Yoshimura MP, Berger C, Chan PH, Wallace DC, Epstein CJ (1995) Dilated cardiomyopathy and neonatal lethality in mutant mice lacking manganese superoxide dismutase. Nat Genet 11:376–381
Melov S, Hertz GZ, Stormo G, Johnson TE (1994) Detection of deletions in the mitochondrial genome of caenorhabditis elegans. Nucleic Acids Res 22:1075–1078
Melov S, Hinerfeld D, Esposito L, Wallace DC (1997) Multi-organ characterization of mitochondrial rearrangements in ad libidum and caloric restricted mice show striking somatic mitochondrial rearrangements with age. Nucleic Acids Res 25:974–982
Melov S, Schneider JA, Day BJ, Hinerfeld D, Coskun P, Mirra SS, Crapo JD, Wallace DC (1998) A novel neurological phenotype in mice lacking mitochondrial manganese superoxide dismutase. Nat Genet 18:159–163
Mizumo Y, Ohta S, Tanaka M, Takamiya S, Suzuki K, Sato T, Oya H, Ozawa T, Kagawa Y (1989) Deficiencies in complex I subunits of the respiratory chain in Parkinson's disease. Biochem Biophys Res Commun 163:1450–1455
Moriya M, Grollman AP (1993) Mutation in the *mut* gene of *Escherichia coli* enhance the frequency of targeted GC → TA transversions induced by a single 8-oxoguanine residue in single-stranded DNA. Mol Gen Genet 239:72–76
Oexel K, Zwirner A (1997) Advanced telomere shortening in respiratory chain disorders. Hum Mol Genet 6:905–908
Orr WC, Sohal RS (1994) Extension of life-span by overexpression of superoxide dismutase and catalase in *Drosophila melanogaster.* Science 263:1128–1130
Park JW, Ames B (1988) 7-Methylguanine adducts in DNA are normally present in high levels and increase on aging: analysis by HPLC with electrochemical detection. Proc Natl Acad Sci USA 85:7467–7470
Pikò L, Hougham AJ, Bulpitt KJ (1988) Studies of sequence heterogeneity of mitochondrial DNA from rat and mouse tissues: evidence for an increased frequency of deletions/additions with aging. Mech Ageing Dev 43:279–293
Powell JR, Caccone A, Amato GD, Yoon C (1986) Rates of nucleotide substitution in *Drosophila* mitochondrial DNA and nuclear DNA are similar. Proc Natl Acad Sci USA 83:9090–9093
Prakash L (1975) Repair of pyrimidine dimers in nuclear and mitochondrial DNA of yeast irradiated with low doses of ultraviolet light. J Mol Biol 98:781–795
Prowse KR, Greider CW (1995) Developmental and tissue-specific regulation of mouse telomerase and telomere length. Proc Natl Acad Sci USA 92:4818–4822
Richter C, Park JW, Ames B (1988) Normal oxidative damage to mitochondrial and nuclear DNA is extensive. Proc Natl Acad Sci USA 85:6465–6467
Shapira AHV, Cooper JM, Dexter D, Clark JB, Jenner P, Marsden CD (1990) Mitochondrial complex I deficiency in Parkinson's disease. J Neurochem 54:823–827
Shoffner JM, Brown MD, Torroni A, Lott MT, Cabell MF, Mirra SS, Beal MF, Yang C-C, Gearing M, Savo R, Watts RL, Juncos JL, Hansen LA, Crain BJ, Fayad M, Reckord CL, Wallace DC (1993) Mitochondrial DNA variants observed in Alzheimer disease and Parkinson disease patients. Genomics 17:171–184
Smith JR, Pereira-Smith O (1996) Replicative senescence: implication for in vivo aging and tumor suppression. Science 273:63–67
Sohal RS, Weindruch R (1996) Oxidative stress, caloric restriction, and aging. Science. 273:59–63
Soon NW, Hinton DR, Cortopassi G, Arnheim N (1992) Mosaicism for a specific somatic mitochondrial DNA mutation in adult brain. Nat Genet 2:318–323

Suganuma N, Itagawa T, Nawa A, Tomoda Y (1993) Human ovarian aging and mitochondrial DNA deletion. Horm Res 39 (Suppl 1):16–21

Sugiyama S, Hattori K, Hayakawa M, Ozawa T (1991) Quantitative analysis of age-associated accumulation of mitochondrial DNA deletions in human hearts. Biochem Biophys Res Commun 180:894–899

Tanaka M, Gong J-S, Zhang J, Yoneda M, Yagi K (1998) Mitochondrial genotype associated with longevity. Lancet 351:185–186

Tanhauser SM, Laipis PJ (1995) Multiple deletions are detectable in mitochondrial DNA of aging mice. J Biol Chem 270:24769–24775

Tomkinson AE, Bonk RT, Kim J, Barfeld N, Linn S (1990) Mammalian mitochondrial endonuclease activities specific for ultraviolet-irradiated DNA. Nucleic Acids Res 18:929–935

Trounce I, Byrne E, Marzuki S (1989) Decline in skeletal muscle mitochondrial respiratory chain function: possible factor in aging. Lancet 1:637–639

Vauwter L, Brown WM (1986) Nuclear and mitochondrial DNA comparisons reveal extreme rate variation in the molecular clock. Science 234:194–196

Wallace DC, Stugard C, Murdock D, Schurr T, Brown MD (1997) Ancient mtDNA sequences in the human nuclear genome. a potential source of errors in identifying pathogenic mutations. Proc Natl Acad Sci USA 94:14900–14905

Yang JH, Lee H-C, Lin KJ, Wei Y-HA (1994) Specific 4977 bp deletion of mitochondrial DNA in human aging skin. Arch Dermatol Res 286:386–390

Yu C-E, Oshima J, Fu Y-H, Wijsman EM, Hisama F, Alisch R, Matthews S, Nakura J, Miki T, Ouais S, Martin GM, Mulligan J, Schellenberg GC (1996) Positional cloning of Werner's syndrome gene. Science 272:258–262

CHAPTER 17

Roles of Mitochondria in Apoptosis 17

B. Mignotte[1] and G. Kroemer[2]

Contents

1 Programmed Cell Death Is a Fundamental Process During Development

Programmed cell death appears as a very early event in the course of evolution, limiting the size cellular populations and eliminating some undesirable cells (Ellis et al. 1991). This process is fundemental for the development of multicellular organisms, in

[1] CNRS UPRES-A 8087, Université de Versailles/Saint-Quentin, 78035 Versailles, France
[2] CNRS UPR420, IRC, 94800 Villejuif, France

the course of which many embryonic cells die. Programmed cell death and proliferation helps determine the size and form of organs, as well as the functional maturation of some systems. During the development of limbs, cell profileration and differentiation allow the appearance and the growth of limb buds, while the morphogenesis of fingers and toes invokes the death cells initially located in interdigital positions. During the development of the nervous system, neurons that do not reach their target are eliminated by a process of programmed cell death (Thompson 1995). This death allows the establishment of functions of the nervous system by playing on its plasticity. The functional maturation of the immune system also involves massive programmed cell death. The clones of self-reactive T lymphocytes are eliminated by a process of programmed cell death. In other cases, structures whose physiological role is only transitory are eliminated, for example during the metamorphosis of amphibians (tail of tadpoles) and insects (intersegmental muscles).

2
Apoptosis Is a Type of Programmed Cell Death Whose Control Is Altered in Several Pathologies

What is the mechanism of programmed cell death? How is it controlled? Different types of programmed cell death can be defined according to morphological criteria (Clarke 1990; Schwartz et al. 1993). Apoptosis, defined in 1972 (Kerr et al. 1972), is the most frequently described type of programmed cell death. Apoptosis is often opposed to necrosis, which is a cellular death that results generally from massive aggressions of the external medium. A dysfunction of apoptosis or of its control can be implicated in various pathologies (Thompson 1995). A defect in the cell death program could be responsible for some autoimmune diseases (Tan 1994). It is also implicated in oncogenesis. Indeed, apoptosis plays a fundamental role in different steps of carcinogenesis (Williams 1991; Bursch et al. 1992). For example, in the course of hepatocarcinogenesis, initiated cells have an apoptocic activity that compensates the intense proliferative activity of the other cells. It results in elimination of most of the intiated cells. Apoptosis can also be observed at other stages of tumor progression as well as in other types of cancers. The control of cell profileration by mitogens and apoptosis is always active in neoplastic cells and even in some malignant tumors with, however, an imbalance between the two phenomena. Some oncogens and oncosuppresor genes are now known for their action on apoptosis. The p53 oncosuppressor induces apoptosis in several transformed cell lines (Yonish-Rouach et al. 1991; Shaw et al. 1992) and the *c-myc* gene als seems to be involved in this process (Evan et al. 1992; Shi et al. 1992). On the other hand, the *bcl-2* oncogene can block apoptosis in different situations, and this capacity is probably involved in the genesis of lymphomas and many cancers (Hockenberry et al. 1990; Sentman et al. 1991; Strasser et al. 1991).

Other pathologies can be due to an ectopic induction of apoptosis. Some neurodegenerative diseases, during which a massive death is observed, could involve apoptosis. This could be the case for Alzheimer and Parkinson diseases and for the amyotrophic lateral sclerosis (Thompson 1995). This same phenomenon could also be involved in the pathogenesis of AIDS where it would lead to the elimination of T lymphocytes and the to immunodefeciency (Gougeon and Montagnier 1993; Martin 1993; Ameisen et al. 1995).

3 Programmed Cell Death Is Genetically Controlled

A genetic approach with the nematode *Caenorhabditis elegans*, has shown that condemned cells express genes that participate actively in apoptosis *(ced-3, ced-4)*. If the coordinated expression of these genes is disturbed, cells do not die. A gene, *ced-9*, capable of blocking apoptosis has also been found. *Ced-9* is the homologue of *bcl-2* which belongs to a family of genes that inhibits or stimulates apoptosis. *Ced-3* is the homologue of a family of genes of cystein proteases (caspases), which the gene *ICE*, which codes for the enzyme for conversion of interleukin-1β, belongs to (Table 1). More recently, a homologue of *Ced-4*, called Apaf-1, has also been identified (Zou et al. 1997). These observations emphasize that this process appears early during evolution. The mode of action of genes of the *bcl-2* family is poorly understood and the substrates of the caspases really implicated in apoptosis remain to be ascertained (Kumar 1995; Guénal et al. 1997a).

In mammals, many other genes have been described as having a negative or positive action on apoptosis. Some of these genes are known to be involved in the control

Table 1. Programmed cell death (PCD) in nematodes and mammals is controlled by homologous genes. Mammalian caspases act either during the activation or the execution phase of PCD. In mammals, some members of the Bcl-2 related proteins are death antagonists while others are death agonists

C. elegans	Mammals
CED-3	*Activation caspases:* caspase-1 (ICE), -4 (ICH-2), -6 (Mch2), -8 (MACH/FLICE) ... *Execution caspases:* caspase-2 (ICH1), -3 (CPP32), -4 (ICH-2), -7 (ICE-LAP3) ...
CED-4	Apaf-1
CED-9	*Anti-apoptotic:* Bcl-2, Bcl-x_L, Bcl-w, Bfl-1, Brag-1, Mcl-1, A1, NR13 ... *Pro-apoptotic:* Bax, Bak, Bcl-x_S, Bad, Bik, Hrk ...

Table 2. Many viral gene products act as antagonists of pro-apoptotic cellular proteins

Pro-apoptotic cellular proteins	Anti-apoptotic viral proteins
p53	Adenovirus E1B 55k SV40 large T HPV E6
FADD (adatator)	v-FLIPs
Caspases of the ICE/Ced-3 family	Cowpox virus Crm A Baculovirus p35
Pro-apoptotic members of the Bcl-2 family	Epstein-Barr virus BHRF1 LMW5HL Adenovirus E1B 19k

of cellular proliferation, this being in agreement with the idea that death by apoptosis allows compensation of cellular multiplication and that it is involved in protection against cancer. Furthermore, many viral gene products are inhibitors of apoptosis by acting as antagonists of cellular proteins implicated in apoptosis (Table 2). This observation shows that these different viruses have developed, in the course of evolution, different strategies to avoid the death of the infected cell before the end of the viral cycle (Gillet and Brun 1996).

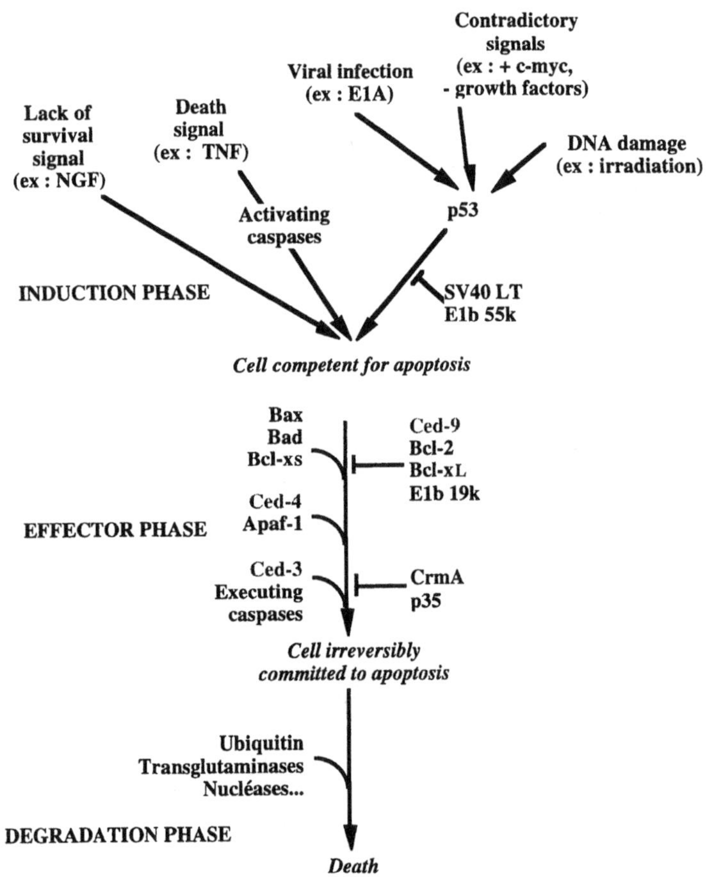

Fig. 1. Simplified model of events occuring during apoptosis. Several induction pathways (lack of survival signal, death signal, viral infection, contradictory signals for proliferation or DNA damage) lead to a series of events similar for all apoptotic deaths. First there is an induction phase leading to an effector (or execution) phase during which only a few morphological changes are visible, and then a degradation phase during which the morphological changes of apoptosis become evident

4 The Different Paths of Induction of Apoptosis Seem to Converge on a Common Effector Phase

Figure 1 presents a simplified model of events occurring during apoptosis. Three phases can be defined. The induction phase can proceed through various steps depending on the cellular type and the nature of the induction signal: default in survival signals, death signals (TGF-β, TNF-α), viral infection, contradictory signals for profileration (example: presence of c-myc and absence of growth factor), damage in DNA. During apoptosis induced by CD95-like surface receptors, an activation of caspases is observed during the induction phase. At the end of this induction phase, cells undergo a series of physiological modifications (effector phase) that is going to lead the irreversibly to death. Then, cells are committed to the degradation phase where they present morphological (condensation of the cell, nucleus and chromatin) and molecular (fragmentation of the DNA in oligonucleosomal fragments) characteristics of apoptosis. In some cases, the phagocytosis of apoptotic cells occurs before these modifications become perceptible. Products of *ced-3, ced-4* and *ced-9* as well as of their mammalian homologues intevene during the effector phase. Several proteins involved in the phase of degradation (ubiquitin, transglutaminase, nuclease) have also been identified.

5 A Drop in the Mitochondrial Membrane Potential Is a Universal and Early Event of Apoptosis

What is the place of mitochondria in apoptosis? The mitochondrial membrane potential ($\Delta\Psi_m$) can be estimated with fluorochromes that incorporate into mitochondria as a function of their potential, according to the Nernst equation (Table 3). The commitment to apoptosis is accompanied by a drop in the $\Delta\Psi_m$ which occurs before the fragmentation of DNA in oligonucleosomal fragments. In the case of apoptosis of fibroblasts immortalized by a temperature sensitive mutant of SV40 LT antigen, where apoptosis is medicated by the p53 protein (Zheng et al. 1994), this drop in $\Delta\Psi_m$ is responsible for a defect in maturation of mitochondrial proteins synthesized in the cytoplasm (Mignotte et al. 1990), cessation of mitochondrial translation and an uncoupling of oxidative phosphorylation (Vayssière et al. 1994). The drop in $\Delta\Psi_m$ is detectable whatever the apoptotic induction signal; physiological (absence of growth factor, glucocorticoids, TNF) or non-physiological (irradiation, chemotherapy; Petit et al. 1995; Zamzami et al. 1995; Marchetti et al. 1996; Sidoti-de Fraisse et al. 1998). Therefore, it seems that the drop in $\Delta\Psi_m$, an event that can be slowed down by cyclosporine A, is a universal characteristic which accompanies apoptosis (Kroemer et al. 1995). These data show, on the one hand, that nuclear fragmentation is a late event compared with the drop in $\Delta\Psi_m$ and, on the other hand, that this drop marks the point of no return of a cell condemned to die. Measurement of $\Delta\Psi_m$ allows identification of cells in "pre-apoptosis", including circulating T lymphocytes from human immunodeficiency virus-1 (HIV-1) carriers (Macho et al. 1995).

Table 3. Some fluorochromes used for the study of mitochondrial and nuclear alterations during apoptosis

Fluorochrome	Specificity
Annexine V	Binds to phosphatydilserine upon its exposure to the outer leaflet of the plasma membrane
Rhodamine 123 3,3' Dihexyloxacarbocyanine iodide (DIOC$_6$(3)) 5,5'6,6'-Tetrachloro-1,1',3,3'-tetraethylbenzimidazolcarbocy anine iodide (JC-1)	Accumulates in mitochondria as a function of the mitochondrial membrane potential
Dihydroethidine (HE)	Reacts with reactive oxygen species (mainly superoxide anion), and is converted to ethidium bromide
Dichlorofluorsceine-diacetate (DCFH-DA)	Reacts with reactive oxygen species (mainly hydrogen peroxide), and is converted to dichlorofluorescein
Monochlorobimane	Forms a fluorescent GST catalyzed adduct with intracellular GSH
Nonyl acridine orange (NAO)	Binds stoichiometrically to cardiolipins of the inner mitochondrial membrane
cis-Parinaric acid	Peroxidation of membranous lipids
Propidium iodide (PI)	Membranes from viable cells are not permeable to this fluorochrome that intercalates in DNA
dUTP-FITC dUTP-biotine	Incorporates in DNA breaks by the action of DNA polymerase or terminal desoxyribonucleotidyl transferase

A kinetic analysis of splenocyte, T cell hybridoma and pre-B cell apoptosis realized with the help of hydroethidine (Table 3) shows reactive oxygen species (ROS) production after the drop in $\Delta\Psi_m$ (Zamzami et al. 1995a). This step can be selectively inhibited by rotenone (inhibitor of the complex I of the respiratory chain) and by ruthenium red (inhibitor to the transporter of calcium of the inner mitochondrial membrane). The increased production of ROS seems to be accompanied by local effects on the cardiolipins of the inner mitochondrial membrane. The alteration of the cardiolipins is detectable by labelling cells with the probe 10-nonyl-acridine orange (Table 3) which binds in a stoichiometric manner to the intact cardiolipins of the inner mitochondrial membrane. At this stage, cells possess a normal size and a plasma membrane whose permeability is not altered. These changes are visible only in final step, coinciding with a diminution of cell size, an increase of membrane permeability and also degradation of the nuclear DNA (Kroemer et al. 1995). This last step can be delayed by compounds that eliminate free radicals, which suggests the implication of ROS in the execution of the apoptotic process.

6 Direct Interventions on Mitochondrial Permeability Transition Modulate Apoptosis

The permeability transition (PT), phenomenon inhibited by cyclosporine A, is characterized by the opening of mitochondrial megachannels that allow permeability to compounds of molecular weight <1500 Da, a reduction in $\Delta\Psi_m$ and the arrest of ATP synthesis (Bernardi et al. 1992; Petronilli et al. 1994). Interestingly, permeability transition has properties of self-amplification. Indeed, the drop of the $\Delta\Psi_m$, that is linked to depletion of non-oxidized glutathione (Macho et al. 1997) and that result from the opening of the PT pores would increase the permeability transition in a retrograde manner (Ichas et al. 1997). We have therefore proposed that the opening of the PT pore may constitute an irreversible state of the effector phase of apoptosis and could account for the apparent synchronization in the drop in $\Delta\Psi_m$ which takes place simultaneously in all the mitochondria of one cell (Kroemer et al. 1995). The molecular composition of these megachannels is not entirely known. The peripheral benzodiazepin receptor, which has recently been implicated in protection against ROS (Carayon et al. 1996), and the translocase of adenine nucleotides (ANT) are components of this channel. Indeed, among compounds described as being able to regulate the PT are two specific inhibitors to the ANT: atractyloside and bongkrekic acid (BA). These two moleculres could favor different conformations of the ANT: a closed conformation induced by BA and cyclosporine and an open conformation induced by atractyloside. The drop in $\Delta\Psi_m$, induced by the protoporphyrine IX, a ligand specific to the peripheral benzodiazepin receptor (which binds to ANT), entails apoptosis (Zamzami et al. 1996a). In contrast, N-methyl Val-4-cyclosporin A (a derivative of cyclosporine that is not an immunosuppressor) and bongkrekic acid prevent the drop in mitochondrial potential and the consequent fragmentation of DNA. These data show that the induction of PT provokes apoptosis while its inhibition provides a protection against it. These results sustain the hypothesis that permeability transition is involved in the drop in $\Delta\Psi_m$ observed during the effector phase of apoptosis (Kroemer et al. 1995; Petit et al. 1996; Mignotte and Vayssière 1998).

7 Mitochondria Can Control Nuclear Apoptosis

Direct alterations of mitochondria can induce apoptosis (Hartley et al. 1994; Wolvetang et al. 1994). The links between mitochondrial perturbations and nuclear alterations can be studied by means of an acellular system where purified nuclei and purified mitochondria are put together (Newmeyer et al. 1994). Such a system allows study of reciprocal and direct effects of one organelle on another and characterization at the biochemical level of the factors involved. The result of such a manipulation has shown that normal mitochondria have no effect on isolated cell nuclei, while the processing of mitochondria with compounds capable of induce the permeability transition (PT) is sufficient to provoke nuclear apoptosis (condensation of the chromatin and fragmentation of the DNA; Zamzami et al. 1996b). A strict correlation between induction of the PT and nuclear apoptosis has been observed by using a variety of known inductors of the PT such as atractyloside, pro-oxidants, calcium, protonophores and substances that provoke linkage of thiol groups such as diamide. These compounds, which have no direct effect on nuclei in the absence of mitochondria, confer pro-apoptotic properties upon mitochondria. The pro-apoptotic character (induction of nuclear apoptosis) of mitochondria treated with atractyloside is altered by inhibitors of the PT such as bongkrekic acid, cyclosporine A and compounds like monochlorobimane which block the cross-linking of the thiols. Cyclosporine A can be replaced by its non-immunosuppressor analogue, N-methyl Val-4-cyclosporine A, which shows that its inhibitory effect on PT and nuclear apoptosis is independent of its calcineurine activity. These results suggest the implication of the PT in the regulation of apoptosis induced via the mitochondria.

Finally, mitochondria isolated from apoptotic cells in vivo are capable of inducing nuclear apoptosis in the acellular system. Indeed, the induction of mouse hepatocyte apoptosis in vivo by a combination of D-galactosamine and lipopolysaccharide entails the reduction of $\Delta\Psi_m$. The mitochondria isolated from these cells provoke apoptosis of HeLa cell nuclei in vitro. A similar result has been obtained with mitochondria isolated from cells spleen processed with dexamethasone. These results show that mitochondria can effectively control nuclear apoptosis (Zamzami et al. 1996b). The mitochondria that undergo the PT would liberate a protein capable of inducing nuclear apoptosis. Some workers suggest that cytochrome c, by dissociating from the mitochondria, activates proteases implicated in nuclear apoptosis (Liu et al. 1996; Kluck et al. 1997; Yang et al. 1997). Other results suggest that the "apoptogenic" protein derived from the mitochondria is a protein of approximately 50 kDa called AIF (Apoptosis Inducing Factor ; Susin et al. 1996).

8
Bcl-2 Inhibits Permeability Transition and the Release of Cytochrome c and AIF in the Cytoplasm

Data concerning Bcl-2 underline the role of mitochondria in apoptosis (Kroemer 1997; Reed 1997; Mignotte and Vayssière 1998). Bcl-2, and other related proteins, are localized in intracellular membranes and specially in the external membrane of mitochondria (Krajewski ete al. 1993; Nakai et al. 1993; Nguyen et al. 1993; Akao et al. 1994; de Jong et al. 1994; Gonzalez-Garcia et al. 1994; Hickish et al. 1994; Janiak et al. 1994; Yang et al. 1995). Anti-apoptotic properties of Bcl-2 are affected in absence of its segment of anchorage to membranes (Tanaka et al. 1993; Nguyen et al. 1994; Zhu et al. 1996). However, it has been shown that *bcl-2* can block apoptosis of cells devoid of mitochondrial DNA ($\rho°$ cells; Jacobson et al. 1993). These experiments show that in cells which have undergone complex selection, and which possess mitochondria having $\Delta\Psi_m$ close to the normal (Skowronek et al. 1992; Marchetti et al. 1996; Sidoti-de Fraisse et al. 1998), the anti-apoptotic activity of *bcl-2* can always be exerted. However, it remains possible that the anti-apoptotic effect of *bcl-2* is linked to an activity which is always present in the mitochondria of $\rho°$ cells or that the anti-apoptotic power is exerted at several levels. It has effectively been shown that Bcl-2, in contrast to inhibitors of caspases, blocks the drop in $\Delta\Psi_m$ which occurs during cellular death induced by respiratory inhibitors (Shimizu and Eguchi 1996).

The importance of the structural location of Bcl-2 in nuclei or mitochondria in relation to its anti-apoptotic function has been studied. These organelles have been purified from hybridomas of T cells transfected by *bcl-2* (Zamzami et al. 1996b). In acellular experiments, the treatment with atractyloside has shown that, in contrast to mitochondria purified from control cells, mitochondria purified from cells transfected by *bcl-2* do not provoke nuclear apoptosis. On the contrary, nuclei purified from cells transfected by *bcl-2* show condensation of chromatin and fragmentation of DNA when they are put together to control mitochondria treated with atractyloside. Furthermore, *bcl-2* inhibits the induction of permeability transition (Decaudin et al. 1997), while *bax* promotes it (Pastorino et al. 1998). These results show that, even if Bcl-2 also intervenes during later events (Guénal et al. 1997b), at least a part of its inhibitory activity on apoptosis is exerted by acting on the mitochondrial permeability transition.

Moreover, the structure of a protein of the Bcl-2 family (Bcl-x_L) has been recently established (Muchmore et al. 1996). It is reminiscent of bacterial toxins, especially the diphtheria toxin, which form a pH-sensitive transmembrane channel. Furthermore, the pro-apoptotic Bax protein can form channels (Antonsson et al. 1997), as reported also for the anti-apoptotic proteins Bcl-x_L (Minn et al. 1997) and Bcl-2 (Schendel et al. 1997). However, the intrinsic properties of Bax and those of Bcl-x_L and Bcl-2 reveal differences. The channel-forming activity of Bcl-x_L and Bcl-2 is observed at highly acidic pH while Bax forms channels at a wide range of pHs including at pH 7, that found in cells. Furthermore, Bcl-2 can block the pore-forming activity of Bax. These results strengthen the hypothesis that these proteins are components of the mitochondrial PT pores. Bax might promote cell death by allowing efflux of ions and small molecules across the mitochondrial membranes, thus triggering permeability transition, while Bcl-2 might counteract this effect.

It has been shown that Bcl-2 inhibits the release of cytochrome c (Kluck et al. 1997; Yang et al. 1997) and of AIF (Susin et al. 1996) outside the mitochondria. Thus, one level of action of Bcl-2 is to control the efflux of proteins from the mitochondrial intermembrane space through the outer membrane and into the cytosol where caspases are found. In the same way, Bcl-x_L inhibits AraC-induced preapoptotic accumulation of cytochrome c in the cytosol (Kim et al. 1997), and cells which overexpress Bcl-x_L fail to accumulate cytosolic cytochrome c or to undergo apoptosis in response to genotoxic stress. These findings support the hypothesis that Bcl-x_L, as well as Bcl-2, protect cells from apoptosis by inhibiting the availability of cytochrome c in the cytosol. The mechanism allowing for the release of AIF and cytochrome c needs further investigation. PT pores, which are thought to connect the mitochondrial matrix to the cytosol, within the contact sites between inner and outer mitochondrial membranes (Beutner et al. 1996) are only permeable to small compounds (molecular mass <1500 Da). Thus PT pores are probably not the structure directly responsible for the efflux of these intermembrane space apoptogenic proteins. Future work will have to determine how AIF and cytochrome c leave mitochondria. Opening of the PT pore could be just the first step in a cascade of events causing an increase in the permeability of the outer mitochondrial membrane. Some results suggest that this permeability increase is due to mechanistic disruption secondary to an increase in matrix volume (Scarlett and Murphy 1997; Van der Heiden et al. 1997).

9
Bcl-2 Family Proteins Relocalize Cellular Proteins to Mitochondrial Membranes

Bcl-2 has alos been found to interact with several other cellular proteins that do not belong to the Bcl-2 related proteins family. These proteins include Nip1, Nip2, and Nip3, the function of which is unknown (Boyd et al. 1994), the GTPase R-ras p23 (Fernandez and Bischoff 1993), Raf-1 (Wang et al. 1994; Ali et al. 1997), BAG-1 (Takayama et al. 1995), the cellular prion protein (PrP; Kurschner and Morgan 1995), the p53 binding-protein p53-BP2 (Naumovski and Cleary 1996), the protein phosphatase calcineurin (Shibasaki et al. 1997) and the mitochondrial membrane protein carnitine palmitoyltransferase I (Paumen et al. 1997). At least some of these interactions could reflect the ability of Bcl-2 to relocalize cellular proteins to mitochondrial membranes.

Indeed, a connection has been made between members of the CED-9/Bcl-2 family and caspases. Mutations that reduce or eliminate CED-9 activity also disrupt its ability to bind CED-4 (Spector et al. 1997). CED-9 (and Bcl-x_L) was found to interact with and inhibit the function of CED-4. Furthermore, CED-4 can simultaneously interact with CED-3 (Irmler et al. 1997), (or mammalian caspases; Chinnaiyan et al. 1997b). Thus, CED-9 might control *C. elegans* cell death by binding to and regulating CED-4 and CED-3 activities. A similar mechanism could also exist in mammalian cells. Bcl-x_L interacts with caspase-1 (Chinnaiyan et al. 1997a) and inhibits the function of CED-4 (Chinnaiyan ete al. 1997b). Furthermore, the human protease activating factor (Apaf-1), which participates in the cytochrome c-dependent activation of caspase-3, contains a caspase-recruitment domain (CARD). Altogether, these results suggest that Apaf-1, like the nematode CED-4, acts by activating caspases and could provide the physical link between Bcl-2 related proteins and caspases (Zou et al. 1997). Since Bax and Bik can disrupt the association between CED-9 (or Bcl-xL) and CED-4 (Chinnaiyan et al. 1997b), it is tempting to speculate that Bcl-x_L, and possibly other members of the Bcl-2 protein family, inhibit apoptosis by maintaining the procaspases/Apaf-1 complexes associated with mitochondrial membranes, and that Bax and Bik permit the activation of procaspases by dissociating the complexes.

10
Mitochondria Participate in Apoptosis Signalling

Recently, it has appeared that mitochondria, in addition to their role during the effector phase of apoptosis, also intervene, in some cases, during the initiation phase of apoptosis. Much of the available data suggest the hypothesis that when ROS are involved in apoptosis signalling, they are the consequence of an impariment of the mitochondrial respiratory chain (Schulze-Osthoff et al. 1992; France-Lanord et al. 1997; Gudz et al. 1997; Quillet-Mary et al 1997; Sidoti-de Fraisse et al. 1998). What is the mechanism leading to the accumulation of mitochondrial ROS? Much of the available data lead to the hypothesis that ROS increases are also the consequence of an impairment of the mitochondrial respiratory chain (Schulze-Osthoff et al. 1992; France-Lanord et al. 1997; Gudz et al. 1997; Quillet-Mary et al. 1997). Indeed, it was shown that an upstream inhibition, with chemical compounds acting on complex I (Schulze-Osthoff et al. 1992; Quillet-Mary et al. 1997), or an elimination of the electron transfer chain by depletion of mtDNA prevent ROS accumulation and consequently protect cells against apoptosis induced by ceramide (Quillet-Mary et al. 1997) or TNF-α (Sidoti-de Fraisse et al. 1998). The ubiquinone site in complex III appears as the major site of mitochondrial ROS production as this site catalyzes the conversion of molecular oxygen to superoxide anion which can lead to the formation of other potent ROS such as hydrogen peroxide and hydrogen radicals. Such a model is supported by the observed potentiation of cell death processes in ROS-dependent apoptosis when electron flow was inhibited distal to the ubiquinone pool. However, the origin of these electron flow disturbances is not clear. The recently observed efflux of cytochrome c from mitochondria to cytosol in the early phases of apoptosis could provide a clue to this question (Liu et al. 1996; Kluck et al. 1997; Yang et al. 1997). Indeed, the release of cytochrome c must lead to a breakdown of the mitochondrial electron flow downstream of the ubiquinone site, which in turn would result in an

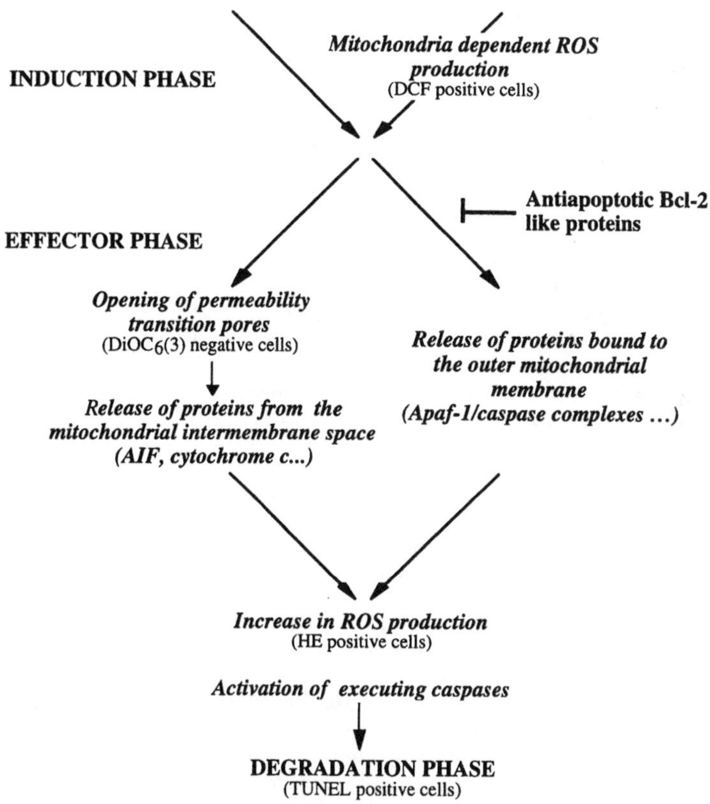

Fig. 2. Roles of mitochondria during apoptosis. First some apoptosis-inducing signals involve an increased mitochondrial ROS accumulation. Second, after the induction phase, mitochondria participate in the activation of executing caspases responsible for the morphological changes in two ways. On the one hand, Ced-4/Ced-3 (Apaf-1/caspase) complexes dissociate from intracellular membranes and are released in the cytosol. On the other hand, the opening of PT pores is a first step of a cascade of events causing an increase in the permeability of the outer mitochondrial membrane, an efflux of proteins as AIF or cytochrome c, an increase in ROS production and an activation of executing caspases

increased generation of ROS. Such a model is supported by the described correlation between loss of cytochrome c and respiratory failure in a Fas-induced apoptosis model (Krippner et al. 1996).

11 Conclusions

In conclusion, mitochondria are involved in the decision of cells to survive or not at several levels (Fig. 2). At a first level, mitochondria can contribute to apoptosis signalling, as shown in TNF-α- or ceramide-induced cell death during which increased mitochondrial ROS production appears as an early event of the induction phase. At a second level, mitochondria are involved in the control of the activation of the cell death machinery by docking at their surface, via Bcl-2 family proteins, execution caspases or by sequestering caspase activators as AIF or cytochrome c in the intermembrane space.

References

Akao Y, Otsuki Y, Kataoka S, Ito Y, Tsujimoto Y (1994) Multiple subcellular localization of Bcl-2: detection in nuclear outer membrane, endoplasmic reticulum membrane, and mitochondrial membranes. Cancer Res 54:2468–2471

Ali ST, Coggins JR, Jacobs HT (1997) The study of cell death proteins in the outer mitochondrial membrane chemical cross-linking. Biochem J 325:321–324

Ameisen JC, Estaquier J, Idziorek T, De Bels F (1995) The relevance of apoptosis to AIDS pathogenesis. Trends Cell Biol 5:27–32

Antonsson B, Conti F, Ciavatta A, Montessuit S, Lewis S, Martinou I, Bernasconi L, Bernard A, Mermod JJ, Mazzei G, Maundrell K, Gambale F, Sadoul R, Martinou JC (1997) Inhibition of Bax channel-forming acitivity by Bcl-2. Science 277:370–372

Bernardi P, Vassanelli S, Veronese P, Colonna R, Szabo I, Zoratti M (1992) Modulation of the mitochondrial permeability transition pore. Effect of protons and divalent cations. J Biol Chem 267:2934–2939

Beutner G, Ruck A, Riede B, Welte W, Brdiczka D (1996) Complexes between kinases, mitochondrial porin and adenylate translocator in rat brain resemble the permeability transition pore. FEBS Lett 396:189–195

Boyd JM, Malstrom S, Subramanian T, Venkatesh LK, Schaeper U, Elangovan B, D'Sa EC, Chinnadurai G (1994) Adenovirus E1B 19 kDa and Bcl-2 proteins interact with a common set of cellular proteins. Cell 79:341–351

Bursch W, Oberhammer F, Schulte-Hermann R (1992) Cell death by apoptosis and its protective role against disease. Trends Pharmacol Sci 13:245–251

Carayon P, Portier M, Dussossoy D, Bord A, Petitpretre G, Canat X, Le Fur G, Casellas P (1996) Involvement of peripheral benzodiazepine receptors in the protection of hematopoietic cells against oxygen radical damage. Blood 87:3170–3178

Chinnaiyan AM, Chaudhary D, O'Rourke K, Koonin EV, Dixit VM (1997a) Role of CED-4 in the activation of CED-3. Nature 388:728–729

Chinnaiyan AM, O'Rourke K, Lane BR, Dixit VM (1997b) Interaction of CED-4 with CED-3 and CED-9: a molecular framework for cell death. Science 275:1122–1126

Clarke PGH (1990) Developmental cell death: morphological diversity and multiple mechanisms. Anat Embryol (Berl) 181:195–213

De Jong D, Prins FA, Mason DY, Reed JC, van Ommen GB, Kluin PM (1994) Subcellular localization of the bcl-2 protein in malignant and normal lymphoid cells. Cancer Res 54:256–260

Decaudin D, Geley S, Hirsch T, Castedo M, Marchetti P, Macho A, Kofler R, Kroemer G (1997) Bcl-2 and Bcl-X_L antagonize the mitochondrial dysfunction preceding nuclear apoptosis induced by chemotherapeutic agents. Cancer Res 57:62–67

Ellis RE, Yuan JY, Horvitz HR (1991) Mechanisms and functions of cell death. Annu Rev Cell Biol 7:663–698

Evan GI, Wyllie AH, Gilbert CS, Littlewood TD, Land H, Brooks M, Waters CM, Penn LZ, Handcock DC (1992) Induction of apoptosis in fibroblasts by c-myc protein. Cell 69:119–128

Fernandez SM, Bischoff JR (1993) Bcl-2 associated with the ras-related protein R-ras p23. Nature 366:274–275

France-Lanord V, Brugg B, Michel PP, Agid Y, Ruberg M (1997) Mitochondrial free radical signal in ceramide-dependent apoptosis: a putative mechanism for neuronal death in Parkinson's disease. J Neurochem 69:1612–1621

Gillet G, Brun G (1996) Viral inhibition of apoptosis. Trends Mircobiol 4:312–317

Gonzalez-Garcia M, Perez-Ballestero R, Ding L, Duan L, Boise LH, Thompson CB, Nunez G (1994) bel-X_L is the major *bcl-x* mRNA from expressed during murine development and its product localizes to mitochondria. Development 120:3033–3042

Gougeon ML, Montagnier L (1993) Apoptosis in AIDS. Science 260:1269–1270

Gudz TI, Tserng KY, Hoppel CL (1997) Direct inhibition of mitochondrial respiratory chain complex III by cell-permeable ceramide. J Biol Chem 272:24154–24158

Guénal I, Risler Y, Mignotte B (1997a) Down regulation of actin genes precedes microfilament network disruption and actin cleavage during p53 mediated-apoptosis. J Cell Sci 110:489–495

Guénal I, Sidoti-de Fraisse C, Gaumer S, Mignotte B (1997b) Bcl-2 and Hsp27 act at different levels to suppress programmed cell death. Oncogene 15:347–360

Hartley A, Stone JM, Heron C, Cooper JM, Schapira AH (1994) Complex I inhibitors induce dose-dependent apoptosis in PC12 cells: relevance to Parkinson's disease. J Neurochem 63:1987–1990

Hickish T, Robertson D, Clarke P, Hill M, di Stefano F, Clarke C, Cunningham D (1994) Ultrastructural localization of BHRF1: an Epstein-Barr virus gene product which has homology with bcl-2. Cancer Res 54:2808–2811

Hockenberg D, Nunez G, Milliman C, Schreiber RD, Korsmeyer SJ (1990) Bcl-2 is an inner mitochondrial protein that blocks programmed cell death. Nature 348:334–336

Ichas F, Jouaville LS, Mazat JP (1997) Mitochondria are excitable organelles capable of generating and conveying electrical and calcium signals. Cell 89:1145–1153

Irmler M, Hofmann K, Vaux D, Tschopp J (1997) Direct physical interaction between the *Caenorhabditis elegans* "death proteins" CED-3 and CED-4. FEBS Lett 406:189–190

Jacobson MD, Burne JF, King MP, Miyashita T, Reed JC, Raff MC (1993) Bcl-2 blocks apoptosis in cells lacking mitochondrial DNA. Nature 361:365–369

Janiak F, Leber B, Andrews DW (1994) Assembly of Bcl-2 into microsomal and outer mitochondrial membranes. J Biol Chem 269:9842–9849

Kerr JFR, Wyllie AH, Currie AR (1972) Apoptosis: a basic biological phenomenon with wide-ranging implications in tissue kinetics. Br J Cancer 26:239–257

Kim CN, Wang X, Huang Y, Ibrado AM, Liu L, Fang G, Bhalla K (1997) Overexpression of Bcl-X(L) inhibits Ara-C-induced mitochondrial loss of cytochrome c and other perturbations that activate the molecular cascade of apoptosis. Cancer Res 57:3115–3120

Kluck RM, Bossy-Wetzel E, Green DR, Newmeyer DD (1997) The release of cytochrome c from mitochondria: a primary site for Bcl-2 regulation of apoptosis. Science 275:1132–1136

Krajewski S, Tanaka S, Takayama S, Schibler MJ, Fenton W, Reed JC (1993) Investigations of the subcellular distribution of the bcl-2 oncoprotein: residence in the nuclear envelope, endoplasmic reticulum, and outer mitochondrial membranes. Cancer Res 53:4701–4714

Krippner Q, Matsuno-Yagi A, Gootlieb RA, Babior BM (1996) Loss of function of cytochrome c in Jurkat cells undergoing Fas-mediated apoptosis. J Biol Chem 271:21629–21636

Kroemer G (1997) The proto-oncogene Bcl-2 and its role in regulating apoptosis. Nat Med 3:614–620

Kroemer G, Petit PX, Zamzami N, Vaysière JL, Mignotte B (1995) The biochemistry of programmed cell death. FASEB J 9:1277–1287

Kumar S (1995) ICE-like proteases in apoptosis. Trends Biochem Sci 20:198–202

Kurschner C, Morgan JI (1995) The cellular prion protein (PrP) selectively binds to Bcl-2 in the yeast two-hybrid system. Mol Brain Res 30:165–168

Liu X, Kim CN, Yang J, Jemmerson R, Wang X (1996) Induction of apoptotic program in cell-free extracts: requirement for dATP and cytochrome c. Cell 86:147–157

Macho A, Castedo M, Marchetti P, Aguilar JJ, Decaudin D, Zamzami N, Girard PM, Uriel J, Kroemer G (1995) Mitochondrial dysfunctions in circulating T lymphocytes from human immunodeficiency virus-1 carriers. Blood 86:2481–2487

Macho A, Hirsch T, Marzo I, Marchetti P, Dallaporta B, Susin SA, Zamzami N, Kroemer G (1997) Glutathione depletion in an early and calcium elevation is a late event of thymocyte apoptosis. J Immunol 158:4612–4619

Marchetti P, Susin SS, Decaudin D, Gamen S, Castedo M, Hirsch T, Zamzami N, Naval J, Senik A, Kroemer G (1996) Apoptosis-associated derangement of mitochondrial function in cells lacking mitochondrial DNA. Cancer Res 56:2033–2038

Martin SJ (1993) Programmed cell death and AIDS. Science 262:1355–1356

Mignotte B, Vayssière JL (1998) Mitochondria and apoptosis. Eur J Biochem 252:1–15

Mignotte B, Larcher JC, Zheng DQ, Esnault C, Couland D, Feunteun J (1990) SV40-induced cellular immortalization: phenotypic changes associated with the loss of proliferative capacity in a conditionally immortalized cell line. Oncogene 5:1529–1533

Minn AJ, Velez P, Schendel SL, Liang H, Muchmore SW, Fesik SW, Fill M, Thompson CB (1997) Bcl-x_L forms an ion channel in synthetic lipid membranes. Nature 385:353–357

Muchmore SW, Sattler M, Liang H, Meadows RP, Harlan JE, Yoon HS, Nettesheim D, Chang BS, Thompson CB, Wong Sl, Ng SC, Fesik SW (1996) X-ray and NMR structure of human Bcl-X_L, an inhibitor of programmed cell death. Nature 381:335–341

Nakai M, Takeda A, Cleary ML, Endo T (1993) The bcl-2 protein is inserted into the outer membrane but not into the inner membrane of rat liver mitochondria in vitro. Biochem Biophys Res Commun 196:233–239

Naumovski L, Cleary ML (1996) The p53-binding protein 53BP2 also interacts with Bcl-2 and impedes cell cycle progression at G2/M. Mol Cell Biol 16:3884–3892

Newmeyer DD, Farschon DM, Reed JC (1994) Cell-free apoptosis in Xenopus egg extracts: inhibition by Bcl-2 and requirement for an organelle fraction enriched in mitochondria. Cell 79:353–364

Nguyen M, Millar DG, Yong VW, Korsmeyer SJ, Shore GC (1993) Targeting of bcl-2 to the mitochondrial outer membrane by a COOH-terminal signal anchor sequence. J Biol Chem 268:25265–25268

Nguyen M, Branton PE, Walton PA, Oltvai ZN, Korsmeyer SJ, Shore GC (1994) Role of membrane anchor domain of Bcl-2 in suppression of apoptosis caused by E1B-defective adenovirus. J Biol Chem 269:16521–16524

Pastorino JG, Chen ST, Tafani M, Snyder JW, Farber JL (1998) The overexpression of bax produces cell death upon induction of the mitochondrial permeability transition. J Biol Chem 273:7770–7775

Paumen MB, Ishida Y, Han H, Muramatsu M, Eguchi Y, Tsujimoto Y, Honjo T (1997) Direct interaction of the mitochondrial membrane protein carnitine palmitoyltransferase I with Bcl-2. Biochem Biophys Res Commun 231:523–525

Petit PX, Lecoeur H, Zorn E, Dauguet C, Mignotte B, Gougeon ML (1995) Alterations in mitochondrial structure and function are early events of dexamethasone-induced thymocytes apoptosis. J Cell Biol 130:157–167

Petit PX, Susin SA, Zamzami N, Mignotte B, Kroemer G (1996) Mitochondria and programmed cell death: back to the future. FEBS Lett 396:7–13

Petronilli V, Nicolli A, Costantini P, Colonna R, Bernardi P (1994) Regulation of the permeability transition pore, a voltage-dependent mitochondrial channel inhibited by cyclosporin A. Biochim Biophys Acta 1187:255–259

Quillet-Mary A, Jaffrezou JP, Mansat V, Bordier C, Naval J, Laurent G (1997) Implication of mitochondrial hydrogen peroxide generation in ceramide-induced apoptosis. J Biol Chem 272:21388–21395

Reed JC (1997) Double identity for proteins of the Bcl-2 family. Nature 387:773–776

Scarlett JL, Murphy MP (1997) Release of apoptogenic proteins from the mitochondrial intermembrane space during the mitochondrial permeability transition. FEBS Lett 418:282–286

Schendel SL, Xie Z, Montal MO, Matsuyama S, Montal M, Reed JC (1997) Channel formation by antiapoptotic protein Bcl-2. Proc Natl Acad Sci USA 94:5113–5118

Schulze-Osthoff K, Bakker AC, Vanhaesebroeck B, Beyaert R, Jacob WA, Fiers W (1992) Cytotoxic activity of tumor necrosis factor is mediated by early damage of mitochondrial functions. Evidence for the involvement of mitochondrial radical generation. J Biol Chem 267:5317–5323

Schwartz LM, Smith SW, Jones MEE, Osborne BA (1993) Do all programmed cell deaths occur via apoptosis? Proc Natl Acad Sci USA 90:980–984

Sentman CL, Shutter JR, Hockenbery D, Kanagawa O, Korsmeyer SJ (1991) Bcl-2 inhibits multiple forms of apoptosis but not negative selection in thymocytes. Cell 67:879–888

Shaw P, Bovey R, Tardy S, Sahli R, Sordat B, Costa J (1992) Induction of apoptosis by wild-type p53 in a human colon tumor-derived cell line. Proc Natl Acad Sci USA 89:4495–4499

Shi Y, Glynn JM, Guilbert LJ, Cotter TG, Bissonnette RD, Green DR (1992) Role for c-myc in activation-induced apoptotic cell death in T cell hybridomas. Science 257:212–214

Shibasaki F, Kondo E, Akagi T, McKeon F (1997) Suppression of signalling through transcription factor NF-AT by interactions between calcineurin and Bcl-2. Nature 386:728–731

Shimizu S, Eguchi Y (1996) Bcl-2 blocks loss of mitochondrial membrane potential while ICE inhibitors act at a different step during inhibition of death induced by respiratory chain inhibitors. Oncogene 13:21–29

Sidoti-de Fraisse C, Rincheval V, Risler Y, Mignotte B, Vayssière JL (1998) TNF-α activates at least two apoptotic signaling cascades. Oncogene 17:1639–1651

Skowronek P, Haferkamp O, Rodel G (1992) A fluorescence-microscopic and flow cytometric study of HeLa cells with an experimentally induced respiratory deficiency. Biochem Biophys Res Commun 187:991–998

Spector MS, Desnoyers S, Hoeppner DJ, Hengartner MO (1997) Interaction between the *C. elegans* cell-death regulators CED-9 and CED-4. Nature 385:653–656

Strasser A, Harris AW, Cory S (1991) Bcl-2 transgene inhibits T-cell death and perturbs thymic self-censorship. Cell 67:889–899

Susin SA, Zamzami N, Castedo M, Hirsh T, Marchetti P, Macho A, Daugas E, Geuskens M, Kroemer G (1996) Bcl-2 inhibits the mitochondrial release of an apoptogenic protease. J Exp Med 184:1–111

Takayama S, Sato T, Krajewski S, Kochel K, Irie S, Millan JA, Reed JC (1995) Cloning and functional analysis of BAG-1: a novel Bcl-2-binding protein with anti-cell death activity. Cell 80:279–284

Tan EM (1994) Autoimmunity and apoptosis. J Exp Med 179:1083–1086

Tanaka S, Saito K, Reed JC (1993) Structure-function analysis of the Bcl-2 oncoprotein. Addition of a heterologous transmembrane domain to portions of the Bcl-2 beta protein restores function as a regulator of cell survival. J Biol Chem 268:10920–10926

Thompson CB (1995) Apoptosis in the pathogenesis and treatment of disease. Science 267:1456–1462

Van der Heiden MG, Chandel NS, Williamson EK, Schumaker PT, Thompson CB (1997) Bcl-xL regulates the membrane potential and volume homeostasis of mitochondria. Cell 91:627–637

Vayssière JL, Petit PX, Risler Y, Mignotte B (1994) Commitment to apoptosis is associated with changes in mitochondrial biogenesis and acitvity in cell lines conditionally immortalized with simian virus 40. Proc Natl Acad Sci USA 91:11752–11756

Wang HG, Miyashita T, Takayama S, Sato T, Torigoe T, Krajewski S, Tanaka S, Hovey LR, Troppmair J, Rapp UR, Reed JC (1994) Apoptosis regulation by interaction of Bcl-2 protein and Raf-1 kinase. Oncogene 9:2751–2756

Williams GT (1991) Programmed cell death: apoptosis and oncogenesis. Cell 65:1097–1098

Wolvetang EJ, Johnson KL, Krauer K, Ralph SJ, Limnane AW (1994) Mitochondrial respiratory chain inhibitors induce apoptosis. FEBS Lett 339:40–44

Yang J, Liu X, Bhalla K, Kim CN, Ibrado AM, Cai J, Peng TI, Jones DP, Wang X (1997) Prevention of apoptosis by Bcl-2: release of cytochrome c from mitochondria blocked. Science 275:1129–1132

Yang T, Kozopas KM, Craig RW (1995) The intracellular distribution and pattern of expression of Mcl-1 overlap with, but are not identical to, those of Bcl-2. J Cell Biol 128:1173–1184

Yonish-Rouach E, Resnitzky D, Lotem J, Sachs L, Kimchi A, Oren M (1991) Wild-type p53 induces apoptosis of myeloid leukaemic cells that is inhibited by interleukin-6. Nature 352:345–347

Zamzami N, Marchetti P, Castedo M, Decaudin D, Macho A, Petit PX, Mignotte B, Kroemer G (1995a) Sequential reduction of mitochondrial transmembrane potential and generation of reactive oxygen species in early programmed cell death. J Exp Med 182:367–377

Zamzami N, Marchetti P, Castedo M, Zanin C, Vayssière JL, Petit PX, Kroemer G (1995b) Reduction in mitochondrial potential constitutes an early irreversible step of programmed lymphocyte death in vivo. J Exp Med 181:1661–1672

Zamzami N, Marchetti P, Castedo M, Hirsch T, Susin SA, Masse B, Kroemer G (1996a) Inhibitors of permeability transition interfere with the disruption of the mitochondrial transmembrane potential during apoptosis. FEBS Lett 384:53–57

Zamzami N, Susin SA, Marchetti P, Hirsch T, Gomez-Monterrey I, Castedo M, Kroemer G (1996b) Mitochondrial control of nuclear apoptosis. J Exp Med 183:1533–1544

Zheng DQ, Vayssière JL, Lecoeur H, Petit PX, Spatz A, Mignotte B, Feunteun J (1994) Apoptosis is antogonized by large T antigens in the pathway to immortalization by polyomaviruses. Oncogene 9:3345–3351

Zhu W, Cowie A, Wasfy GW, Penn LZ, Leber B, Andrews DW (1996) Bcl-2 mutants with restricted subcellular location reveal spatially distinct pathways for apoptosis in different cell types. EMBO J 15:4130–4141

Zou H, Henzel WJ, Liu X, Lutschg A, Wang X (1997) Apaf-1, a human protein homologous to *C. elegans* CED-4, participates in cytochrome c-dependent activation of caspases-3. Cell 90:405–413

CHAPTER 18

The Triiodothyronine Mitochondrial Pathway

18

C. WRUTNIAK, P. ROCHARD, F. CASAS, and G. CABELLO

Contents

1 Introduction

Numerous data show that mitochondrial activity is hormonally regulated. Besides the influence of α-agonists, glucagon and vasopressin, several studies show that glucocorticoid, thyroid hormone, vitamin D3 and peroxisome proliferators stimulate the activity of this organelle. Interestingly, thyroid hormone, glucocorticoid, peroxisome proliferators and vitamin D3 receptors all belong to a common nuclear receptor superfamily. These data suggest that such receptors could have a particular importance in the regulation of mitochondrial activity. However, the mechanisms involved

INRA, Laboratoire de Différenciation Cellulaire et Croissance, Unité d'Endocrinologie Cellulaire, 2, place Viala, 34060 Montpellier Cedex 1, France in this endocrine influence remain poorly understood. The action of thyroid hormone, especially triiodothyronine (T3), upon the organelle has been studied, leading to the proposition of a direct T3 mitochondrial pathway. In this paper, we present numerous data in agreement with this hypothesis, with special reference to recent work from our laboratory identifying a T3 binding protein in the mitochondrial matrix.

2 The Nuclear T3 Pathway

It is generally assumed that the fundamental influence of thyroid hormone involves a nuclear pathway. In a pioneering study, Tata et al. (1963) first established that thyroid hormone stimulates the activity of ARN-polymerases I and II, thus focusing the interest of numerous thyroid scientists upon nuclear thyroid hormone receptors. As a consequence, several studies reported the presence of nuclear high affinity T3 binding proteins (KA: 10^9 M^{-1}). Additional studies also suggested the occurrence of several nuclear T3 receptors (Casanova et al. 1984; Ichikawa and De Groot 1987).

Whereas all attempts to purify these T3 binding proteins remained unsuccessful, characterization of the v-erb A oncogen in the avian erythroblastosis virus genome (A.E.V.) unexpectedly led to the identification of nuclear T3 receptors. In 1986, two groups simultaneously demonstrated that proteins encoded by two different c-erb A genes (α and β), cellular counterparts of the v-erb A oncogen, displayed a nuclear location. In addition, they bound T3 with an affinity similar to that previously recorded for native nuclear receptors (Weinberger et al. 1986; Sap et al. 1986).

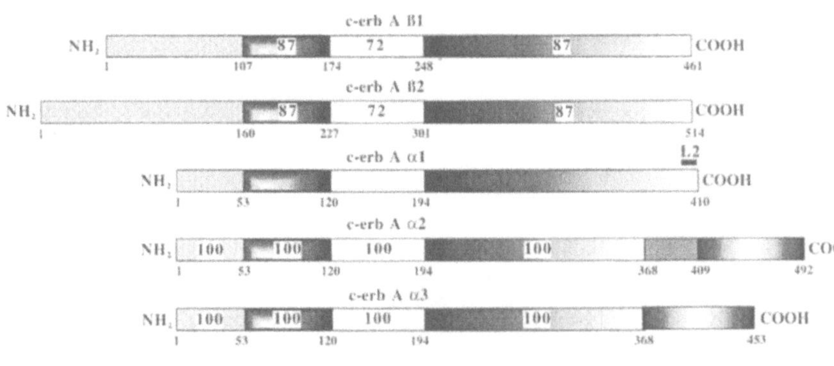

Fig. 1. Comparison of c-erb A proteins. The c-erb A β gene encodes c-erb A β1 and c-erb A β2 receptors. In some species, the c-erb A α gene encodes the c-erb A α1 receptor and c-erb A α2 and c-erb A α3 proteins which do not bind T3 due to the absence of a short C-terminal domain (L2) needed for hormone binding. The first and last amino acids of each functional domain are given below each protein. *Numbers* given in each functional domain provide the percent of amino acid identity in relation to c-erb A α1. Homologies lower than 15% are not mentioned. ▒ NH2 terminal domain containing a T3-independent transactivation domain. ■ DNA binding domain. □ Hinge region containing a nuclear localization signal anc co-repressor interacting sequences. ■ Multifunctional domain containing dimerization, T3 binding and co-activators interacting sequences. Regions 409–492 (α2) and 368–453 (α3) are identical

This discovery leads to a spectacular progression in the knowledge of the nuclear T3 pathway. Two messengers, c-erb Aβ1 and c-erb Aβ2, are produced by alternative splicing from the c-erb A β gene. The corresponding proteins only differ by their NH2 terminus (Fig. 1). Whereas c-erb Aβ1 is expressed in practically all tissues, c-erb A β2 expression is restricted to the pituitary. In addition, three messengers are also produced by alternative splicing from the c-erb A α gene. The α-encoded proteins differ by their C-terminus (Fig. 1). The α2 isoform identified in the rat, mouse and man, and the α3 isoform (rat, mouse) do not bind T3 because they lack a small functional domain located in the α1 C-terminus, which plays an essential role for T3 binding. In addition, c-erb A α2 binds to the same nucleotidic sequence as c-erb A α1, with a reduced affinity. However, as the α2 isoform is devoid of positive transcriptional activity, it is generally assumed that this particular variant could antagonize c-erb A α1 and β receptor's activity.

C-erb A receptors are T3 dependent transcription factors which bind as dimeric complexes on specific nucleotidic sequences (T3REs) generally located in target gene promoters. In the absence of ligand, they are associated with transcriptional repressors (Horlein et al. 1995; Chen and Evans 1995) which inhibit gene expression. T3 binding induces a change in receptor conformation leading to a dissociation of transcriptional repressors. Simultaneously, it allows the exposure of a functional domain of the receptor which interacts with co-activators (Fig. 2; for review, Horwitz et al. 1996). In these conditions, the T3 activated receptor strongly stimulates target gene expression. However, through some particular TREs (negative T3REs), the liganded receptor inhibits transcription.

T3REs generally comprise two half-sites. The nucleotidic sequence of each of them displays some variations around the AGGTCA consensus motif. Today, three groups of T3REs have been characterized: palindromic sequences (AGGTCA TGACCT), direct repeats (AGGTCA nnnn AGGTCA in which n is a variable nucleotide) and inverted palindromes (TGACCT nnnnnnn AGGTCA). The half-sites are not generally perfectly

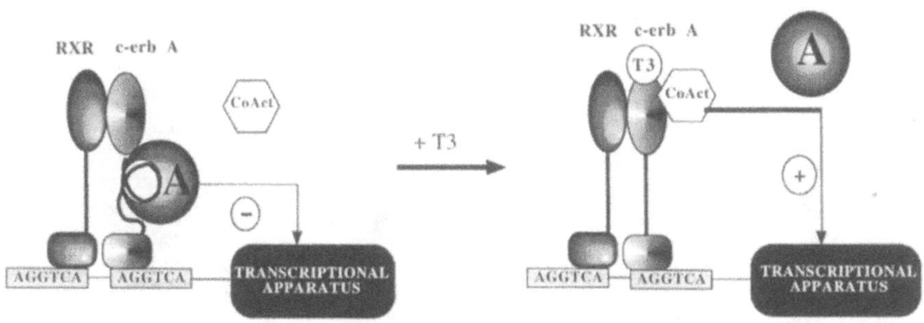

Fig. 2. In contrast to the glucocorticoid receptor, sequestrated in the cytosol with heat shock proteins (hsp 90, hsp 70 et hsp 56), in the absence of the hormone, nuclear T3 receptors are translocated into the nucleus and bind as homodimeric or heterodimeric complexes upon T3REs sequences identified on the promoter of target genes, independently of hormone presence. Without ligands, these receptors are complexed with transcriptional inhibitors *(A)* repressing the basal transcription rate (NCor ou SMRT). T3 binding to its receptors induces a conformational change leading to: (1) dissociation of transcriptional inhibitors; (2) exposure of a functional domain of the receptor, allowing direct interactions with several transcriptional coactivators (CoAct) inducing bridging with the transcriptional apparatus and probably modifying the structure of chromatin, thus stimulating transcription of the target gene

symetric or identical and a number of significant nucleotidic variations could be observed in each of them. Consequently, in relation to this variability, classification of such sequences as T3RE needs the demonstration that they are recognized by c-erb A proteins and that they drive T3 activation of gene transcription. More complex T3REs, including multiple half-sites, have been also identified, in particular in the GH gene promoter.

Nuclear T3 receptors generally bind to DNA as homodimeric or heterodimeric complexes. All major heterodimerization partners of c-erb A belong to the subgroup II of the nuclar receptors superfamily (Forman and Samuels 1990): retinoic acid, 9-cis-retinoic acid (retinoid X), vitamin D3 receptors, as well as peroxisome proliferator-activated receptor (PPAR), considered for a time as an orphan receptor. COUP-TF (chicken ovalbumin upstream promoter-transcription factor), an orphan receptor belonging to the superfamily subgroup III (Forman and Samuels 1990), has also been proposed as a heterodimerization partner.

Today, numerous reports suggest that the retinoic x receptor (RXR) is absolutely required for the transcriptional activity of c-erb A (Perlmann et al. 1993; Piedrafita et al. 1995). However, more recent data obtained using RXR free-vertebrate cells clearly indicate that the functional importance of this partner could be restricted to the induction of transcription trough direct repeat sequence 4 (DR4) T3REs (Cassar-

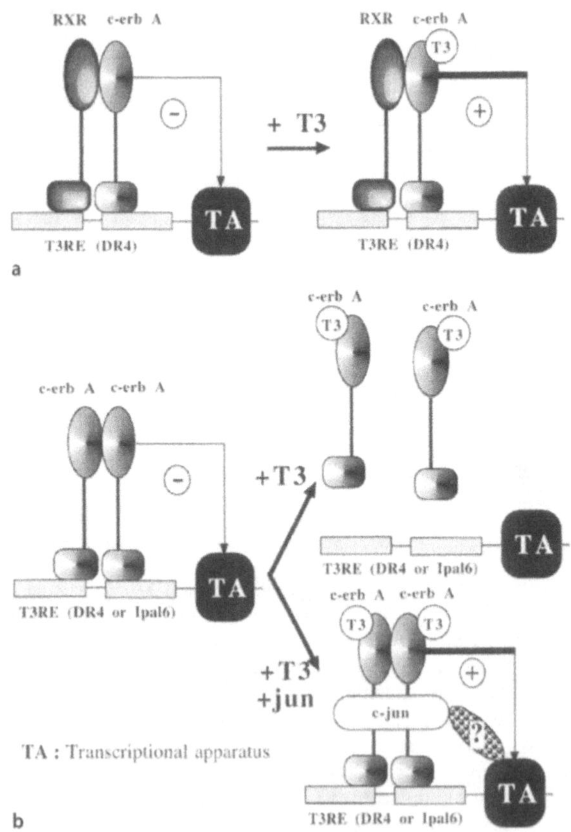

Fig. 3. Without triiodothyronine, the T3 receptor binds to several kinds of response elements as homodimeric (**b**) or heterodimeric complexes involving in particular RXR (**a**); under these conditions, it represses basal transcription of the target gene (see Fig. 2). In the presence of T3, the heterodimer stimulates transcription at least through a DR4-T3RE (**a**). In contrast, T3 inhibits homodimer binding to DR4 and IPal 6-T3REs (**b**), thus only inducing an abrogation of transcriptional repression by the unliganded complexes (restoration of the basal transcription rate). However, when c-jun expression is elevated, this oncoprotein probably stabilizes homodimer binding to T3REs in conjunction with a partner not yet identified, and induces a T3-dependent transcriptional activation through DR4 and IPal 6-T3REs. (Cassar-Malek et al. 1996b)

Malek et al. 1996a, b). Heterodimerization with the other partners could differentially modulate the T3 transcriptional activity upon its different target genes, and, in some cases, could induce a shift in the set of stimulated genes (Bogazzi et al. 1994). Several reports indicate that the c-erb A homodimer binding to DNA is inhibited by T3, and the transcriptional activity of the homodimer is still debated (Yen et al. 1992; Andersson et al. 1992). However, recent data indicate that expression of the c-jun oncoprotein restores a full transcriptional activity to the liganded complex, probably by stabilizing homodimer binding to DNA (Cassar-Malek et al. 1996a; Fig. 3).

Lastly, besides their direct transcriptional activity, liganded c-erb A T3 receptors repress the activity of the AP-1 (jun/fos) complex (Desbois et al. 1991; Zhang et al. 1991). Although the molecular basis of this influence is not clearly understood, it appears that RXR expression is required to induce the functionality of this interaction (Cassar-Malek et al. 1996a).

3 The T3 Mitochondrial Pathway

Besides the well-documented nuclear pathway, numerous data demonstrate that T3 exerts influence upon mitochondrial activity. In a recent review, we have classified such influence into three groups in relation to their latency periods, probably reflecting the involvement of different mechanisms (Wrutniak and Cabello 1996).

3.1 The Early Thyroid Hormone Influence

Sterling's group first demonstrated that T3 very quickly stimulates mitochondrial oxygen consumption and ATP synthesis (Sterling et al. 1977, 1980). This influence was detected after a latency period as short as 15 min in vivo and 2 min using isolated mitochondria. More recently, Sterling and Brenner (1995) reported in vivo that T3 also stimulates mitochondrial translocase activity in less than 15 min. In addition, these early actions were not affected by inhibition of nuclear or mitochondrial protein synthesis. Therefore, such influences display three features ruling out the involvement of T3 nuclear receptors: (1) a very short latency period; (2) occurrence in the absence of nuclear influence (isolated mitochondria); (3) insensitivity to protein synthesis inhibitors. These features strongly support the hypothesis that an extra-genomic pathway is involved in the early T3 mitochondrial influence.

Although several more or less convincing hypotheses have been proposed, these influences remain poorly understood. One of them involves changes in mitochondrial Ca influx, on the following basis. In less than 1 min, T3 stimulates Ca^{2+} entry into the cell (Segal 1990). In addition, glucagon and vasopressine exert a similar action and, as T3, induce an early stimulation of mitochondrial activity (Soboll and Sies 1989). Lastly, a stimulation of mitochondrial calcium influx potentiates dehydrogenase activities (Hansford 1991). Such a mechanism could satisfactorily explain the early T3 influences. However, in this hypothesis it is also assumed that a T3-induced increase in cell calcium influx is related to a rise in the Ca^{2+} mitochondrial pool. Unfortunately, this assumption disagrees with the observation of Murphy (1992) showing that, in fact, T3 reduces the calcium mitochondrial uptake in relation to a decreased mitochondrial membrane potential.

Hardy and Mowbray (1992) proposed another mechanism, which was quite unexpected. According to these workers, T3 induces the ADP-ribosylation of a 11 kDa inner membrane protein, before increasing oxygen consumption. In addition, they provide evidence that the early T3 mitochondrial influences are repressed by an ADP-ribosylation inhibitor (Thomas and Mowbray 1987). In a more recent study (Mowbray and Hardy 1996), it is suggested that the mitochondrial translocator could be the target of this ADP-ribosylation. However, such assumption remains to be experimentally confirmed.

Lastly, Goglia et al. (1994) observed that diiodothyronines (3.3'-T2 and 3,5-T2, physiological T3 derivates obtained by deiodination of the hormone) bind to purified cytochrome oxidase, induce conformation changes and activate this enzyme located in the inner membrane. The occurrence of such a mechanism in vivo or in isolated mitochondria remains to be confirmed, but this hypothesis is in agreement with the observation of mitochondrial diiodothyronines-specific binding sites (Lanni et al. 1994).

3.2
The Rapid Thyroid Hormone Influence

Two T3 influences, occuring in few hours, have been described: (1) a reduction in the mitochondrial membrane potential; (2) a stimulation in the expression of the mitochondrial genome.

3.2.1
Influence on the Mitochondrial Membrane Potential

The passive proton leak occurring across the mitochondrial inner membrane is one of the parameters regulating the membrane potential of the organite. In rat hepatic cells, the proton leak represents 22% of the control of mitochondrial oxygen consumption (Brown 1992).

The respiratory state IV is characterized by an uncoupling of oxidative phosphorylation in which mitochondrial oxygen consumption is essentially affected by the proton leak across the inner membrane. In these conditions, T3 increases the inner membrane permeability to protons of isolated mitochondria (Harper et al. 1993); this influence in turn reduces the membrane potential of the organite and consequently induces a stimulation of oxygen consumption (Hafner et al. 1988). Changes in inner membrane permeability are detected in less than 12 h after T3 administration. Moreover, they are not affected by actinomycin D (Horrum et al. 1992). These data obtained using isolated mitochondria are in agreement with experiments involving isolated hepatocytes in which mitochondrial respiration status is intermediary between states III and IV (Gregory and Berry 1991).

The T3 influence on inner membrane permeability to protons has been partly explained by an increase in the area of the mitochondrial inner membrane (Brand et al. 1992). However, changes induced by T3 in the compositon and saturation of fatty acids included in membrane phospholipids appear to be a major factor (Brand et al. 1992). In particular, cardiolipin content, a phospholipid specifically present in mitochondria, is reduced by hypothyroidy and increased by hyperthyroidy (Paradies and Ruggiero 1989, 1990); in turn, cardiolipin influences cytochrome oxidase activity, as well as the function of several mitochondrial carriers, including mitochondrial translocator. Therefore, such a mechanism could be involved in the stimulation of mitochondrial activity by T3.

More recent data provide another interesting way to explain the influence of T3 upon the mitochondrial membrane potential. UCP (mitochondrial uncoupling protein) is involved in uncoupling of oxidative phosphorylation, and as a consequence, in the regulation of the mitochondrial membrane potential. Cloning of two UCP-related proteins (UCP 2 and 3) showed the existence of a UCP family. In addition, whereas UCP1 is highly specific of brown adipose tissue, UCP3 is mainly expressed in muscle (Larkin et al. 1997) and UCP 2 is found in many tissues. These observations emphasize that at least one UCP-related protein probably occurs in each cell type. As T3 stimulates UCP-1 (Cassard-Doulcier et al. 1994; Guerra et al. 1996; Rabelo et al. 1996), UCP-2 (Lanni et al. 1997) and UCP-3 expression (Larkin et al. 1997), a part of the rapid thyroid hormone mitochondrial influence could be initiated at the nuclear level by stimulation of UCP synthesis.

3.2.2
Influence on Mitochondrial Genome Transcription

Numerous data demonstrated that thyroid hormone induces a rise in the amount of mitochondrial mRNAs by stimulating mitochondrial genome transcription (Gadaleta et al. 1972; Mutvei et al. 1989). These results are in agreement with the observation that T3 increases the steady state levels of mtTFA mRNAs. MtTFA is a transcription factor of the mitochondrial genome encoded by a nuclear gene (Fisher and Clayton 1988). Therefore, the hormone could regulate the synthesis of a mitochondrial transcription factor at the nuclear level.

However, the fact that T3 could increase the amounts of mitochondrial mRNAs with a latency period lower than 5 min in isolated mitochondria (Martino et al. 1986) is not in agreement with the hypothesis of an exclusive T3 nuclear pathway. This last observation raises the question of the occurrence of a possible mitochondrial T3 receptor, acting as a T3-dependent transcription factor.

3.3
The Long-Term Thyroid Hormone Influence

Among the delayed influences of thyroid hormone, the best established is probably the stimulation of mitochondriogenesis occurring after a latency period longer than 24 h (Gustafsson et al. 1965; Jakovilcic et al. 1978).

Proteins encoded by nuclear genes are largely predominant in the mitochondria. However, a reduced number (13 proteins) encoded by the mitochondrial genome are absolutely required for the functionality of the respiratory chain located in the inner membrane. The importance of this small number of proteins of mitochondrial origin is well demonstrated by the occurrence of mitochondrial cytopathies induced by mutations or deletions of the mitochondrial genome. Therefore, mitochondrial biogenesis needs a coordinated activation of nuclear and mitochondrial gene expression.

As previously discussed, T3 could directly activate the expression of nuclear genes encoding mitochondrial proteins by the way of c-erb A receptors. For instance, the β-F1 ATPase subunit is the product of a nuclear gene directly regulated by T3 (Izquierdo and Cuevas 1993). Similarly, despite the fact that regulation at the transcriptional level is not clearly demonstrated, the steady state level of cytochrome c1 mRNA, product of a nuclear gene, is upregulated by T3 (Joste et al. 1989). Lastly, as previously reported in this review, the steady-state level of mtTFA mRNA is positively regulated by the hormone (Garstka et al. 1994). The increase in the synthesis of mtTFA, able to stimulate mitochondrial genome expression after a few hours, could therefore sustain the rapid influence of a possible mitochondrial T3-dependent transcription factor previously hypothesized in this review.

4 First Evidence of the Occurrence of T3 Mitochondrial Receptors

Some observations suggest the question of the occurrence of a direct T3 mitochondrial pathway: (1) the observation of early influences of this hormone, insensitive to protein synthesis inhibitors and occurring in isolated mitochondria and (2) a rapid stimulation of the mitochondrial genome expression by T3. The existence of a T3 mitochondrial receptor was first proposed by Sterling's group. These workers discovered rapid and important mitochondrial T3 uptake (Sterling et al. 1984). Therefore, whereas it was generally assumed that T3 exerts its cellular action at the nuclear level, mitochondria are a major T3 storage compartment. In addition, three different groups, working independently, showed the presence of specific T3 binding sites in mitochondria (Sterling and Milch 1975; Goglia et al. 1981; Hashizume and Ichikawa 1982).

5 Truncated c-erb Aα1 Proteins are Located in Mitochondria (Wrutniak et al. 1995)

A photoaffinity labelling technique (PAL-T3), previously used to demonstrate heterogeneity of T3 nuclear receptors (Casanova et al. 1984) was used on highly purified mitochondrial preparations obtained from rat liver. Two specific T3-binding proteins were detected: a minor 28 kDa protein located in the inner membrane, and therefore displaying a molecular weight and a localization in agreement with Sterling's data, and a major 43 kDa protein located in the mitochondrial matrix.

Several indications led to the search for c-erb A-related proteins in mitochondria. Firstly, transfection of a cDNA enabling the synthesis of the T3 nuclear receptor c-erb Aα1 (47 kDa) into fibroblasts also induced the expression of a lower-sized c-erb A protein (Sap et al. 1986). Secondly, truncated c-erb Aα1 proteins, synthetized by the use of internal AUG observed in the c-Erb Aα1 messenger were identified in avian cells (Bigler and Eisenman 1988; Bigler et al. 1992). In addition, some of them displayed an extra-nuclear localization.

Western blot experiments using two antibodies raised against non-overlapping sequences of the c-erb A hormonal binding domain identified two c-erb Aα1-related proteins in mitochondria. The respective amounts, molecular weights (43 and 28 kDa) and location (matrix and inner membrane) of these proteins were similar to those obtained for PAL-T3 binding proteins. Moreover, after radioactive PAL-T3 labelling of mitochondrial proteins, an antibody raised against c-Erb Aα specifically immunoprecipitated a radioactive PAL-T3 labelled 43 kDa protein. The same experiment did not show positive data for the 28 kDa protein, but thus may be due to the very low amounts recorded in the mitochondrion.

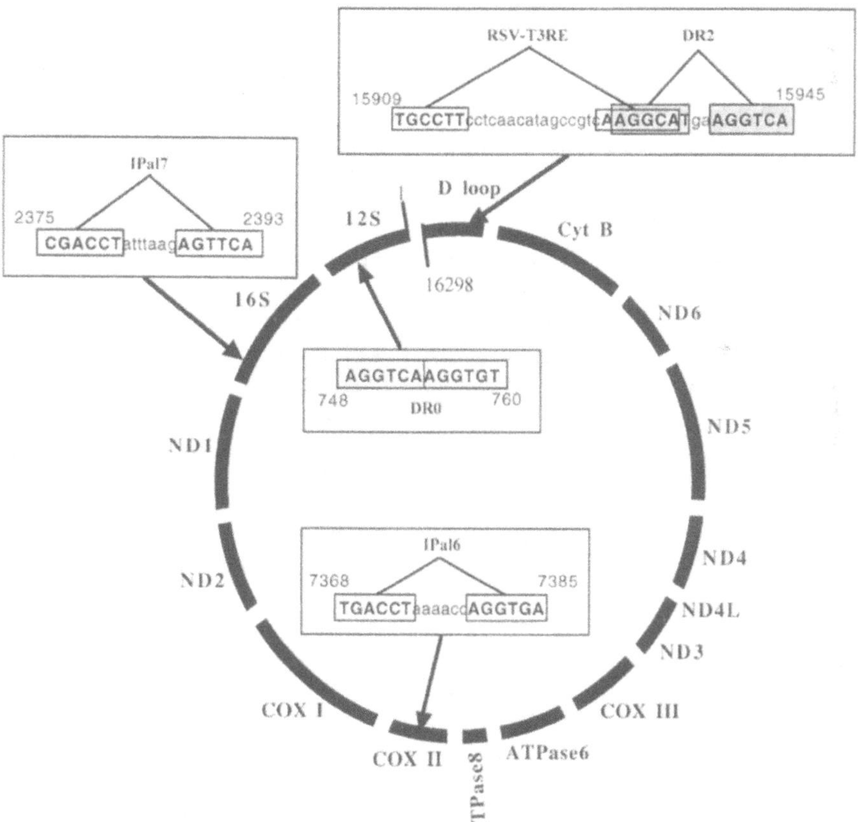

Fig. 4. Localization on the rat mitochondrial genome of nucleotidic sequences displaying strong homologies with nuclear T3REs previously described. *DR0:* direct repeat without spacing between the two half-sites of the T3RE. *DR2:* direct repeat with two nucleotides spacing between the two half-sites of the T3RE. *Ipal 6* and *7:* inverted palindromes with 6 or 7 nucleotides spacing. *RSV-T3RE:* T3RE identified on the Rous sarcoma virus (RSV) genome. *Numbers* provided behind the sequences give positions of the first and last nucleotides of each T3RE. (According to Gadaleta et al. 1989)

A 43 kDa c-erb Aα1 protein corresponding to an N-terminal deleted form of the T3 nuclear receptor was overexpressed in CV1 cells stably transfected. This experiment directly demonstrated by a cytoimmunofluorescence approach that this protein was specifically adressed to mitochondria (Wrutniak et al. 1995). Therefore, it seems that the c-erb Aα gene simultaneously encodes a T3 nuclear receptor and a T3 binding protein imported into the mitochondrion.

The 43 kDa protein (p43) probably possesses the DNA binding domain of the nuclear receptor c-erb Aα1. As expected, EMSA experiments established that p43 specifically binds to nucleotidic probes corresponding to nuclear T3REs. In addition, several T3REs-related sequences were identified in the rat mitochondrial genome, two of them located in the D-loop which comprises the promoters of this genome (Fig. 4). The demonstration that p43 specifically binds to at least four of this sequences suggests that this protein could be the first hormone-dependent transcription factor identified in the mitochondrion (Wrutniak et al. 1994, 1995; Casas et al., submitted). This hypothesis is in agreement with the first results of in organello transcription obtained in the laboratory.

These data establish the occurrence of a truncated form of the T3 nuclear receptor c-erb Aα1 in the mitochondrial matrix, probably involved in the rapid stimulation of the mitochondrial genome expression by T3.

6 Influence of the c-erb Aα1 Truncated Protein Upon Mitochondrial Activity

The observation of a tissue-specific expression of p43 gave some interesting insights concerning the physiological importance of this protein (Wrutniak et al. 1995). First, the intramitochondrial amounts of this protein were positively related to the organite amounts in several tissues. In particular, very important amounts of p43 were recorded in brown adipose tissue mitochondria, thus suggesting an involvement of this protein in the regulation of mitochondriogenesis, a T3-regulated process. Second, p43 was not detected in mitochondria from adult rat brain, a tissue insensitive to the thermogenic action of thyroid hormone, thus suggesting an involvement of this protein in the T3 thermogenic action. A similar approach could not be used for p28, taking in to account the low levels recorded in the mitochondrion.

Interestingly, p43 overexpression in CV1 cells induced a strong stimulation of mitochondrial activity assessed by rhodamine 123 uptake and cytochrome oxidase activity (COX). However, citrate synthase and malate dehydrogenase activities were only moderately affected (Wrutniak et al. 1994, 1995). Interestingly, in contrast to COX, all subunits of these two last enzymatic complexes are encoded by nuclear genes. These results, demonstrating the involvement of p43 in the regulation of mitochondrial activity, are also in agreement with a specific action of this protein on mitochondrial genome expression.

Overall, this work confirms the existence of a direct T3 mitochondrial pathway. The involvement of p43, a probable transcription factor of the mitochondrial genome, in the rapid and long-term T3 influences is conceivable. It remains to establish the mechanisms involved in the early effects of the hormone. In addition, it is probably interesting to understand how a protein having a nuclear localization signal could be imported into the mitochondrion. Last, the absence of a c-erb Aβ form of the T3 nuclear receptor into the organite (Wrutniak et al. 1995) is the first clear demonstration of a functional difference between the two forms of the T3 receptors.

7 Involvement in the Regulation of Cell Differentiation

The T3 influence upon muscle development is well established. This hormone stimulates myoblast terminal differentiation by accelerating myoblast withdrawal from the cell cycle (Marchal et al. 1993, 1995). The T3 nuclear receptor c-erb Aα1 and a stimulation of the cAMP pathway have been involved in the myogenic influence of the hormone (Cassar-Malek et al. 1994; Marchal et al. 1996). In addition, the T3 mitochondrial pathway could be another mechanism implicated in this effect: p43 overexpression in avian myoblasts induces a strong potentiation of terminal differentiation (Rochard et al. 1996a; Wrutniak et al. 1998).

In fact, the involvement of mitochondrial activity in myogenesis regulation was not unexpected. For instance, transient changes in the organelle activity occur just before the induction of terminal differentiation. These changes are abrogated by an experimental block of differentiation (Rochard et al. 1996b). Moreover, expression of the v-erb A oncogene clearly influences mitochondrial activity in avian myoblasts (Rochard et al. 1996b) before exerting its myogenic effect (Cassar-Malek et al. 1995). Lastly, an inhibitor of the mitochondrial membrane potential (FCCP) or of the mitochondrial protein translation (chloramphenicol) strongly inhibits myoblast differentiation (Korohoda et al. 1993; Rochard et al. 1996a; Wrutniak et al. 1998).

Some mechanisms able to explain the involvement of mitochondrial activity in the regulation of myogenesis have been proposed. In particular, the decrease in membrane potential induced by FCCP stimulates the transcriptional activity of the AP-1 complex (jun/fos), considered as a potent repressor of myogenesis (Rochard et al. 1996a). In addition, inhibition of mitochondrial protein synthesis by chloramphenicol strongly and selectively decreases expression of myogenin, a myogenic factor involved in the induction of myoblast differentiation (Rochard et al. 1996a). These observations demonstrate the importance of mitochondrial activity for myogenic differentiation. In addition, previous studies performed on erythrocytic or neuronal cells already suggested that a mitochondrial involvement in the regulation of cell differentiation may not be restricted to these three cell types (Kaneko et al. 1988; Vayssière et al. 1992). More generally, these data demonstrate that mitochondrial activity influences the expression of nuclear genes regulating cell differentiation, and therefore underline the existence of reciprocal relationships between the nucleus and mitochondria.

8 The Mitochondria: A Direct Target of Receptors Belonging to the Nuclear Receptor Superfamily?

Besides the influence of thyroid hormone reviewed in this chapter, numerous studies also report the influence of glucocorticoids (Van Itallie 1992), vitamin D3 (Chou et al. 1995) or of peroxisome proliferator (El Kebbaj et al. 1995; Cai et al. 1996) upon mitochondrial activity. All these substances act via transcription factors belonging to the nuclear receptor superfamily.

The identification of a truncated c-erb Aα1 protein in the mitochondrial matrix raises the more general problem of the presence of other members of this superfamily in the organite. Although its intramitochondrial localization has not been defined, the occurrence of the glucocorticoid receptor in mitochondria has already been reported. In addition, glucocorticoid responsive elements have been identified in the mitochondrial genome (Demonacos et al. 1995).

The simultaneous presence of mtTFA and hormone-dependent transcription factors in mitochondria is interesting. It could be proposed that, as in the nucleus, the interaction of hormone receptors with proteins of the transcriptional apparatus induces an endocrine stimulation of mitochondrial genome expression. Is mt-TFA one of these proteins in the organite?

9 Conclusions

Besides the well-known T3 nuclear pathway, the occurrence of a direct T3 mitochondrial pathway, supported by numerous data, remains highly controversial. However, an objective analysis of the literature underlines the probable occurrence of T3 extragenomic influences inducing, for instance, an early stimulation of mitochondrial activity. In addition, some authors suggest the presence of a mitochondrial T3-dependent transcription factor. Without characterization, the observation of specific T3 binding sites in the organite does not definitively establish the reality of such a pathway.

Recently, we have characterized a truncated c-erb Aα1 protein in the organite. This protein (p43) binds to specific sequences of mitochondrial DNA and is probably a T3-dependent transcription factor of the mitochondrial genome. Its overexpression stimulates the general activity of the organite and, unexpectedly, myoblast differentiation. These results allow better understanding of some T3 influences. Using its mitochondrial and nuclear receptors, the hormone could induce a coordinated stimulation of the expression of nuclear genes encoding mitochondrial proteins and of the mitochondrial genome. Such an activity suggests that T3 could be a major coordinator of the expression of the two genomes, thus explaining the important role of this hormone in the regulation of mitochondriogenesis.

All these data clearly support the hypothesis of the occurrence of a direct mitochondrial pathway, mediated by a mitochondrial c-erb A protein. In addition, this pathway seems to be involved in the regulation of cell differentiation, and consequently in the regulation of the expression of some nuclear genes. Therefore, the nucleus/mitochondria relationships demonstrate a reciprocity in which the role of mitochondria has probably been largely underestimated.

Lastly, other members of the receptor superfamily appear to be located in the organite, such as the glucocorticoid receptor. This suggests that the mitochondrion is probably a direct target of hormonal receptors previously believed to display a strict nuclear localization.

We have yet to understand the intramitochondrial mechanisms of action of these receptors and particularly the relationships with mtTFA, a constitutive transcription factor of the mitochondrial genome. Similarly, p43, as the glucocorticoid receptor, displays a nuclear localization signal. Under these conditions, what are the processes involved in their mitochondrial import? Are other members of the nuclear receptor superfamily located in the organite? The molecular tool provided by the characterization of p43 could help answer some of these questions in the near future.

Acknowledgements. The work of the laboratory presented in this review was supported by grants from the Institut National de la Recherche Agronomique (INRA), Association Française Contre les Myopathies (AFM) and Association de Recherche contre le Cancer (ARC). P.R. and F.C. are recipients of fellowships from Ministère de la Recherche et de l'Enseignement (MRE) and INRA and Direction Générale de l'Enseignement et de la Recherche (DGER) respectively.

References

Andersson ML, Nordström K, Demczuk S, Harbers M, Vennström B (1992) Thyroid hormone alters the DNA binding properties of chicken thyroid hormone receptors α and β. Nucleic Acids Res 20:4803–4810

Bigler J, Eisenman RN (1988) C-erb A encodes multiple protein in chicken erythroid cells. Mol Cell Biol 8:4155–4161

Bigler J, Hokanson W, Eisenman RN (1992) Thyroid hormone receptor transcriptional activity is potentially autoregulated by truncated forms of the receptor. Mol Cell Biol 12:2406–2417

Bogazzi F, Hudson LD, Nikodem VM (1994) A novel heterodimerization partner for thyroid hormone receptor. Peroxisome proliferator-activated receptor. J Biol Chem 269:11683–11686

Brand MD, Steverding D, Kadenbach B, Stevenson PM, Hafner RP (1992) The mechanism of the increase in mitochondrial proton permeability induced by thyroid hormones. Eur J Biochem 206:775–781

Brown GC (1992) Control of respiration and ATP synthesis in mammalian mitochondria and cells. Biochem J 284:1–13

Cai Y, Nelson BD, Li R, Luciakova K, DePierre JW (1996) Thyromimetic action of the peroxisome proliferators clofibrate, perfluorooctanoic acid, and acetylsalicylic acid includes changes in mRNA levels for certain genes involved in mitochondrial biogenesis. Arch Biochem Biophys 325:107–112

Casanova J, Horowitz ZD, Copp RP, McIntyre WR, Pascual A, Samuels HH (1984) Photoaffinity labeling of thyroid hormone nuclear receptors: influence of *n*-butyrate and analysis of the half-lives of the 47000 and 57000 molecular weight receptor forms. J Biol Chem 259:12084–12091

Cassard-Doulcier AM, Larose M, Matamala JC, Champigny O, Bouillaud F, Ricquier D (1994) In vitro interactions between nuclear proteins and uncoupling protein gene promoter reveal several putative transactivating factors including Etsl, retionid X receptor, thyroid hormone receptor, and a CACCC box-binding protein. J Biol Chem 269:24335–24342

Cassar-Malek I, Marchal S, Altabef M, Wrutniak C, Samarut J, Cabello G (1994) V-erb A stimulates quail myoblast differentiation in a T3 independent, cell-specific manner. Oncogene 9:2197–2206

Cassar-Malek I, Marchal S, Rochard P, Casas F, Wrutniak C, Samarut J, Cabello G (1996a) Induction of c-Erb A-AP-1 interactions and c-Erb A transcriptional activity in myoblasts by RXR. Consequences for muscle differentiation. J Biol Chem 271:11392–11399

Cassar-Malek I, Marchal S, Rodier A, Rochard P, Casas F, Wrutniak C, Cabello G (1996b) Functional interactions between the T3 nuclear receptor c-Erb Aα1, c-Jun and MyoD are probably involved in the regulation of myoblast differentiation. Ann Endocrinol 57 (Suppl 4):58

Chen JD, Evans RM (1995) A transcriptional co-repressor that interacts with nuclear hormone receptors. Nature 377:397–404

Chou SH, Hannah SS, Lowe KE, Norman AW, Henry HL (1995) Tissue-specific regulation by vitamin D status of nuclear and mitochondrial gene expression in kidney and intestine. Endocrinology 136:5520–5226

Demonacos C, Djordjevic-Markovic R, Tsawdaroglou N, Sekeris CE (1995) The mtiochondrion as a primary site of action of glucocorticoids: the interaction of the glucocorticoid receptor with mitochondrial DNA sequences showing partial similarity to the nuclear glucocorticoid responsive elements. J Steroid Biochem Mol Biol 55:43–55

Desbois C, Aubert D, Legrand C, Pain B, Samarut J (1991) A novel mechanism of action for v-erb A: abrogation of the inactivation of the transcription factor AP-1 by retinoic and thyroid hormone receptors. Cell 67:731–740

El Kebbaj MS, Cherkaoui Malki M, Latruffe N (1995) Effects of peroxisome proliferators and hypolipenic agents on mitochondrial inner membrane linked D-3-hydroxybutyrate dehydrogenase (BDH). Biochem Mol Biol Int 35:65–77

Fisher RP, Clayton DA (1988) Purification and characterization of human mitochondrial transcription factor 1. Mol Cell Biol 8:3496–3509

Forman BM, Samuels HH (1990) Interaction among a subfamily of nuclear receptors: the regulatory zipper model. Mol Endocrinol 4:1293–1301

Gadaleta G, Pepe G, De Candia G, Quagliariello C, Sbisà E, Saccone C (1989) The complete nucleotide sequence of the *Rattus novegicus* mitochondrial genome: cryptic signals revealed by comparative analysis between vertebrates. J Mol Evol 28:497–516

Gadaleta MN, Barletta A, Galdarazzo M, De Leo T, Saccone C (1972) Triidothyronine action on RNA synthesis in rat liver mitochondria. Eur J Biochem 30:376–381

Garstka HL, Fäcke M, Escribano JR, Wiesner RJ (1994) Stoichiometry of mitochondrial transcripts and regulation of gene expression by mitochondrial transcription factor A. Biochem Biophys Res Commun 200:619–626

Goglia F, Lanni A, Barth J, Kadenbach B (1994) Interaction of diiodothyronines with isolated cytochrome c oxidase. FEBS Lett 346:295–298

Goglia T, Torresani J, Bugli P, Barletta A, Liverini G (1981) In vitro binding of triiodothyronine to rat liver mitochondria. Pflugers Arch 390:120–124

Gregory RB, Berry MN (1991) The administration of triiodothyronine to rats results in a lowering of the mitochondrial membrane potential in isolated hepatocytes. Biochim Biophys Acta 1133:89–94

Guerra C, Roncero C, Porras A, Fernandez M, Benito M (1996) Triiodothyronine induces the transcription of the uncoupling protein gene and stabilizes its mRNA in fetal rat brown adipocyte primary cultures. J Biol Chem 271:2076–2081

Gustafsson R, Tata JR, Lindberg O, Ernster L (1965) The relationship between the structure and activity of rat skeletal muscle mitochondria after thyroidectomy and thyroid treatment. J Cell Bol 26:555–578

Hafner RP, Nobes CD, McGown AD, Brand MD (1988) Altered relationship between protonmotive force and respiration rate in non-phosphorylating liver mitochondria isolated from rats of different thyroid hormone status. Eur J Biochem 178:511–518

Hansford RG (1991) Dehydrogenase activation by Ca^{2+} in cells and tissues. J Bionerg Biomemb 23:823–854

Hardy DL, Mowbray J (1992) The rapid response of isolated mitochondrial particles to 0.1 nM tri-iodothyronine correlates with the ADP-ribosylation of a single inner-membrane protein. Biochem J 283:849–854

Harper ME, Ballantyne JS, Leach M, Brand MD (1993) Effects of thyroid hormones on oxidative phosphorylation. Biochem Soc Trans 21:785–792

Hashizume K, Ichikawa K (1992) Localization of 3,5,3'-triiodothyronine receptor in rat mitochondrial membrane. Biochem Biophys Res Commun 106:920–926

Horlein AJ, Naar AM, Heinzel T, Torchia J, Gloss B, Kurokawa R, Ryan A, Kamei Y, Soderstrom M, Glass CK, Rosenfeld MG (1995) Ligand-independent repression by the thyroid hormone receptor mediated by a nuclear receptor co-repressor. Nature 377:397–404

Horrum MA, Tobin RB, Ecklung RE (1992) The early triiodothyronine-induced changes in state IV respiration is not regulated by the proton permeability of the mitochondrial inner membrane. Biochem Int 28:813–821

Horwitz KB, Jackson TA, Bain DL, Richer JK, Takimoto GS, Lung T (1996) Nuclear receptor coactivators and corepressors. Mol Endorinol 10:1167–1177

Ichikawa K, De Groot LJ (1987) Thyroid hormone receptors in a human hepatoma cell line: multiple receptor forms on isoelectric focusing. Mol Cell Endocrinol 51:135–143

Izquierdo JM, Cuezva JM (1993) Thyroid hormones promote transcriptional activation of the nuclear gene coding for mitochondrial β-F1-ATPase in rat liver. FEBS Lett 323:109–112
Jakovilcic S, Swift HH, Gross NJ, Rabinowitz R (1978) Biochemical and stereological analysis of rat liver mitochondria in different thyroid states. J Cell Biol 77:887–901
Joste V, Goitom Z, Nelson BD (1989) Thyroid hormone regulation of nuclear-encoded mitochondrial inner membrane polypeptides of the liver. Eur J Biochem 184:255–260
Kaneko T, Watanabe T, Oishi M (1988) Effect of mitochondrial protein synthesis inhibitors on erythroid differentiation of mouse erythroleukemia (Friend) cells. Mol Cell Biol 8:3311–3315
Korohoda W, Pietrzkowski Z, Reiss K (1993) Chloramphenicol, an inhibitor of mitochondrial protein synthesis, inhibits myoblast fusion and myotube differentiation. Folia Histochem Cytobiol 31:9–13
Lanni A, Moreno M, Horst C, Lombardi A, Goglia F (1994) Specific binding sites for 3,3'-diiodo-L-tyhronine (3,3'-T2) in rat liver mitochondria. FEBS Lett 351:237–240
Lanni A, De Felice M, Lombardi A, Moreno M, Fleury C, Ricquier D, Goglia F (1997) Induction of UCP2 mRNA by thyroid hormones in rat heart. FEBS Lett 418:171–174
Larkin S, Mull E, Miao W, Pittner R, Albrandt K, Moore C, Young A, Denaro M, Beaumont K (1997) Regulation of the third member of the uncoupling protein family, UCP3, by cold and thyroid hormone. Biochem Biophys Res Commun 240:222–227
Marchal S, Cassar-Malek I, Pons F, Wrutniak C, Cabello G (1993) Triiodothyronine influences quail myoblast proliferation and differentiation. Biol Cell 78:191–197
Marchal S, Cassar-Malek I, Magaud JP, Rouault JP, Wrutniak C, Cabello G (1995) Stimulation of avian myoblast differentiation by triiodothyronine: possible involvement of cAMP pathway. Exp Cell Res 220:1–10
Marchal S, Cassar-Malek I, Rodier A, Wrutniak C, Cabello G (1996) Mécanismes moléculaires impliqués dans l'activité myogénique de la triiodothyronine (T3). Médecine/Sciences 12:1065–1076
Martino G, Covello C, De Giovanni R, Filipelli R, Pitrelli G (1986) Direct in vitro action of thyroid hormones on mitochondrial RNA-polymerase. Mol Biol Rep 11:205–211
Mowbray J, Hardy DL (1996) Direct thyroid hormone signalling via ADP-ribosylation controls mitochondrial nucleotide transport and membrane leakiness by changing the conformation of the adenine nucleotide transporter. FEBS Lett 394:61–65
Murphy MP (1992) Calcium uptake by liver mitochondria form hypothyroid rats is inhibited in vitro by triiodothyronine. Biochem Int 27:1019–1026
Mutvei A, Kuzela S, Nelson BD (1998) Control of mitochondrial transcription by thyroid hormone. Eur J Biochem 180:235–240
Paradies G, Ruggiero FM (1989) Decreased activity of pyruvate translocator and changes in the lipid composition in heart mitochondria from hypothyroid rats. Arch Biochem Biopyhys 269:595–602
Paradies G, Ruggiero FM (1990) Stimulation of phosphate transport in rat-liver mitochondria by thyroid hormones. Biochim Biophys Acta 1019:133–136
Perlmann T, Rangarjian PN, Umesono I. Evans RM (1993) Determinants for selectivity of RAR and TR recognition of direct repeat HREs. Genes Dev 7:1411–1422
Piedrafita FJ, Bendik I, Ortiz MA, Pfahl M (1995) Thyroid hormone receptor homodimers can function as ligand-sensitive repressors. Mol Endocrinol 9:563–578
Rabelo R, Reyes C, Schifman A, Silva JE (1996) Interactions among receptors, thyroid hormone response elements, and ligands in the regulation of the rat uncoupling protein gene expression by thyroid hormone. Endocrinology 137:3478–3487
Rochard P, Cassar-Malek I, Marchal S, Casas F, Rodier A, Wrutniak C, Cabello G (1996a) L'activité mitochondriale est impliquée dans la régulation de la différenciation musculaire. VIéme Colloq sur les maladies neuromusculaires 21–25 Oct, Versailles
Rochard P, Cassar-Malek I, Marchal S, Wrutniak C, Cabello G (1996b) Changes in mitochondrial activity during avian myoblast differentiation: influence of triiodothyronine or v-erbA expression. J Cell Physiol 168:239–247
Sap J, Munoz A, Damm K, Goldberg Y, Ghysdael J, Leutz A, Beug H, Vennström B (1986) The c-erb A protein is a high affinity receptor for thyroid hormones. Nature 324:242–244

Segal J (1990) In vivo effect of 3,5,3'-triiodothyronine on calcium uptake in several tissues in the rat: evidence for a physiological role of calcium as the first messenger for the prompt action of thyroid hormone at the level of the plasma membrane. Endocrinology 126:17–24

Soboll S, Sies H (1989) Effects of hormones on mitochondrial processes. Methods Enzymol 174:118–130

Sterling K, Brenner MA (1995) Thyroid hormone action: effect of triiodothyronine on mitochondrial adenine nucleotide translocase in vivo and in vitro. Metabolism 44:193–199

Sterling K, Milch PO (1975) Thyroid hormone binding by a component of mitochondrial membrane. Proc Natl Acad Sci USA72:3225–3229

Sterling K, Milch PO, Brenner MA, Lazarus JH (1977) Thyroid hormone action: the mitochondrial pathway. Science 197:996–999

Sterling K, Brenner MA, Sakurada T (1980) Rapid effect of triiodothyronine on the mitochondrial pathway in rat liver in vivo. Science 210:340–343

Sterling K, Campbell GA, Taliadouros GS, Nunez EA (1984) Mitochondrial binding of triiodothyronine (T3). Demonstration by electron microscopic radioautography of dispersed liver cells. Cell Tissue Res 236:321–325

Tata JR, Ernster L, Lindberg O, Arrhenius E, Pedersen S, Hedman R (1963)The action of thyroid hormones at the cell level. Biochem J 86:408–428

Thomas WE, Mowbray J (1987) Evidence for ADP-ribosylation in the mechanism of rapid thyroid hormone control of mitochondria. FEBS Lett 223:279–283

Van Itallie MC (1992) Dexamethasone treatment increases mitochondrial RNA synthesis in rat hepatoma cell line. Endocrinology 130:567–576

Vayssière JL, Cordeau-Lossouarn L, Larcher JC, Basseville M, Gros F, Croizat B (1992) Participation of the mitochondrial genome in the differentiation of neuroblastoma cells. In Vitro Cell Dev Biol 28A:763–772

Weinberger C, Thompson CC, Ong ES, Lebo R, Gruol DJ, Evans RM (1986) The c-erb A gene encodes a thyroid hormone receptor. Nature 324:641–646

Wrutniak C, Cabello G (1986) La voie d'action mitochondriale directe de la triiodothyronine: mythe or realité? Médecine/Sciences 12:475–484

Wrutniak C, Cassar-Malek I, Marchal S, Rochard P, Dauça M, Fléchon J, Samarut J, Ghysdael J, Cabello G (1994) In search for the mitochondrial T3 receptor: a 43 kDa c-erbAα1 related protein is located in the matrix and affects mitochondrial activity. J Endocrinol Invest 17:15

Wrutniak C, Cassar-Malek I, Marchal S, Rascle A, Heusser S, Keller JM, Fléchon J, Dauça M, Samarut J, Ghysdael J, Cabello G (1995) A 43-kDa protein related to c-erb A α1 is located in the mitochondrial matrix of rat liver. J Biol Chem 270:16347–16354

Wrutniak C, Rochard P, Casas F, Fraysse A, Charrier J, Cabello G (1998) Physiological importance of the T3 mitochondrial pathway. Ann N Acad Sci 839:93–100

Yen PM, Darling DS, Carter RL, Forgione M, Umeda PK, Chin WW (1992) Triiodothyronine (T3) decreases binding to DNA by T3-receptor homodimers but not receptor-auxiliary protein heterodimers. J Biol Chem 267:3565–3568

Zhang XK, Wills KN, Husmann N, Hermann T, Pfahl M (1991) Novel pathway for thyroid hormone receptor action through interaction with jun and fos oncogene activities. Mol Cell Biol 11:6016–6025

CHAPTER 19

Cytoskeleton – Mitochondrial Interactions 19

J.-F. LETERRIER[1] and M. LINDÉN[2]

Contents

1 Distribution of Mitochondria in the Cytoplasm

The direct relationship between mitochondrial distribution and cell metabolism has been reviewed previously (Bereiter-Hahn and Vöth 1994). Their localization in the cytoplasm depends first on the local energetic requirements of other organelles and second on the presence of specialized subcellular domains in differentiated cells, such as the synapses of neurons (Gotow et al. 1991), the muscular fibrils of skeletal muscles (Rappaport and Samuel 1988; Stromer and Bendayan 1990), the flagellar bodies of spermatozoids (Olson and Winfrey 1990), and oocytes in which the segregation of mitochondria determines the future cleavage plans of the egg (Klymkowsky 1995). These particular situations are the result of specific interactions between mitochondria and the cytoskeleton of differentiated cytoplasmic domains (Fig. 1).

2 Saltatory Motion of Mitochondria

The localization and intracellular transport of organelles occur through stable and/or transient interactions with the cytoskeleton. These phenomena have mainly been explored in axons where no major protein synthesis occurs, allowing the measure-

[1] U298 INSERM, CHRU, 49033, Angers, France
[2] Arrhenius Laboratory, Stockholm University, Stockholm, Sweden

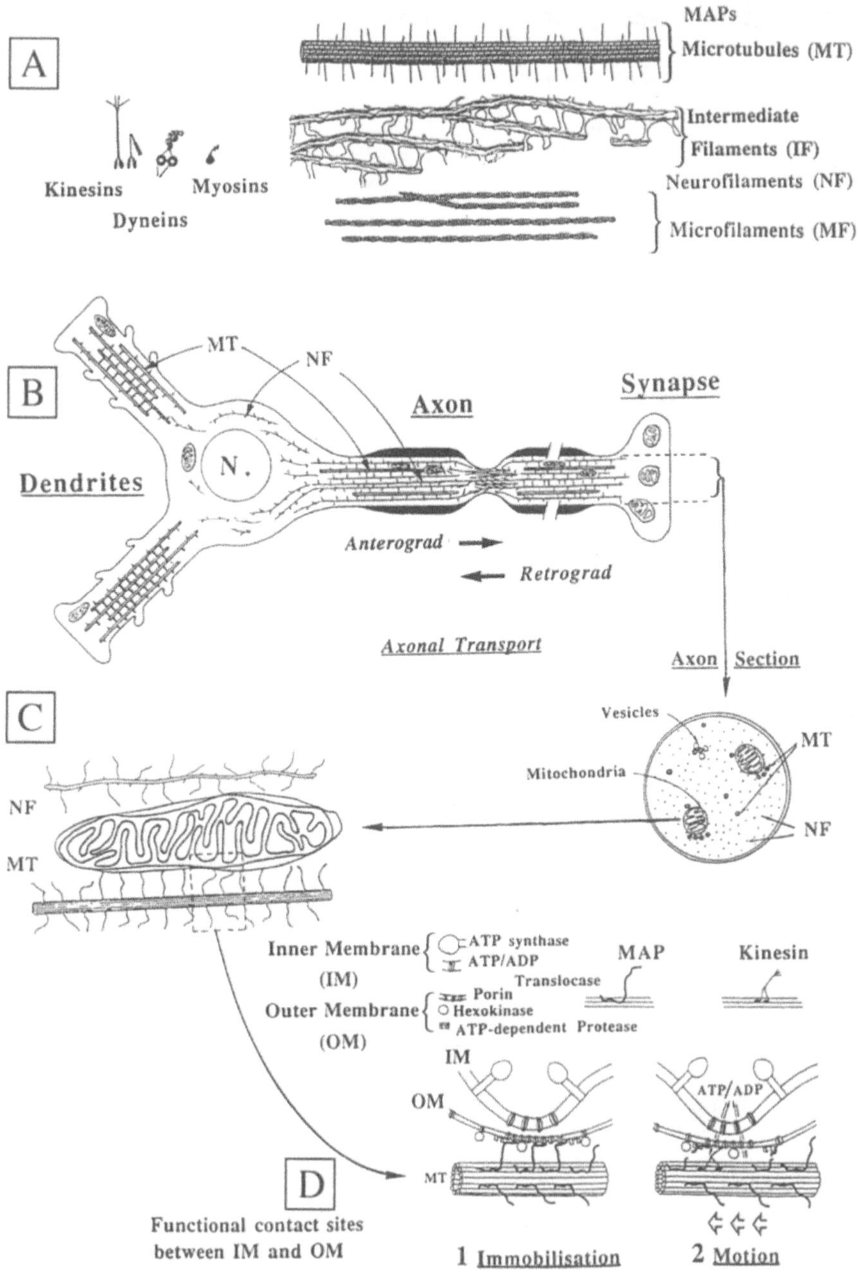
A
MAPs
Microtubules (MT)
Intermediate
Filaments (IF)
Neurofilaments (NF)
Microfilaments (MF)
Kinesins
Dyneins
Myosins
B
MT
NF
Axon
Synapse
Dendrites
N.
Anterograd
Retrograd
Axonal Transport
Axon Section
C
Vesicles
MT
Mitochondria
NF
NF
MT
Inner Membrane (IM)
ATP synthase
ATP/ADP Translocase
Outer Membrane (OM)
Porin
Hexokinase
ATP-dependent Protease
MAP
Kinesin
IM
OM
MT
ATP/ADP
D
Functional contact sites between IM and OM
1 Immobilisation
2 Motion

ment of intracellular motion along considerable distances (Burgoyne 1991; Hirokawa 1998). Although mitochondria do not exhibit an active motility in all cell types, their distribution depends on their anchorage to the cytoskeleton. This occurs through physical contacts between the mitochondrial outer membrane and the three types of cytoskeletal polymers (microtubules, MT; actin filaments or microfilaments, MF; intermediate filaments, IF; Martz et al. 1984; Eckert 1986; Forman 1987; Carmo-Fonseca and David-Ferreira 1990; Price et al. 1991; Soltys and Gupta 1992; Couchmann and Rees 1992; Drubin et al. 1993; Morris and Hollenbeck 1993, 1995). The three major polymer species, MT, IF, and MF which form the cytoskeleton are present in the majority of eukaryotic cells with some exceptions such as red blood cells (Bershadsky and Vasiliev 1988; Fuchs and Weber 1994). These polymers are associated with structural proteins whose functions range from the catalytic stabilisation or fragmentation of each type of labile polymer (MT and MF) to the organization of polymers into networks and their anchorage to various membranous compartments. These proteins contain two or more identical or distinct binding sites to other proteins, located on distinct subdomains of the molecule (Bershadsky and Vasiliev 1988). A second group of cytoskeleton-associated proteins is constituted by large families of molecular motors of the dynein (driving motion to the – end of MT), kinesin (driving motion to

Fig. 1. Scheme of interactions between mitochondria and cytoskeleton in neurons. **A** Elements of the neuronal cytoskeleton. To the wall of *MT* (25-nm tube) are bound MAPs and motors whose lateral projections determine the spacing between *MT* and other adjacent cytoplasmic organelles. *NF* (11 nm) are the most abundant *IF* species. Lateral projections of *NF* subunits are similar to that of MAPs on *MT*. *MF* (7 nm) are associated with many proteins, regulating their polymerization, reticulation in networks and anchorage to membranes. Molecular motors associated with *MT* drive motion towards the + end (kinesins) or the – end of *MT* (dyneins) while myosins are associated with *MF*. The hydrolysis of ATP by globular heads generates the transport of cargoes bound to the other end of the molecule, through elementary jumps of the motor heads along the polymer. **B** Organization of the neuronal cytoskeleton. Dendrites contain *MT* bundles of opposite polarities, associated with a specific class of MAPs (MAP2). *MT* of identical polarity (+ end towards the synapse) characterize axons, as well as their MAPs (Tau, MAP1b). *NF*, rare in dendrites, are organized into bundles in axons. *MF* (not shown) are underneath the plasma membrane of most cytoplasmic compartments, and associated with *MT* and *NF* in dendrites and axons. Structural crossbridges between *MT*, *NF* and *MF* (MAPs, sidearms of *NF*, actin-binding proteins) are regulated by phosphorylation, proteolysis or Ca ions. Fast axonal transport occurs along *MT* through the kinesin and dynein motors, and in local association with MF and myosins. Mitochondria are transported at a lower velocity than that of other organelles, as a consequence of their saltatory motion, in strict association with the *MT* domains frequently organized in ribbons of 3–4 *MT* in the immediate proximity of the mitochondrial membrane (scheme of axonal section). **C** Local interactions between mitochondria and axonal cytoskeletal polymers. Multiple crossbridges between the mitochondrial membrane and the lateral projections of *NF* and *MT* must be disrupted (by phosphorylation/dephosphorylation, proteolysis) to allow the function of bound motors. The mitochondrial motion requires the coordination of such events between domains distant from each other on the membrane surface. **D** Membranous domains specialized in contacts between neuronal mitochondria and the cytoskeleton. Hypotheses for the mechanisms of saltatory motion: Porin is concentrated in restricted domains of the outer mitochondrial membrane, through which nucleotides and other small molecules exchange between the cytoplasmic and the intramitochondrial media. **1** Temporary immobilization of mitochondria through crossbridges with *MT* via structural *MAPs*. **2** A local high *ATP* concentration in the vicinity of the crossbridge induces the degradation of mitochondria-bound *MAPs* by an *ATP*-dependent protease. The binding of kinesin to sites close to that for *MAPs* in porin-rich domains, stimulated by *ATP*, induces mitochondrial motion the + end of MT) and myosin (associated with MF) types (Hirokawa 1998; Mermall et al. 1998). The structural and dynamic interactions between the three cytoskeletal polymers through their structurally associated proteins and the molecular motors determine a highly plastic, integrated network all through the cytoplasmic space, which mediates functional communications between specialized domains of the plasma membrane and the other intracellular compartments. Detailed observations of mitochondrial behaviour in axons of neurons, where an intense intracellular traffic occurs, demonstrate that two fundamental characteristics distinguish mitochondrial motility from that of other intracellular organelles: (1) the axonal transport of mitochondria is a saltatory motion, including long arrested phases between motion episodes either to the direction of the synaptic terminal (anterograd transport) or to the cell body (retrograd transport). A consequence of this saltatory motion is a net velocity of the axonal transport of mitochondria towards the synaptic end of the axon that is much slower than that of other membranous organelles (Martz et al. 1984; Forman 1987; Morris and Hollenbeck 1995). (2) Mitochondria bind simultaneously to the cytoskeletal polymers through several domains of the outer membrane, at a distance from each other, which mediate either stable or dynamic interactions with one or several MT (Martz et al. 1984).

Important morphological deformations of mitochondria occur during the various phases of saltatory motion, which suggest a possible direct contribution of the mitochondrial physiology to this process (Martz et al. 1984). Thus, a major particularity of the dynamic behaviour of mitochondria is their apparent motile autonomy, consecutive to multiple interactions with the cytoskeleton via several membranous specialized domains coordinated with a simultaneous change in the organelle shape. This is in strong contrast with the longitudinal motion of secretory vesicles along a single MT (Weiss et al. 1986). Mitochondrial motility does not involve IF (Morris and Hollenbeck 1995), to which no specific molecular motor has yet been found associated. The saltatory motion of mitochondria could result from alternate crossbridging between the mitochondrial membrane and the molecular motors associated with MT or MF, on one hand, and with (non-motor) structural proteins associated with the three types of polymers, on the other.

3 Cytoskeletal Proteins Which Interact Specifically with Mitochondria

3.1 MT Proteins

Mitochondria are predominantly associated with MT in most cells. The association of tubulin isoforms to isolated mitochondria has been reported in vitro (Bernier-Valentin et al. 1983; Hargreaves and Avila 1985). The possibility of specific interactions between tubulin subunits and mitochondria in vivo under physiological conditions is supported by the finding that a contracture in cardiac myocytes induces the depolymerization of MT followed by the sequestration of β tubulin at the surface of the outer mitochondrial membrane (Saetersdal et al. 1990). The structural MT-associated fibrous proteins (MAPs), lacking motor activity, consist of a domain that binds to the tubulin wall of MT and a domain projecting away from the MT axis (Mandelkow and

Mandelkow 1995). This lateral decoration of MT by MAPs, abundant in neurons, mediate interactions between MT and the adjacent cytoplasmic organelles, by forming non-covalent crossbridges, regulated by the phosphorylation level of MAPs. High molecular weight MAPs of MT from the central nervous system are bound to mitochondria in situ, and are found in purified outer mitochondrial membranes (Leterrier et al. 1994; Lindén et al. 1989a). Two binding sites of high and medium affinity exist on the membrane of intact mitochondria, while one site of intermediate affinity is conserved in purified outer membrane fragments. The binding of MAPs to mitochondria is neither affected by the presence of tubulin, nor the polymerization status of MT. MAPs mediate stable crossbridges between MT and mitochondria in vitro, similar in ultrastructure to those observed in situ (Leterrier et al. 1991). The topography of crossbridges between mitochondria and MT suggests that the membrane-binding domain of MAP2 is located on the N-terminal lateral sidearm of MAP2, since the tubulin-binding site is restricted to a relatively small C-terminal domain of MAP2 (Mandelkow and Mandelkow 1995). Although tau proteins, the second group of MAPs present in purified MT, have been suggested to bind (weakly) to mitochondrial membranes in vitro (Jancsik et al. 1989), this finding is not consistent with investigations in situ (Migheli et al. 1988). MAPs-mediated crossbridges between MT and mitochondria in vitro are static. However, mitochondrial motion in situ involves molecular motors. Recent investigations suggest the lack of involvement of dynein in mitochondrial distribution and motility (Burkhardt et al. 1997). A major neuronal kinesin isoform was suggested to be associated with mitochondria in situ (Elluru et al. 1995) and in purified mitochondria (Jellali et al. 1994). However, this molecular motor seems to be preferentially involved in the axonal transport of endoplasmic reticulum tubulovesicles (Feiguin et al. 1994). The discovery by Nangaku et al. (1994) of a non-conventional (monomeric) kinesin that is specifically associated to mitochondria in situ and which induces their motion in vitro supports the hypothesis that each type of organelle is targeted by binding to a particular subset of motors (Hirokawa 1998). The finding of a specific kinesin isoform exclusively associated with the positioning of mitochondria during mitosis (Pereira et al. 1997) is in agreement with this hypothesis. However, the evidence that the knockout of the major neuronal isoform results in the selective sequestration of mitochondria in cell bodies of axons demonstrates that this motor is required for the export of mitochondria in axons (Tanaka et al. 1998). From these data altogether, it could be inferred that the anterograd motion of mitochondria on MT involves either two distinct kinesin subtypes working simultaneously, or that each of the kinesin species is required sequentially for (1) the specific addressing of mitochondria from the cell body to the axon and (2) their transport to the axon terminal.

3.2 MF Proteins

Mitochondria are more frequently found anchored to MT, but their dynamic association with actin MF is an alternate mechanism to their active distribution, recently identified in neurons. The demonstration by Kuznetsov et al. (1992) that an active motion occurs on actin networks in isolated axonal extrudates from squid suggests complementary roles of MT and MF in the intracellular motility of organelles. Morris and Hollenbeck (1995) have established that mitochondria move in neurites of cultured neurons in association with either MT or MF, following anterograde and retrograde velocities that are characteristic of each type of cytoskeletal element, while NF do not participate in mitochondrial translocation. In a previous work, the same authors showed the direct involvement of the actin cytoskeleton in the maintenance of a high mitochondrial density in the proximal domain of the growth cone, a dynamic localization which disappears under the arrest of neuritic growth (Morris and Hollenbeck 1993). The participation of actin networks in the dynamic organization of mitochondria in the cytoplasm has been documented in yeast with the use of actin mutants (Drubin et al. 1993; Smith et al. 1995). The binding of actin filaments to yeast mitochondria in vitro is regulated by ATP and is inhibited by the previous saturation of actin filaments with myosin heads, suggesting a direct interaction between actin subunits and the mitochondrial outer membrane (Lazzarino et al. 1994). On the other hand, the exclusive localization of an actin isoform in the immediate vicinity of the mitochondrial membrane in muscle fibers of the mouse diaphragm (Pardo et al. 1983), suggests unknown interactions between this particular actin isoform and the immobile mitochondria of this tissue, similar to the mitochondrial binding of β-tubulin in heart myocytes (Saetersal et al. 1990). If this evidence points to a direct contact between actin (monomeric or polymeric forms) and the mitochondrial membrane, the specific characteristics of the MF-dependent motion of mitochondria (Morris and Hollenbeck 1995) strongly suggests the involvement of MF-associated motors (non-muscular myosins). The existence of a mitochondria-specific cytoplasmic myosin has been recently established (Simon et al. 1995). In addition, experimental indications exist which suggest also the possibility of interactions between mitochondria and MF through actin-binding proteins other than myosins: the annexins V and VI which bind to phospholipid membranes in a Ca^{2+}-dependent manner, are associated with both the cytoskeleton and mitochondria in myocardium (Sun et al. 1993), and fodrin, a protein involved in anchoring actin filaments to membranes, is regularly found in purified fractions of brain outer mitochondrial membrane (Leterrier et al. 1994).

3.3 Interactions Between Mitochondria and IF

Although the interactions between mitochondria and either MT or MF involve primarly their active motion among other putative less dynamic functions, interactions between mitochondria and IF remain enigmatic, since these cytoskeletal polymers are not associated with the cytoplasmic motility of organelles (Fuchs and Weber 1994). With regard to the main known function of IF in the organization and the mechanical resistence of the cytoplasm to stresses, a presumed function of IF-mitochondria interactions could be their immobilization in precise localizations within the cytoplasm for the maintenance of an appropriate distribution of energy sources specific of the tissue needs, as initially suggested by Eckert (1986). The occurrence of physical contacts between IF and the outer mitochondrial membrane has been shown in various differentiated tissues such as in smooth muscle cells (Stromer and Bendayan 1990) and in neurons in which regular crossbridges between axonal mitochondria and NF were clearly identified by Hirokawa (Burgoyne 1991). The phenotype of desmin knockout mice includes morphological alterations of muscular mitochondria, suggesting that the localization and possibly the functional integrity of mitochondria, which are normally associated with the desmin-containing Z line of myofibrils depends on their interaction with this muscle-specific IF type (Li et al. 1996). The in vitro analysis of interactions between IF and mitochondria was studied with NF and brain mitochondria (a mixture of mitochondria from neuronal and glial origins) which regulate the formation of networks between NF in an ATP-dependent manner, while hepatic mitochondria are inefficient, suggesting the presence of binding sites for NF specific for neuronal mitochondria (Leterrier et al. 1991). Yaffe et al. identified an immunolgically IF-related protein in yeast, Mdm1p, the mutants of which impair nuclear and mitochondrial transfer in reproduction buds (Fisk and Yaffe 1997). These observations raised the possibility that IF-related proteins contribute to the regulation of mitochondrial division. IF-like proteins were also found in purified brain mitochondria, where they are differentially distributed between the outer and inner mitochondrial membranes (Leterrier et al. 1994). The evidence for a homology between the sequence of the N-terminal domain of the type II-8 keratin and that of the signal sequence targeting proteins to the mitochondrial membranes (Quellet et al. 1988) suggests similarly the possibility that IF proteins might insert in mitochondrial membranes, and is supported by evidence for the phosphorylation of cytokeratin subunits in the hepatic inner mitochondrial membrane (Gorlach et al. 1995).

4 Mitochondrial Proteins Interacting with the Cytoskeleton

The control of mitochondrial morphology depends on its cytoplasmic surrounding. The finding that an outer membrane protein, MMM1, involved in the maintenance of mitochondrial shape, is also necessary for the segregation of yeast mitochondria in daughter cells following mitosis, suggests that both functions are linked (Burgess et al. 1994). The possibility that the MMM1 protein functions by association with a cytoskeleton frame was suggested by Burgess et al. (1994) in the light of indirect evidence. The in vitro approach of interactions between MT and brain mitochondria demonstrates that the binding of MAP2 to outer mitochondrial membrane fragments induces changes in the physicochemical behaviour of porin through its purification procedure with 2% Triton X 100 and NaCl (Lindén et al. 1998a, b). Porin is a channel-forming protein in the outer mitochondrial membrane involved in the voltage-dependent exchange of small molecules between the cytoplasm and the intramitochondrial space (Lindén et al. 1982a; Mannella 1997). Porin has been identified as the binding site for the cytoplasmic enzyme hexokinase to the outer mitochondrial membrane (Lindén et al. 1982b; Fiek et al. 1982) and is thought to interact with the inner membrane nucleotide transporter protein carrying ADP and ATP (nucleotide translocase) in transient contact domains between the outer and the inner mitochondrial membranes (Kottke et al. 1988; Adams et al. 1991; BeltrandeloRio and Wilson 1992). The nearly complete inhibition of MAP2 binding to intact mitochondria by porin antibodies (Leterrier et al. 1994) and the selective trapping of porin on a MAP2 affinity column (Lindén and Karlsson 1996) demonstrate the direct involvement of porin in MAP2-binding sites. However, porin is likely to be associated with other membrane proteins in these sites, as two other unknown MAP2-binding proteins are copurified with porin through its extraction from purified outer membrane fragments (Lindén et al. 1989b; Leterrier et al. 1994; Lindén and Karlsson 1996). Furthermore, MAP2 binding sites contain an ATP-dependent MAP2 protease which could be involved in the regulatory mechanisms of stable crossbridges between mitochondria and MT (Leterrier et al. 1994). Crossbridges formed by the binding of lateral sidearms of the two high molecular weight NF subunits (NF-H and NF-M) occur on sites on the outer mitochondrial membrane that are either common or close to that for MAP2, since the binding of MAPs is partially inhibited by the addition of pure NF subunits (Leterrier et al. 1991). These experimental indications suggest that porin domains on the surface of neuronal mitochondria may bind either NF (sidearms of NF-H and NF-M) or MT-associated proteins (sidearms of MAPs). Evidence for inhibition by MAPs of the binding of the major neuronal molecular motor kinesin isoform to mitochondria further supports the possibility that kinesin also binds to porin domains (Leterrier et al. 1992). The demonstration that the mitochondrial surface is organized in domains of high density of porin, hexokinase, or bound cytoskeletal poteins (Leterrier et al. 1994) opens an interesting functional hypothesis in which such porin-rich domains of the outer mitochondrial membrane are priviledged sites for crossbridging with the neuronal cytoskeletal elements MT and NF.

From the direct observation of the dynamic behaviour of mitochondria in axons (Martz et al. 1984; Forman 1987; Morris and Hollenbeck 1993, 1995), it may be suggested that mitochondrial physiology contributes to the coordination of simultaneous interactions of the organelle with distinct cytoskeletal polymers through limited sites at a distance from each other on the outer mitochondrial membrane. If the active motion of small spherical organelles (secretory vesicles) could depend on their interaction with MT through one single molecular motor (Ashkin et al. 1990), the translocation of large and elongated mitochondria implies the temporal coordination between several anchoring sites of the organelle to the cytoskeleton polymers (Martz et al. 1984). The saltatory motion of mitochondria, which includes arrested phases alternating with active motion phases in either direction, might be based on the alternate contribution of stable and dynamic crossbridges with MT, MF or IF. If such interactions were to occur through the binding of cytoskeleton associated proteins on specialized membrane domains such as those described in vitro for brain mitochondria (Leterrier et al. 1994), regulatory mechanisms should exist locally within these domains, allowing the release of mitochondria from structural, immobile crossbridges (mediated by the MAPs sidearms of MT or the NF sidearms), giving room for the action of molecular motors (MT dependent kinesins, MF-dependent myosins). The existence of such specialized domains on endoplasmic reticulum membranes has been reported recently, which mediate the formation and the transport of tubulovesicular structures along MT via kinesin molecules (Allan and Vale 1994). A similar organization of the mitochondrial surface is likely. The evidence that a specific monomeric kinesin is associated with the mitochondria (Nangaku et al. 1994) further implies the organization of its membrane binding domains into functional clusters containing several motor molecules, as shown recently by Hancock and Howard (1998). Thus, the specific motile behaviour of mitochondria (deformations coordinated with saltatory motion) might depend first on the presence of (porin-rich?) domains containing sites for the anchorage of structural cytoskeleton-associated proteins and second on the presence of specific monomeric motors concentrated into clusters on the mitochondrial outer membrane for achieving the active translocation of the organelle. This hypothesis, originating from studies with brain mitochondria

(Leterrier et al. 1994), may be extended to mitochondria from other tissues, as the occurrence of porin-rich domains was also demonstrated on the surfaces of isolated heart mitochondria (Konstantinova et al. 1995).

References

Adams V, Griffin L, Towbin J, Gelb B, Worley K, McCabe ERB (1991) Porin interaction with hexokinase and glycerol kinase: microcompartmentation at the outer mitochondrial membrane (Review). Biochem Med Metab Biol 45:271–291

Allan V, Vale R (1994) Movement of membrane tubules along microtubules in vitro: evidence for specialized sites of motor attachment. J Cell Sci 107:1885–1897

Ashkin A, Schütze K, Dziedzic JM, Euteneuer U, Schliwa M (1990) Force generation of organelle transport measured in vivo by an infrared laser trap. Nature 348:346–348

BeltrandeloRio H, Wilson J (1992) Interaction of mitochondrially bound rat brain hexokinase with intramitochondrial compartments of ATP generated by oxidative phosphorylation and creatine kinase. Arch Biochem Biophys 299:116–124

Bereiter-Hahn J, Vöth M (1994) Dynamics of mitochondria in living cells: shape changes, dislocation, fusion, and fission of mitochondria. Microsc Res Tech 27:198–219

Bernier-Valentin F, Aunis D, Rousset B (1983) Evidence for tubulin-binding sites on cellular membranes: plasma membranes, mitochondrial membranes and secretory granule membranes. J Cell Biol 97:209–216

Bershadsky AD, Vasiliev JM (1988) Cytoskeleton. Plenum Press, New York

Burgess SM, Delannoy M, Jensen RE (1994) MMM1 encodes a mitochondrial outer membrane protein essential for establishing and maintening the structure of yeast mitochondria. J Cell Biol 126:1375–1391

Burgoyne RD (ed) (1991) The neuronal cytoskeleton. Wiley-Liss, New York

Burkhardt JK, Echeverri C, Nilsson T, Vallee RB (1997) Overexpression of the dynamitin (p50) subunit of the dynactin complex disrupts dynein-dependent maintenance of membrane organelle distribution. J Cell Biol 139:469–484

Carmo-Fonseca M, David-Ferreira JF (1990) Interactions of intermediate filaments with cell structures. Electron Microsc Res 3:115–141

Couchman JR, Rees DA (1992) Organelle-cytoskeleton relationships in fibroblasts: mitochondria, golgi apparatus, and endoplasmic reticulum in phases of movement and growth. Eur J Cell Biol 27:47–54

Drubin DG, Jones HD, Wertman KF (1993) Actin structure and function: roles in mitochondrial organization and morphogenesis in budding yeast and identification of the phalloidin-binding site. Mol Cell Biol 4:1277–1294

Eckert BS (1986) Alteration of the distribution of intermediate filaments in PtK1 cells by acrylamide. II. Effect on the organization of cytoplasmic organelles. Cell Motil Cytoskeleton 6:15–24

Elluru RG, Bloom GS, Brady ST (1995) Fast axonal transport of kinesin in the rat visual system: functionality of kinesin heavy chain isoforms. Mol Biol Cell 6:1–40

Feiguin F, Ferreira A, Kosik KS, Caceres A (1994) Kinesin-mediated organelle translocation revealed by specific cellular manipulations. J Cell Biol 127:1021–1039

Fiek C, Benz R, Roos N, Brdiczka D (1982) Evidence for identity between the hexokinase binding protein and the mitochondrial porin in the outer membrane of rat liver mitochondria. Biochim Biophys Acta 688:429–440

Fisk HA, Yaffe MP (1997) Mutational analysis of Mdm1p function in nuclear and mitochondrial inheritance. J Cell Biol 138:485–494

Forman DS (1987) Axonal transport of mitochondria. In: Smith RS, Bigby MA (eds) Axonal transport. Liss, New York, pp 155–163

Fuchs E, Weber K (1994) Intermediate filaments: structure, dynamics, function and disease. Annu Rev Biochem 63:345–382

Gorlach M, Meyer HE, Eisermann B, Soboll S (1995) cAMP-dependent phosphorylation of cytokeratin in hepatic inner mitochondrial membrane. Biol Chem Hoppe Seyler 376:51–55

Gotow T, Miyaguchi K, Hashimoto PH (1991) Cytoplasmic architecture of the axon terminal: filamentous strands specifically associated with synaptic vesicles. Neuroscience 40:587–598
Hancock WO, Howard J (1998) Processivity of the motor protein kinesin requires two heads. J Cell Biol 140:1395–1405
Hargreaves AJ, Avila J (1985) Localization and characterization of tubulin-like proteins associated with brain mitochondria: the presence of a membrane-specific isoform. J Neurochem 45:490–496
Hirokawa N (1998) Kinesin and dynein superfamily proteins and the mechanism of organelle transport. Science 279:519–526
Jancsik V, Filliol D, Felter S, Rendon A (1989) Binding of microtubule-associated proteins (MAPs) to rat brain mitochondria: a comparative study of the binding of MAP2, its microtubule-binding and projection domains, and tau proteins. Cell Motil Cytoskeleton 14:372–381
Jellali A, Metz-Boutigue MH, Surgucheva I, Jancsik V, Schwartz C, Filliol D, Gelfand VI, Rendon A (1994) Structural and biochemical properties of kinesin heavy chain associated with rat brain mitochondria. Cell Motil Cytoskeleton 28:79–93
Klymkowksy MW (1995) Intermediate filament organization, reorganization, and function in the clawed frog *Xenopus*. Curr Top Dev Biol 31:455–486
Konstantinova SA, Mannella CA, Skulachev VP, Zorov DB (1995) Immunoelectron microscopy study of the distribution of porin on outer membranes of rat heart mitochondria. J Bionerg Biomembr 27:93–99
Kottke M, Adams V, Reisinger I, Bremm G, Bosh W, Sandri G, Panfili E (1988) Mitochondrial boundary membrane contact sites in the brain: points of hexokinase and creatine kinase location and control of Ca transport. Biochim Biophys Acta 935:807–832
Kuznetsov SA, Langford GM, Weiss DG (1992) Actin-dependent organelle movement in squid axoplasm. Nature 356:722–725
Lazzarino DA, Boldogh I, Smith MG, Rosand J, Pon LA (1994) Yeast mitochondria contain ATP-sensitive, reversible actin-binding activity. Mol Cell Biol 5:807–818
Leterrier JF, Eyer J, Weiss DG, Lindén M (1991) In vitro studies of the physical interactions between neurofilaments, microtubules and mitochondria isolated from the central nervous system. In: Paillotin G (ed) Conf. Proc. on The living cell in four dimensions, American Institute of Physics, Vol. 226, pp 91–105
Leterrier JF, Lindén M, Kuznetsov SA, Weiss DG (1992) Effects of nucleotides and microtubule-associated proteins (MAPs) on the interactions of kinesin with rat brain mitochondria. Eur J Cell Biol (Suppl) 36:49
Leterrier JF, Rusakov DA, Nelson BD, Lindén M (1994) Interactions between brain mitochondria and cytoskeleton: evidence for specialized outer membrane domains involved in the association of cytoskeleton-associated proteins to mitochondria in situ and in vitro. J Microsc Res Tech 27:233–261
Li Z, Colucci-Guyon E, Pincon-Raymond M, Merieskay M, Pournin S, Paulin D, Babinet C (1996) Cardiac lesions and skeletal myopathy in mice lacking desmin. Dev Biol 175:362–366
Lindén M, Gellefors P, Nelson BD (1982a) Purification of a protein having pore forming activity from the rat liver mitochondrial outer membrane. Biochem J 208:77–82
Lindén M, Gellerfors P, Nelson BD (1982b) Pore protein and the hexokinase-binding protein from the outer membrane of rat liver mitochondria are identical. FEBS Lett 141:189–192
Lindén M, Nelson BD, Leterrier JF (1989a) The specific binding of the microtubule-associated protein 2 (MAP2) to the outer membrane of rat brain mitochondria. Biochem J 261:167–173 (Erratum: Biochem J 262:1002)
Lindén M, Nelson BD, Loncar D, Leterrier JF (1989b) Studies on the interactions between mitochondria and the cytoskeleton. J Bionerg Biomembr 21:507–518
Lindén M, Karlsson G (1996) Identification of porin as a binding site for MAP2. Biochem Biophys Res Commun 218:833–836
Mannella CA (1997) Minireview: on the structure and gating mechanisms of the mitochondrial channel, VDAC. J Bionerg Biomembr 29:525–531
Mandelkow E, Mandelkow EM (1995) Microtubules and microtubule-associated proteins. Curr Opin Cell Biol 7:72–81

Martz D, Lasek RJ, Brady ST, Allen RD (1984) Mitochondrial motility in axons: membranous organelles may interact with the force generating system through multiple surface binding sites. Cell Motil Cytoskeleton 4:89–101

Mermall V, Post PL, Mooseker MS (1998) Unconventional myosins in cell movement, membrane traffic, and signal transduction. Science 279:527–533

Migheli A, Butler M, Brown K, Shelanski ML (1988) Light and electron microscope localization of the microtubule-associated tau protein in rat brain. J Neurosci 8:1846–1851

Morris RL, Hollenbeck PJ (1993) The regulation of bidirectional mitochondrial transport is coordinated with axonal outgrowth. J Cell Sci 104:917–927

Morris RL, Hollenbeck PJ (1995) Axonal transport of mitochondria along microtubules and F-actin in living vertebrate neurons. J Cell Biol 13:1315–1326

Nangaku M, Sato-Yoshitake R, Okada Y, Noda Y, Takemura R, Yamazaki H, Hirokawa N (1994) KF1B, a novel microtubule plus end-directed monomeric motor protein for transport of mitochondria. Cell 79:1209–1220

Olson GE, Winfrey VP (1990) Mitochondria-cytoskeleton interactions in the sperm midpiece. J Struct Biol 103:13–22

Quellet T, Levac P, Royal A (1988) Complete sequence of the mouse type-II keratin EndoA: its amino-terminal region resembles mitochondrial signal peptides. Gene 70:75–84

Pardo JV, Pittenger FM, Craig SWW (1983) Subcellular sorting of isoactins: selective association of γ actin with skeletal muscle mitochondria. Cell 32:1093–1103

Pereira AJ, Dalby B, Stewart RJ, Doxsey SJ, Goldstein LSB (1997) Mitochondrial association of a plus end-directed microtubule motor expressed during mitosis in *Drosophila.* J Cell Biol 136:1081–1090

Price RL, Lasek RJ, Katz MJ (1991) Microtubules have special associations with smooth endoplasmic reticula and mitochondria in axons. Brain Res 540:209–216

Rappaport L, Samuel JL (1988) Microtubules in cardiac myocytes: In: Bourne GH, Frielander M, Jeon KW (eds) International review of cytology 13. Academic Press, London, pp 101–143

Saetersdal T, Greve G, Dalen H (1990) Association between beta-tubulin and mitochondria in adult isolated heart myocytes as shown by immunofluorescence and immunoelectron microscopy. Histochemistry 95:1–10

Simon VR, Swayne TC, Pon LA (1995) Actin-dependent mitochondrial motility in mitotic yeast and cell-free systems: identification of a motor activity on the mitochondrial surface. J Cell Biol 130:345–354

Smith MG, Simon VR, O'Sullivan H, Pon LA (1995) Organelle-cytoskeletal interactions: actin mutations inhibit meiosis-dependent mitochondrial rearrangement in the budding yeast *Saccharomyces cerevisiae.* Mol Biol Cell 6:1381–1396

Soltys BJ, Gupta RS (1992) Interrelationships of endoplasmic reticulum, mitochondria, intermediate filaments, and microtubules; a quadruple immunofluorescence labeling study. Biochem Cell Biol 70:1174–1186

Stromer MH, Bendayan M (1990) Immunocytochemical identification of cytoskeletal linkages to smooth muscle cell nuclei and mitochondria. Cell Motil Cytoskeleton 17:11–18

Sun J, Bird CH, Salem HH, Bird P (1993) Association of annexin V with mitochondria. FEBS Lett 329:79–83

Tanaka Y, Kanai Y, Okada Y, Nonaka S, Takeda S, Harada A, Hirokawa N (1998) Targeted disruption of mouse conventional kinesin heavy chain, kif5B, results in abnormal perinuclear clustering of mitochondria. Cell 93:1147–1158

Weiss DG, Keller F, Gulden J, Maile W (1986) Towards a new classification of intracellular particle movements based on quantitative analysis. Cell Motil Cytoskeleton 6:128–135

CHAPTER 20

NADH Brain Determination by Micromeasurements of a Laser-Induced Fluorescence in the Rat

20

S. MOTTIN[1], P. LAPORTE[1], and R. CESPUGLIO[2]

Contents

1 Introduction

In 1962, Chance et al. asked how NADH (nicotinamide adenine dinucleotide), an intracellular compound of paramount importance, could be measured in vivo, and this still remains a topical issue. This is particularly true for the approaches dealing with mitochondrial functions at the level of organs like the brain (Mottin 1997b), the liver (Sato et al. 1995) and the myocardium (Sholtz et al. 1995). Rapid voltammetric methods (100 V/s) for NADH measurements reach concentrations around 1 μmol l^{-1} (Kuhr 1993). In vivo, this compound is not accessible by voltammetry or microdialysis since it is located in the intracellular compartment.

Moreover, are we looking at free or bound NADH measurements? The soluble-free form of this compound is not bound to another molecule like a dehydrogenase or a receptor and is present in solution in two types of configurations, opened and closed.

[1] TSI Laboratory, CNRS UMR 5516, 42023 St. Etienne Cedex 02, France; MOTTIN@UNIV-ST-ETIENNE-fr.
[2] INSERM U480, Cl. Bernard University, 69373 Lyon Cedex 08, France

The closed configuration predominates in aqueous solution. According to the fact that the equilibrium of the NADH enzymatic reaction is often displaced in favour of the oxidised form of the substratum and that the optimal efficiency of a dehydrogenase is in a redox balanced position, it is possible to write, with K as a constant (brackets are used rather than ordinary parentheses because the concept of the homogeneous concentration has to be replaced by a heterogeneous distribution):

$$\frac{\{\mathrm{Ox}\}\ \{\mathrm{Free\ NADH}\}}{\{\mathrm{Red}\}\ \{\mathrm{Free\ NAD^+}\}} = \frac{K_0}{\{H^+\}} = K$$

and the evaluation of the redox state

$$\frac{\{\mathrm{Free\ NADH}\}}{\{\mathrm{Free\ NAD^+}\}}$$

is obtained by measuring

$$\frac{\{\mathrm{Ox}\}}{\{\mathrm{Red}\}} \quad \text{and K.}$$

Finally, the choices of enzyme and redox couple are also very important for determining the mitochondrial redox state. In this respect, two enzymatic systems were originally selected from rat hepatic cells as being the most relevant for this purpose (Krebs 1967; Siesjo 1978): (1) glutamate-2-acetoacetate; and (2) β-hydroxybutyrate dehydrogenase. Several other enzymatic systems like lactate/pyruvate, malate/oxaloacetate and glycerophosphate/dihydroxy-acetone phosphate have been also tested. The main difficulty attached to these methods lies in the discrepancy existing between the redox values obtained by direct and those obtained by non-direct measurements (Siesjo 1978): direct methods do not reflect the redox state of the free forms since large quantities of pyridine nucleotides are in a reduced form and linked to the constitutive compounds of the tissue and the concentration ratio is different between the cytosol and mitochondria.

It is therefore evident that measurements carried out within biopsies do not reflect the redox states attached to individual compartments. After examination of the various difficulties attached to the measurement of NADH in vivo, it appears that direct procedures of measurement, notably in vivo, remain the most promising. Continuous measurements of this compound in the unanaesthetised animal over several weeks are now possible by new biophotonic methods.

2 NADH Measurement

Despite the great diversity of living species, they have a common characteristic, i.e. ATP (adenosine triphosphate), which is necessary for their survival. In cells, the production of this compound is ensured by the metabolism of energetic substrata which is achieved by glycolysis (anaerobic conditions), followed by lactic fermentation (anaerobic conditions) or oxidative phosphorylation (aerobic conditions). This last step includes the respiratory chain located in the internal part of the mitochondrial membrane. The energy supplied by the electrochemical gradient existing at this level is partly recovered by the ATP-synthase and converted into ATP. These processes can be summarised as follows:

1. NADH production: $SH_2 + NAD^+ \rightarrow S + NADH + H^+$ (SH_2 = reduced substrata)
2. Aerobic NADH consumption (Pi = inorganic phosphate):

$$NADH + H^+ + 3\ ADP + 3\ Pi + 1/2\ O_2 \rightarrow NAD^+ + H_2O + 3\ ATP$$

In a first approximation, within a given cellular medium and without distinction between the free or linked forms of NADH, it is possible to consider that $\{NADH\} + \{NAD^+\} = C$ (C being a constant). The measurement of the variations occuring in NADH fluorescence without changes in quantum efficiency allows the evaluation of the ratio:

$$R = \frac{\{NADH\}}{\{NAD+\}} \text{ which varies according to } (\times\ \varepsilon\ Jo;\ 1C).$$

The NADH measurement thus gives relevant information about the mitochondrial redox state.

2.1 Photophysical Properties of NAD(P)H

With the model of Gaussian sum (a, b, c: a × exp(–((b-x)/c)2)), the adsorption spectrum of free NADH in water (pH 7) presents two peaks at 260 and 339 nm (for the b parameter), respectively, 13 9060 and 5860 $(mol\ l^{-1})^{-1}\ cm^{-1}$ for an, and for c: 17.8 and 24.2 nm. The peak at 339 nm is linked to the chromophore, 1,4-dihydronicotinamide since the adenine part adsorbs only up to 300 nm (peak 260 nm). NAD^+ presents only a single peak corresponding to the peak at 260 nm. In neutral media, where the excitation concerns these two peaks, the fluorescence emission occurs in a spectral band centred around 460 nm (Wolfbeis and Schulman 1985). A fluorescence emission is obtained only with NADH since NAD^+ is not fluorescent under the above conditions, except in strong basic solutions. When NADH is linked to enzymatic proteins, the maximum of the excitation, emission wavelengths and fluorescence decay times can change considerably. In this respect, extreme values from 325/4040 to 365/478 nm have been measured (Wolfbeis and Schulman 1985). The time necessary to achieve the fluorescence measurement following a very short light impulse is τ_F and the speed constant of the fluorescence measured is: $k_F = 1/\tau_F$. Acording to the equation $K_F = k_R + k_{NR}$, where the speed constants k_R and k_{NR} are related to the radiative and non-radiative phenomena, respectively, $t_R = 1/k_R$ would then be the fluorescence time without desecxitation phenomena. The lifetime of the fluorescence measured is the consequence of spontaneous molecular emissions as well as of competitive phenomena resulting from the heterogeneity of the fundamental state (static inhibition) and/or the excited state (dynamic inhibition). These competitive phenomena contribute negatively to the quantum efficiency, the fluorescence intensity and the decay time. The experimental NADH decay times are in the range of 0.2–4 ns.

The question "are the measurements of free or bound NADH?" can be answered by TRF (time-resolved fluorescence). The forms of NADH are under many types of configurations which can be separated into two parts: bound or unbound, with heterogeneous modes of interactions and, in each, the two opposite molecular forms: opened and closed with a "continous" or multimodal distribution. The "unbound" closed configuration predominates in vitro aqueous solution and has a decay time of 0.2 ns, and the opened, 0,7 ns. The decay time of the bound forms can increase to 4 ns. So the measurement of decay times can give information on the molecular heterogeneity distribution.

2.2
NADH Measurement In Vivo

Fluorescence measured from endogenous NADH from mitochondrial rat hepatocyte preparations is found to be about ten times more intense than that of an NADH solution. Under the same conditions, the NADPH signal is found to be only two to four times higher (Estabrook 1962). The increment in fluorescence observed in mitochondrial preparations compared with NADH or NAD(P)H solutions might depend, in the first instance, on the presence of other different compounds.

In the case of brain slice preparations, Riepe et al. (1996) have recently substantiated the correlations existing between intramitochondrial NADH and the tissular fluorescence measured at 460 nm under 337-nm excitation. The decrease in fluorescence observed after addition of amytal (an inhibitor of complex I) or sodium-cyanide (an inhibitor of complex IV), as well as its reversibility after secondary addition of a mitochondrial uncoupler, argue in favour of the fact that the 460-nm fluorescence is mainly dependent upon NADH.

The first spectrophotometric and spectrofluorometric measurements on isolated mitochondria were achieved in 1956 (Chance and Williams 1956). The same techniques have been used since for studies on skeletal muscles (Dubosc et al. 1987), liver (Chance et al. 1965a; Kobayashi et al. 1971a), kideny (Kobayashi et al. 1971b), heart (Chance et al. 1965b; Renault 1987) and brain (Chance et al. 1962; Jobsis 1971; Mayevsky et al. 1974; Harbig et al. 1976; Kramer and Pearlstein 1979; Mayevsky and Chance 1982). Several approaches have also been focused on direct measurements of NADH in vivo, since the mitochondrial energetic metabolism is closely linked to oxygen availability and under ex-vivo conditions this aspect might be deficient. In this respect, brain studies have been carried out in anaesthetised animals without an FO (fiber optic) for guiding the light, but through direct illumination of the superficial cortex. These approaches were solely focused on the study of the redox processes in order to determine the influence on the NADH fluorescence of: (1) oxygen (Chance et al. 1962; Mayevsky et al. 1974; Harbig et al. 1976; Kramer and Pearlstein 1979; Mayevsky and Chance 1982); (2) glucose (Bryan and Jobsis 1983, 1986; Uematsu et al. 1989); (3) various ions (Mayevsky and Chance 1982); (4) anaesthetics (Chance et al. 1962; Nowicki et al. 1987). Emission spectra recorded on cerebral slices or on intact organs (Chance et al. 1962; Harbig et al. 1976; Riepe et al. 1996; Mottin et al. 1997b) were found close to those obtained with NADH in vitro or in mitochondrial preparations.

Despite the fact that cerebral fluorescence is correlated with NADH contents (Jobsis 1971), several physiological and physical factors have also been suspected for independently modifying the in vivo measurements (Mayevsky 1982). These factors include variations in the absorption and the reflection/diffusion of the tissue, also termed oxymetric and haemodynamic artefacts; mechanical artefacts produced by breathing and cardiac flutters.

2.3 Methodological Difficulties of In Vivo NADH Measurement

The NADH and NAD(P)H coenzymes are present in all forms of life (0.01–0.2 mg/g). In the brain, the level of the NADH/NAD^+ couple generally dominates the NADPH/$NADP^+$ one by a factor of 8 to 24. NAD^+ is reported to be predominant over NADH by a factor of 2 to 4, while the reserve is found for $NADP^+$ (MacIlwain 1959). Brain concentrations of pyridine nucleotides are high, around 0.3 mmol/g of tissue. Finally, it must also be noted that the NADH/NAD^+ redox couple is mainly located in the internal membrane of the mitochondria while the other couple is present in the cytosol. No NAD(P)H could be measured in the cerebral spinal fluid (Mottin 1994). The autofluorescence of the cerebral tissue is correlated with the intramitochondrial content in NADH (Chance et al. 1962; Jobsis 1971; Welsh and Rieder 1978; Paschen et al. 1981; Dora et al. 1984; Mayevsky 1984; Eng 4t al. 1989; Riepe et al. 1996) and this is due to the facts that: (1) NADH absorbs strongly at 340 nm and fluoresces in the range of 400–530 nm; (2) in grey matter (Welsh and Rieder 1978) or in the cortex (Jobsis 1971), the increase in the 450–460 nm autofluorescence is correlated with the endogenous variations of NADH; (3) the variations occurring in the 460–530 nm autofluorescence are inversely correlated with glucose and ATP levels (Paschen et al. 1981); (4) the 460–530 nm autofluorescence is correlated with the histological staining of the mitochondria (Eng et al. 1989); and (5) there is a link between the 440–460 nm fluorescence and the endogenous content in NADH (Riepe et al. 1996).

Despite the above arguments, opposite variations occurring in autofluorescence have been reported with analogous conditions of stimulation. This aspect, obviously, has been conroversial. However, when comparing the different experimental approaches, it appears that the reproducibility of the results is also closely dependent on the physiological conditions of the animals as well as on the redox state of the preparation just before the beginning of the experiment. This last point indeed seems the determinant, since the redox state existing before stimulation appears to determine the direction of the NADH variations (increase or decrease) even when identical stimulations are applied. In this sense, the habituation of the animals to the experimental conditions is very important, since basal levels of pyruvate or lactate can be significantly altered (Shimizu et al. 1966; Reich et al. 1972). The different anaesthetics used are also a source of difficulty. All these aspects show that for NADH in vivo measurements, the experimental protocol used and the physiological state of the animal at the beginning of each experiment are important (Mottin et al. 1997a, b). These aspects, already suspected by Dora et al. (1984), questioned the methodology which was originally developed by Chance et al. in 1962 for NADH measurements.

2.3.1
Oxymetric Artefacts

The difference between the absorption spectra of oxy-haemoglobin and deoxy-haemoglobin, particularly in the 350–450 nm window, can entail noticeable modification of the excitation and emission light. Chance et al. (1962), on the basis of the results obtained with cerebral perfused slices, attributed only a small role to this phenomenon. This aspect, however, cannot be generalised to other organs (Mayevsky 1984) and the importance of anoxia on fluorescence has been emphasised (Kramer and Pearlstein 1979).

2.3.2
Haemodynamic Artefacts

For organs like the liver, kidney, heart and, to a minor extent, the brain, several authors have shown that variations of arterial pressure significantly impair the intensity of the emitted fluorescent signal (Chance et al. 1962; Schnitger et al. 1965; Jobsis 1971; Kobayashi et al. 1971a, b). The essential perturbation, termed "haemodynamic artefact", comes from modifications in the light distribution and absorption (excitation and emission) related to the variation occurring in the blood flow. In this respect, the primary increase appearing in the fluorescence intensity of NADH following a decrease in the blood flow could result from changes in the light distribution and absorption in the 330–370 nm window for the excitation and in the 420–540 nm window for emission. Regarding the cerebral tissue, a fluorescence increase of 20% has been determined within perfused slices depleted of red blood cells. In this experimental situation, the low amplitude of the variations observed is surprising since with renal tissue under the same experimental conditions, a much more important effect is obtained (Chance et al. 1962; Kobayashi et al. 1971a, b). Finally, it has also been documented that a large and sudden increase in the blood flow in the area being measured produces a fall in the fluorescent signal (Chance et al. 1962).

2.3.3
Corrections

According to the above considerations, the main artefact influencing the fluorescent signal is generated by variations in the blood flow occurring within the analysed area (Kobayashi et al. 1971a) and according to the low magnitude reported with preparations in non-physiological situations, it appears likely that under true physiological conditions (unanaesthetised and freely moving animals, used only after 10 days of habituation) the spectral modifications dependent on NADPH as well as those linked to blood pigments (haemoglobins), mitochondrial chromophores (cytochromes and flavines) and cellular or conjunctive tissues can be neglected (Mottin et al. 1997b). Generally, when corrections of the oxymetric and haemodynamic artefacts are applied, they correspond to the ratio (or a substraction) between the fluorescent signal and the values of reflection (Harbig et al. 1976; Kramer and Pearlstein 1979). Finally, the mechanical artefacts mentioned previously can now be easily avoided by using optical fibers design which allows, under freely moving conditions, a perfect mechanical concordance between the brain tissue and the sensor.

3
Brain NADH Measurement Throughout the Rat Sleep-Wake Cycle

Regarding the mammalian sleep-wake cycle, three cardinal states of vigilance are currently described, i.e. waking (W), slow wave sleep (SWS) and paradoxical sleep (PS). These states, defined by means of polygraphic methods associating the electroencephalogram (EEG) and neck muscles electromyogram (EMG), are used routinely. We combined this methodology with that allowing brain NADH measurement, recently perfected in the unanaesthetised freely moving rat (Mottin et al. 1997b). Such an approach focused on the study of the sleep-energy relationships and pertained to a field as yet poorly explored but which is promising regarding PS functions which still remain uncertain (Jouvet 1992). More precisely, the interest in the NADH measurement lies in the fact that it is mainly indicative of the functional level of oxidative phosphorylation and that PS is now reported as an energy-gated state. Here, it must be emphasised that sleep-related studies are among the most complex approaches developed in living animals, since heteroregulations involving energy but also conventional neurotransmitters are needed for a normal sleep-wake cycle (Cespuglio et al. 1990, 1995; Jouvet 1992). Sleep studies also require the achievement of biochemical measurements while the animal is sleeping or waking and, in this respect, direct evaluation of the sleep-wake related NADH variations has never been carried out before. For this purpose, we have developed a first instrumental setup made up of a subnanosecond nitrogen laser and an FO microsensor ($\phi = 200$ μm) with a real time subnanosecond detection. With this setup, measurement of the 460-nm fluorescence can be achieved within 10 s (Mottin et al. 1997).

3.1
Origin of the Fluorescence Measured at 450–460 nm

In the brain, three essential compartments can be outlined, namely the vascular, the intracellular and the extracellular compartments. Mitochondria are located in the intracellular compartment and, at this level, it is the intramitochondrial compartment that is the site of interest. According to our technology, the maximal volume influenced by the 337-nm stimulation is less than 60 nl. Within this volume the three compartments described above are probed. As already discussed, under the physiological conditions of our experiments, the influence of the vascular compartment can be considered as weak. Concerning the haemodynamic or oxymetric artefacts, however, a contribution to the signal cannot be avoided. The density of the vascular bed is thus important and among the structures that we have already investigated, the nRD (raphe dorsalis nucleus) is a nucleus reported as poorly vascularised (300–600 lumens/mm^2; Descarries et al. 1982). But, whatever the structure considered and its level of vascularisation, it is necessary to underline that the variations attached to the vascular compartment remain within the physiological limits of homeostasis.

Concerning the extracellular compartment, we have already suggested that in the fresh cerebrospinal fluid, no traces of fluorescence have been observed (Mottin et al. 1994). According to the above considerations, it therefore appears very likely that the NADH reflected in our measurements is mainly located in the intracellular compartment and in the intramitochondrial part. Finally, we also specified that the NADH fluorescence can be dependent on the nature of the NADH molecular environment. In this way, a variation of the brain tissue autofluorescence could be attributed to a change in the redox state. As regards this aspect, time-resolved measurement of the fluorescence that we performed in the nRD did not exhibit great changes in the decay time (0.9–1 ns measured; Mottin et al. 1997). This indicates that there is not great variation in the quantum efficiency of the chomophores probed. With all the photophysical properties of laser-induced fluorescence of nRD and biochemical studies already discussed, the variation in nRD fluorescence intensity can be correlated with (NADH) and thus the mitochondrial redox state.

3.2
Brain NADH and Sleep-Wake Cycle

Owing to the constraints attached to this type of PS approach, the development of new optical techniques and methods was necessary (Mottin et al. 1997). In Fig. 1, the setup is described. The main requirements are as follows:

1. Integration of each measurement in less than 30 s, since duration of a PS episode in the rat is less than 2 min. Now we currently carry out our measurements in less than 5 s and future developments planned will allow faster measurements (1/30 s).
2. Reduction of the volume probed in order to investigate subareas of the nRD. The present anatomical resolution of our probe is within 200 μm in diameter.

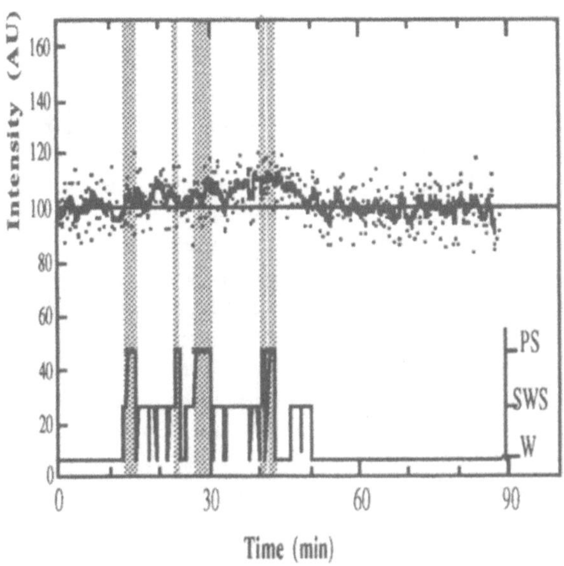

Fig. 1. Variations occuring in the 460 nm fluoresence within the nRD in correlation with the sleep-wake cycle. Mild increases in the NADH fluorescence intensity occur during PS *(grey columns)*. The correlations are illustred by the hypnogram (*bottom* of the figure). *nRD* raphe dorsalis nucleus, *W* waking, *SWS* slow wave sleep, *PS* paradoxical sleep, *AU* arbitrary units

3. Reinforcement of biochemical selectivity. Our efforts in this sense have been aimed at obtaining a triple signature, i.e. absorption wavelength, emission wavelength and decay time of the fluorescence. The selectivity presently obtained is satisfactory and reinforced by the data obtained with specific drugs.
4. Improvement of sensitivity: for NADH measurement in vivo, a sensitivity allowing the detection of weak variations is necessary.
5. Limitation of the excitation light. In this sense a real time single photon counting method has been chosen. The weak power used (less than 50 μW and 1.5 mW/mm^2), however, cannot entail local overheating, vaporisation and strong photochemical phenomena. The invasive character of the interface FO-tissue has been, however, controlled and no glial reaction (verified by post-mortem examination) has been evidenced in the sites probed.
6. More efficient collection of the fluorescence by mono-fiber configurations: the same optical fiber brings the excitation to the brain and guides back the signal collected to the detection setup:
7. Realisation of the experiments under chronic conditions. Our experiments are now generally achieved in freely moving animals. Recordingsessions are performed over 8 to 12 h and sometimes over 48 h. The brain prothesis developed allows maintenance of animals over several weeks. At the end of each experimental session and in order to check the site probed, the animals are sacrificed with a lethal dose of anaesthetic, their brain removed and the slices stained with cresyl violet.

The main results obtained indicate that the intensity of the NADH signal increases druing PS in the cerebral cortex as well as in the nRD area (Mottin et al. 1997). The magnitude of the effect observed, however, is different depending on the sites recorded. In the nRD, a structure known for containing an important depending on the sites recorded. In the nRD, a structure known for containing an important serotoninergic component whose neurons become progressively silent during SWS and PS, the effect obtained is the most important. This aspect is paradoxical since it tends to indicate that the maintenance of the cellular inhibition of the nRD serotoninergic neurons during PS necessitates more energy than the maintenance of an active neuronal discharge during wakefulness.

3.3 Pharmacological Modifications of Brain NADH

Pharmacological modifications can be obtained with a great variety of compounds (see Chep. 5, this Vol.). It is beyond the scope of this chapter to give an overview of the pharmacological experiments of brain NADH fluorescence.

After a lethal dose of anaesthetic and while the breathing and heart beats stop, a large (+70%) and rapid (less than 15 s) increase in the NADH fluorescent signal was obtained (Mottin et al. 1997). Here, it is necessary to emphasise that measurements realised under ex vivo or post-mortem conditions are poorly exploitable for studies related to bioenergetics since they reflect a brain status where all the parameters are extra-physiologic. The short time lapse during which these important changes occur after death is certainly the basis of the controversial data reported in the literature. It also indicates the extreme precariousness of the NADH/NAD$^+$ system and the necessity to perform measurements under authentic physiological conditions. Moreover, our method allows behavioural experiments with freely moving animals.

4 Conclusions and Perspectives

The current knowledge in experimental life sciences and particularly of in vivo functional aspects of the subsystems in living species is often based on the possibility of detecting the static charge or currents yielded by their cellular elements. In this respect, several methodologies can be cited, i.e. electroencephalogram, electromyogram, electrocardiogram, unitary discharge studies, electrochemical approaches, stimulation processes, etc. The possibility, illustrated in this chapter, of using light for information transfer is new and promising. Beyond the uses of electron or photon, the main difficulty of analysis methods in life science is that subsystems [molecular levels, often microheterogeneously bound molecules, at macromolecular scales, isolated mitochondria models (Balaban 1990), "true" mitochondria clustering in isolated cells (Jones 1986), or isolated organs] cannot be summed to reach the organism integrity. In this sense, methods and data reported and related to the sleep-wake cycle of freely moving animals support the pertinence of the coupling of analysis methods with the whole organism. As a general conclusion. our research strategy can also be illustrated by the following quotation of Professor Jouvet: "The problem now is the exploration of cerebral energy. As long as experimental sciences have existed, research constantly stumbles over this wall that one does not know very well how to proceed for technical reasons. There are areas, in physiology, where the progress of knowledge needs new technologies, these are coming" (translated from a French citation; Jouvet 1992).

Acknowledgements. This work was supported by CNRS UMR 5516, INSERM unit U480, 97-99 Rh ône-Alpes region grant. We also thank Miss C. Limoge for improving the English version.

References

Balaban RS (1990) Regulation of oxidative phosphorylation in the mammalian cell. Am J Physiol 258:C377–C389

Bryan RM, Jobsis FF (1983) The cerebral redox state in cats during severe insulin induced hypoglycemia. Brain Res 279:266–270

Bryan RM, Jobsis FF (1986) Insufficient supply of reducing equivalents to the respiratory chain in cerebral cortex during severe insulin-induced hypoglycemia in cats. J Cereb Blood Flow Metab 6(3):286–291

Cespuglio R, Chastrette N, Prevautel H, Jouvet M (1990) Serotonin in hypnogenic factors: functional relationship for sleep induction. In: Inou S, Krueger JM (eds) Endogenous sleep factors. SPB Academic Publishing, The Hague, pp 87–98

Cespuglio R, Marinesco S, Baubet V, Bonnet C, El Kafi B (1995) Evidence for a sleep-promoting influence of stress. Adv Neuroimmunol 5:145–154

Chance B, Williams GR (1956) The respiratory chain and oxydative phosphorylation. Adv Enzymol 17:65–134

Chance B, Cohen P, Jobsis F, Schoener B (1962) Intracellular oxidation-reduction states in vivo. Science 137(3529):499–508

Chance B, Schoener B, Krejci K, Rüssmann W, Wesemann W, Schnitger H, Bücher T (1965a) Kinetics of fluorescence and metabolite changes in rat liver during a cycle of ischemia. Biochem Z 341:325–333

Chance B, Williamson JR, Jamieson D, Schoener B (1965b) Properties and kinetics of reduced pyridine nucleotide fluorescence of the isolated and in vivo rat heart. Biochem Z (341):357–377

Descarries L, Watkins KC, Garcia S, Beaudet A (1982) The serotonin neurons in nucleus raphe dorsalis of adult rat: a light and electron microscope radioautographic study. J Comp Neurol (207):239–254

Dora E, Gyulai L, Kovach AGB (1984) Determinants of brain activation-induced cortical NAD/NADH response in vivo. Brain Res (299):61–72

Dubosc D, Renault G, Polianski J, Muffat-Joly M, Toussaint M, Guérin F, Pocidalo JJ, Fardeau M (1987) NADH measured by laser fluorometry in skeletal muscle in McArdle's disease. N Engl J Med 316(26):1664–1665

Eng J, Lynch RM, Balaban RS (1989) Nicotinamide adenine dinucleotide fluorescence spectroscopy and imaging of isolated cardiac myocytes. Biophys J 55:621–630

Estabrook RW (1962) Fluorometric measurement of reduced pyridine nucleotide in cellular and subcellular particles. Anal Biochem 4:231–245

Harbig K, Chance B, Kovach AGB, Reivich M (1976) In vivo measurement of pyridine nucleotide fluorescence from cat brain cortex. J Appl Physiol 41(4):480–488

Jobsis FF (1971) Intracellular redox changes in functional cerebral cortex. J Neurophysiol 34:735–749

Jones DP (1986) Intracellular diffusion gradients of O_2 and ATP. Am J Physiol 250:C663–675

Jouvet M (1992) Le sommeil et le rêve. Odile Jacob, Paris

Kobayashi S, Nishiki K, Kaede K, Ogata E (1971a) Optical consequences of blood substituion on tissue oxidation-reduction state microfluorometry. J Appl Physiol 31(1):93–96

Kobayashi S, Kaede K, Nishiki K, Ogata E (1971b) Microfluorometry of oxidation-reduction rate of the rat kideny in situ. J Appl Physiol 31(5):693–696

Kramer RS, Pearlstein RD (1979)Cerebral cortical microfluorometry at isobestic wavelengths for correction of vascular artifacts. Science 205(17):693–696

Krebs HA (1967) The redox state of NAD in the cytoplasm and mitochondria of rat liver. Advances in enzyme regulation, vol 5. Pergamon Press, Oxford, pp 409–434

Kuhr WG, Barrett VL, Gagnon MR, Hopper P, Pantano P (1993) Deshydrogenase-modified carbon-fiber microelectrodes for the measurement of neurotransmitter dynamics 1. NADH voltammetry. Anal Chem 65:617–622

MacIlwain H (1959) Biochemistry and the central nervous system. Churchill, London, pp 140–143

Mayevsky A (1984) Brain NADH redox state monitored in vivo by fiber optic surface fluorometry. Brain Res Rev 7:49–68

Mayevsky A, Chance B (1982) Intracellular oxidation-reduction state measured in situ by a multichannel fiber-optic surface fluorometer. Science 217(6):637–540

Mayevsky A, Jamieson D, Chance B (1974) Oxygen poisoning in the unanesthetized brain: correlation of the oxidation-reduction state of pyridine nucleotide with electrical activity. Brain Res 76:481–491

Mottin S, Tran-Minh C, Laporte P, Cespuglio R, Jouvet M (1994) Spectrofluorimetric analytical investigation of ex vivo cerebrospinal fluids and in vivo raphe dorsalis nuclei of the brain of freely moving rats. J Phys IV C4:261–264

Mottin S, Laporte P, Jouvet M, Cespuglio R (1997a) Spectroscopie d'emission du noyau raphe dorsalis chez le rat non anesthésié soux excitation laser UV 300–355 nm. Ann Phys C1 22:179–180

Mottin S, Laporte P, Jouvet M, Cespuglio R (1997b) Brain NADH determination throughout the rat sleep-wake cycle by use of a fiver optic time-resolved fluorescence sensor. Neuroscience 79:683–693

Nowicki JP, Jourdain D, MacKenzie ET (1987) NADH fluorescence in vivo: changes in cerebral oxidative metabolism and perfusion induced by pentobarbital, indomethacin and salicylate. J Cereb Blood Flow Metab 7(3):280–288

Paschen W, Niebuhr I, Hossmann KA (1981) A bioluminescence method for the demonstration of regional glucose distribution in brain slices. J Neurochem 36:513–517

Reich P, Geyer SJ, Karnovsky ML (1972) Metabolism of brain during sleep and wakefulness. J Neurochem 19:487–47

Renault G (1987) Clinical applications of laser fluorometer. Lasers Opton (12):56–59

Riepe MW, Schmalzigaug K, Fink F, Oexle K, Ludolph AC (1996) NADH in the pyramidal cell layer of hippocampal regions CA1 and CA3 upon selective inhibition and uncoupling of oxidative phosphorylation. Brain Res 710:21–27

Sato B, Tanaka A, Mori S, Yanabu N, Kitai T, Tokura A, Inomoto T, Iwata S, Yamaoka Y, Chance B (1995) Quantitative analysis of redox gradient within the rat liver acini by fluorescence images: effects of glucagon perfusion. Biochim Biophys Acta 1268:20–26

Schnitger H, Scholtz R, Bücher T, Lübbers DW (1965) Comparative fluorometric studies on rat liver in vivo on isolated, perfused, hemoglobin-free liver. Biochem Z 341:334–339

Shimizu H, Tabushi K, Hishikawa Y, Kakimoto Y, Kaneko Z (1966) Concentration of lactic acid in rat brain during natural sleep. Nature 212:936–937

Sholtz TD, Laughlin MR, Balaban RS, Kupriyanov VV, Heineman FW (1995) Effect of substrate on mitochondrial NADH, cytosolic redox state, and phosphorylated compounds in isolated hearts. Am J Physiol 268:H82–H91

Siesjo BK (1978) Bain energy metabolism. Wiley, London, pp 187–270

Uematsu D, Greenberg JH, Reivich M, Karp A (1989) Cytosolic free calcium, NAD/NADH redox state and hemodynamic changes in the cat cortex during severe hypoglycemia. J Cereb Blood Flow Metab 9(2):149–155

Welsh FA, Rieder W (1978) Evaluation of in situ freezing of cat brain by NADH fluorescence. J Neurochem 31:299–309

Wolfbeis OS, Schulman SG (1985) The fluorescence of organic natural products. Molecular luminescence spectroscopy, method and applications, part 1, vol 77. Wiley, New York, pp 167–370

Allotopic Expression of Mitochondrial Genes: Further Steps in the Evolution of Eukaryotic Cells and Therapeutic Strategy

21

C. Jacq, M. Corral-Debrinski, and S. Hermann-Le Denmat

Contents

1 Introduction

This chapter will address the questions surrounding the raison d'être of the mitochondrial genome. We will describe how mitochondrial diseases can potentially be managed by strategies currently under development which replace defective mitochondrial genes with a nuclear copy (allotopic expression). These studies have contributed towards a better understanding of the evolution of the mitochondrial genome and the mechanisms of mitochondrial protein import.

Laboratoire de Génétique Moléculaire, CNRS URA 1302, Ecole Normale Supérieure, 46 rue d'Ulm, 75230 Paris Cedex 05, France

1.1
The Mitochondrial Genome, an Indispensable Fossil

Mitochondrial biogenesis and function are controlled by about 600 nuclear genes and a dozen protein-coding genes located in the mitochondrial genome, In both man (and in the yeast *Saccharomyces cerevisiae*), the mitochondrial genome codes for 13 (and 8) polypeptide chains and 22 (and 24) tRNAs and the 12S and 16S ribosomal RNAs. Although few in number, these mitochondrial genes play an essential role in mitochondrial respiration. Their mutation results in serious diseases, especially of the central nervous system, and has been linked to cellular ageing (Wallace 1993; Lestienne and Bataille 1994). The absence of mitochondrial transfection vectors precludes the use of classical gene therapy to correct these genetic defects. The efficiency of DNA transfer from the nucleus to mitochondria is 10^5 times lower than in the oppsite direction (Thorsness and Fox 1990) making the use of "gentle" methods for DNA delivery to mitochondria extremely difficult. Several alternative methods have therefore been devised, one of which couples the DNA molecule with a mitochondrial transport signal, thereby enabling mitochondrial import of DNA, at least in vitro (Seibel et al. 1995).

The approach we will discuss here mimics the natural evolution of the mitochondrial genome, which consists of copying mitochondrial genes into nuclear DNA. In fact, considering the widely accepted hypothesis of the endosymbiotic evolution of mitochondria, the large majority of bacterial genes from which mitochondria originated have been transferred to the nuclear genome. This phenomenon very often results in the complete disappearance of the mitochondrial gene, undoubtedly due to the still poorly understood advantage conferred by the nuclear copy. Several questions concerning this evolution merit further discussion within the scope of our project. First, why were not all mitochondrial genes transferred to the nucleus? This would have saved the cell the costly job of having to come up with an entire set of replication, transcription and translation machinery for the mitochondrion. Second, why is it that in all organisms studied so far, the mitochondrial genome is always found to contain, in addition to genes coding for ribosomal RNAs, the same two genes: those coding for cytochrome b and for cytochrome oxidase subunit 1 (COX1)? One way to answer these questions is to try to continue the unfinished process of mitochondrial gene transfer to the nucleus. Such an approach requires that both epistemological and technological barriers be overcome. In a first step, we can briefly summarize what is already known about how evolutionary tinkering has come up with some solutions.

2
Mitochondrial Protein Transport: A Key Mechanism of Gene Transfer to the Nucleus

Copies of mitochondrial genes, which no doubt integrated at random into the nuclear genome where they are now dispersed, must have acquired additional information so that their translation products would be correctly imported by mitochondria. Our current understanding of mitochondrial protein import is based largely on in vitro studies of proteins with low hydrophobicity imported into isolated mitochondria. Although these studies have unquestionably provided valuable data on certain key points, as discussed below, it is important to keep in mind that the in vivo control of mitochondrial protein import may be more complex, particularly in the case of hydrophobic proteins, and that further events such as coupling between translation and import may occur.

2.1
Signal Sequences

Most proteins translated in the cytoplasm for import into mitochondria contain an N-terminal signal peptide which is cleaved after passage through the inner membrane. As many as 25% of randomly generated peptides can act as signal peptides, a high percentage, explainable by the fact that simply an amphiphilic helix with a particular distribution of positve charges is all that is needed for interaction with the membrane receptor and mitochondrial import. The relationship between the signal sequence and the protein to be imported is obscure. However, it is known that *Neurospora crassa* ATPase subunit 9, the most hydrophobic protein known to be imported, has the longest signal sequence: 66 amino acids as opposed to the usual signal sequence length of 20 to 35 amino acids.

2.2
Membrane Receptors and Protein Channels

The main mechanisms of protein translocation into mitochondria are no doubt common to most organisms (Lill and Neupert 1996) and they show significant analogy with other translocation mechanisms (Schatz and Dobberstein 1996). The basic elements of this translocation machinery are outlined in Fig. 1.

A small number of proteins at the mitochondrial surface play an essential role in the recognition of precursors for import. Thus, the protein complexes Tom20p-Tom22p and Tom37p-Tom70p interact respectively with the amphiphilic signal peptide and its psoitive charges and with the protein precursors themselves. Some of these proteins, such as Tom22p and Tom70p, are buried in the membrane and form part of the import channel with Tim23, Tim17 and Tim44 (in association with mHSP70p) present in the inner membrane. The main driving force for translocation undoubtedly stems from the repeated complexing of the mHSP70p chaperone with the denatured parts of the protein to be imported (molecular ratchet model).

2.3
Chaperones and Protein Structure

In most membrane systems, translocation can only occur if the proteins are at least partly denatured. Most proteins, especially hydrophobic ones, must be complexed with chaperones in order to retain a certain degree of denaturation. Some chaperones are specific, such as the mitochondrial import-stimulation factor MSF which interacts with proteins to be transported in or through the membrane. The MSF-precursor complex then interacts with the Tom37p-Tom70p receptor at the mitochondrial surface (Hachiya et al. 1995).

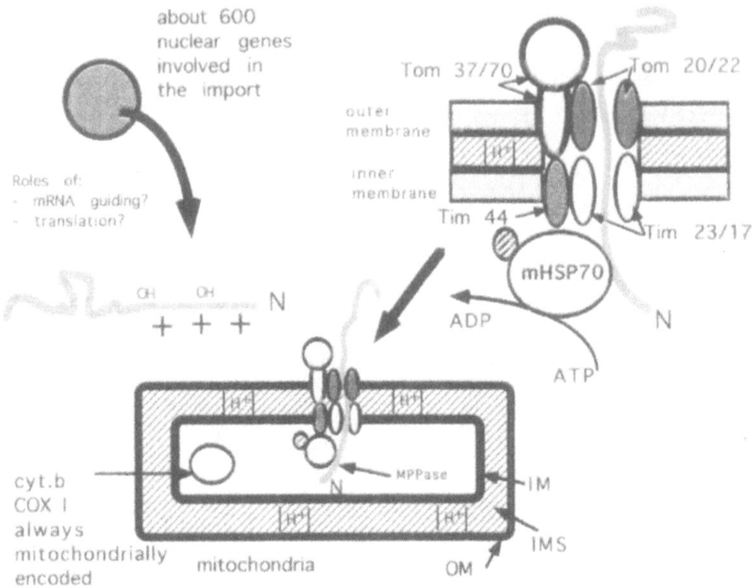

Fig. 1. General outline of mitochondrial protein import. The essential features involved in the mitochondrial import of cytoplasmic preproteins are described. The majority of mitochondrial proteins are coded in the nucleus and imported into mitochondria. The specific processes by which these mRNA might be guided to mitochondria are not well understood (Lithgow et al. 1997). The signal peptide and the denatured form of the preprotein are recognized by receptors on the outer membrane (Tom20/22 and 37/70). Aften it penetrates the mitochondrial matrix, the preprotein interacts with the chaperone mHSP70 and the signal peptide is cleaved by a specific protease (MPPase). Repeated interaction of mHSP70 with the denatured parts of the protein is an essential feature of the translocation mechanism. (Hachiya et al. 1995; Lill and Neupert 1996; Schatz and Dobberstein 1996)

3 Allotopic Expression of Mitochondrial Genes: Can We Continue the Long March of Evolution?

The data discussed above together with preliminary studies (Banroques et al. 1986; Gearing and Nagley 1996) have shown that after its sequence has been adapted to the universal code (see below), a mitochondrial gene can be transferred to the nucleus while guiding the localization of its translation product to the mitochondrion. This is referred to as allotopic expression.

3.1 Adaptation of a Mitochondrial Gene to the Nuclear Genome

It is most certainly due to their different origins that the nuclear and mitochondrial genomes follow different rules of gene expression. First, transcription and translation initiation and termination are controlled by different signals which must be taken into account for allotopic expression of a mitochondrial gene. Furthermore, the genetic codes are different. For example, in the case of yeast there are three types of codons which correspond to different amino acids in mitochondria and in the cytoplasm. Therefore, base changes must be introduced in the gene to be transferred in order for the final protein sequence to remain unchanged. As an example, nuclear transfer of the cytochrome b gene would require 18 base changes. Recent in vitro mutagenesis techniques have made it possible to considerably increase the efficiency of introducing multiple changes in a sequence by PCR amplification of the newly synthesized strand (Claros et al. 1996). The sequence can then even be adapted according to the codon bias in the two translation systems or restriction sites introduced to facilitate manipulation of the gene. The entire modified gene can then be placed in a nuclear vector. In the case of yeast, two types of vectors exist: multicopy vectors (2 μm) or single copy vectors containing a centromeric sequence. This new nuclear gene can therefore be integrated in the desired position of the nuclear genome by homologous recombination with adjacent sequences determined by the choice of insertion site. An important feature of this approach is the availability of reliable screening methods to ensure proper functioning of the gene. The most direct method consists of complementing a mitochondrial deficiency of the gene under study. For the yeast *S. cerevisiae*, mutant strains for most mitochondrial genes are available. The protein to be imported can also be conjugated with a marker activity or specific epitope to follow the final localization of the protein.

3.2
Case of Hydrophobic Proteins

It is already known that most genes which remained in the mitochondrial genome during evolution have a greater average hydrophobicity than those which were transferred to the nucleus. This notion that protein hydrophobicity is a limiting factor in the gene transfer process is strengthened by the fact that for cytochrome b and cytochrome oxidase subunit 1, by far the two most hydrophobic proteins of the mitochondrion, the corresponding genes have never been found in a functional form in the nuclear genome. The recent discovery of nuclear pseudogenes for cytochrome b (Collura and Stewart 1995) indicates that the gene transfer itself is not limiting. Systematic studies of allotopic expression of different parts of the cytochrome b gene have helped to define the limiting step in this process. In these studies (Claros et al. 1995), different combinations fof eight transmembrane helices of cytochrome b were produced in the nucleus in the form of hybrids with a marker protein used to detect the presence of the hybrid in mitochondria by gtenetic complementation (Fig. 2). It was shown that although each individal transmembrane helix could be imported into mitochondria, a combination of two or three of some helices was nonpermissive for mitochondrial import of the marker protein. These observations show that the average hydrophobicity value, called mesohydrophobicity (Claros et al. 1995) determines whether a protein can be imported. More conclusively, these studies indicate that the entire cytochrome b protein cannot be imported into mitochondria and agree with the evolutionary bias in the population of genes retained in the mitochondrion.

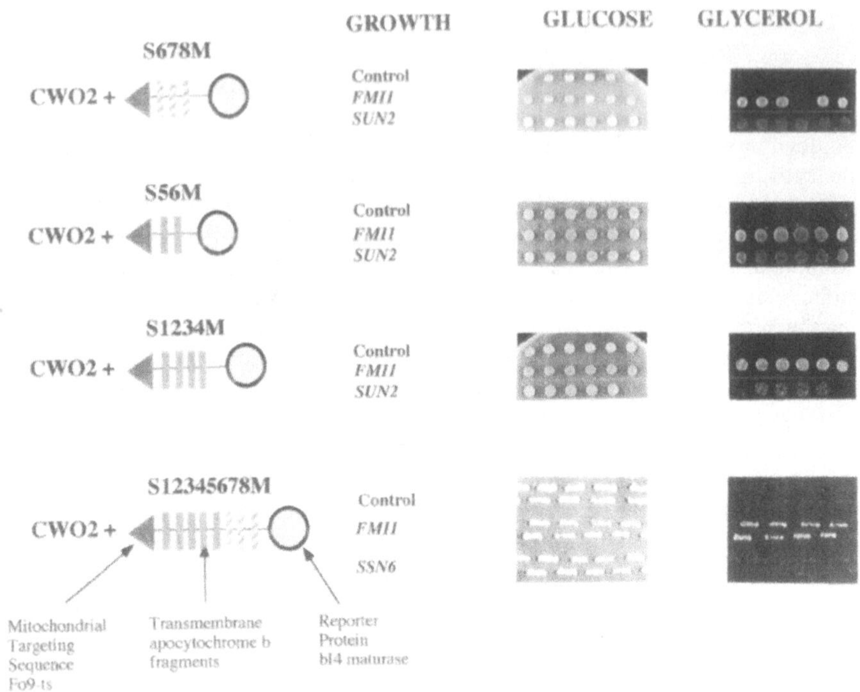

Fig. 2. Mitochondrial import of chimeric proteins with different hydrophobicities and effect of overexpression of certain nuclear genes. A family of isologous proteins with increasing hydrophobicity was constructed to analyze the effect of hydrophobicity on protein import. To do so, nuclear plasmids were constructed harboring chimeric genes coding for (1) a mitochondrial signal sequence *(triangles);* (2) a variable number of transmembrane segments of apocytochrome b *(rectangles);* and (3) RNA maturase *(circles)* which, if it enters the mitochondrion, can complement certain mitochondrial deficiencies. In this case, the mutant strain can again grow on glycerol. Upon overexpression of certain nuclear genes, some proteins which were too hydrophobic to complement the mitochondrial deficiency of the CW02 strain alone, became permissive for growth of this strain on glycerol. Thus, overexpression of the FMI 1 gene allows mitochondrial import of RNA maturase, even when the latter is conjugated with all the transmembrane helices of cytochrome b

3.3 Development of a Method for Mitochondrial Import of Hydrophobic Proteins

From both a fundamental and applied standpoint, it is important to understand how we can design tools to allow mitochondrial import of hydrophobic proteins. In other words, it is possible to find conditions in the cell which allow us to overcome the limits established during evolution? The following methodology is being developed as one potential approach to this problem.

In this method, *S. cerevisiae* is used as the model organism due to the wealth of genetic tools it offers for answering this type of question. Furthermore, all evidence suggests that the protein signalling mechanism is highly conserved between yeast and man, thereby further justifying this choice of a model.

The main objective of this approach is to identify genes which play a specific role in mitochondrial import of hydrophobic proteins. Figure 2 shows the nuclear gene constructs used. The coresponding proteins have different degrees of hydrophobicity due to the presence of a variable number of transmembrane helices from cytochrome b. These hydrophobic helices are directed towards the mitochondrion via the mitochondrial signal sequence used for a hydrophobic peptide, the *N. crassa* ATPase subunit 9. Mitochondrial import in vivo is detected by a reporter gene coding for RNA maturase which is essential for intron splicing in the mitochondrial COX1 and cytochrome b genes. An absence of RNA maturase, which itself is coded by a mitochondrial gene, therefore results in a respiration defect and an inability of the coresponding yeast strain to grow on a substrate such as glycerol. In such a strain, mitochondrial import of an RNA maturase artifically coded in the nuclear genome can be detected by its capacity to restore a wild-type phenotype (Banroques et al. 1986). RNA maturase is particularly suited as a marker of mitochondrial import because it is active in very low amounts, making this assay highly sensitive.

These tools form the basis of a direct approach to discover new genes controlling mitochondrial import of hydrophobic proteins. In fact, by using a construct leading to synthesis of a chimeric protein sufficiently hydrophobic to prevent its import into mitochondria, we can directly screen (by identifiying strains capable of growing on glycerol) for genetic environments permissive for mitochondrial import of these hydrophobic proteins. These genetic alteraitons can be induced by mutagenesis of the nuclear genome or by use of a multicopy vector to overexpress a yeast gene bank. This latter approach will be used to isolate genes that can facilitate the import of hydrophobic proteins.

3.4
Preliminary Results

In this manner, nine different genes have been isolated by transforming a suitable yeast strain (see above) with a yeast genomic bank constructed with a multicopy vector. These overexpressed genes can, to a varying degree, change the extent of mitochondrial import of hydrophobic proteins. Some of these gene products have a limited effect an only act on constructs with low hydrophobicity. Such constructs may act indirectly by altering the quantitiy of hybrid protein through an action on transcription or on RNA or protein stability, etc. Other gene products, in contrast, enable mitochondrial import of proteins with hydrophobicity similar to that of cytochrome b, as confirmed by direct biochemical analysis of mitochondrial contents.

This is the case for the FMI 1 (facilitating mitochondrial import 1) gene which is described in Fig. 2. Although these results are preliminary, it is interesting to note that among these genes are several new genes with no known function, which were not known a fortiori to be involved in mitochondrial transport. Second, these genes include essential genes whose presence is indispensable for cell function and whose knock-out is lethal to the cell. Third, some are genes for which homologs exist in the human genome (although their function is obviously unknown). Fourth, several of these genes code for RNA-binding proteins, thus reviving the stimulating idea that the directed transport of mRNA out of the nucleus might contribute to the fidelity of protein targeting to mitochondria (Lithgow et al. 1997).

4 Conclusion and Perspectives

The therapy of mitochondrial diseases, especially the mitochondrial myopathies, requires the development of tools to better control the activity of the mitochondrial genome. Since it is virtually impossible to deliver DNA to mitochondria, it is necessary to transfer mitochondrial genes to the nucleus and import their translation products, many of which are very hydrophobic, back into mitochondria. Yeast, which offers a wealth of well-understood genetic tools, is a perfect model organism for such an approach. Several new nuclear genes whose translation products apparently play a role in in vivo mitochondrial import process have already been isolated. These genes, most of which have human homologs, are curently being characterized. A full understanding of the mechanisms of mitochondrial biogenesis is clearly important for the management of mitochondrial diseases and can also provide an essential tool for gene therapy of mitochondrial cytopathies.

References

Banroques J, Delahodde A, Jacq C (1986) A mitochondrial RNA maturase gene transferred to the yeast nucleus can control mitochondrial mRNA splicing. Cell 46:837–844

Claros M, Perea J, Shu Y, Samatey FA, Popot JL, Jacq C (1995) Limitations to the in vivo import of hydrophobic proteins into yeast mitochondria. The case of a cytoplasmically synthesized apocytochrome b. Eur J Biochem 228:762–771

Claros M, Perea J, Jacq C (1996) Allotopic expression of yeast mitochondrial maturase to study mitochondrial import of hydrophobic proteins. Methods Enzymol 264:389–403

Collura RV, Stewart C (1995) Insertions and duplications of mtDNA in the nuclear genomes of Old World monkeys and hominoids. Nature 378:485–489

Gearing DP, Nagley P (1996) Yeast mitochondrial ATPase subunit 8, normally a mitochondrial gene product expressed in vitro and imported back into the organelle. EMBO J 5:3651–3655

Hachiya N, Mihara K, Suda K, Horst M, Schatz G, Lithgow T (1995) Reconstitution of the initial steps of mitochondrial protein import. Nature 376:705–709

Lestienne P, Bataille N (1994) Mitochondrial DNA alterations and genetic diseases: a review. Biomed Pharmacother 48:199–214

Lill R, Neupert W (1996) Mechanisms of protein import across the mitochondrial outer membrane. Trends Cell Biol 6:56–61

Lithgow T, Cuezva JM, Silver PA (1997) Highways for protein delivery to the mitochondria. TIBS 22:110–113

Schatz G, Dobberstein B (1996) Common principles of protein translocation across membranes. Science 271:1519–1526

Seibel P, Trappe J, Villani G, Klopstock T, Papa S, Reichman H (1995) Transfection of mitochondria: strategy towards a gene therapy of mitochondrial diseases. Nucelic Acids Res 23:10–17

Thorsness PE, Fox TD (1990) Escape of DNA from mitochondria to the nucleus in Saccharomyces cerevisiae. Nature 346:376–379

Wallace DC (1993) Mitochondrial diseases: genotype versus phenotype. Trends Genet 9:128–133

CHAPTER 22

Import of tRNA into Yeast Mitochondria: Experimental Approaches and Possible Applications

22

I. A. TARASSOV[1], N. S. ENTELIS[1,2], and R. P. MARTIN[1]

Contents

1 Introduction

Mitochondria contain hundreds of polypeptides, only a minor part of which are encoded in the mitochondrial DNA and synthesized in the organelle. The majority of mitochondrial proteins are coded for by nuclear genes and are addressed to the organelle in the form of precursors from the cytoplasm. The mechanism of pre-protein import has been studied in detail, and a number of components of targeting and translocation across the mitochondrial membranes have been identified (Kübrich et al. 1995). Proteins are not the only type of macromolecules to be mitochondrially imported. In fact, several RNAs of nucleo-cytoplasmic origin were identified as asso-

[1] UPR 9005 CNRS, 15 rue René Descartes, 67084 Strasbourg, France
[2] Dept. Molecular Biology, Moscow State University, Moscow 119899, Russia

Table 1. Nuclear DNA – coded mitochondrial tRNAs

Species	No. of imported tRNAs	Evidence of import
Plants		
Phaseolus vulgaris	>8	Hybridization
Solanum tuberosum	>11	Hybridization, import in vivo
Triticum vulgaris	>8	Sequence
Zea mais	>8	Sequence
Marchantia polymorpha	2	Sequence
Chlamydomonas reinhardtii	19	Sequence
Fungi		
Saccharomyces cerevisiae	1	Hybridization, sequence, import in vitro and in vivo
Protozoans		
Tetrahymena pyriformis	26	Hybridization
Paramecium aurelia	19	Sequence
Plasmodium falciparum	totality	Sequence
Trypanosoma brucei	totality	Hybridization, sequence, import in vivo
Leishmania tarentolae	totality	Hybridization, sequence, import in vivo and in vitro

ciated with mitochondria: 5S rRNA, the RNA component of MRP RNase, RNase P RNA and numerous tRNAs (Schneider 1994). tRNA import can be considered as the best documented. Mitochondrially imported tRNAs were found in plants, yeast and protozoans (Table 1). The number of imported tRNAs varies among organisms and ranges between a single in the yeast *S. cerevisiae* and the totality of mitochondrial tRNAs in trypanosomatids. In some protozoans and in plants, the number of imported tRNAs is intermediate and species-specific (see Chap. 23, this Vol.). Studies of import mechanisms are very recent and rely on the development of model systems in vivo for the trypanosomes (Hauser and Schneider 1995; Schneider 1996; Lima and Simpson 1996), Tetrahymena (Rusconi and Cech 1996a,b), plants (Small et al. 1992) and yeast (Entelis et al. 1998) and, on the other hand, in vitro, for the trypanosomatids (Mahapatra et al. 1994; Adhya et al. 1997) and yeast (reviewed in: Tarassov and Martin 1996). This chapter describes experimental approaches for studying import mechanisms in yeast.

2 A Single Mitochondrially Imported Yeast tRNA

Mitochondrial import of tRNA in yeast was described nearly 20 years ago. Comparison of tRNA populations before and after hybridization to mitochondrial DNA revealed the presence of a single tRNA present in the mitochondrial inner compartment which does not hybridize with the mitochondrial DNA. This tRNA was identified as a nuclear DNA-coded $tRNA^{Lys}_{CUU}$, further referred to as tRK1 (Martin et al. 1977, 1979). tRK1 is unequally distributed between the cytoplasm (>95%) and the

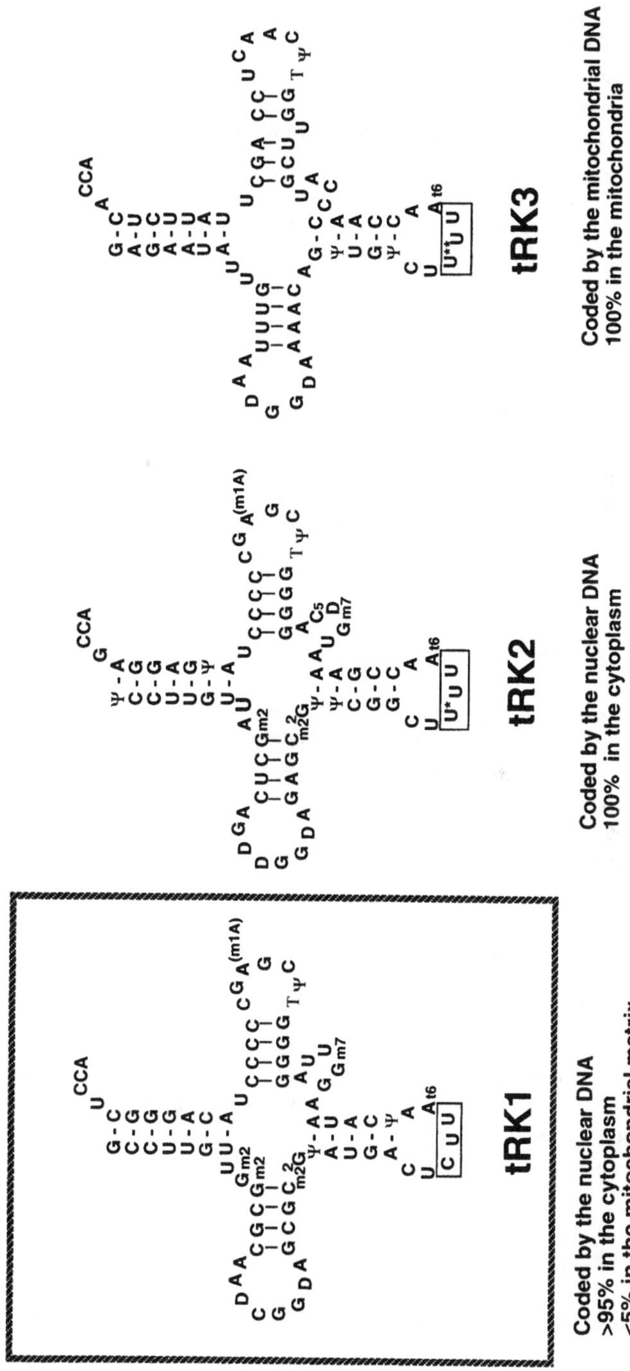

Fig. 1. Cloverleaf structures of the three yeast lysine isoacceptor tRNAs. *tRK1*, $tRNA^{Lys}_{CUU}$, *tRK2* cytoplasmic $tRNA^{Lys}_{U*UU}$, *tRK3* mitochonodrial $tRNA^{Lys}_{U**UU}$. *U** is for 5-[(methylamino)-methyl]-2-thiouridine, *U*** is for 5-(carboxymethylaminomethyl)-uridine

mitochondrial matrix (<5%; Entelis et al. 1996). Yeast cells contain two additional lysine isoacceptors, $tRNA^{Lys}{}_{mem5S2UUU}$ (tRK2), also nuclear DNA encoded but found only in the cytoplasm, and $tRNA^{Lys}{}_{cmnm5UUU}$ (tRK3) encoded by mitochondrial DNA and associated exclusively with mitochondria (Fig. 1). tRK1 is the major lysine acceptor: the nuclear genome contains 14 identical copies of the corresponding gene and its cellular concentration is 2– 3 times higher than that of tRK2. Although only a minor portion of tRK1 is imported into the mitochondrion, its mitochondrial concentration is comparable with that of other mitochondrial tRNAs coded by the organellar DNA. The mitochondrial function of tRK1 remains unclear. The $cmnm^5U$ base in the "wobble" position of the tRK3 anticodon can recognize both A and G in the third position of lysine coding codons (AAA and AAG). This suggests that this host mitochondrial tRNA is, in principle, sufficient for mitochondrial protein synthesis (Yokoyama et al. 1985; Martin et al. 1990). One hypothesis, based on a complementarity of tRK1 and several intron-exon junctions in mitochondrial mRNAs, suggests its participation in splicing (Soidla 1983). Another hypothesis is that tRK1 takes part in reverse transcription or DNA replication in mitochondria (Martin et al. 1979). Finally, one cannot exclude its participation in organellar protein synthesis and some of our results, described herein, favor such as possibility.

3 Experimental Systems to Study Mitochondrial Import

Several experimental tests were developed to study tRK1 import. Both in vivo and in vitro approaches were used. Selectivity of tRK1 import was used as the most important criterion of evaluation of the various assays.

3.1 Electroporation of Intact Yeast Cells

Electroporation of intact yeast cells was optimized to efficiently introduce 5'-end [^{32}P]-labelled tRNAs into the cytoplasm, and after 30-min incubation of the cells in culture medium, they were subfractionated to see if labelled tRNAs were associated with mitochondria (Tarassov and Entelis 1992). Among nearly 20 various RNAs tested, only tRK1 was found in the mitochondria. These results suggested that electroporated cells conserved the import selectivity of living cells and also were the first proof that tRK1 mitochondrial import can be reproduced in a model system. Unfortunately, electroporation has not quantitatively reproducible results, perhaps due to a rapid degradation of the exogenous tRNA in the cytoplasm.

3.2 Import into Isolated Mitochondria

An in vitro import system was developed to selectively introduce 5'-end [^{32}P]-labelled tRK1 into isolated yeast mitochondria (Tarassov and Entelis 1992). The main features of the import assay are listed in the Table 2. This import required the presence of ATP,

Table 2. Conditions of tRK1 import into isolated yeast mitochondria

Import assay (100 μl)
- Purified mitochondria (100 μg of mitochondrial protein)
- ATP (1 mM)
- ATP generation system (pyruvate kinase; phosphoenol pyruvate)
- [^{32}P]-tRNA (tRK1) 5'-end labelled (3 pmol, 10^5 cpm)
- Import directing proteins (S100 from pre-MSK overexpressing cells[a], 100 μg)
- L-lysine (0.1 mM), salts, proteinase inhibitors

Detection of the imported tRNA
- Treatment of the mitochondria with the nuclease cocktail[b] and re-purification of the mitochondria
- Hypotonic shock and generation of mitoplasts (mitochondria without the outer membrane)
- Treatment of the mitoplasts with the nuclease cocktail[b]
- Extraction of the mitochondrial RNA by hot phenol-SDS treatment
- Denaturing gel electrophoresis and autoradiography

[a] Tom20 – deficient yeast cells overexpresing MSK (Tom20 is the mitochondrial pre-protein receptor).
[b] RNAse A, micrococcal nuclease and phosphodiesterase.

of the electrochemical potential (ΔΨ) across the mitochondrial membranes, proteins associated with the mitochondrial outer membrane and soluble cytosolic proteins. This approach was used to search for import determinants in tRK1, because in vitro synthesized tRK1 transcripts can be imported under the same conditions as natural tRK1 (Entelis et al. 1996)

3.3
In Vivo Import of Mutant tRNAs

The third type of assay was to construct mutant tRK1 genes, to express them in yeast cells and to detect mutant tRNAs by oligonucleotide-probe hybridization in various subcellular compartments. This approach has two major problems: (1) several mutant genes do not direct efficient transcription of the corresponding tRNA in vivo; and (2) the quantitation of the import efficiencies of mutant tRNAs are complicated by competition with the normally imported tRK1. Therefore, this approach was mainly used to validate the most important conclusions of in vitro assays (Entelis et al. 1998).

4
Import Directing Proteins

Biochemical, energetic and temporal parameters of tRK1 import in vitro share similarity with those of pre-protein import (Söllner et al. 1991; Tarassov and Entelis 1992; Glick et al. 1992). The main differences are in the need of cytoplasmic proteins in the import assay. For this reason we tested if there were common components of these two targeting pathways, and also tried to identify cytoplasmic factors specific for tRNA import.

4.1
Pre-Protein Receptors

The mechanisms of pre-protein mitochondrial import have been described in detail (Söllner et al. 1991; Glick et al. 1992; Höhfeld and Hartl 1994; Kübrich et al. 1995). Pre-proteins possess, in general, a N-terminal signal sequence of 25–50 amino acids which is recognized by mitochondrial outer membrane receptors and is then inserted into a multiprotein complex GIP (general insertion pore). These receptors are exposed on the mitochondrial outer surface and are sensitive to protease treatment. Pre-proteins are then addressed to the mitochondrial matrix via a GIP-independent complex localized in the inner membrane. This process requires ATP hydrolysis and uses the $\Delta\Psi$ potential across the inner membrane. Finally, the signal peptide is cleaved off by a specific peptidase in the matrix and the protein is refolded by interaction with the mitochondrial hsp60 chaperon.

Like pre-protein import, tRK1 import in vitro depends on temperature and ATP hydrolysis (Tarassov and Entelis 1992). Elimination of outer membrane receptors by mild protease treatment of mitochondria, as well as dissipation of $\Delta\Psi$ by phosphorylation uncouplers, inhibitits import (Tarassov et al. 1995a). Partial depletion of ATP and dissipation of $\Delta\Psi$ by oligomycine (Söllner et al. 1992) arrests tRK1 translocation and blocks it on the outer membrane in the form of a ribonucleoproteic complex sensitive to protease treatment. Formation of such a complex is reversible: addition of ATP allows the completion of tRK1 import even without restoration of $\Delta\Psi$ (Tarassov et al. 1995a).

Analysis of the mutants of pre-protein import apparatus components directly proved its participation in tRK1 import. Two outer membrane receptors were tested, Tom20 and Tom70. Mitochonria depleted of Tom20 (purified from *tom*20 cells) failed to import tRK1, while Tom70-depleted organelles (from the *tom70* strain) retained the capacity to import tRK1 with the same efficiency as mitochondria from wild-type cells (Tarassov et al. 1995a). This signifies that Tom20 is implicated in tRK1 import or in the import of a protein which is necessary for tRK1 import. Depletion of mitochondria with Tim44, an essential component of pre-protein import localized in the inner membrane, also leads to an inhibition of tRK1 import, while its gradual depletion in vivo results in a decrease in the mitochondrial concentration of tRK1 (Tarassow et al. 1995a). These experiments clearly demonstrate that this translocator also plays a significant role in tRK1 import.

We suggest that tRK1 is co-translocated with a mitochondrially imported pre-protein which is recognized by the Tom20 receptor and is then imported in the mitochondrial matrix via interaction with the Tim44 protein.

4.2
Aminoacyl-tRNA Synthetases

The hypothesis suggesting that mitochondrial aminoacyl-tRNA synthetases (aaRS) can participate in mitochondrial import of tRNAs was first proposed more than 20 years ago by Suyama and Hamada (1976). These enzymes, essential for organellar protein synthesis, were considered as ideal tRNA cariers, because all of them are imported and their main feature is to selectively recognize tRNAs with different aminoacylation specifities. Since the single imported tRNA species in yeast is an acceptor of lysine, we tested if lysyl-tRNA synthetases are indeed involved in tRK1 import (Tarassov et al. 1995b).

S. cerevisiae possesses two distinct LysRS, the cytoplasmic enzyme (KRS) and the mitochondrial one (MSK). KRS and MSK are well characterized both biochemically and genetically (Gatti and Tzagoloff 1991; Martinez and Mirande 1992). The two proteins are encoded by distinct nuclear genes sharing homology with other lysRS. KRS has a cytoplasmic localization, aminoacylates tRK1 and tRK2, but is not able to aminoacylate tRK3, while MSK is mitochondrially imported, aminoacylates tRK3, but cannot aminoacylate tRK1.

Two protein fractions are needed to direct import of tRK1 in vitro, one of these contains KRS and the other one, MSK activities (Tarassov and Entelis 1992). KRS-depleted protein extracts fail to direct import of the deacylated form of tRK1 but direct import of its aminoacylated form. Furthermore, the import-directing capacity of various mutant forms of KRS is correlated with their aminoacylation activities. These results suggest that aminoacylation of tRK1 is a prerequisite for its import (Tarassov et al. 1995b). On the other hand, MSK is required for in vitro import of both aminoaclated and deacylated forms of tRK1. This remains true in vivo, since a MSK version with an N-terminal deletion, which cannot be imported and accumulates in the cytoplasm, does not direct tRK1 import. Finally, although MSK is not able to aminoacylate tRK1, it can form a stable ribonucleoproteic complex with this tRNA, but only if either KRS is also present or tRK1 is aminoacylated. These data suggest that tRK1 is co-imported with the precursor of MSK (pre-MSK).

5 Determinants of Import Selectivity

5.1 In Vitro Assays

To study import determinants within the tRK1 molecule, we introduced a panel of replacements in the original sequences of the imported and non-imported lysine-tRNAs, which correspond to domains or individual residues differing between these two isoacceptors (Entelis et al. 1998). The different mutant transcripts were tested for import, aminoacylation and binding to pre-MSK. Experiments with in vitro import of T7 transcripts showed, that the anticodon region of tRK1 contains determinants for its import selectivity. In fact, replacement of the entire anticodon arm of tRK1 by that of tRK2 totally blocked the import, both in vitro and in vivo. The most important position in this region appears to be the wobble position of the anticodon (C_{34}), since introduction of C_{34} into a tRK2 transcript directs its efficient import in vitro. Position 34 is not only important by itself but so is its context, since no significant loss of import efficiency was found with a $tRK1_{UUU}$ transcript. It is possible that in tRK2, the uridine modification (5-[(methylamino)-methyl]-2-thiouridine, mnm^5s^2U) in position 34 may act as an anti-determinant for import, since a tRK2 transcript is weakly imported while natural (fully modified) tRK2 is not. Therefore, the yeast mitochondrial import apparatus seems to be able to discriminate between mnm^5s^2U and C in positon 34 of the tRNA molecule, but not (or, at least, with a much lesser efficiency) between unmodified U and C in this position.

A second domain of the tRK1 molecule which might be important for its import selectivity is the acceptor stem, since replacement of residues 67–69 ot tRK1 by those of tRK2 blocks the import. However, this effect can be reversed by additional changes in position 72 and 73, which restore the import.

As we stressed above, pre-MSK can form a stable complex with tRK1, but only with its aminoacylated form. We therefore hypothesized that aminoacylation is essential for binding to pre-MSK, which in turn would serve as a carrier for mitochondrial transport of the tRNA. In fact, pre-MSK binding is indeed of critical importance for the import of mutant tRNA transcripts since all transcripts having high pre-MSK binding capacities are well imported whereas those showing low binding are poorly imported. However, comparison of aminoacylation capacities of mutant tRNAs showed that aminoacylation is needed for pre-MSK binding and mitochondrial import of only a portion of the mutant tRNAs. We suggest that the role of aminoacylation in import is to induce a conformational change in tRK1 in order to facilitate its binding to pre-MSK. As a matter of fact, the role of aminoacylation as a factor of conformational change in tRNA was recently described for the tRNA-like domain of several viral RNAs (Giegé 1996). It was shown that the role of aminoacylation is to induce productive interaction of the tRNA-like domain with replication factors, while the nature of the amino acid at the 3'-terminus of the viral RNA is not curcial. If we accept the idea that aminoacylation of tRK1 facilitates its binding to pre-MSK, how then can some $tRNA^{Lys}$ variants bind to pre-MSK in their deacylated form? It is worth noting that these transcripts either contain mutations in the acceptor stem or have an altered tertiary interaction. We suggest that some particular alterations in the core of the tRK1 L-shape structure or in the acceptor helix might induce conformational changes which can mimic the effect of aminoacylation on tRK1 and allow pre-MSK binding.

5.2
In Vivo Assays

The above results showed that a U to C change in the first position of the anticodon of a tRK2 transcript conferred efficient import in our in vitro system. To test if this is also the case in vivo, we constructed a version of the tRK2 gene with the anticodon CUU ($tRK2_{CUU}$) and expressed it from a centromeric plasmid, pTRK2-CUU (Entelis et al. 1998). The construct was made on the basis of one of the seven chromosomal tRK2 gene copies. All the tRK2 gene copies contain an identical intron of 23 bases. The U_{34} to C change should not interfere with correct folding of the pre-TRNA, but replaces a weak U:G base pair by a stron C:G pair. The mutated gene was found to be expressed in vivo and gave rise to a tRNA molecule with a slightly lower electrophoretic mobility than the natural tRK2. The altered mobility might be due to a difference in base modification (absence of mnm^5s^2U in position 34) but not to the absence of intron splicing, since the unspliced precursor should have a significantly lower gel mobility and would not have been detected with the oligonucleotide probe used (which corresponds to the anticodon-arm and variable loop of $tRK2_{CUU}$). Cell fractionation and subsequent hybridization with a $tRK2_{CUU}$-specific oligonucleotide probe revealed that a small fraction of the mutated tRNA is associated with the mitochondrial matrix. At $tRK2_{UUU}$-specific probe, used as a negative control, did not detect the import. The level of expression of $tRK2_{CUU}$ was comparable to that of tRK2 and approximately two times lower than that of tRK1. The amount of imported $tRK2_{CUU}$ is about ten-times lower than that of imported tRK1, which suggests an approximately five-times lower import efficiency for the $tRK2_{CUU}$. The fact that a single base change in the anticodon of a normally non-imported $tRNA^{Lys}$ results in its import in vivo is in agreement with the in vitro results and demonstrates that residue C_{34} is important for import selectivity. However, the weakness of the import signal of $tRK2_{CUU}$ indicates that the first position of the anticodon is not the only position affecting import efficiency.

To verify if changes in the tRK1 anticodon arm that block the import in vitro have the same effect in vivo we constructed a tRK1 gene variant containing the tRK2 anticodon-arm in the tRK1 backbone. Although the mutant tRNA was well expressed in vivo it was not found to be imported into mitochondria. This result is in agreement with the in vitro import assays and confirms that the anticodon arm of tRK1 contains residues critical for import selectivity.

A similar analysis was done to test bases of the acceptor stem of the imported tRNAs for their input in import efficiency (Kazakowa et al. 1999) This study was done on the basis of the weakly imported $tRK2_{CUU}$ variant (see above). In fact, replacement of the first base pair of tRK2 (Ψ:A) by (G:C) and replacement of the discriminator base (position 73) G → U resulted in a well-expressed version which was imported in vivo with an efficiency comparable to that of the naturally imported tRK1. These results are in agreement with the in vitro assays and are in favor of the importance of the acceptor stem.

5.3 Import of Nicked tRK1 Transcripts

It seems clear that at least two regions of the tRNA molecule are involved in determination of import selectivity, the anticodon and the acceptor stem. On the other hand, several mutations that should disturb the overall tRNA L-shape also have an inhibiting effect on the import in vitro (in vivo such tRNAs are unstable and thus cannot be tested). To verify if the import apparatus recognizes the "sequence motif" or the 3D-structure is important, we analyzed the import of nicked tRK1 transcripts (Entelis et al. 1998).

It has been shown previously that polynucleotides cross-linked to a signal peptide for pre-protein targeting into mitochondria can be imported into isolated mitochondria (Vestweber and Schatz 1989; Seibel et al. 1995). Likewise, if we assume that tRK1 import involves formation of a covalent link with its carrier protein (presumably pre-MSK), then folding of the tRNA should not be required for its mitochondrial translocation. To verify this hypothesis we tested TRK1 molecules assembled from two distinct transcripts, long (48-bases) 5'- and short (28 bases) 3'-moieties, annealed one to another by in vitro import assays. The 3'-moiety of tRK1 was found to direct mitochondrial import of the 5'-portion of the tRNA, which alone is not imported. Therefore, our results are clearly in favor of the importance of the tRNA cloverleaf structure, and suggest translocation of the tRNA as a folded molecule.

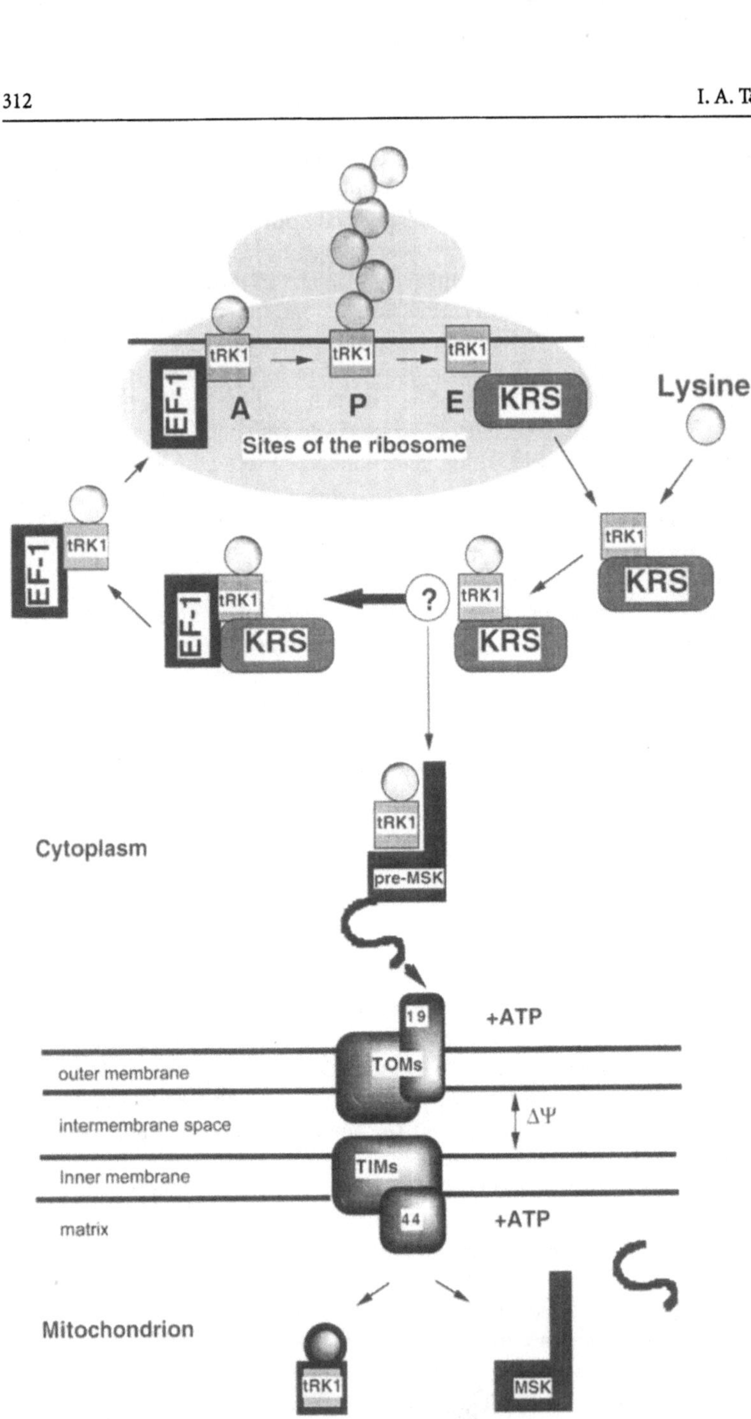

Fig. 2. Hypothetical mechanism of tRK1 mitochondrial import. *KRS* cytoplasmic LysRS, *pre-MSK* cytoplasmic precursor of the mitochondrial LysRS; *EF-1* elongation factor EF1, *TOMs* pre-protein receptors of the mitochondrial outer membrane, *TIMs* pre-protein translocators of the mitochondrial inner membrane (see the text for details)

6 Possible Mechanisms of Import

The results described above permitted us to propose a model (Tarassov et al. 1995b; Tarassov and Martin 1996) of RNA-protein interactions leading to mitochondrial import of tRK1 in yeast (Fig. 2). This model suggests that tRK1 is aminoacylated by KRS and is adressed thereafter either towards the cytoplasmic translation apparatus (via interaction with EF-1 elongation factor), or towards mitochondria (via association with pre-MSK). tRK1-pre-MSK complex is then recogized by the Tom20 receptor of the mitochondrial outer membrane. The two possibilities remain credible: either the tRNA is co-imported with pre-MSK across the pre-protein import channel, or it penetrates into the organelle using a tRNA-specific channel. Taking into account the importance of the Tim44 translocator, the co-import hypothesis seems more attractive.

The proposed mechanism would mean that tRK1, once aminoacylated, would be competed for by cytoplasmic translation factors and mitochondrial import factors (pre-MSK and, probably, other proteins not yet identified). Taking into account a low concentration of the cytoplasmic precursor of MSK compared with translation factors, this model explains an unequal distribution of tRK1 between the cytoplasm (95%) and mitochondria (5%). Recent results demonstrate that inhibition of cytoplasmic protein synthesis leads to a temporary activation of tRK1 import in vivo, and that in vitro, tRK1 import is dependent on pre-MSK concentration (Tarassov, unpubl.).

The model of co-import has a serious problem, because mitochondrial import of pre-proteins implies unfolding, at least partially, of the polypeptide during translocation across mitochondrial membranes (Höhfeld and Hartls 1994; Stuart et al. 1994). Import of tRK1 as an RNP complex with pre-MSK suggests that this enzyme possesses an RNA-binding domain which is not unfolded during the import. On the other hand, crystallographic studies of aaRS-tRNA complexes clearly demonstrate that large surfaces of the protein molecule are involved in interaction with the tRNA (Moras 1993). Such interaction must be destroyed by unfolding of the 3D structure of the protein. Therefore, we think that the link between pre-MSK and tRK1 is different from that of classical interaction between a tRNA and an aaRS. This idea is supported by several observations (Tarassov et al. 1995b): (1) although pre-MSK can aminoacylate tRK3 (its cognate mitochondrial tRNA), it is not able to direct import of this tRNA in vitro; (2) pre-MSK forms stable RNP complexes only with the aminoacylated form of tRK1; (3) a tertiary complex can be formed including tRK1 and both pre-MSK and KRS. It is therefore possible that pre-MSK-tRK1 interaction is locally restricted and the interacting domain(s) of the pre-protein are not unfolded during translocation.

Beside the two aaRS, at least one additional factor is required for tRK1 import. This conclusion is based on the fact that the two purified aaRS cannot direct tRK1 import in vitro in the absence of another fraciton of cytosolic proteins. There factor(s) are not yet identified. It may be that these are chaperones which help the imported tRNA to adopt a structure which facilitates the translocation across the mitochondrial membranes. Such an "RNA-chaperone" activity has been described for several hnRNPs and for HIV nucleocapsid proteins (Hershlag 1995).

Are the mechanisms of mitochondrial import of tRNAs found in yeast conserved in evolution? Although current experimental data are not sufficient to answer this question definitively, tRNA import data on trypanosomatids lets us suggest that other mechanisms could be selected in different organisms. In fact, poor selectivity of import as well as efficient import of non-aminoacylable tRNAs in trypanosomes does not favor participation of aaRS in import (Schneider 1994; Hauser and Schneider 1995). Furthermore, recent in vitro data suggest that import in trypanosomatids does not require any cytosolic factors and can use an RNA-specific channel localized in the mitochondrial membrane (Adhya et al. 1997). On the other hand, in plants, tRNA mitochondrial import is far more selective, and there are data implicating aaRS, which suggest possible similarities with the yeast system (see Chapt. 23, this Vol.).

7
Is It Possible to Use tRNA Import for the Therapy of Mitochondrial Diseases?

Mutations in the mitochondrial genome are associated with a number of neurodegenerative and muscular human diseases. Several of these pathologies are caused by point mutations in tRNA genes. Two diseases of this type which are better characterized are MERRF (myoclonic epilepsy with ragged red fibers) associated with a point mutation in the $tRNA^{Lys}$ gene and MELAS (mitochondrial myopathy, encephalopathy with lactic acidosis and stroke-like episodes), associated with a mutation in the $tRNA^{Leu}$ gene (Shoffner et al. 1990; Kobayashi et al. 1990). Other diseases, like several cardiomyopathies, certain cases of diabetes or syndromes CIPO and CPEO are also linked with mitochondrial tRNA mutations (Lestienne 1992; Wallace 1994). Some of these mutations result in a severe decrease of concentration of the mutated tRNA and consecutive deficiencies in mitochondrial protein synthesis (Enriquez et al. 1995; Flierl et al. 1997; Hao and Moraes 1997). Therefore, it seems attractive to complement these deficiencies by tRNAs expressed from nuclear genes and imported from the cytoplasm.

For many years, it was thought that mitochondrial tRNA import did not exist in mammals. However, it has been recently suggested that one cytoplasmic lysine isoacceptor is imported in the opossum, *Marsupialis domesticus* (S. Pääbo, pers. comm.). Furthermore, there are data demonstrating association of other nucleus-encoded RNAs with mammalian mitochondria: 5S rRNA (Yoshionari et al. 1994, Magalhaes et al. 1998), the RNA component of MRP endonuclease (Li et al. 1994) and the RNA component of the mitochondrial RNase P (G. Attardi, pers. comm.). These results favor the idea that human mitochondria are able to import cytoplasmic RNAs. Our knowledge of the mechanisms of yeast mitochondrial tRNA import can enable us to develop an artifical system of tRNA import in human cells, thus developing a new gene therapy approach to mitochondrial diseases. Very recently, we have shown that the yeast $tRNA^{Lys}_{CUU}$, tRK1, is selectively imported in isolated human mitochondria, but only in the presence of yeast import-directing proteins (Tarassov and Martin, unpubl.). This result signifies that import factors which are used for tRK1 import in yeast are absent in human cells and let us hope that co-expression of the genes coding for these proteins with the genes coding for tRNA(s) to be imported in the nucleus will allow the creation of an artificial import system in human cells.

References

Adhya S, Ghosh T, Das A, Bera SK, Mahapatra S (1997) Role of an RNA-binding protein in import of tRNA into *Leishmania* mitochondria. J Biol Chem 272:21396–21402

Enriquez JA, Chomyn A, Attardi G (1995) MtDNA mutation in MERRF syndrome causes defective aminoacylation of $tRNA^{Lys}$ and premature translation termination. Nat Gent 10:47–55

Entelis NS, Krasheninnikov IA, Martin RP, Tarassov IA (1996) Mitochondrial import of a yeast cytoplasmic $tRNA^{Lys}$: possible roles of aminoacylation and modified nucleosides in subcellular partitioning. FEBS Lett 384:38–42

Entelis NS, Kieffer S, Kolesnikova OA, Martin RP, Tarassov IA (1998) $tRNA^{Lys}$ structural requirements for its import into yeast mitochondria. Proc Natl Acad Sci USA 95:2338–2843

Flierl A, Reichmann H, Seibel P (1997) Pathophysiology of the MELAS 3243 transition mutation. J Biol Chem 272:27189–27196

Gatti DL, Tzagoloff A (1991) Structure and evolution of a group of related aminoacyl-tRNA synthetases. J Mol Biol 218:557–568

Giegé R (1996) Interplay of tRNA-like structure from plant viral RNAs with partners of the translation and replication machineries. Proc Natl Acad Sci USA 93:12078–12081

Glick BS, Beasley EM, Schatz G (1992) Protein sorting in mitochondria. Trends Biochem Sci 17:453–459

Hao H, Moracs CT (1997) A disease-associated G5703A mutation in human mitochondrial DNA causes a conformational change and a marked decrease in steady-state levels of the mitochondrial $tRNA^{Asn}$. Mol Cell Biol 17:6831–6837

Hauser R, Schneider A (1995) tRNAs are imported into mitochondria of *Trypanosoma brucei* independently of their genomix context and genetic origin. EMBO J 14:4212–4220

Herschlag D (1995) RNA chaperones and the RNA folding problem. J Biol Chem 270:20871–20874

Höhfeld J, Hartl U (1994) Post-translational protein import and folding. Curr Opin Cell Biol 6:499–509

Kazakova HA, Entelis NS, Martin RP, Tarassov IA (1999) The aminoacceptor stem of the yeast $tRNA^{Lys}$ contains determinants of mitochondrial import selectivity. FEBS Lett 442, 193–197

Kobayashi Y, Momoi MY, Tominagha K, Momoi T, Nihei K, Yanagishawa M, Kagawa Y, Ohta S (1990) A point mutation in the mitochondrial $tRNA^{Leu}$ (UUR) gene in MELAS (mitochondrial myopathy, encephalopathy, lactic acidosis and soke-like episodes). Biochem Biophys Res Commun 173:816–822

Kübrich M, Dietmeier K, Pfanner N (1995) Genetic and biochemical dissection of the mitochondrial protein-import machinery. Curr Genet 27:393–403

Lestienne P (1992) Mitochondrial DNA mutations in human diseases: a review. Biochimie 74:123–130

Li K, Smagula CS, Parsons WJ, Richardson JA, Gonzales M, Hagler HK, Williams RS (1994) Subcellular partitioning of MRP RNA assessed by ultrastructural and biochemical analysis. J Cell Biol 124:871–882

Magalhes PJ, Andreu AL, Schon EA (1998) Evidence for the presence of 5S rRNA in mammalian mitochondria. Mol Biol Cell 9:2375–2382

Mahapatra S, Ghosh T, Adhia S (1994) Import of small RNAs into *Leishmania* mitochondria in vitro. Nucleic Acids Res 22:3381–3386

Martin RP, Schneller JM, Stahl AJC, Dirheimer G (1977) Study of yeast mitochondrial tRNAs by two-dimensiopnal electrophoresis: characterization of isoaccepting species and search for imported cytoplasmic tRNA. Nucleic Acids Res 4:3497–3510

Martin RP, Schneller JM, Stahl AJC, Dirheimer G (1979)Import of nuclear deoxyribonucleic acid coded lysine-accepting transfer ribonucleic acid (anticodon C-U-U) into yeast mitochondria. Biochemistry 18:4600–4605

Martin RP, Sibler AP, Gherke CW, Kuo K, Edmonds CG, McClosley JA, Dirheimer G (1990) 5-[[(Carboxymethyl)amino]methyl]uridine is found in the anticodon of yeast mitochondrial tRNAs recognizing two-codon families ending in a purine. Biochemistry 29:956–959

Martinez R, Mirande M (1992) The polyanion-binding domain of cytoplasmic Lys-tRNA synthetase from *Saccharomyces cerevisiae* is not essential for cell viability. Eur J Biochem 207:1–11

Moras D (1993) Structural aspects and evolutionary implications of the recognition between tRNAs and aminoacyl-tRNA synthetases. Biochimie 75:651–657

Rusconi CP, Cech T (1996a) Mitochondrial import of only one of three nuclear-encoded glutamine tRNAs in *Tetrahymena thermophila.* EMBO J 15:3286–3295

Rusconi CP, Cech TR (1996b) The anticodon is the signal sequence for mitochondrial import of glutamine tRNA in *Tetrahymena.* Genes Dev 10:2870–2880

Schneider A (1994) Import of RNA into mitochondria. Trends Cell Biol 4:282–286

Schneider A (1996) Cytosolic yeast $tRNA^{His}$ is covalently modified when imported into mitochondria of *Tyrpanosoma brucei.* Nucleic Acid Res 24:1225–1228

Seibel P, Trappe J, Villani G, Klopstock T, Papaa G, Richmann H (1995) Transfection of mitochondria: strategy towards a gene therapy of mitochondrial DNA diseases. Nucleic Acids Res 23:10–17

Shoffner J, Lott M, Lezza AMS, Seibel P, Ballinger SW, Wallace DC (1990) Myoclonic epilepsy and ragged-red fiber disease (MERRF) is associated with a mitochondrial DNA $tRNA^{Lys}$ mutation. Cell 61:931–937

Sibler AP, Dirheimer G, Martin RP (1986) Codon reading patterns in *Saccharomyces cerevisiae* mitochondria based on sequences of mitochondrial tRNAs. FEBS Lett 194:131–138

Small I, Maréchal-Drouard L, Masson J, Pelletier G, Cosset A, Weil H-H, Dietrich A (1992) In vivo import of a normal or mutagenized heterologous transfer RNA into the mitochondria of transgenic plants: towards novel ways of influencing mitochondrial gene expression? EMBO J 11:1291–1296

Soidla TR (1983) Possible role of the mitochondrially-imported $tRNA^{Lys}{}_1$ in mitochondrial RNA splicing in yeast. Mol Biol Russia 17:1154–1162

Söllner T, Rassow J, Pfanner N (1991) Analysis of mitochondrial protein import using translocation intermediates and specific antibodies. Methods Cell Biol 34:345–358

Stuart RA, Cyr DM, Craig EA, Neupert W (1994) Mitochondrial molecular chaperones: their role in protein translocation. Trends Biochem Sci 19:87–92

Suyama Y, Hamada J (1976) Imported tRNA: its synthetase as a probable transport protein. In: Bücher T, Neupert W, Sebald W, Werner S (eds) Genetics and biogenesis of chlorooplasts and mitochondria. Elsevier/North-Holland, Amsterdam, pp 763–774

Tarassov IA, Entelis NS (1992) Mitochondrially imported cytoplasmic $tRNA^{Lys}$(CUU) of *Saccharomyces cerevisiae*: in vivo and in vitro targeting systems. Nucleic Acids Res 20:1277–1281

Tarassov IA, Entelis NS, Martin RP (1995a) An intact protein translocation machinery is required for mitochondrial import of a yeast cytoplasmic tRNA. J Mol Biol 245:315–323

Tarassov IA, Entelis NS, Martin RP (1995b) Mitochondrial import of a lysine-tRNA in yeast is mediated by cooperation of cytoplasmic and mitochondrial lysyl-tRNA synthetases. EMBO J 14:3461–3471

Tarassov IA, Martin RP (1996) Mechanisms of tRNA import into yeast mitochondria: an overview. Biochimie 78:502–510

Vestweber D, Schatz G (1989) DNA-protein conjugates can enter mitochondria via the protein import pathway. Nature 338:170–172

Wallace DC (1994) Mitochondrial DNA sequence variation in human evolution and disease. Proc Natl Acad Sci USA 91:8739–8736

Yokoyama S, Watanabe T, Murao K, Ishikura H, Yamaizumi Z, Nishimura S, Miyazama T (1995) Molecular mechanism of codon recognition by tRNA species with modified uridine in the first position of the anticodon. Proc Natl Acad Sci USA 82:4905–4909

Yoshionari S, Koike T, Yokogawa T, Nishigawa K, Ueda T, Miura K, Watanabe K (1994) Existence of nuclear encoded 5S rRNA in bovine mitochondria. FEBS Lett 338:137–143

CHAPTER 23

The Import of Cytosolic tRNA into Plant Mitochondria 23

L. Maréchal-Drouard[1], A. Dietrich[1], H. Mireau[2], N. Peeters[2], and I. Small[2]

Contents

1 Introduction

A consideration of the sparse genetic information carried by mitochondrial genomes illustrates why these organelles are so dependent on the rest of the cell, as most of the proteins necessary for their function are encoded by the nuclear genome and imported from the cytosol. However, certain protein components of the respiratory chain complexes and (in some cases) a few ribosomal proteins are synthesized by the mitochondrial translation machinery. Hence, the mitochondrial translation system must include a full set of transfer RNAs (tRNAs) to ensure the incorporation of the usual 20 amino acids into the polypeptides being synthesized. Vascular plant mitochondrial genomes contain at most 12 to 14 functional "native" tRNA genes deriving from the genome of the original endosymbiotic bacterium (Sangaré et al. 1990; Weber-Lotfi et al. 1993; Ceci et al. 1996; Unseld et al. 1997). These genes are obviously insufficient to cover the needs of the mitochondrial translation system. To get round

[1] Institut de Biologie Moléculaire des Plantes, UPR A0406 du CNRS, Université Louis Pasteur, 12 rue de Général Zimmer, 67084 Strasbourg Cedex, France
[2] Station de Génétique et d'Amélioration des Plantes, INRA, Route de St-Cyr, 78026 Versailles Cedex, France

this problem, plant mitochondria use tRNAs that have to be imported from the cytosol (Dietrich et al. 1992; Kumar et al. 1996). This specific transport of nucleic acids across the double mitochondrial membrane occurs in many diverse organisms, including plants, some fungi and protozoa (Schneider 1994), and it can be expected that a sufficient understanding of the tRNA import mechanism will permit the development of a strategy for gene therapy of human cytopathies such as MELAS or MERRF provoked by mutations in mitochondrial tRNA genes. The general idea would be to complement these mutations by importing into human mitochondria functional tRNAs transcribed from the nuclear genome.

2 The Evidence in Favour of tRNA Import

The first evidence for the existence of nuclearly encoded tRNAs in plant mitochondria was indirect: it was not possible to find a complete set of tRNA genes in the mitochondrial genomes that had been studied. Subsequently, more direct evidence was obtained by the discovery of mitochondrial tRNAs with identical primary sequence to their cytosolic counterparts and which hybridized only to nuclear DNA, and not to the mitochondrial genome (Dietrich et al. 1992). Direct proof for the existence of a mechanism for importing nuclearly encoded tRNAs into plant mitochondria was finally obtained by experiments with transgenic plants (Small et al. 1992). The nuclear gene from bean coding for $tRNA^{Leu}$(C*AA) was used to transform potato protoplasts, from which transgenic plants were subsequently generated. The use of a hybridization probe specific for the bean tRNA then demonstrated that this tRNA was present in the cytosol and in the mitochondria of the transgenic potato plants. By directed mutagenesis, four nucleotides (TCGA) could be inserted into the transgene at the position of the anticodon without affecting the mitochondrial import of the tRNA.

3 The Population of Imported tRNAs

Different techniques have been employed with the aim of identifying the nuclearly encoded tRNAs present in plant mitochondria and studying the import mechanism (Maréchal-Drouard et al. 1995). It turned out that the import process is highly selective; in general only cytosolic tRNAs that have no equivalent encoded in the mitochondrial genome are imported.

Systematic studies with numerous specific hybridization probes were carried out on different plant species, such as potato *(Solanum tuberosum)*, maize *(Zea Mays)*, wheat *(Triticum aestivum)* and larch (*Larix* x *leptoeuropaea;* Joyce and Gray 1989; Maréchal-Drouard et al. 1990; Kumar et al. 1996). Comparison of the mitochondrial tRNA populations of larch, potato, sunflower *(Helianthus annuus)*, wheat and maize revealed important differences in the number and identity of the tRNAs imported, sometimes even between phylogenetically related plants. Transfer RNA^{Ala}, $tRNA^{Leu}$ and $tRNA^{Thr}$ were found to be imported into mitochondria in all the plants tested. Other tRNAs were found to be always encoded by the mitochondrial genome: $tRNA^{Asp}$, $tRNA^{Glu}$, $tRNA^{Ile}$(L * AU), $tRNA^{fMet}$ and $tRNA^{Tyr}$. It is worth noting, however, that there is a clear tendency throughout the evolution of higher plants towards the loss of the "native" tRNA genes. The bryophyte *Marchantia polymorpha* (a liverwort) contains 29 native mitochondrial tRNA genes, but only a dozen or so remain in the mitochondrial genome of vascular plants (Weber-Lotfi et al. 1993; Kumar et al. 1996; Unseld et al. 1997).

4
General Considerations About the Import Mechanism

The comparison of the sequences of a number of plant cytosolic tRNAs and nuclear tRNA genes, corresponding to imported or non-imported tRNAs, has revealed no conserved sequence or structural motif specific to the tRNAs which are targeted to mitochondria, either in the tRNAs themselves, or in the flanking regions of their genes (Ramamonjisoa et al. 1998). This contrasts with the situation in the protozoan *Leishmania,* where tRNA import seems to involve a conserved sequence present in the D-arm of many imported tRNAs (Mahapatra et al. 1998). The apparent absence of conserved import signals in plants, together with the differences in the number and identity of the tRNAs imported in different plant species, suggests that each imported tRNA is recognized specifically by one or several factors. It is hard to imagine a specific import channel for each imported tRNA, therefore the hypothesis which has generally been favoured is that there exists a series of specific carrier proteins capable of directing tRNAs to the import system. It is tempting to believe that in such a mechanism tRNAs can subsequently pass through the protein import channel. The mitochondrial aminoacyl-tRNA synthetases, which are all encoded in the nuclear genome and imported into mitochondria, are plausible candidates as specific carrier proteins. This idea was reinforced by the following observation: when a tRNA or a group of isoacceptor tRNAs is imported, it seems that the same aminoacyl-tRNA synthetase is used in the cytosol and in the mitochondria. In contrast, the tRNAs encoded by the mitochondrial genome are aminoacylated by enzymes specific to mitochondria which generally do not recognize the corresponding cytosolic tRNAs.

5
The Cytosolic and Mitochondrial Forms of Several Plant Aminoacyl-tRNA Synthetases Are Encoded by the Same Genes

In higher plants, the aminoacyl-tRNA synthetases required in the three cellular compartments where protein synthesis takes place (i.e. the cytosol, the mitochondria and the plastids) are all encoded by nuclear genes. This poses questions about the relationships and similarities between aminoacyl-tRNA synthetases from different compartments. Just as the $tRNA^{Leu}$ partition between the cytosol and the mitochondria, it seems, based on biochemical and immunological data, that the same leucyl-tRNA synthetase is present in the cytosol and in mitochondria, at least in bean *(Phaseolus vulgaris;* Maréchal-Drouard et al. 1988; P. Guillemaut, L. Maréchal-Drouard, A. Dietrich, unpubl. results). A more detailed study was carried out in the case of the *Arabidopsis thaliana* alanyl-tRNA synthetase (AlaRS) encoded by the ALATS gene. Immunological, biochemical and molecular analyses have shown that this gene is bifunctional, as it codes for both the cytosolic and the mitochondrial form of the enzyme, depending on which initiation codon is used for tranlsation (Mireau et al. 1996). Given that the mitochondrial $tRNA^{Ala}$ is transcribed from a nuclear gene and imported from the cytosol in all higher plants studied to date, we find once again a situation where the same tRNA and the same aminoacyl-tRNA synthetase are present in the cytosol and in the mitochondria. Several other *Arabidopsis* genes encoding cytosolic/mitochondrial aminoacyl-tRNA synthetases that aminoacylate mitochondrially imported tRNAs were shown to have a similar structure and presumably similar modes of expression (Small et al. 1998). It is logical in this context to consider the possibility of a co-import of the tRNAs with the corresponding aminoacyl-tRNA synthetases.

6
Interaction with the Aminoacyl-tRNA Synthetase Is Probably Necessary But Not Sufficient for Import of a tRNA into Plant Mitochondria

In the absence of a suitable in vitro system based on isolated plant mitochondria, the work aimed at testing the implication of the aminoacyl-tRNA synthetases in the mechanism of mitochondrial tRNA import has essentially relied on in vivo experiments analyzing the import of heterologous tRNAs into mitochondria in transgenic plants.

As judged by aminoacylation assays and suppression tests in vivo (Fig. 1), the alteration of U_{70} to C_{70} in plant $tRNA^{Ala}$, which generates a normal "Watson-Crick" G_3:C_{70} base-pair, destroys the specific interaction between plant $tRNA^{Ala}$ and AlaRS (Carneiro et al. 1994). Constructs carrying the wild-type (U_{70}) or mutant (C_{70}) form of *Arabidopsis thaliana* $tRNA^{Ala}$ were used to transform tobacco protoplasts. Whereas both forms of $tRNA^{Ala}$ were present in cytosolic fractions from the corresponding transgenic tobacco plants, only the wild-type form, which is recognized by the AlaRS, was found in mitochondria (Fig. 1; Dietrich et al. 1996a). Thus, a single nucleotide change is sufficient to abolish both aminoacylation and import of $tRNA^{Ala}$, suggesting that recognition by the AlaRS is necessary for import of the tRNA into plant mitochondria.

As AlaRS appeared to be implicated in mitochondrial import of $tRNA^{Ala}$ in plants, we felt it would be interesting to express this enzyme in an organism in which mitochondria do not naturally import this tRNA. The idea was to see whether the presence of this plant enzyme with its mitochondrial targeting presequence would be sufficient to induce import of cytosolic $tRNA^{Ala}$. Hence, the mitochondrial form of the *Arabidopsis thaliana* AlaRS was expressed in the fungi *Saccharomyces cerevisiae* and *Schizosaccharomyces pombe*, and shown to be correctly imported and accumulated in mitochondria. Despite this, neither the endogenous cytosolic $tRNA^{Ala}$, nor the *Arabidopsis* $tRNA^{Ala}$, co-expressed with the AlaRS, were imported (H. Mireau, A, Dietrich, L. Maréchal-Drouard , A. Cosset, S. Kauffmann-Brubacher, T. Fox, I. Small, submitted). We conclude that the AlaRS is probably not sufficient to support import of $tRNA^{Ala}$, and that the import mechanism must involve one or several other factors present in plants.

An experiment carried out with transgenic plants gave a similar result. Potato plants were supplied with an extra aminoacyl-tRNA synthetase, capable both of being imported into mitochondria and of recognizing a cytosolic tRNA that is not normally

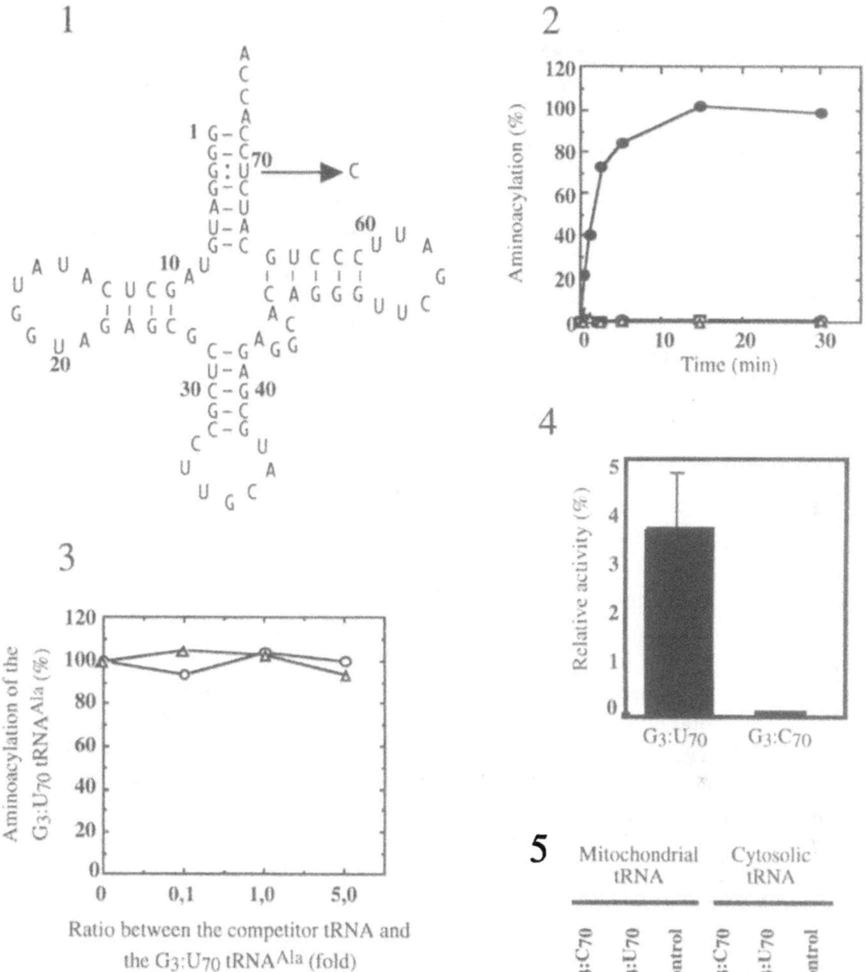

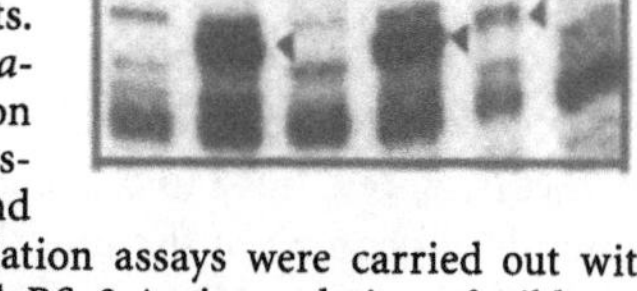

Fig. 1. The *Arabidopsis thaliana* tRNA^{Ala}/AlaRS system, a model for the study of mitochondrial tRNA import in plants. **1** Secondary structure of tRNA^{Ala}(UGC) from *Arabidopsis thaliana.* The *arrow* indicates the mutation studies at position 70. **2** Aminoacylation kinetics of in vitro transcripts corresponding to wild-type tRNA^{Ala} (g3:u70; *filled circles*) and mutant tRNA^{Ala} (G_3:C_{70}; *open triangles*). The aminoacylation assays were carried out with saturating amounts of partially purified bean cytosolic AlaRS. **3** Aminoacylation of wild-type tRNA^{Ala} (G_3:U_{70}) transcripts (3 μM) by limiting amounts of partially purified bean cytosolic AlaRS in the presence of increasing amounts of transcripts corresponding to mutant tRNA^{Ala} (G_3:C_{70}; *open triangles*), or to wild-type tRNA^{Phe} from *Arabidopsis thaliana (open circles)*. **4** Amber suppressor activity of tRNA^{Ala}(CUA) in vivo. Tobacco protoplasts were electroporated with a β-glucuronidase (GUS) marker gene containing a premature amber stop codon, together with a gene coding for an amber suppressor derivative of tRNA^{Ala} (CUA anticodon) with either a G_3:U_{70} or a G_3:C_{70} base-pair. GUS activity was calculated relative to the activities obtained by electroporation of the same batch of protoplasts with the wild-type GUS gene. **5** RNase protection analysis of total and mitochondrial tRNA extracted from transgenic tobacco plants transformed with the gene encoding the wild-type *Arabidopsis* tRNA^{Ala} (G_3:U_{70}), with the gene encoding the mutant tRNA^{Ala} (G_3:C_{70}), or with an equivalent plasmid lacking the tRNA gene (control). The *arrows* indicate the band specific to the tRNA^{Ala} derived from the transgenes

imported. The idea again was to see whether this could induce import of the tRNA. The cytosolic aspartyl-tRNA synthetase (AspRS) from the yeast *Saccharomyces cerevisiae* supplied with a plant mitochondrial tareting presequence was chosen for the experiment, as the $tRNA^{Asp}$ present in potato mitochondria is mitochondrially encoded and potato cytosolic $tRNA^{Asp}$ is a good substrate for the yeast enzyme. The cytosolic $tRNA^{Asp}$ could not be detected in the mitochondrial fraction of transgenic plants accumulating the yeast AspRS in mitochondria. Import of the AspRS did not, therefore, lead to import of its substrate tRNA (Dietrich et al. 1996b).

In conclusion, it seems reasonably clear that recognition by an aminoacyl-tRNA synthetase imported into plant mitochondria is not sufficient to permit efficient import of a tRNA. This is reminiscent of the situation described for yeast (Saccharomyces cerevisiae); the precursor of the mitochondrial lysyl-tRNA synthetase (LysRS), although essential for $tRNA^{Lys}$(CUU) mitochondrial import, is not by itself sufficient for the process (Chap 22, this Vol.).

7 Mitochondrial Import of tRNAs in Other Organisms

As we have already mentioned, the import of tRNAs into mitochondria is not limited to the plant kingdom. Similar observations were reported for yeast, higher plants, *Tetrahymena pyriformis*, trypanosomes and *Leishmania* (Martin et al. 1979; Suyama 1986; Maréchal-Drouard et al. 1988; Simpson et al. 1989; Hancock and Hajduk 1990). The experimental approaches which provided direct proof of tRNA import fall into two types: in vivo techniques involving transgenic material, which have been used with plants and several protozoa (*Leishmania, Trypanosoma, Tetrahymena;* Small et al. 1992; Chen et al. 1994; Hauser and Schneider 1995; Dietrich et al. 1996a; Lima and Simpson 1996; Rusconi and Cech 1996), and in vitro import of RNA into isolated mitochondria, as demonstrated for yeast (Tarassov et al. 1995; Chap 22, this Vol. and *Leishmania* (Adhya et al. 1997).

The number of tRNAs imported differs significantly between these organisms. For example, yeast mitochondria import a single tRNA, $tRNA^{Lys}$(CUU) (Martin et al. 1979), whereas mitochondria from trypanosomes and *Leishmania* import all their tRNAs (Simpson et al. 1989; Hancock and Hajduk 1990; Lye et al. 1993). Other organisms fall between these two extrems: in *Marchantia polymorpha* (a bryophyte), only a few tRNAs are imported (Akashi et al. 1998), whereas higher plant mitochondria import from one third to one half of their tRNAs (e.g. Kumar et al. 1996) and the mitochondrial genome of the unicellular alga *Chlamydomonas reinhardtii* contains only three tRNA genes (Michaelis et al. 1990). In *Tetrahymena pyriformis,* about three-quarters of the mitochondrial tRNAs are nuclearly encoded (Suyama 1986), more than 19 tRNAs are likely to be imported into *Paramecium aurelia* mitochondria (Pritchard et al. 1990) and the minimal mitochondrial genome of the parasite *Plasmodium falciparum* (6 kpb) contains no tRNA genes at all (Feagin 1994). Finally, the mitochondrial DNA of the chytridiomycete fungus *Spizellomyces punctatus* apparently encodes only 8 tRNAs, implying that about two-thirds of the mitochondrial tRNAs are probably imported in this organism (Laforest et al. 1997).

In conclusion, the above observations indicate that in yeast, plants and some protozoa *(Tetrahymena, Paramecium)* the tRNA import process must be selective because only certain nuclearly encoded tRNAs are of nuclear origin, and even heterologous tRNAs can be importet in vivo (Hauser and Schneider 1995). This may reflect fundamental differences in the import mechanisms between the two groups of organisms.

8 Possible Models for the Mechanism of tRNA Import

Two different mechanisms for tRNA import have been proposed: a co-import of tRNA with specific protein factors via the protein import channel and a direct import of tRNA after recognition by a membrane-bound receptor.

The first model is based on the fact that the protein import channel is the only currently known transport system capable of carrying molecules as big as a tRNA. Indeed, in yeast, import of $tRNA^{Lys}$(CUU) requires an intact protein import system and involves soluble protein factors, including both the cytosolic LysRS and the precursor to the mitochondrial LysRS (Tarassov et al. 1995; Chap. 22, this Vol.). In plants, the in vivo experiments described above using wild-type and non-aminoacylatable forms of $tRNA^{Ala}$ also suggest the necessity of some sort of interaction between the tRNA to be imported and the corresponding aminoacyl-tRNA synthetase. However, as mentioned earlier, factors other than aminoacyl-tRNA synthetases appear to be required, both in yeast and in plants.

The direct import model, at first sight a rather simpler hypothesis, may apply only to trypanosomatids. RNA import in *Leishmania* initially involves binding of the RNA to a specific receptor present on the outer mitochondrial membrane before the passage into the mitochondrial matrix. A potential receptor has been identified (Adhya et al. 1997). The fact that trypanosome mitochondria can import a non-aminoacylatable $tRNA^{Tyr}$ precursor carrying an 11-nucleotie intron (Schneider et al. 1994) suggests that aminoacyl-tRNA synthetases are not involved in this case.

It is clear that, even if it is plausible that the tRNA import mechanisms differ between plants and yeast on one hand and trypanosomatids on the other, each of these models requires further confirmation. The factors involved and the different transport steps need to be more precisely defined.

9 Conclusions and Perspectives

The elucidation of the tRNA import mechanism(s) could have far-reaching practical applications.

The first of these, in particular in plants, could be to permit the manipulation of mitochondrial gene expression by targeting other RNAs (antisense RNAs or ribozymes) into these organelles. Yeast remains the only organism for which mitochondrial transformation is routine, and the possibility of targeting transcripts from nuclear transgenes to mitochondria would provide a much-needed alternative for other organisms.

The second and more immediate application could be in the medical field. The characterization of the different key components of the tRNA import system in different organisms could suggest ways of treating patients suffering from encephalomyopathies such as MERRF or MELAS that are due to mutations in mitochondrial tRNA genes. In principle, by inducing mitochondrial import of nuclear-encoded tRNA in human MELAS or MERRF cells, one should be able to functionally complement the mutated tRNA. This is feasible only if the MELAS or MERRF phenotype is due to a lack of functional tRNA and not due to an RNA processing problem, for example. The primary defect in the MERRF syndrome appears to be premature termination of translation at lysine codons, presumably because of a deficiency in amnioacylated $tRNA^{Lys}$ (Enriquez et al. 1995). More recently, a mutation in the $tRNA^{Leu}$(UUR) gene associated with the MELAS syndrome (Goto et al. 1990), was shown to be complemented by a mutation in the second $tRNA^{Leu}$ in the human mitochondrial genome that allowed this tRNA to read UUR codons (El Meziane et al. 1998). These data strongly suggest that the major defect in both MERRF and MELAS is a lack of functional tRNA ($tRNA^{Lys}$ and $tRNA^{Leu}$, respectively) for translation. All higher plants import $tRNA^{Leu}$ and some import $tRNA^{Lys}$. We feel, therefore, that research on tRNA import in plants holds out considerable hope that both of these syndromes can be "cured", at least in cell culture and in the longer term perhaps by gene therapy. First, however, we need to define all the factors involved in the import process.

Acknowledgements. We would like to thank all our collaborators who have taken part in these studies: A Cosset, AM Duchêne-Louarn, G Green, S Kauffmann-Brubacher, D Ramamonjisoa, G Souciet, H Wintz (Strasbourg) and VTC Carneiro, N Choisne, R Kumar, D Lancelin (Versailles). This work has been financed by the Institut National de la Recherche Agronomique (INRA), the Centre National de la Recherche Scientifique (CNRS) and the Université Louis Pasteur (Strasbourg).

References

Adhya S, Ghosh T, Das A, Bera SK, Mahapatra S (1997) Role of an RNA-binding protein in import of tRNA into *Leishmania* mitochondria. J Biol Chem 272:21396–21402

Akashi K, Takenaka M, Yamaoka S, Suyama Y, Fukuzawa H, Ohyama K (1998) Coexistence of nuclear DNA-encoded $tRNA^{Val}$(AAC) and mitochondrial DNA-encoded $tRNA^{Val}$(AAC) in mitochondria of a liverwort *Marchantia polymorpha*. Nucleic Acids Res 26:2168–2172

Carneiro VTC, Dietrich A, Maréchal-Drouard L, Cosset A, Pelletier G, Small I (1994) Characterization of some major identitiy elements in plant alanine and phenylalanine transfer RNAs. Plant Mol Biol 26:1843–1853

Ceci L, Veronico P, Gallerani R (1996) Identification and mapping of tRNA genes on the *Helianthus annuus* mitochondrial genome. DNA Seq 6:159–166

Chen DT, Shi X, Suyama Y (1994) In vivo expression and mitochondrial import of normal and mutated $tRNA^{Thr}$ in *Leishmania*. Mol Biochem Parasitol 64:121–133

Dietrich A, Weil JH, Maréchal-Drouard L (1992) Nuclear-encoded transfer RNAs in plant mitochondria. Annu Rev Cell Biol 8:115–131

Dietrich A, Maréchal-Drouard L, Carneiro V, Cosset A, Small I (1996a) A single base change prevents impot of cytosolic $tRNA^{Ala}$ into mitochondria in transgenic plants. Plant J 10:913–918

Dietrich A, Small I, Cosset A, Weil JH, Maréchal-Drouard L (1996b) Editing and import: strategies for providing plant mitochondria with a complete set of functional transfer RNAs. Biochimie 78:530–538

El Meziane A, Lehtinen SK, Hance N, Nijtmans LGJ, Dunbar D, Holt IJ, Jacobs HT (1998) A tRNA suppressor mutation in human mitochondria. Nat Genet 18:350–353

Enreiquez JA, Chomyn A, Attardi G (1995) MtDNA mutation in MERRF syndrome causes defective aminoacylation of $tRNA^{Lys}$ and premature translation termination. Nat Genet 10:47–55

Feagin JE (1994) The extrachromosomal DNAs of apicomplexan parasites. Annu Rev Microbiol 48:81–104
Goto Y, Nonaka I, Horai S (1990) A mutation in the $tRNA^{Leu}$(UUR) gene associated with the MELAS subgroup of mitochondrial encephalomyopathies. Nature 348:651–653
Hancock K, Hajduk SL (1990) The mitochondrial tRNAs of *Trypanosoma brucei* are nuclear-encoded. J Biol Chem 265:19208–19215
Hauser R, Schneider A (1995) tRNAs are imported into mitochondria of *Trypanosoma brucei* independently of their genomic context and genetic origin. EMBO J 14:4212–4220
Joyce PBM, Gray MW (1989) Chloroplast-like transfer RNA genes expressed in wheat mitochondria. Nucleic Acids Res 17:5461–5476
Kumar R, Maréchal-Drouard L, Akama K, Small I (1996) Striking differences in mitochondrial tRNA import among different plant species. Mol Gen Genet 252:404–411
Laforest MJ, Roewer I, Lang BF (1997) Mitochondrial tRNAs in the lower fungus *Spizellomyces punctatus:* tRNA editing and UAG "stop" codons recognized as leucine. Nucelic Acids Res 25:626–632
Lima BD, Simpson L (1996) Sequence-dependent in vivo importation of tRNAs into the mitochondrion of *Leishmania tarentolae.* RNA 2:429–440
Lye DF, Chen DT, Suyama Y (1993) Selective import of nuclear-encoded tRNAs into mitochondria of the protozoan *Leishmania tarentolae.* Mol Biochem Parasitol 58:233–246
Mahapatra S, Ghosh S, Kanti Bera S, Ghosh T, Das A, Adhya S (1998) The D arm of $tRNA^{Tyr}$ is necessary and sufficient for import into *Leishmania* mitochondria in vitro. Nucleic Acids Res 26:2037–2041
Maréchal-Drouard L, Weil JH, Guillemaut P (1988) Import of several tRNAs from the cytoplasm into the mitochondria in bean *Phaseolus vulgaris.* Nucleic Acids Res 16:4777–4788
Maréchal-Drouard L, Guillemaut P, Cosset A, Arbogast M, Weber F, Weil JH, Dietrich A (1990) Transfer RNAs of potato *(Solanum tuberosum)* mitochondria have different genetic origins. Nucleic Acids Res 18:3689–3696
Maréchal-Drouard L, Small I, Weil J-H, Dietrich A (1995) Transfer RNA import into plant mitochondria. Methods Enzymol 260:310–327
Martin RP, Schneller JM, Stahl A, Dirheimer G (1979)Import of nuclear deoxyribonucleic acid coded lysine accepting transfer ribonucleic acid (anticodon CUU) into yeast mitochondria. Biochemistry 18:4600–4605
Michaelis G, Vahrenholtz C, Pratje E (1990) Mitochondrial DNA of *Chlamydomonas reinhardtii:* the gene for apocytochrome b and the complete functional map of the 15.8 kb DNA. Mol Gen Genet 223:211–216
Mireau H, Lancelin D, Small I (1996) The same *Arabidopsis* gene encodes both cytosolic and mitochondrial alanyl-tRNA synthetases. Plant Cell 8:1027–1039
Pritchard AE, Seihlâmer JJ, Mahalingam R, Sable CL, Venuti SE, Cummings DJ (1990) Nucleotide sequence of the mitochondrial genome of *Paramecium.* Nucleic Acids Res 18:173–180
Ramammonjisoa D, Kaufmann S, Choisne N, Maréchal-Drouard L, Green G, Wintz H, Small I, Dietrich A (1998) Structure and expression of several bean *(Phaseoulus vulgaris)* nuclear transfer RNA genes: relevance to the process of tRNA import into plant mitochondria. Plant Mol Biol 36:613–625
Rusconi CP, Cech TR (1996) The anticodon is the signal sequence for mitochondrial import of glutamine tRNA in *Tetrahymena.* Genes Dev 10:2870–2880
Sangaré A, Weil JH, Grienenberger JM, Fauron C, Lonsdale D (1990) Localization and organization of tRNA genes on the mitochondrial genomes of fertile and male-sterile lines of maize. Mol Gen Genet 223:224–232
Schneider A (1994) Import of RNA into mitochondria. Trends Cell Biol 4:282–286
Schneider A, Martin J, Agabian N (1994) A nuclear-encoded tRNA of *Trypanosoma brucei* is imported into mitochondria. Mol Cell Biol 14:2317–2322
Simpson AM, Suyama Y, Dewes H, Campbell DA, Simpson L (1989) Kinetoplastid mitochondria contain functional tRNAs which are encoded in nuclear DNA and also contain small minicircle and maxicircle transcripts of unknown function. Nucleic Acids Res 17:5427–5445
Small I, Maréchal-Drouard L, Masson J, Pelletier G, Cosset A, Weil JH, Dietrich A (1992) In vivo import of a normal or mutagenized heterologous transfer RNA into the mitochondria of transgenic plants: towards novel ways of influencing mitochondrial gene expression? EMBO J 11:1291–1296

Small I, Wintz H, Akashi K, Mireau H (1998) Two birds with one stone: genes that encode products targeted to two or more compartments. Plant Mol Biol 38:265–277

Suyama Y (1986) Two dimensional polyacrylamide gel electrophoresis analysis of *Tetrahymena* mitochondrial tRNA. Curr Genet 10:411–420

Tarassov IA, Entelis N, Martin RP (1995) Mitochondrial import of a cytoplasmic lysine-tRNA in yeast is mediated by cooperation of cytoplasmic and mitochondrial lysyl-tRNA synthetases. EMBO J 14:3461–3471

Unseld M, Marienfeld JR, Brandt P, Brennicke A (1997) The mitochondrial genome of *Arabidopsis thaliana* contains 57 genes in 366 924 nucleotides. Nat Genet 15:57–61

Weber-Lotfi F, Maréchal-Drouard L, Folkerts O, Hanson M, Grienenberger JM (1993) Localization of tRNA genes on the *Petunia hybrida* 3704 mitochondrial genome. Plant Mol Biol 21:403–407

Plant Cytoplasmic Male Sterility: A Mitochondrial Pathology and Its Biotechnological Application

24

S. Litvak[1], M. Hernould[2], E. Zabaleta[1], V. Blanc[1], D. Begu[2], I. Kurek[3], A. Breiman[3], X. Jordana[4], A. Mouras[1,2], and A. Araya[1]

Contents

1 Introduction

While the molecular origin of mitochondrial human pathologies started to be known in some detail at the end of the 1980s, plant mitochondrial mutations were known several decades before (Rogers and Edwardson 1952). In this review we describe the molecular origin of a mitochondrial mutation observed in plants leading to the emer-

[1] EP 630, CNRS-Université Victor Segalen-Bordeaux 2, 1 rue Camille Saint Saëns, 33077 Bordeaux Cedex, France
[2] Laboratoire de Biologie Cellulaire et Biotechnologie Végétale, Université Victor Segalen-Bordeaux 2, Avenue des Facultés, 33405 Talence, France
[3] Department of Botany, University of Tel-Aviv, Tel-Aviv, Israel
[4] Departamento de Genética Molecular y Microbiologia, Facultad de Ciencias Biológicas, P. Universidad Católica de Chile, P.O.Box 114-D, Santiago, Chile

gence of cytoplasmic male sterility (CMS). We will focus the general description on maize and petunia. CMS can be defined as a phenotype produced by the incompatibility between nuclear and mitochondrial genomes of a given plant. The CMS phenotype may be relied to have high gene recombination, which plays an important role in modeling the plant mitochondrial genome. Mitochondrial gene recombination may lead to the formation of chimeric genes. Chimeric proteins having an altered mitochondrial function, may lead to the emergence of the CMS phenotype. The male sterility trait is used to obtain hybrids seeds carrying favourable agronomic traits.

We have used an experimental approach to mimic the CMS. This approach involves the targeting of a modified mitochondrial protein into the organelle which may interfere with normal mitochondrial function. For this purpose, we have used a novel process of genetic expression: RNA editing. Editing in plant mitochondria RNA involves the transformation of some cytidine residues into uridines by a deamination mechanism. RNA editing may result in amino acid changes and/or the emergence of new stop or initiation codons. Transgenic plants carrying an unedited gene, fused to a mitochondrial targeting peptide were constructed. After integration in the nuclear genome, the expression of "unedited" proteins induced male sterile phenotype in transgenic plants. Fertility restoration of sterile plants was obtained by crossing the latter with plants carrying the same gene in the antisense orientation. Other laboratories have obtained male sterile plants by expressing a bacterial ribonuclease (RNase) called barnase in pollen-producing tissues. The nuclease activity led to the loss of pollen production. Restoration in this case was obtained by crossing the male sterile plants with plants carrying the gene for a Barnase inhibitor.

1.1 Plant Male Sterility

Spontaneous or induced male sterility is a plant pathology characterized by the absence or abortion of pollen. Male sterility may have two origins: (1) nuclear, when the genetic locus is in the nuclear genome; in this case this trait is inherited as a classical Mendelian trait. The allele involved in this phenotype is usually recessive. Nuclear sterile mutants, spontaneous or induced, have been described in nearly 200 plants. (2) Cytoplasmic, maternal inheritance explains that this mutation is called cytoplasmic male sterility (CMS). This type of mutation is linked to mitochondrial function. Other mitochondrial mutations leading to abnormal growth and stripes in the leaf characterize a pathology named "non-chromosomal stripe" (NCS; Feiler and Newton 1987). NCS mutations, concerning several mitochondrial genes, have been found exclusively in the case of corn. An important difference in these two plant mitochondrial pathologies lies in the fact that in CMS plants all molecules of mtDNA are identical (homoplasmy) while in the case of NCS only a fraction of the mtDNA molecules carries the mutation (heteroplasmy). This chapter will be focused on plant CMS and how artificial male sterility can be induced by using experimental approaches related to genetic engineering.

2
Plant Mitochondrial DNA (mtDNA)

The plant organellar genome is more heterogeneous in size and complex in organization than that of fungi and mammals (Gray 1990). The size of mitochondrial plant genomes studied until now vary from 200 to 2500 kb, while the size range of fungi mitochondrial genomes is between 17–78 kb and around 16 kb in mammals.

Plant mtDNA can be described as a group of molecules having different sizes and forms composed of a large circular molecular (master circle) containing direct or inverted repeated sequences. These sequences are involved in homologous recombination events which generate either subgenomic DNA ("loop-out" mechanism), or an inversion of the sequence involved in the recombination process ("flip-flop" mechanism). In the case of corn mtDNA (570 kb), the presence of five direct repeated sequences leads to the emergence of multiple sub-genomic molecules (Lonsdale et al. 1984). The higher the number of repeated sequences, the higher the complexity of plant mtDNAs. This rule is not universal since in *Brassica hirtha* (207 kb) no subgenomic circles are found, probably due to the absence of repeated sequences (Palmer and Herbon 1987) while in *Marchantia polymorpha* subgenomic circles are absent in spite of the presence of several repeated sequences. In addition to the master circle and the subgenomic mtDNA molecules, small linear and circular plasmid-like molecules have been described without a defined genetic function, since their loss does not affect the plant phenotype (Pring et al. 1977; Palmer et al. 1983).

2.1
Coding Capacity of the Mitochondrial Genome

The complexity and size of plant mitochondrial genomes raise several questions: is the whole genome functional, as in mammals? Why is there such a wide range of genomic sizes between different species? Are there structural genes responsible for specific functions in the plant mtDNA?

To know whether the large size of plant mtDNA is linked to a greater coding capacity, studies based on the conservation of the sequences of mitochondrial genomes during evolution have been undertaken (Palmer 1990). The complete mtDNA sequence of *Marchantia polymorpha* and *Arabidopsis thaliana* outlines the organization and genetic content of two very distinct plant mtDNAs (Oda et al. 1992; Unseld et al. 1997).

Plant mtDNA codes for some of the polypeptides of the respiratory chain, the ATP synthase and ribosomes. Practically all mitochondrial genes homologous to those from mammals and yeast are present in the mitochondrial genomes of higher plants and in the moss *M. polymorpha:* (eg. respiratory chain complex I, III, IV and ATP synthase, some tRNAs and 18S and 26S rRNA). Plant mitochondrial genomes also have some genes not found in the mtDNA from other organisms: genes coding for proteins of the small *(rps)* and large *(rpl)* ribosomal subunits, genes of the respiratory chain complex II, proteins from the cytochrome-c biogenesis, the small rRNA 5S and a number of *orfs* of unknown function.

Some *orfs*, found in plant mtDNA introns, code for RNA maturases. The latter enzymes play a crucial role in the splicing and mobility of introns as shown in the yeast

Saccharomyces cerevisiae (Carignani et al. 1983). In certain plants, new *orfs*, orginated by recombination events, have been identified. They contain fragments of sequences issued from functional genes. The most studied of these *orfs* are the corn *T-urf13*, petunia *pcfS* and *orfB* from *Brassica.*

The tRNA population encoded in plant mtDNA is not sufficient to translate all codons in the mRNAs; some of the tRNAs have to be imported from the nucleus (Maréchal-Drouard et al. 1993).

The total coding capacity is not directly correlated with the size of the genome, since *A. thaliana* codes for 57 genes in a 367-kb mtDNA while *M. polymorpha* codes for 81 genes in a 187-kb molecule. This situation may be linked to genetic variability in different species. It is a possible explanation that the different genetic composition of mitochondria from different organisms is related to the biological history of each species. Part of the mitochondrial genetic information has been transferred to the nuclear genome according to the endosymbiotic hypothesis (Nugent and Palmer 1991).

2.2
Genetic Expression of Plant mtDNA

Replication, transcription and translation processes are poorly understood in plant mitochondria. The initiation of transcription has been studied in some detail, allowing the description of a mitochondrial promoter consensus sequence (Mulligan et al. 1991; Newton et al. 1995). The discovery of RNA editing (see below) confirmed unambiguously that plant mitochondria use the universal genetic code while previously some amino acids were supposed to be coded by non-universal codons.

3
Cytoplasmic Male Sterility (CMS) and the Origin of This Phenotype

Plant mitochondria and chloroplasts have their own genome which is maternally inherited. Cytoplasmic male sterility or CMS phenotype was linked to morphological and functional damage of the mitochondrial compartment. This conclusion was strongly supported by the use of somatic hybridization and genetic approaches (Belliard et al. 1978; Hanson et al. 1985). Protoplast fusion experiments showed that chloroplasts were not linked to the CMS trait. On the contrary, the comparison of mtDNA from fertile and CMS plants of the same species indicated significant differences (Belliard et al. 1979). The mutation analysis of the traits involved in the reversion of the CMS phenotype constituted strong support for a link between CMS and the mitochondrial genetic elements (Hanson et al. 1989).

The contribution of mitochondria in cell energy production is essential in non-chlorophyllien organs as well as in growing tissues. It seems obvious that non-lethal mtDNA mutation(s) leading to a decrease in cell energy production should be characterized by a particular phenotype. These mutants are very interesting for studying the mitochondrial function of higher plants.

3.1
Agricultural Importance of CMS

The production of hybrid plants is very important in modern agriculture, both in terms of increasing the yield and improving the quality of the product (hybrid vigor). However, the cross between two different plants is often limited due to the hermaphrodite trait of most plants; fertilisation resulting mainly from a self-pollination process. A drastic way to facilitate the production of hybrid plants is based on the blockage of self-pollination by selectively destroying either genetically, chemically (gametocides) or physically the male organs or the male gametes. Thus, it is important to obtain a CMS phenotype to produce hybrid seeds since it prevents self-pollination without using the above-mentioned methods. Crossing a CMS plant (female partner or recipient) with a male fertile plant (male partner or pollen donor) leads to a male-sterile progeny. The CMS phenotype is inherited as a maternal (non-Mendelian) trait allowing the transfer of the CMS from the female parent to the hybrid plant. The contribution of the male parent is restricted almost exclusively to nuclear traits (Laughnan and Gabay-Laughnan 1983).

The use of CMS in plant breeding programs was possible only when genes involved in fertility restoration, called Rf, were available. In the case of maize, the discovery of the Rf genes showed the diversity of CMS and led to the classification of maize cytoplasms C, T and S as a function of the fertility restoration genes (Vedel et al. 1994). CMS is characterized by some deleterious effects, like abnormal development of the male organs (anthers), a diminished function of these organs, etc. Cytological studies of the anther tissue showed a degeneration of the microspore feeder layer. The cases of petunia and maize will be described in some detail.

3.2
CMS in Maize

The mtDNA of the CMS lines of *Zea mays* called T (Texas), C (Charrua) and S (USDA), differ clearly from each other, when analyzed by restriction fragment length polymorphism as shown in Fig. 1a. Specific differences between these cytoplasms at the level of the genetic expression products have been also described (Hanson 1991).

The molecular basis of the CMS in maize CMS-T line, which has been the best characterized so far, started with the isolation of genes differently transcribed in CMS-T and in fertile plants. This approach allowed the localization of a locus called T-URF 13 carrying an *orf* present only in CMS-T mtDNA (Dewey et al. 1986, 1987). The CMS-T chimeric gene *(T-urf13)* is originated by multiple intramolecular recombination events implying mitochondrial genes. This chimeric gene has 115 amino acids corresponding to a polypeptide of 13-kDa molecular weight as shown in Fig. 1b *(T-urf 13)*. It has 88 codons of the 3' flanking region of the *atp6* gene and 18 codons orginated from a sequence of the mitochondrial 26 S rRNA. The nature of nine other codons is unknown. The transcription of *T-urf13* is controlled by a region homologous to that regulating the transcription of the gene *atp6*. The *T-urf13* gene is transcribed and translated in all the organs of the maize CMS-T but it is absent in fertile lines. The protein URF13 is found associated with the mitochondrial membrane fraction (Dewey et al. 1987). Fertility restoration in CMS-T plants can be obtained

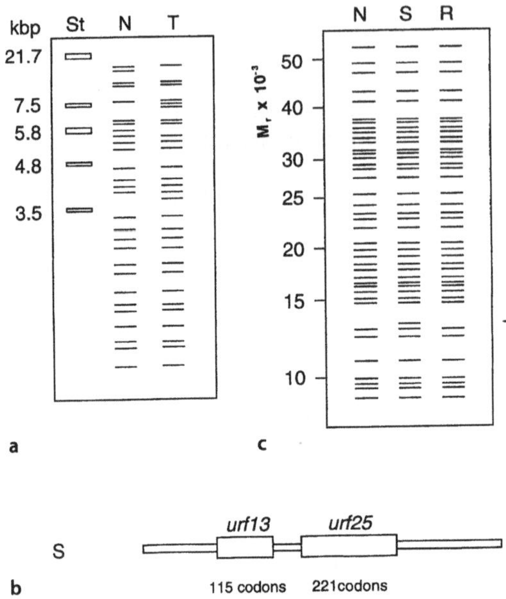

Fig. 1. **a** Schematic profile of the Eco R1 restriction nuclease pattern of mtDNA from normal fertile corn *(N)* and CMS-T corn *(T)*. *Left lane* shows the bacteriophage λ DNA profile digested with Hind III as marker *(St)*. **b** Scheme of the mtDNA CMS-T region carrying the 13 kDa open reading frame (T-urf13). **c** In organello protein synthesis products labeled with [^{35}S]-methione in normal male-fertile lines *(N)*, CMS-T *(S)* and lines restored to fertility *(R)*. The *arrow* indicates the PAGE-SDS migration of the 13-kDa protein specific to the CMS-T line

through the action of two nuclear restoring genes, Rf1 and Rf2. The presence of Rf1 in restored CMS-T plants, as seen in Fig. 1c, leads to a strong reduction of the level of T-URF13 protein in the organelle.

A striking property of the CMS-T maize, concerns the increased sensitivity to a fungal toxin produced by a pathogenic agent *Bipolarys maydis* toxin or the pesticide methomyl. The action of the *B. maydis* toxin on mitochondria produces a dissipation of the membrane potential and the consequent leakage of Ca^{+2} and NAD^{+} leads to the uncoupling of the respiratory chain. In the 1960s and 1970s the extensive use of CMS-T maize (more than 85% of the cultivated area in USA) for the production of hybrid plants produced a dramatic loss of the harvest due to the effect of the fungal toxin. The 1969 and 1970 epidemics showed dramatically the genetic vulnerability observed when a single, uniform plant cytoplasm is widely used.

The expression of T-URF13 in *Escherichia coli* increased the sensitivity of the bacteria to the action of methomyl or *B. maydis* toxin. In the presence of these agents, bacterial growth was reduced, and the ion permeability of the bacterial membrane was increased. A swelling phenomenon and the inhibition of respiration was also observed as described in CMS-T mitochondria (Dewey et al. 1988). The expression and targeting of the protein T-URF13 in *S. cerevisiae* generates a similar phenotype in the presence of methomyl to that obtained in CMS-T mitochondria (Glab et al. 1990).

The main anomaly observed concerns the feeder layer of the pollen grains, the tapetum. A degeneration of the tapetum and middle layer of the anther occurs after meiosis. A crucial question is how to explain the tissue-specific effect in plants showing the CMS phenotype? The expression of the *T-urf13* gene is found in all cells while the noxious effect of T-URF13 protein is tissue specific, i.e., restricted to the tapetum in anthers. Two hypotheses have been proposed: the first involves the anther-specific expression of a gene which, in the presence of the protein T-URF13 in the CMS-T plants, should produce the same effect as the one observed in the presence of the *B. maydis* toxin. The second one is related to the concept of threshold, widely used in human mitochondrial pathologies. The latter involves the idea that a minimum of energy is necessary for the development of a given tissue. Thus, in the case of CMS-T maize, the amount of energy produced, although lower than that of fertile plants, is enough to insure the development of the plant vegetative organs. In contrast, the development of organs involved in sexual reproduction, with a high respiratory rate, would be strongly affected.

3.3
CMS in Petunia

The CMS phenotype in *petunia* is also associated with a mitochondrial genome rearrangement generating the chimeric gene *S-pcf* (Young and Hanson 1987). *S-pcf* results from the fusion of the 35 first codons of *atp9* with sequences corresponding to two exons of *coxII* and an unidentified reading frame *(urfS)*. *S-pcf* is under the control of sequences which regulate the transcription of *atp9*. Immunological analysis of *urfS* allowed the identification of a 25-kDa protein present in the CMS plants and absent in fertile lines (Nivinson and Hanson 1989). The 25-kDa protein is localized in the membrane and soluble mitochondrial fractions. As in corn *T-urf14*, expression of *S-pcf* in *E. coli* showed that the PCF protein induced morphological changes in the bacteria and arrested growth in a liquid medium (Hanson 1991). Fertility restoration is due to a unique nuclear gene which markedly decreases the expression of the *S-pcf* gene.

3.4
Strategies Used for the Production of the Male-Sterile Phenotype

With the aim of producing hybrid plants, several groups have tried to develop new systems allowing the production of male-sterility using a plant transformation strategy. A system based on the specific expression of the ribonuclease gene from *Bacillus amyloliquefaciens* (barnase) in the anther tissues has been described (Mariani et al. 1990). Barnase expression in tapetum cells leads to the hydrolysis of cellular RNAs, thus avoiding pollen production. Restoration to fertility was obtained by using a specific inhibitor of RNase called "barstar". The co-expression of genes coding for barnase and barstar proteins in tapetal cells results in the formation of an inactive enzyme/inhibitor complex, thus restoring male fertility of the offspring (Mariani et al. 1992).

Another strategy to create CMS plants is based on the anther-specific expression of the enzyme callase (b-1,3-glucanase) at an early stage of microsporogenesis (Worral et al. 1992). Callase gene is naturally expressed in the tapetum at the end of meiosis, and the enzyme is excreted in the anther loculi allowing the release of microspores from their callose envelope. The tobacco callase gene is expressed under the control of an anther-specific promoter which operates at the beginning of meiosis. Transgenic plants expressing callase in anthers produce abnormal pollen. No fertility restoration approach is available for this strategy.

Another approach to producing male-sterile plants is based on the inhibition of the flavonoid pigment synthesis at the step catalyzed by chalcone synthase (CHS; Van der

Meer et al. 1992). Chalcone biosynthesis was blocked by using an antisense RNA complementary to CHS mRNA.

The strategies mentioned above to obtain male-sterile plants are based either on molecule degradation or the inactivation of cellular functions in plant organs involved in pollen production. They do not take into account the mitochondrial function. We have produced male-sterile tobacco plants by making use of a post-transcriptional step in genetic expression: RNA editing. Introducing into the mitochondrion a protein which has not undergone RNA editing ("unedited protein") produced an artificial form of male-sterility, described in further detail in this chapter. These results suggest that failing of the RNA editing process may be one of the physiological causes of CMS in plants. However, our own comparative studies of the degree of RNA editing in fertile, male-sterile and restored lines of wheat have not shown significant differences (Kurek et al. 1997).

4
RNA Editing in Higher Plants

The word "RNA editing" was used for first time a dozen years ago, when the insertion and deletion of uridine residues in the mitochondrial RNAs of trypanosomes was discovered. RNA editing has also been found in mitochondria and chloroplasts from land plants, in the mitochondria of some fungi, in the nucleo-cytoplasmic compartments from animal cells and in the genomes of some viruses (Table 1).

The RNA editing process involves the modification of residues in RNA or the insertion and/or deletion of nucleotides in messenger RNA (for reviews see Araya et al. 1994; Hanson et al. 1996; Smith et al. 1997; Stuart et al. 1997). An important consequence of the discovery of RNA editing is that the genomic sequence does not coincide necessarily with the primary structure of the corresponding protein.

RNA editing in the organelles of higher plants involves change of some C residues to U. Moreover, the discovery of RNA editing in plant mitochondria showed the use of the universal genetic code in these organelles. RNA editing involves preferentially coding regions of plant mitochondrial transcripts. and less frequently non-coding sequences like introns (Knoop et al. 1991; Zanlungo et al. 1995). The rare editing events found in introns, tRNAs and rRNAs lead generally to a correction of mispaired stem regions (for references see the RNA editing reviews mentioned above). In mRNAs, when the C residue edited to U is placed at the first or the second codon position, the change leads to an amino acid substitution, while changes in the third codon position does not change the nature of the amino acid. In a few cases, the C-to-U changes originate initiation codons (Chapdelaine and Bonen 1991), and most frequently, termination codons. It is interesting to point out that U-to -C changes can also be found, although they are extremely rare.

RNA editing is observed in all land plants tested except in the moss *Marchantia polymorpha*, suggesting that all bryophytes had lost the RNA editing process (Malek et al. 1996). However, recent reports show that RNA editing is present in other bryophytes. It remains to be established why *M. polymorpha* does not perform this process. No RNA editing is found in algae.

Table 1. Different types of RNA editing

Type	Compartment/Organism	Mechanism and possible cofactors
I. Insertion/deletion		
U insertion/deletion	Kinetoplastids *(Trypanosoma, Leishmania, Crithidia)*	Cleavage, ligation guide RNAs ...
C insertion (rarely U, AA CU, GU, GC insertions)	Mitochondria, *Physarum polycephalum*, mRNAs, rRNAs and tRNAs, *Paramyxoviruses*	Unknown Viral polymerase Slippery transcription
A insertion	*Ebola viruses*	Viral polymerase. Slippery transcription
3' mRNA poly A synthesis	Mitochondria, *Vertebrates*	Cleavage/TATase action
II. Base modification or replacement		
C to U	*Land plants*, mitochondria, mRNAs, tRNAs, rRNAs	C-deamination
C to U	Chloroplasts, *land plants*, mRNAs	Unknown
C to U	*P. polycephalum*	Unknown
C to U	*Mammals*, mRNAs (Apo-B)	Mooring sequence, Cytidine deaminase (APOBEC-1)
C to U	*Mammals and marsupials*	Unknown
U to C	*Land plants* (mitochondria, chloroplasts), mRNA	Transamination (?)
A to I	*Vertebrates*, mRNAs, (glutamic and serotonin receptor subunits)	DRADA and A-deamination
A to I	*Hepatitis delta virus* (HDV)	A-deamination of the HDV antigenome
A to G	*Drosophila*, 4f-rnp	A to I deamination (?)
U to A	*Humans*, alpha-galactosidase mRNA (Phe-to Tyr)	Unknown
C to A, A to G U to G, U to A	*Acanthamoeba castellani*, mitochondria, tRNAs	Nucleotide replacement (?)

4.1 Biochemical Mechanism of RNA Editing in Plants

RNA editing involving a C-to-U change may take place via different pathways (1) RNA cleavage followed by cytidine release, uridine incorporation and RNA ligation; (2) cytidine deamination; (3) transamination; or (4) transglycosylation reactions.

The insertion/deletion of base modification RNA editing processes are very different in terms of the biochemical mechanisms. This model needs several enzymatic activities able to catalyze the addition or deletion of uridine residues (for references see: Kable et al. 1996; Smith et al. 1997). In the case of plants, changes are less radical and the length of the edited transcripts are not modified.

Using an in vitro system from wheat mitochondria, we found that the RNA editing process proceeds by a C-to-U deamination step in plant mitochondria (Araya et al. 1992; Blanc et al. 1995). These results are in good agreement with the in organello experiments previously described showing that the α-phosphate is retained during C-to-U conversion in plant mitochondrial RNAs (Rajasekhar and Mulligan 1993). A similar in vitro approach using mitochondrial extracts obtained from pea seedlings was described by Yu and Schuster (1995). In summary, although all the details of the reaction involved in plant mitochondrial RNA editing have not been elucidated, this process seems biochemically similar to that described in animals for the editing of apoB (Teng et al. 1993) and glutamic acid and serotonin receptors in the central nervous system (Melcher et al. 1995).

4.2 What is the Biological Implication of RNA Editing?

The functional consequences of RNA editing, specially in non-coding regions, have not been explained and must be considered with caution. It can be postulated that RNA editing upstream of coding regions and/or the creation of initiation codons may modify the sequence or secondary structure required for protein synthesis. RNA editing events in introns may increase the similarity of the putative secondary structure towards the consensus motif necessary for splicing of group II introns (Michel et al. 1989).

The main outcome of RNA editing in plant organelles would be the increase of the homology between the edited transcript products and their counterpart found in organisms other than plants. RNA editing, in this case, may be considered as a correction process leading to the synthesis of functional proteins. In the case of *coxII*, the homologous proteins from other species have a cysteine residue which is essential for activity. In wheat *coxII*, this cysteine is absent and emerges only after editing of mRNA (Covello and Gray 1990). Similarly, editing of wheat *atp9* and potato apocytochrome b *(cob)* increases the hydrophobic character of the protein, thus probably favouring their integration in the mitochondrial membrane (Araya et al. 1993; Zanlungo et al. 1993). The hypothesis that RNA editing is necessary for the production of functional proteins is the basis of our strategy to produce male-sterile plants (see below).

5 Artifical Production of Male Sterile Plants

Several strategies have been used to produce male-sterile plants. None of these procedures has been focused on modfying the mitochondrial function which plays a key role in the emergence of the CMS phenotype.

The approach developed in our laboratory tried to mimic the physiological situation observed in CMS. This is a difficult task since direct mitochondrial transformation is hard to obtain given the absence of mitochondrial selection markers.

5.1
Construction of Male-Sterile Transgenic Plants Using an Unedited Gene

CMS is linked to a mitochondrial dysfunction in pollen producing tissues. Our hypothesis was that the introduction into mitochondria of a modified (non-functional) mitochondrial protein, should interfere with the multimeric respiratory chain complex whence the protein originates, and produce a mitochondrial dysfunction. Our aim was to interfere with the function of tissues having a great need of energy production like the tapetum in anthers. For this purpose, we transformed tobacco cells with an unedited gene that became integrated in the nucleus. No RNA editing has been reported in the plant nucleus.
We have chosen *atp9*, whose function is crucial in the production of ATP. The wheat *atp9* transcript encodes a short protein where RNA editing involves the change of five amino acid residues and the creation of a stop codon shortening the protein by six residues (Bégu et al. 1990). The *atp9* transcript is efficiently edited in vivo since less than 5% of *atp9* mRNA is partially edited (Kurek et al. 1997). Moreover, the ATP9 protein is encoded in animal and fungal mitochondria in the nuclear genome. Thus, import of ATP9 from the cytoplasm to the mitochondria should not be an obstacle.

The scheme in Fig. 2 shows transformation of tobacco cells with the unedited *atp9* gene, fused to a mitochondrial transit peptide, under the control of a strong promoter (Hernould et al. 1993). Transgenic plants carrying the "unedited" protein were phenotypically normal, but they were severely affected in their male fertility. Cytological analysis of anthers showed important modifications and degeneration of the tapetal cells. In contrast, control plants expressing the edited transgene produced

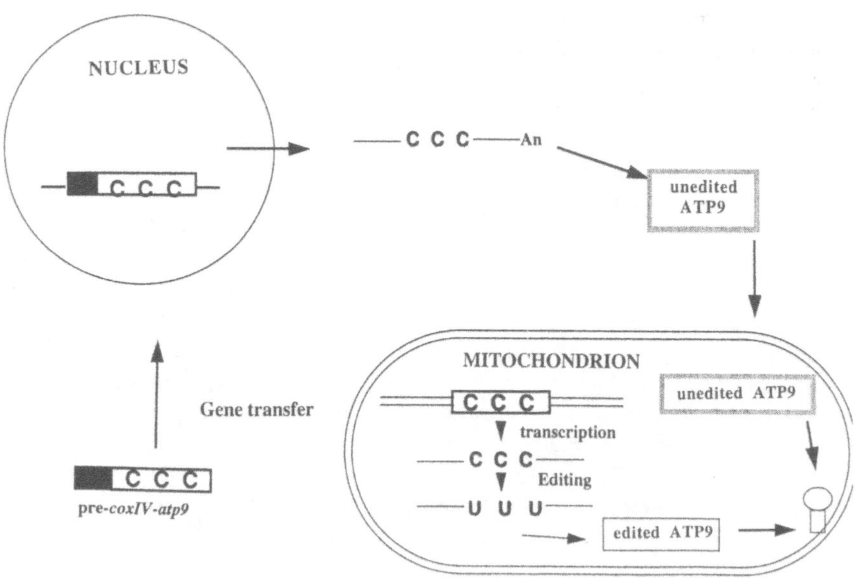

Fig. 2. Schematic representation of the strategy used in our laboratory to produce artificial transgenic male-sterile plants using an unedited mitochondrial gene. Restoration was obtained using an approach similar to that described in this scheme except that the unedited gene was expressed in the antisense orientation

functional pollen, were fertile and did not show any morphological anther alterations (Hernould et al. 1998). Molecular analysis showed that the chimeric *atp9* transgene was transcribed and translated. The ATP9 protein, either edited or unedited, was synthesized in the cytoplasm and imported into the mitochondria. Genetic analyses showed that the transgene was transmitted as a Mendelian character. Thus, the introduction of a modified mitochondrial gene and the targeting of the protein to the mitochondria was able to produce pollen abortion in the anthers of transgenic plants. This approach represents a new method to create artificial male-sterility in plants.

5.2 Recovering the Fertility of Artificially Created Male-Sterile Plants

The agronomical interest in creating male-sterile plants is linked with the possibility of obtaining hybrids with restored fertility. The use of barnase to generate male sterility is a drastic experimental approach that elegantly substitutes hand emasculation. Such a strategy needs the use of anther-specific expression of the RNase gene to avoid general lethality. After cell transformation, protein expression and plant regeneration, the active nuclease destroyed the pollen-producing organs (Mariani et al. 1990). To restore fertility, the sterile plants were crossed with transgenic plants expressing a specific inhibitor of barnase (Mariani et al. 1992). Our approach to obtain male sterility did not induce cell destruction, but aims to interfere with the physiological process of energy production. Our approach to restore fertility was based on the antisense strategy (Zabaleta et al. 1996). Our hypothesis was that crossing male-sterile plants with plants expressing the unedited *atp9* gene in the antisense orientation should lead to inhibiton of the synthesis of the "unedited" protein which caused the male sterility. As predicted, when we crossed a plant producing a high level of antisense *atp9* mRNA with a male-sterile plant expressing unedited *atp9* mRNA, male fertility was restored. In all cases, the restoration to male fertility was associated with 100% loss of unedited *atp9* mRNA. Very recently, a related strategy was described where the repression of the expression of the 55 kDa, NADH-binding subunit of complex I was disturbed by antisense expression in transgenic potato plants. Antisense effect led to a dramatic decrease of pollen production and a lower male fertility (Heisert et al. 1997).

Acknowledgements. Work in our laboratory was supported by the CNRS, the University of Bordeaux 2, the Human Frontiers Science Program Organization (HFSPO) (RG-437/94 M); the European Union (CI1+-CT93-0058), the Conseil Général d'Aquitaine, the French Ministère de l'Agriculture et de la Pêche, The Pôle GBM Aquitaine (L97083), the France-Israel Program of Biotechnological Applications to Agriculture. The authors acknowledge the expert technical assistance of Ms. Christina Calmels and Mrs. Evelyne Sargos, VB benefited from a pre-doctoral fellowship from the Minisère de la Recherche et de la Technologie, EZ benefited from a post-doctoral fellowship from the Ministère Français des Affaires Etrangères.

References

Araya A, Domec C, Bégu D, Litvak S (1992) An in vitro system for the editing of ATP synthase subunit 9 mRNA using wheat mitochondrial extract. Proc Natl Acad Sci USA 89:1040–1044

Araya A, Bégu D, Graves PV, Hernould M, Litvak S, Mouras A, Suharsono S (1993) Of RNA editing and cytoplasmic male sterility in plants. In: Brennicke A, Kück U (eds) Plant mitochondria. VCH Chemie, Weinheim, pp 83–91

Araya A, Bégu D, Litvak S (1994) RNA editing in plants. Physiol Plant 91:543–550

Bégu D, Graves PV, Domec C, Arselin G, Litvak S, Araya A (1990) RNA editing of wheat mitochondrial ATPase subunit 9: direct protein and cDNA sequencing. Plant Cell 2:1238–1290

Belliard G, Pelletier G, Vedel F, Quetier F (1978) Morphological characteristics and chloroplast DNA distribution in different cytoplasmic parasexual hybrids of *Nicotiana tabacum*. Mol Gen Genet 165:231–237

Belliard G, Vedel F, Pelletier G (1979)Mitochondrial recombination in cytoplasmic hybrids of *Nicotiana tabacum* by protoplast fusion. Nature 28:401–403

Blanc V, Litvak S, Araya A (1995) RNA editing in wheat mitochondria proceeds by a deamination mechanism. FEBS Lett 373:56–60

Carignani G, Groudinsky O, Frezza D, Schiavon E, Bergantino E, Slonimski P (1983) An mRNA maturase is encoded by the first intron of the mitochondrial gene for the subunit I of cytochrome oxidase in *S. cerevisiae*. Cell 35:733–742

Chapdelaine Y, Bonen L (1991) The wheat mitochondrial gene for subunit I of NADH dehydrogenase complex: a transplicing model for this gene-in-pieces. Cell 65:465–472

Covello PS, Gray MW (1990) RNA sequence and the nature of the Cu_A-binding site in cytochrome c oxidase. FEBS Lett 268:5–7

Dewey RE, Levings CS III, Timothy DH (1986) Novel recombination in the maize mitochondrial genome produces a unique transcriptional unit in the Texas male sterile cytoplasm. Cell 44:439–444

Dewey RE, Timothy DH, Levings CS III (1987) A mitochondrial protein associated with male sterility in the T cytoplasm of maize. Proc Natl Acad Sci USA 84:5374–5378

Dewey RE, Siedow JN, Timothy DH, Levings CS III (1988) A 13 kDa maize mitochondrial protein in *E. coli* confers sensitivity to *Bipolaris maydis* toxin. Science 239:293–295

Feiler HS, Newton KJ (1987) Altered mitochondrial gene expression in the nonchromosomal stripe 2 mutant of maize. EMBO J6:1535–1539

Glab N, Wise RP, Pring DR, Jacq C, Slominski P (1990) Expression in *S. cerevisiae* of a gene associated with cytoplasmic male sterility from maize: respiratory dysfunction and uncoupling. Mol Gen Genet 223:24–32

Gray MW (1990) Origin and evolution of mitochondrial DNA. Annu Rev Cell Biol 5:25–50

Hanson MR (1991) Plant mitochondrial mutations and male sterility. Annu Rev Genet 25:461–486

Hanson MR, Rothenberg M, Boeshore ML, Nivison HT (1985) Organelle segregation and recombination following protoplast fusion: analysis of sterile cytoplasms. In: Zaitlin M, Day P, Hollaender A (eds) Biotechnology in plant science. Academic Press, New York, pp 129–144

Hanson MR, Pruitt KD, Nivison HT (1989) Male sterility loci in plant mitochondrial genomes. Oxf Surv Plant Mol Cell Biol 6:61–85

Hanson MR, Sutton CA, Lu B (1996) Plant organelle gene expression: altered by RNA editing. Trends Plant Sci 1:57–64

Heisert V, Rasmusson AG, Thick O, Brennicke A, Grohman L (1997) Antisense repression of the mitochondrial NADH-binding subunit of complex I in transgenic potato plants affects male fertility. Plant Sci 127:61–69

Hernould M, Suharsono S, Litvak S, Araya A, Mouras A (1993) Male-sterility induction in transgenic tobacco plants with an unedited *atp9* mitochondrial gene from wheat. Proc Natl Acad Sci USA 90:2370–2374

Hernould M, Suharsono, Zabaleta E, Carde JP, Litvak S, Araya A, Mouras A (1998) Impairment of tapetum and mitochondria in engineered male-sterile tobacco plants. Plant Mol Biol 36:499–508

Kable ML, Seuwert SD, Heidmann S, Stuart K (1996) RNA editing: a mechanism of gRNA-specified uridylate insertion into precursor mRNA. Science 273:1189–1195

Knoop V, Schuster W, Wissinger B, Brennicke A (1991) Transplicing integrates an exon of 22 nucleotides into the nad5 mRNA in higher plant mitochondria. EMBO J 10:3483–3493

Kurek I, Ezra D, Bégu D, Erel N, Litvak S, Breiman A (1997) Studies on the effect of nuclear background and tissue specificity on RNA editing of the mitochondrial ATP synthase subunits α, 6 and 9 in fertile and cytoplasmic male sterile (CMS) wheat. Theor Appl Genet 95:1305–1311

Laughnan JR, Gabay-Laughnan G (1983) Cytoplasmic male sterility in maize. Annu Rev Genet 17:27–48

Lonsdale DM, Hodge JP, Fauron CMR (1984) The physical map and organization of the mitochondrial genome from fertile cytoplasm of maize. Nucleic Acids Res 12:9249–9261

Malek O, Lättig K, Hiesel R, Brennicke A, Knoop V (1996) RNA editing in bryophytes and a molecular phylogeny of land plants. EMBO J 15:1403–1411

Maréchal-Drouard L, Weil JH, Dietrich A (1993) tRNA and rRNAs genes in plants. Annu Rev Plant Physiol Plant Mol Biol 44:13–32

Mariani C, de Beuckeleer M, Truettner J, Leemans J, Goldberg RB (1990) Induction of male sterility in plants by chimaeric ribonuclease gene. Nature 347:737–741

Mariani C, Gossele V, de Beuckeleer M, de Block M, Goldberg RB, de Greef W, Leemans J (1992) A chimaeric ribonuclease inhibitor gene restores fertility to male sterile plants. Nature 257:384–387

Melcher T, Maas S, Higuchi M, Keller W, Seeburg PH (1995) Editing of a-amino-3-hydroxy-5-methylisoxazole-4-propionic acid receptor GluR-B pre-mRNA in vitro reveals site-selective adenosine to inosine conversion. J Biol Chem 270:8566–8570

Michel F, Umesono K, Ozeki H (1989) Comparative and functional anatomy of group II catalytic introns – a review. Gene 82:5–30

Mulligan RM, Leon P, Walbot V (1991) Transcriptional and post-transcriptional regulation of maize mitochondrial gene expression. Mol Cell Biol 11:533–543

Newton KJ, Winberg B, Yamato K, Lupold S, Stern DB (1995) Evidence for a novel mitochondrial promoter preceding the cox2 gene of perennial teosintes. EMBO J 14:585–593

Nivinson HT, Palmer JD (1991) RNA-mediated transfer of the gene coxII from the mitochondrion to the nucleus during flowering plant evolution. Cell 66:473–481

Oda K, Yamato K, Ohta E, Nakamura Y, Takemuru M, Nozato N, Akashi K, Kanegue T, Ogura Y, Kohshi T, Okyama K (1992) Gene organization deduced from the complete sequence of liverwort *Marchantia polymorpha* mitochondrial DNA: a primitive form of plant mitochondria genome. J Mol Biol 223:1–7

Palmer JD (1990) Contrasting modes and tempos of genome evolution in land plant organelles. Trends Genet 6:115–120

Palmer JD, Herbon LA (1987) Unicircular structure of the *Brassica hirta* mitochondrial genome. Curr Genet 11:565–570

Palmer JP, Shields CR, Cohen DB, Orton TJ (1983) An unusual mitochondrial DNA plasmid in the genus *Brassica*. Nature 301:725–728

Pring DR, Levings CS III, Huww L, Timothy DM (1977) Unique DNA associated with mitochondria in the S type-cytoplasm of male sterile maize. Proc Natl Acad Sci USA 74:2904–2908

Rajesakhar VK, Mullligan RM (1993) RNA editing in plant mitochondria: α-phosphate is retained during C-to-U conversion in mRNAs. Plant Cell 5:1843–1852

Rogers SJ, Edwardson JR (1952) The utilization of cytoplasmic male sterile inbreds in the production of corn hybrids. Agron J 44:8–13

Smith HC, Gott JM, Hanson MR (1997) A guide to RNA editing. RNA 3:1105–1123

Stuart K, Allen TE, Heideman S, Seiwert SD (1997) RNA editing of kinetoplast protozoa. Microbiol Mol Biol Rev 61:105–120

Teng B, Burant CF, Davidson NO (1993) Molecular cloning of an apolipoprotein B messenger RNA editing protein. Science 260:1816–1819

Unseld M, Marienfeld JF, Brandt P, Brennicke A (1997) The mitochondria genome of *Arabidopsis thaliana* contains 57 genes in 366 924 nucleotides. Nat Genet 15:57–61

Van der Meer I, Stam ME, van Tunen AJ, Mol JNM, Stuitje AR (1992) Antisense inhibition of flavonoid synthesis in *Petunia* anthers results in male sterility. Plant Cell 4:253–262

Vedel F, Pla M, Vitart V, Gutierrez S, Chétrit P, De Paepe R (1994) Molecular basis of nuclear and cytoplasmic male sterility in higher plants. Plant Physiol Biochem 32:601–618

Worrald D, Hird DL, Hodge R, Paul W, Draper J, Scott R (1992) Premature dissolution of the microsporocyte callose wall causes male sterility in transgenic tabacco. Plant Cell 4:759–771

Young EG, Hanson MR (1987) A fused mitochondrial gene associated with CMS is developmentally regulated. Cell 50:41–49

Yu W, Schuter W (1995) Evidence for a site-specific cytidine deamination reaction involved in C to U RNA editing in plant mitochondria. J Biol Chem 270:18227–18233

Zabaleta E, Mouras A, Hernould M, Suharsono S, Araya A (1996) Transgenic male-sterile plant induced by an unedited *atp9* gene is restored to fertility by inhibiting its expression with antisense RNA. Proc Natl Acad Sci USA 93:11259–11263

Zanlungo S, Bégu D, Quiñones V, Araya A, Jordana X (1993) RNA editing of apocytochrome b *(cob)* transcripts in mitochondria from two genera of plants. Curr Genet 24:344–348

Zanlungo S, Quiñones V, Holuigue L, Moenne A, Jordana X (1995) Splicing and editing of rps 10 transcripts in potato mitochondria. Curr Genet 27:565–571

CHAPTER 25

Histopathology of Skeletal Muscle Mitochondria

25

N. B. Romero[1], M. Coquet[2], and H. Carrier[3]

Contents

1 Introduction

The muscle biopsy can often be one of the most decisive steps in the diagnosis of several mitochondrial diseases, even when the patients do not exhibit any apparent clinical symptom of muscular involvement. The technique is particularly efficient in the case of respiratory chain deficiencies. The histological examination of this muscle biopsy is perhaps the best technique for visualizing rare muscle fibers which exhibit striking mitochondrial anomalies. Indeed, in biochemical studies which are generally performed with muscle homogenates, it is difficult to distinguish differences between normal and diseased tissue when, for example, 10% of the fibers are damaged.

[1] Laboratoire de Pathologie Musculaire, Institut de Myologie, Hôpital Pitié-Salpêtrière, 83 Bd de l'Hôpital, 75651 Paris, France
[2] Laboratoire d'Anatomie Pathologique et de Neurologie, CHR Pellegrin, Université de Bordeaux II, 33076 Bordeaux Cedex, France
[3] Groupe de Recherches en Pathologie Neuromusculaire, INSERM U499, Faculté de Médecine Laënnec, 12 rue Guillaume Paradin, 69372 Lyon Cedex 08, France

2
Morphological Anomalies

2.1
Accumulation of Abnormal Mitochondria

Abnormal mitochondria were first reported in muscle fibers by Luft et al. (1962). Electron microscopy then allowed the further description of aggregations of abnormal mitochondria under the sarcolemma of type 1 fibers, their abnormal ultrastructure and the presence of dense or paracrystallin inclusions (Gruner 1963; Shy et al. 1966; Price et al. 1967). One can see examples of normal or pathological mitochondria which are enlarged, containing stacking of cristae (Fig. 1a), or paracrystallin inclusions (Fig. 1b). Engel and Cunningham (1963) have shown that the modified Gomori trichrome stain could reveal the proliferation of these abnormal mitochondria by

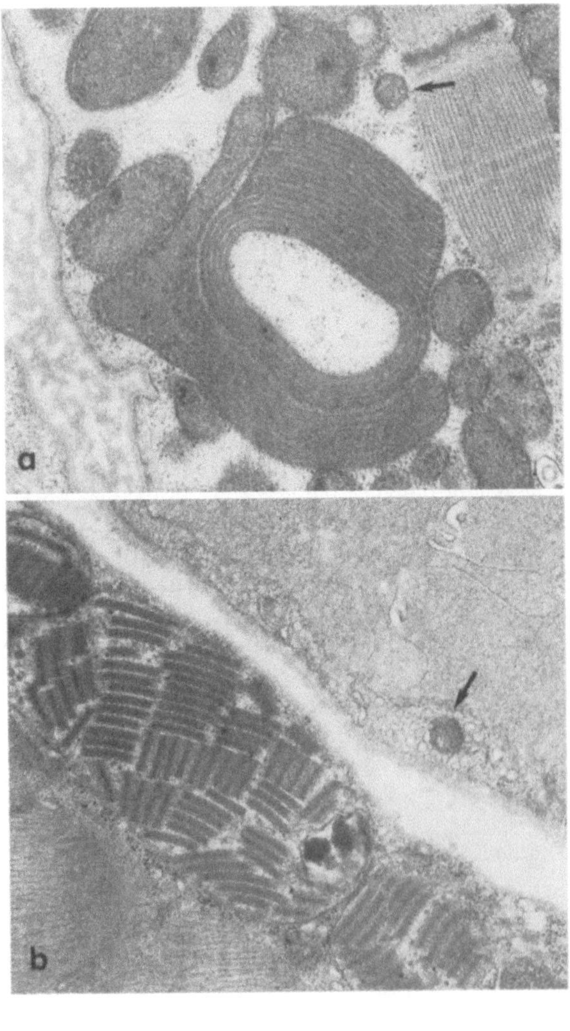

Fig. 1. Electron microscopy: pathologic mitochondria with hyperplasia and stacking of cristae from a MELAS patient **(a)**, and giant mitochondrion containing about 75 paracrystallin inclusions inside the cristae **(b)**. The *arrow* indicates a normal mitochondrion in an endothelial cell. (x 15 000)

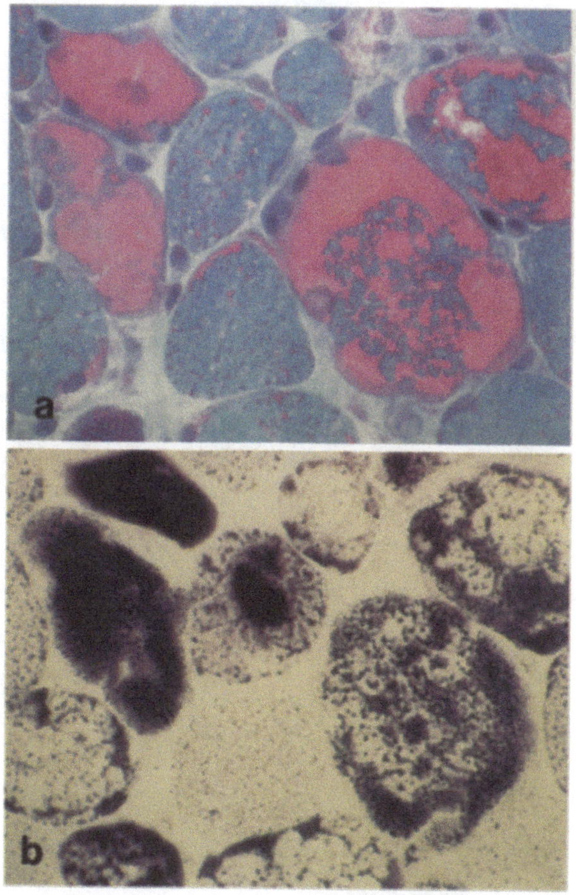

Fig. 2. Serial cross sections exhibiting RRF, characterized by the presence of sub-sarcolemmal and intermyofibrillar aggregates of abnormal mitochondria which take a red staining with the modified trichrome Gomori technique (**a**) and a (dark) staining for the SDH reaction (**b**)

using light microscopy and suggested the term of "ragged red fiber" (RRF) to describe them.

The RRF (Fig. 2a) is a characteristic histological feature of skeletal muscle fibers presenting mitochondrial DNA (mtDNA) deletions, duplications or mitochondrial tRNA point mutations. RRF are exceptionally found in the muscle of patients presenting mutations in mtDNA genes encoding proteins. They can also be observed in muscles of patients with respiratory chain deficiencies with a sporadic or presumed autosomic mode of inheritance.

RRF are occasionnally present in a variety of other muscle pathologies, such as chronic polymyositis (Chou 1972), or in aged individuals (Müller-Hocker 1992). It cannot be ascertained at the present time whether or not they are a specific hallmark of deficient mitochondria. They have also been described in cases where respiratory chain impairment is secondary to drug treatment such as in AZT-treated patients (Dalakas et al. 1994).

In the muscle of adult patients, RRF are generally not very numerous, usually representing 3 to 10% of the fibers. In babies or children a complete absence of RRF can occur (Fig. 3a), even though the respiratory chain deficiency can be found by bio-

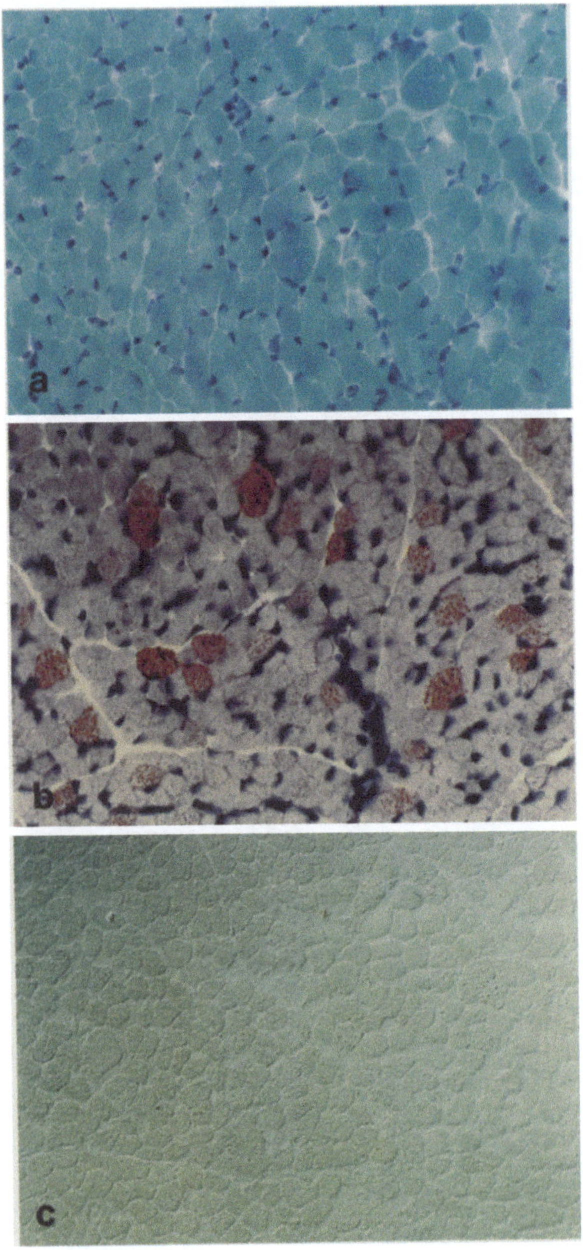

Fig. 3. Newborn muscle cross section exhibiting a severe COX deficiency, with absence of mitochondrial aggregates, as shown by the trichrome Gomori technique **(a)**, with many muscular fibers containing lipid storage **(b)** and with complete absence of COX activity **(c)**

chemical techniques. It is probably because the pathophysiological processes which lead to RRF formation are not achieved at the time of the biopsy. In such cases, the presence of other abnormalities can sometimes be detected (see below, COX negative fibers). Different muscles of the same patient do not contain the same RRF density. For instance, the deltoid muscle usually exhibits more RRF than the quadriceps (Romero et al. 1991).

2.2
Lipid Storage

Storage of neutral lipids within type 1 muscle fibers is occasionally associated with mitochondrial respiratory deficiencies and RRF. However, it can sometimes be the only morphological anomaly in newborn and children evoking a mitochondrial deficiency (Fig. 3b; Romero et al. 1996). Lipid storage can be revealed by oil red O, black or red Sudan stains (Dubowitz 1989). Electron microscopy shows that the lipid droplets are located in close contact with the outer membrane. Lipid storage is most often observed in muscles with a beta-oxidation deficiency (Tyni et al. 1996), and a carnitine deficiency (Morand et al. 1979; Carrier and Berthillier 1980). It is very rare in palmityl-carnitine transferase II deficiency.

3
Histoenzymatic Anomalies

3.1
Cytochrome c Oxidase (COX)

COX allows the visualization of complex IV activity in muscle fibers (Selignman et al. 1968). Cytochrome c oxidation is coupled with 3-3' diaminobenzidine (DAB) oxidation which forms a brown precipitate, easily recognized by light microscopy. Type I fibers in which mitochondria are more abundant, are more heavily stained than type 2 fibers (Romero et al. 1990).

In cases of total complex IV deficiency, as for example in fatal infantile myopathy (Di Mauro et al. 1980; Di Mauro and Moraes 1993), the COX reaction is negative in all muscle fibers, including the intrafusal ones. However, in some cases with exclusive muscle COX deficiencies, intrafusal fibers exhibit a residual COX reaction (Zeviani et al. 1985; Oldfors et al. 1989). In other instances, when the complex IV deficiency is partial, the muscle fibers can present a pale diffuse staining (Possekel et al. 1995).

Very often, the complex IV deficiency disclosed in the COX-negative fibers is not specific, but is associated with a general mitochondrial respiratory chain impairment. As a general rule, the COX reaction may be negative not only within the RRF but also in a number of other type 1 fibers. Furthermore, it is not rare that a muscle biopsy exhibiting no RRF presents a significant number of COX-negative fibers.

In addition, the COX activity may vary in successive segments of the same fiber, a normal staining alternating with a weak staining or with an absence of staining (Fig. 4a). A correlation seems to exist between the length of the COX-negative segments and the percentage of deleted mtDNA in the muscle (Romero and Fardeau 1993). This heterogeneous COX activity must be due to differences in the percentage of modified mtDNA (mutation, duplication or deletion) in the various fibers or in different segments of the same fiber (Mita et al. 1989).

When the deficiency of one or several respiratory chain complexes is related to a point mutation in a tRNA gene, the COX reaction is sometimes reinforced under the RRF sarcolemma (Conjard et al. 1997). Finally, it has been suggested that COX abnormal reactions might precede mitochondrial proliferation in individual muscle fibers, giving a mosaic aspect to the muscle cross-section due to the coexistence of

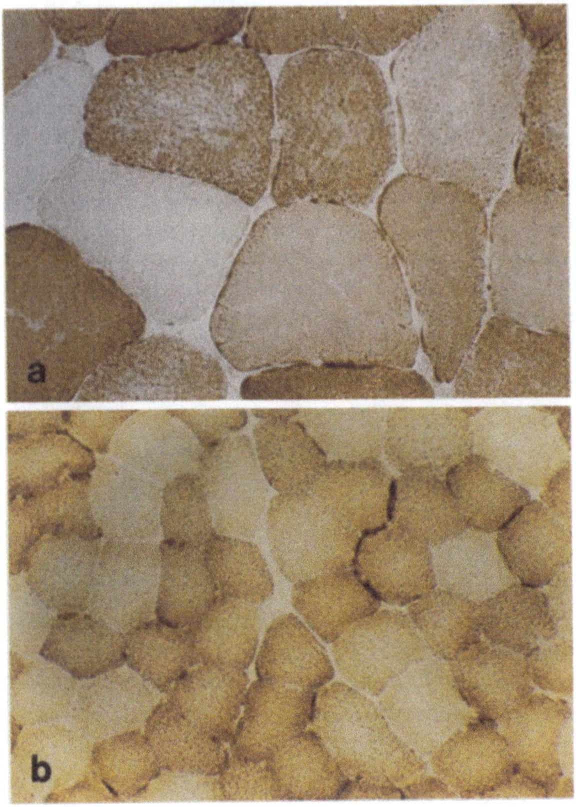

Fig. 4. Muscle serial cross sections with COX-positive and COX-negative fibers (**a**), compared with a control exhibiting a normal COX activity (**b**)

well stained COX-positive fibers and of COX-negative fibers completely devoid of enzymatic activity (Matsuaka et al. 1991).

3.2 Succinate Dehydrogenase (SDH)

In this histoenzymatic reaction, the succinate oxidation is coupled with the reaction of a colorless tetrazolium salt which precipitates into an insoluble dark-blue formazan (Watterberg and Leong 1960). The SDH reaction, as well as the COX reaction is normally more intense in mitochondria-rich type 1 fibers than in type 2 fibers. In respiratory chain deficient muscles, it is strongly reinforced under the RRF sarcolemma where the abnormal mitochondria accumulate. Therefore, the SDH staining coincides with the red staining observed with the trichrome Gomori technique (Fig. 2b).

In contrast, the complete absence of SDH activity has been reported in very rare cases of complex II deficiency. In addition, some SDH negative muscle fibers are also occasionally seen in atypical mitochondrial diseases (Fig. 5). The same fibers are frequently COX-negative but very rarely RRF.

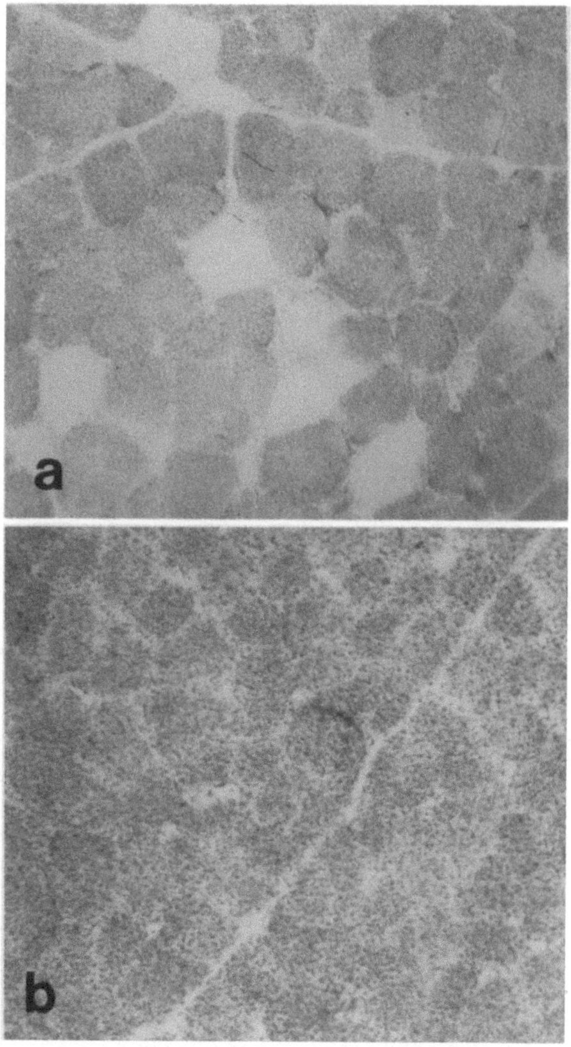

Fig. 5. Cross section exhibiting a mosaic of SDH-positive and SDH-negative fibers (**a**), compared with a control muscle with normal SDH distribution and absence of mitochondrial aggregates (**b**)

3.3
Mitochondrial ATPase Reaction

The mitochondrial ATPase activity has been analyzed by the technique described by Meijer and Vloedman (1980). It is successively compared in the absence of an inhibitor and in the presence of an uncoupler without and with oligomycin. This reaction can therefore give an idea about the state of coupling or uncoupling of the mitochondria.

3.4 Myofibrillar ATPase Reaction

A first myofibrillar ATPase reaction, made at pH 9.4, permits the identification of clear type 1 and dark type 2 fibers. A modified ATPase reaction made by preincubation at pH 4.2 and 4.6 followed by reaction at pH 9.4 identifies the type 1 fibers and the subtype 2A and 2B fibers (Brooke and Kaiser 1969).

In muscle harboring a respiratory chain deficiency, the percentage of type 1 fibers is usually increased. In these slow-twitch, endurant fibers, the oxidative energy metabolism is pre-eminent (Guth and Smaha 1969). It is likely that the plasticity of these fibers is important in developing a compensatory mechanism when the oxidative energy metabolism is deficient.

4 Immunolabelling

The immunolabelling techniques allow the visualization of antigenic sites or epitopes located on molecules present inside the cell or, more precisely, inside the mitochondria. The proteins are recognized by specific monoclonal or polyclonal antibodies that will then react with a secondary antibody coupled to a fluorochrome or to peroxidase. Immunolabelling therefore is the first visual molecular approach for studying the expression of a protein that can be at least partly responsible for mitochondrial enzymatic activity.

Thanks to the availability of antibodies specifically recognizing the various COX subunits, one of us (NR) has been able to investigate the differential expression of the COX subunits in normal and diseased muscles. In normal muscle, type 1 fibers are generally more heavily stained than type 2 fibers by antibodies directed against the COX subunits, as shown above for the COX histoenzymatic reaction (Romero et al. 1996).

In two newborn patients, the selective absence of one of the COX isoenzymes specific to striated muscle, namely subunit VIIa-H suggested a direct relationship between the expression of this tissue-specific subunit and the myopathic and myocardiopathic symptoms that the patients exhibited (Possekel et al. 1995).

A loss of immunoreactivity of different COX subunits was found in children presenting symptoms related to various organs. The pattern suggested a global deficiency of several ubiquitous COX subunits (Possekel et al. 1995; Romero et al. 1996).

In COX-deficient patients, a normal expression of all the COX subunits for which antibodies were available suggested either that the defect impaired a subunit which could not been studied, or that a structural modification occurred in one of the analyzed subunits without modifying epitope recognition by the antibodies (Romero et al. 1993).

In cases of muscles exhibiting a heterogeneous histoenzymatic COX reaction, immunostaining varying in distribution as well as in intensity was found to be in agreement with mtDNA anomalies and the studied protein. Two situations occurred:

1. The expression of COX subunits was normal in all muscle fibers whether they were COX-positive or COX-negative in histoenzymatic analyses.

2. The expression of specific subunits was deficient in the COX-negative segments and was normal in other segments as well in other fibers.

In the muscle of patients exhibiting a qualitative (deletion or mutation) or more seldom a quantitative (depletion) mtDNA deficiency, immunolabelling demonstrated that histoenzymatic COX-negative fibers express the subunits encoded by nuclear DNA normally but express weakly, or not at all, the mtDNA encoded subunits, even when the encoding gene was not impaired. Finally, rare cases exhibiting COX-negative fibers and no immunolabelling of some COX subunits were not associated with a known mtDNA rearrangement (Possekel et al. 1995).

The expression of a large number of proteins which are not respiratory chain components can also be modified in muscles presenting mitochondrial diseases for example, the expression of heat shock proteins which interfere with mitochondrial biogenesis and of enzymes involved in the dismutation (Ohkoshi et al. 1995) and reduction of free radicals are induced or repressed in RRF (unpubl. data).

5 Molecular Histology

Serial cross sections of skeletal muscle alternating in situ hybridizations with suitable staining (COX, SDH, modified Gomori's trichrome) allow a qualitative analysis of the deleted and wild-type mtDNA transcripts and those of some nuclear genes (Fig. 6A–D). This histological method is perhaps more instructive than the biochemical ones performed with muscle homogenates because it allows correlation of transcript expression with RRF, which often represents a tiny part of the muscle fibers.

Studies using in situ DNA/DNA hybridization have demonstrated that deleted mtDNA is overexpressed within RRF (Shoubridge et al. 1990; Collins et al. 1991; Hammans et al. 1992; Moraes et al. 1991; Sciaccio et al. 1994). However, the question of whether wild-type mtDNA is present or not in reduced amounts remains unanswered. The expression of messenger and ribosomal RNA from genes located within or outside mtDNA single deletions has been studied using in situ DNA/RNA hybridization. Studying the messenger transcripts from nuclear genes encoding mitochondria-imported proteins is also interesting for understanding the interactions between the expression of the two genomes (Carrier et al. 1996). Differential studies of normal tRNAs or of tRNAs which present a point mutation by using in situ hybridization are not currently feasible.

Complementary probes to the RNAs from the mitochondrial genes located outside the mtDNA deletions display an overexpression of the messenger and ribosomal transcripts in RRF (Fig. 6A–C; Mita et al. 1989; Shoubridge et al. 1990; Hammans et al. 1992; Moraes et al. 1991; Carrier et al. 1996). This overexpression could be a consequence of either the mitochondrial proliferation in these fibers or that of a decreased degradation of mRNAs if they are not translated. Indeed, the translation can be impaired because many genes encoding the mitochondrial tRNAs are carried away by the mtDNA deletion, preventing a efficient protein synthesis (Nakase et al. 1990), in spite of the ribosomal RNAs' overexpression.

Nuclear encoded transcripts such as that of the mitochondrial ATPase β subunit were also overexpressed within the RRF. Since the translation of β-ATPase occurs in

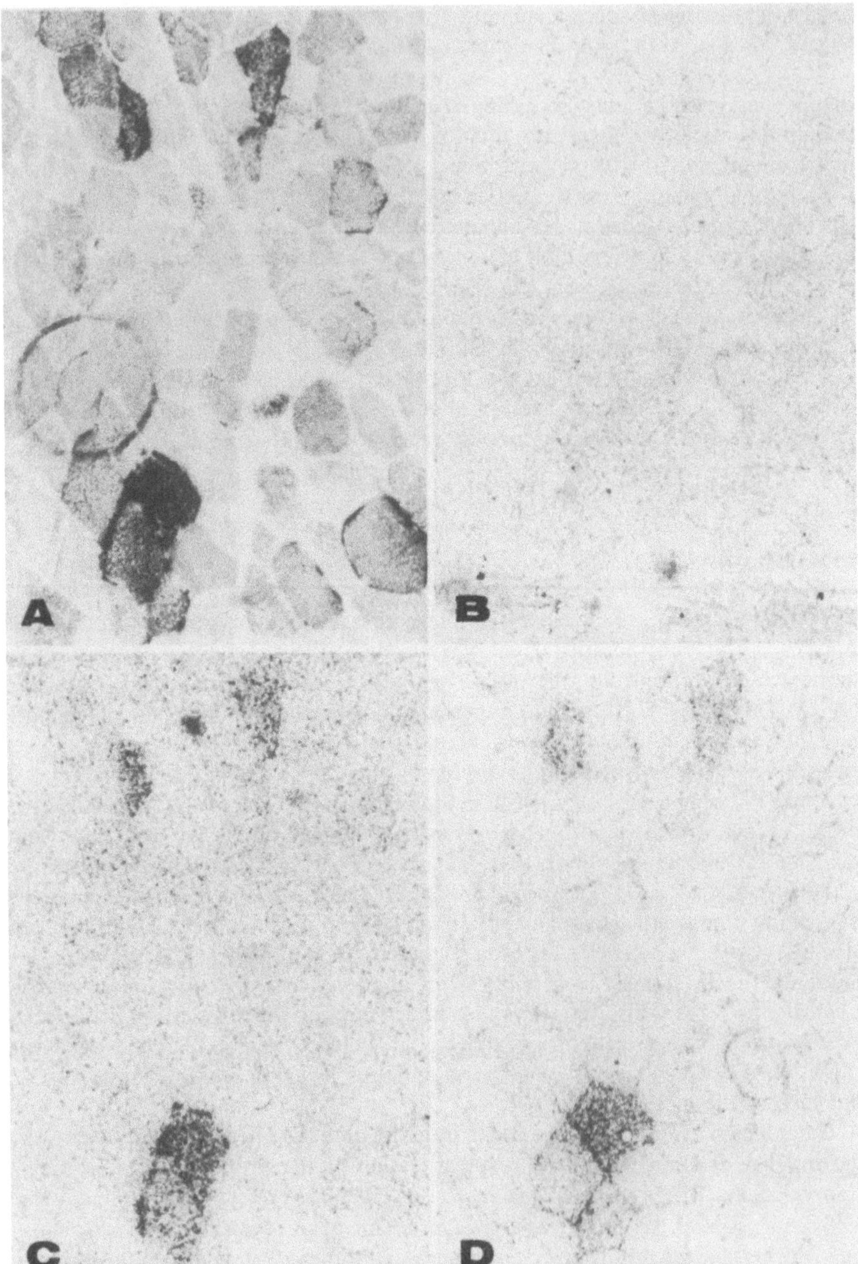

Fig. 6. Serial cross section of a skeletal muscle harboring a 5.5 kb single deletion encompassing ATP8 to ND5 mitochondrial genes. (x 120). **A** Succinic dehydrogenase reaction displays the characteristic oxidative hyperactivity within RRF. **B** In situ hybridization (ISH) of the ND4 transcripts, the gene of which is located within the mtDNA deletion. Their expression is normal in all the fibers. **C** ISH of the COII transcripts, the gene of which is located outside the mtDNA deletion. They are overexpressed in the RRF. **D** ISH of the β-ATPase transcripts from a nuclear gene. They are also overexpressed in the RRF

the cytosol, its transcription cannot be directly related to the amount of mtDNA in the cell (Fig. 6A–D). This result rather suggests the existence of an up-regulation mechanism inducing the simultaneous expression of both mitochondrial and nuclear transcripts which are involved in oxidative phosphorylations, as previously proposed by Heddi et al. (1993) and Haraguchi et al. (1994) who have observed an increase in the amount of β-ATPase, and of ATP/ADP translocator transcripts, in patients muscle biopsies. Whatever its mechanism may be, this up-regulation is restricted to RRF.

In situ hybridization failed to detect increased expression of another nuclear gene transcript which is not directly involved in oxidative phosphorylation, the sarcomeric mitochondrial creatine kinase.

Probes complementary to the RNAs from the mitochondrial genes located within the mtDNA deletions only hybridize the transcripts from the wild-type mtDNA. No decrease in their expression was found in any fiber, including RRF (Shoubridge et al. 1990; Carrier et al. 1996). These results differ from those of previous reports where a decreased expression of wild-type mtDNA transcripts was described in RRF (Mita et al. 1989; Nakamura et al. 1990; Hammans et al. 1992). The existence of an up-regulation mechanism of the transcriptions, as demonstrated above, might explain such an apparently normal expression. Rather it is an overexpression of the wild-type mtDNA transcripts which complements the untranscribed deleted genes within the RRF mitochondria. Such a complementation is likely to be efficient until a threshold in the deleted mtDNA proportion is reached. It is likely to be a high threshold, since the expression of such transcripts was found to be normal in the RRF of a muscle which harbored 80% of deleted mtDNA (Carrier et al. 1996).

Acknowlegements. We thank Dr. C. Godinot for stimulating advice.

References

Brooke MH, Kaiser KK (1969) Some comments on the histochemical characterization of muscle adenosine triphosphatase. J Histochem Cytochem 17:431–435

Carrier H, Berthillier G (1980) Carnitine levels in normal children and adults in patients with diseased muscle. Muscle Nerve 3:326–334

Carrier H, Burt-Pichat B, Flocard F, Guffon N, Mousson B, Dumoulin R, Godinot C (1996) Molecular histology of mitochondrial and nuclear transcripts in the muscle of patients harboring a single mitochondrial DNA deletion. Acta Neuropathol (Berl) 91:104–111

Chou SM (1972) Megaconial mitochondria observed in a case of chronic polymyositis. Acta Neuropathol (Berl) 12:68–89

Collins S, Rudduck C, Marsudi S, Dennett X, Byrne E (1991) Mitochondrial genome distribution in histochemically cytochrome oxidase negative muscle fibers in patients with a mixture of deleted and wild-type mitochondrial DNA. Biochim Biophys Acta 1097:309–317

Conjard A, Martin M, Ferrier B, Durozard D, Carrier H, Baverel G (1997) Increase in oxidative key enzymes in a case of muscle ubiquinol-cytochrome c reductase deficiency. Acta Neuropathol (Berl) 93:592–598

Dalakas MC, Leon-Monzon ME, Bernardini I, Gahl WA, Jay CA (1994) Zidovudine-induced mitochondrial myopathy is associated with muscle carnitine deficiency and lipid storage. Ann Neurol 35:482–487

Di Mauro S, Moraes CT (1993) Mitochondrial encephalomyopathies. Arch Neurol 50:1197–1208

Di Mauro S, Mendel JR, Sahenk Z, Bachman D, Scarpa A, Scofield RM, Reiner C (1980) Fatal infantile mitochondrial myopathy and renal dysfunction due to cytochrome c oxidase deficiency. Neurology 30:795–804

Dubowitz V (1989) Muscle biopsy: a practical approach. Baillière Tindall, London
Engel WK, Cunningham GG (1963) Rapid examination of muscle tissue: an improved trichrome method for rapid diagnosis of muscle biopsy fresh-frozen section. Neurology 13:919–923
Gruner JE (1963) Sur quelques anomalies mitochondriales observées au cours d'affections musculaires variées. C R Soc Biol Paris 157:181–182
Guth L, Smaha FJ (1969) Qualitative differences between actomyosin ATPase of slow and fast mammalian muscles. Exp Neurol 25:138–152
Hammans SR, Sweeney MG, Wicks DA, Morgan-Hughes JA, Harding AE (1992) A molecular genetic study of focal histochemical defects in mitochondrial encephalomyopathies. Brain 115:343–365
Haraguchi Y, Chung AB, Neil S, Wallace DC (1994) Transcriptional control of nuclear genes for the mitochondrial muscle ADP/ATP translocator and the ATP synthase β subunit. J Biol Chem 269:9330–9334
Heddi A, Lestienne P, Wallace DC, Stepien G (1993) Mitochondrial DNA expression in mitochondrial myopathies and coordinated expression of nuclear genes involved in ATP production. J Biol Chem 268:12156–12163
Luft R, Ikkos D, Palmieri G, Ernster L, Afzelius BA (1962) A case of severe hypermetabolism of nonthyroid origin with a defect in the maintenance of mitochondrial respiratory control – a correlated clinical, biochemical and morphological study. J Clin Invest 41:1776
Matsuaka T, Goto Y, Yoneda M, Nonaka I (1991) Muscle histopathology in myoclonus epilepsy with ragged red fibers (MERRF). J Neurol Sci 106:193–198
Meijer AEFH, Vloedman AHT (1980) The histochemical characterization of the coupling state of skeletal muscle mitochondria. Histochemistry 69:217–232
Mita S, Schmidt B, Schon E, Di Mauro S, Bonilla E (1989) Detection of "deleted" mitochondrial genomes in cytochrome c oxidase deficient muscle fibers of a patient with Kearns-Sayre syndrome. Proc Natl Acad Sci USA 86:9509–9513
Moraes CT, Andreetta F, Bonilla E, Shanske S, Di Mauro S, Schon EA (1991) Replication competent human mitochondrial DNA lacking the heavy strand promoter region. Mol Cell Biol 11:1631–1637
Morand P, Despert F, Carrier H, Saudubray JM, Fardeau M, Romieux B, Fauchier C, Combe P (1979) Myopathie lipidique avec cardiomyopathie sévère par déficit généralisé en carnitine. Arch Mal Cœur 5:536–544
Müller-Hocker J (1992) Mitochondria and ageing. Brain Pathol 2:149–158
Nakamura S, Sato T, Hirawake H, Kobayashi R, Fukada Y, Kawamura J, Ujike H, Horai S (1990) In situ hybridization of muscle mitochondrial mRNA in mitochondrial myopathies. Acta Neuropathol (Berl) 81:1–6
Nakase H, Moraes CT, Rizzuto R, Lombes A, Di Mauro S, Schon EA (1990) Transcription and translation of deleted mitochondrial genomes in Kearns-Sayre syndrome: implication for pathogenesis. Am J Genet 46:418–427
Ohkoshi N, Mizuzawa H, Shiraiwa N, Shoji S, Harada K (1995) Superoxide dismutase of muscle in mitochondrial encephalomyopathies. Muscle Nerve 18:1265–1271
Oldfors A, Sommerland H, Holme E, Tulinius M, Kristianss O (1989) Cytochrome c oxidase deficiency in infancy. Acta Neuropathol (Berl) 77:267–275
Possekel S, Lombes A, Ogier de Baulny H, Cheval MA, Fardeau M, Kadenbach B, Romero NB (1995) Immunohistochemical analysis of muscle cytochrome c oxidase deficiency in children. Histochemistry 103:59–68
Price HM, Gordon GB, Munsat TL, Pearson CM (1967) Myopathy with atypical mitochondria in type I skeletal muscle fibers. A histochemical and ultrastructural study. J Neuropathol Exp Neurol 26:474–497
Romero NB, Fardeau M (1993) Morphological approaches to the study of mitochondrial disorders. In: Fejermann N, Chamoles NA (eds) New trends in pediatric neurology. Elsevier, Amsterdam, pp 221–226
Romero NB, Marsac C, Fardeau M, Droste M, Schneyder B, Kadenbach B (1990) Immunohistochemical demonstration of fibre type-specific isozymes of cytochrome c oxidase in human skeletal muscle. Histochemistry 94:211–215

Romero NB, Scanove-Moran C, Lombes A, Tome F, Fardeau M (1991) Variabilité de l'expression morphologique des myopathies mitochondriales: étude comparative entre les muscles deltoïdes et quadriceps. Rev Neurol (Paris) 147:689

Romero NB, Marsac C, Paturneau-Jouas M, Ogier H, Magnier S, Fardeau M (1993) Infantile familial cardiomyopathy due to mitochondrial complexes I and IV associated deficiency. Neuromusc Disord 3(1):31–42

Romero NB, Lombes A, Touati G, Rigal O, Frachon P, Cheval MA, Giraud M, Possekel S, Fardeau M, Ogier de Baulny H (1996) Morphological studies of skeletal muscle in lactic acidosis. J Inherit Metab Dis 19:528–534

Sciaccio M, Bonilla E, Schon EA, Di Mauro S, Moreas CT (1994) Distribution of wild-type and common deletion forms of mtDNA in normal and respiration-deficient muscle fibers from patients with mitochondrial myopathy. Hum Mol Genet 3:13–19

Seibel P, Degoul F, Bonne G, Romero NB, Paturneau-Jouas M, Zeigler F, Eymard B, Fardeau M, Marsac C, Kadenbach B (1991) Genetic, biochemical and pathophysiological characterization of a familial mitochondrial encephalomyopathy (MERRF). J Neurol Sci 105:217–224

Selignman AM, Karnousky MJ, Wasserkrug H, Hanker JS (1968) Non-droplet ultrastructural demonstration of cytochrome oxidase activity with a polymerizing osmiophilic reagent, diaminobenzidine (DAB). J Cell Biol 38:1–14

Shoubridge EA, Karpati G, Hastings KEM (1990) Deletion mutants are functionally dominant over wild-type mitochondrial genomes in skeletal muscle fiber segments in mitochondrial diseases. Cell 62:43–49

Shy GM, Gonatas MK, Perez M (1966) Two childhood myopathies with abnormal mitochondria. I. Megaconial myopathy. II. Pleoconial myopathy. Brain 89:133–158

Tyni T, Majander A, Kalimo H, Rapola J, Pihko H (1996) Pathology of skeletal muscle and impaired respiratory chain function in long-chain 3-hydroxyacyl CoA dehydrogenase deficiency with the G152C mutation. Neuromusc Disord 6:327–337

Wallace DC (1995) Mitochondrial DNA variation in human evolution, degenerative disease, and aging. Am J Hum Genet 57:201–223

Wallace DC, Zheng X, Lott MT, Shoffner JM, Hodge JA, Kelly RI, Epstein CM, Hopkings LC (1988) Familial mitochondrial encephalopathy (MERRF): genetic, pathophysiological, and biochemical characterization of mitochondrial DNA disease. Cell 55:601–610

Watterburg LW, Leong JL (1960) Effect of coenzyme Q10 and menadione on succinic dehydrogenase activity as measured by tetrazolium salt reduction. J Hitochem Cytochem 8:296–300

Zeviani M, Nonaka I, Bonilla E, Okino E, Moggio M, Jones S, Di Mauro S (1985) Fatal infantile mitochondrial myopathy and renal dysfunction caused by cytochrome c oxidase deficiency: immunological studies in a new patient. Neurology 17:414–417

Zheng X, Shoffner JM, Lott MT, Voljavec AS, Krawiecki NS, Winn K, Wallace DC (1989) Evidence in a lethal infantile mitochondrial disease for a nuclear mutation affecting respiratory complexes I and IV. Neurology 39:1203–1209

CHAPTER 26

Enzymatic and Polarographic Measurements of the Respiratory Chain Complexes

26

M. Malgat, T. Letellier, G. Durrieu, and J.-P. Mazat

Contents

EMI 99-29, INSERM, Université Bordeaux 2, 146 rue Léo Saignat, 33076 Bordeaux Cedex, France

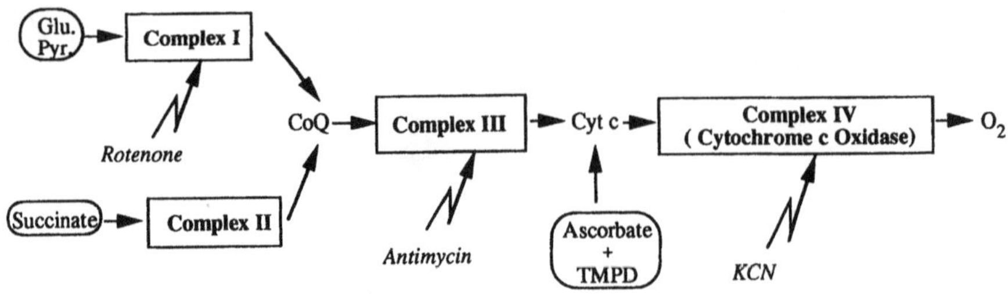

Fig. 1. Respiratory chain

1 Introduction

The different respiratory chain complexes are represented in Fig. 1. Their activity can be measured either separately or in groups, enzymatically (Sect. 3) or by the polarographic measurement of oxygen consumption (Sect. 4). Classically, these measurements are made on isolated mitochondria. The isolation of muscular mitochondria from a small quantity of tissue (around 100 mg) is difficult, so two other methods of preparation may be used: (1) the preparation of a homogenate by potterisation of a biopsy fragment (between 20–100 mg) and (2) the permeabilization of muscular fibers (10 to 100 mg). The former is particularly useful for studying enzymatic activities of isolated complexes. The latter is useful in oxygraphy instead of isolated mitochondria. These techniques are completely described in Section 2.

The determination of control values with confidence intervals is necessary to obtain a precise diagnosis. This is usually done with classical statistic methods. Section 5 presents an analysis of results by the method of principal component analysis which is particularly suited to the problem of mitochondrial pathology diagnosis.

2 Sample Preparations

2.1 Homogenate

Between 20 and 100 mg of muscular biopsy are cut into thin pieces and homogenised in a cold chamber in a potter (in glass) with an isotonic buffer to avoid the breakdown of mitochondria (225 mM mannitol, 75 mM sucrose, 10 mM Tris-HCl, 0.10 mM EDTA, pH 7.2). At this stage, the homogenate is at 10% (w/v) and is centrifuged for 20 min at 650 g. This obtains a "post-nuclear" supernatant (or S1) devoid of tissular fragments and nuclei. Protein concentrations are determined by the procedure of Lowry et al. (1951).

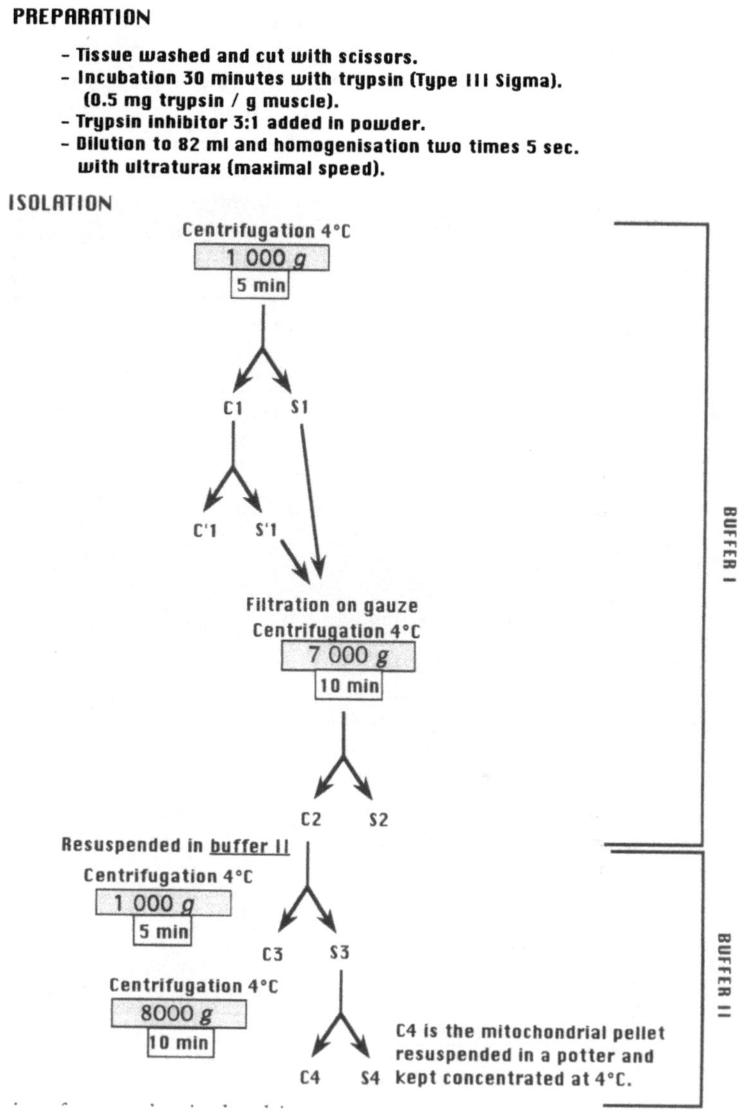

Fig. 2. Preparation of rat muscle mitochondria

2.2 Mitochondria

Muscle mitochondria are isolated by differential centrifugation as described by Morghan-Hughes et al. (1982; Fig. 2). The muscle (about 300 to 500 mg) is collected in the isolation medium I (210 mM mannitol, 70 mM sucrose, 50 mM Tris-HCl (pH 7.4), 10 mM EDTA) and digested by trypsin (0.5 mg/g of muscle) for 30 min. The reaction is stopped by addition of trypsin inhibitor (from soybean; 3 : 1 inhibitor to trypsin) and is homogenized with ultraturax. The homogenate is centrifuged at 1000 g for 5 min.

The supernatant is strained on gauze and recentrifuged at 7000 g for 10 min. The resulting pellet is resuspended in ice-cold isolation medium II (225 mM mannitol, 75 mM sucrose, 10 mM Tris-HCl (pH 7.4), 0.1 mM EDTA) and a new series of centrifugations (1000 g and 7000 g) are performed. The last mitochondrial pellet is resuspended in a minimum volume of isolation medium II in order to obtain a mitochondrial concentration between 20 and 40 mg protein/ml. Protein concentration is estimated by the biuret method using bovine albumin as standard. When there are few mitochondria, the protein concentration can be determined by measuring the fluorescence of tryptophan (excitation: 295 nm; emission: 340 nm) with a dilution of mitochondria in distilled water. The method is also rapid and sensitive (between 1 and 20 µg mitochondrial proteins). Bovine albumin is used as a standard (Dr. Velours, pers. comm.).

2.3 Permeabilised Muscle Fibers

A simple way to monitor the oxygen consumption polarographically is to permeabilize the external membrane of the muscular fibers without affecting the mitochondria. Then it is possible to work with mitochondria in situ as with isolated mitochondria. The method is described in Veksler et al. (1987) and Letellier et al. (1992). The method (Fig. 3) is the following: bundles of muscular fibers between 10 and 20 mg are incubated for 20 min in 2 ml of solution A (10 mM EGTA, 3 mM Mg^{2+}, 20 mM taurine, 0.5 mM dithiothreitol, 20 mM imidazole, 0.1 M K^{+}2-[N-morpholino]ethane sulfonic acid, pH 7.0, 5 mM ATP and 15 mM phosphocreatine) containing saponin 50 µg/ml. The bundles are then washed twice for 15 min each time in solution B (10 mM EGTA,

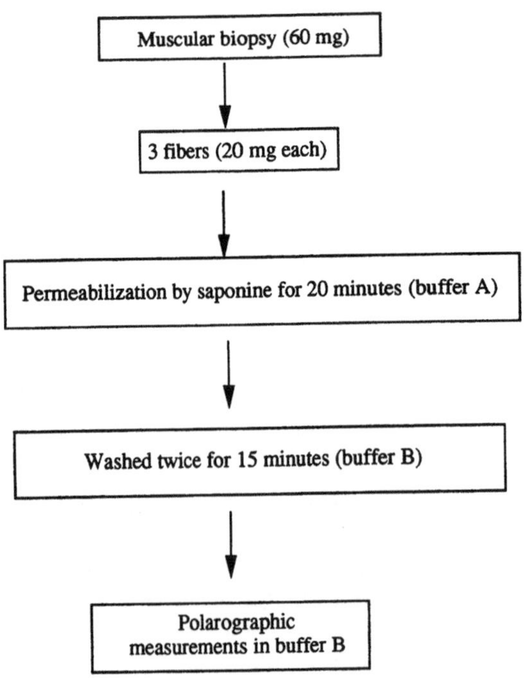

Fig. 3. Permeabilization of muscular and cardiac fibers

3 mM Mg^{2+}, 20 mM taurine, 0.5 mM DTT, 20 mM imidazole, 0.1 M K^+ MES, pH 7.0, 3 mM phosphate and 5 mg/ml fatty acid-free BSA) to remove the saponin. All procedures are carried out at 4 °C with extensive stirring.

The permeabilized fibers can be kept for at least 3 h without loss of respiratory coupling. This allows about six polarographic recordings as described below (Sect. 4).

The biopsies may be preserved at least 24 h in buffer A before utilization for homogenate or permeabilized fibers. No change in enzymatic or polarographic activities is observed after 24 h.

3 Enzymatic Determinations

3.1 Introduction

The enzymatic activities of different respiratory chain complexes are measured by observing the transfer of electrons from a natural or artificial substrate (NADH; succinate; ubiquinol; reduced cytochrome c, TMPD) to a natural or artificial respiratory chain acceptor (O_2; ubiquinone; oxidized cytochrome c). The rate of disappearance or appearance of the reduced acceptor is measured spectrophotometrically.

The results are normalized with the activities of citrate synthase, located in the mitochondrial matrix, by succinate dehydrogenase, located in the inner mitochondrial membrane, and by the cellular protein concentration.

In principle, normalization by citrate activity expresses the activities as a function of the amount of mitochondria. Succinate deshydrogenase, an enzyme involved in the respiratory chain, is also a reliable marker of the number of mitochondria, as experience shows.

Note that measurement of citrate synthase activity with or without triton X-100 (0.1%) can also be used to verify the quality of mitochondria. This activity must be zero or close to zero without triton (integrity of the inner mitochondrial membrane).

3.2 Calculation of Enzymatic Activities

The activities are expressed in nanomoles of transformed substrate (or product) (disappearing or appearing) per minute and by milligrams of proteins (nmol min^{-1} mg^{-1}). ΔOD is the variation of optical density of substrate or the product:

$$\Delta OD = \varepsilon \, \Delta c,$$

where c is expressed in mol/l and ε is expressed in $M^{-1} \cdot cm^{-1}$.

The corresponding concentration in the cuvette is:

$$\Delta c = \Delta OD/\Delta t/\varepsilon \text{ in mol } l^{-1} \text{ min}^{-1}.$$

The variation of the number N of moles is:

$$N = \Delta c \cdot V \cdot 10^{-3} = \frac{\Delta OD/\Delta t}{\varepsilon} \cdot V \cdot 10^{-3} \text{ mol min}^{-1},$$

V being the volume of the cuvette expressed in ml.

The activity/mg protein is $= \frac{N}{Q \cdot 10^{-3}}$ Q = amount protein in the cuvette (μg)

The specific activity is:

$$\frac{\Delta OD/\Delta t}{\varepsilon} \times \frac{V(\text{ml}) \times 10^{-3}}{Q(\mu\text{g}) \times 10^{-3}} = \frac{\Delta OD/\Delta t}{\varepsilon(M^{-1} \cdot \text{cm}^{-1})} \times \frac{V(\text{ml})}{Q(\mu\text{g})} \text{ mol min}^{-1} \text{ mg}^{-1} \text{ protein}$$

or in nmol/min/mg protein.

$$\text{A.S.} = \frac{\Delta OD/\Delta t}{\varepsilon(M^{-1} \cdot \text{cm}^{-1})} \times \frac{V(\text{ml})}{Q(\text{mg})} \times 10^{9} \text{ nmol min}^{-1} \text{ mg}^{-1} \text{ protein}$$

3.3 Pyruvate Dehydrogenase

3.3.1 *Principle*

An original method of pyruvate dehydrogenase (PDH) measurement was described in Chretien et al. (1995) and is based on specific inhibiton of lactate dehydrogenase by oxamate. Indeed, lactate dehydrogenase (LDH) interferes with PDH in reoxidizing $NADH_2$ produced by PDH activity. Unfortunately, due to the amount of LDH, the measurement of PDH activity is only possible for a preparation at least enriched in mitochondria, but not at all on a homogenate prepared as indicated above. However, this measurement is possible if the homogenate is prepared from permeabilized fibers already used for polarographic measurements with substrates other than pyruvate. In fact, it is probable that in the course of permeabilization, most of the lactate dehydrogenase activity leaks out of the cytosol.

3.3.2 *Procedure*

Fibers (around 30–50 mg) are put in medium A (0.3 M mannitol, 10 mM KCl, 5 mM MgCl2, 10 mM KH_2PO_4 and 0.1% triton X-100, pH 7.8) to be washed and to eliminate an excess of substrates. The fibers are centrifuged at 500 *g* for 3 min. The supernatant is eliminated and the fibers are collected in a potter with medium A at 7.5% (w/v). After homogenization, a centrifugation at 200 *g* for 10 min is performed. Protein concentration is estimated in the supernatant by the method of Lowry (1951) using bovine albumin as standard.

The pyruvate dehydroganse activity is spectrophotometrically measured in supernatant according to Chretien et al. (1995), by measuring NAD^+ reduction at 340 nm, in 1 ml of medium A in which are added: 0.5 mM NAD^+, 200 μM CoA, 1 mM cysteine and

200 μM thiamine pyrophosphate. In this medium, 30 mM oxamate are added to inhibit lactate dehydrogenase activity. The homogenate is added (around 5–10 mg fibers or 100–200 μg proteins) and is incubated for 3 min at 37 °C. The reaction is started by addition of 0.5 mM pyruvate. A linear kinetic can be monitored for 2 min. The PDH activity inhibition can be checked by adding 4 mM arsenic (III) oxide. The activity is calculated using the molar extinction coefficient: ε (NADH) = 6220 $M^{-1}cm^{-1}$ at 340 nm.

3.4 Citrate Synthase Activity (CS)

3.4.1 *Prinicple*

Citrate synthase allows the formation of citrate from oxaloacetate and acetyl coenzyme A. Its activity is monitored by coupling with the reaction between the reduced coenzyme A formed in the first reaction and DTNB [5,5'-dithio-bis (2-nitrobenzoic acid)]. DTNB reacts with 2 CoA-SH to form a disulfur group between these two molecules and releases two TNB molecules (yellow). The rate of TNB formation is proportional to the rate of the enzymatic reaction.

$$\text{2 Acetyl CoA + 2 Oxaloacetate} \xrightarrow{\text{CS}} \text{2 Citrate + 2 CoA-SH}$$
$$\text{2 CoA-SH + DTNB} \longrightarrow \text{2 TNB + CoA-S-S-CoA}$$

3.4.2 *Procedure*

Citrate synthase activity is measured by addition of triton X-100 according to Shepard and Garland (1969). The reactive medium (1 ml) includes: 2 mM Tris-HCl, pH 8.0, 0.1 mM DTNB (from a 1 mM solution in 5 mM Tris ph 8.0), 0.4 mM acetyl CoA and 5 to 20 μl of muscle homogenate (25 to 100 μg proteins) or isolated mitochondria (10 to 40 μg mitochondrial proteins) and is incubated in the presence of triton X-100 (1‰). The citrate synthase reaction is then started by addition of 0.5 mM of oxaloacetate. The absorption at 412 nm is monitored for 5 min. A blank is made without homogenate or without mitochondria. The activity is calculated using the molar extinction coefficient of TNB at 412 nm: ε (TNB) = 13600 $M^{-1}cm^{-1}$.

3.5 NADH CoQ Reductase (Complex I)

3.5.1 *Principle*

In the mitochondria, the NADH is oxidized into NAD^+ at the level of complex I. In mammals, the reaction is located on the inner side of the inner mitochondrial membrane. For this reason, to ensure maximal accessibility and fixation of the NADH, the mitochondrial membranes have to be disrupted. Therefore, the homogenate or the mitochondria are frozen and thawed twice in liquid nitrogen. The natural acceptor of complex I electrons is coenzyme Q. Experimentally, complex I can be isolated by blocking complexes III and IV with antimycin and KCN, respectively. NADH oxidation is recorded at 340 nm using the ubiquinone analogue decylubiquinone as electron acceptor. The cytochrome b_5 reductase activity of the outer membrane also oxidises NADH. Its activity is measured in the presence of rotenone, a specific inhibitor of mitochondrial complex I.

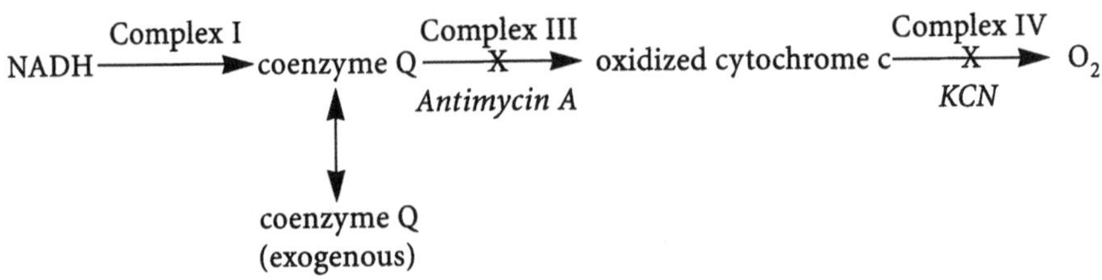

3.5.2
Procedure

The assay medium (35 mM KH_2PO_4 pH = 7.2, 5 mM $MgCl_2$, 2 mM KCN, ph 7.0) is supplemented with defatted BSA (2.5 mg/ml), antimycin (2 µg/ml), 0.1 mM decylubiquinone and 20 to 40 µl homogenate (100–200 µg proteins) or 40–80 µg mitochondrial proteins. After 5 min of incubation at 30 °C, the reaction is started with 0.14 mM NADH. The decrease in O.D. due to the oxidation of NADH is measured at 340 nm.

3.5.3
Measurement of Activity

Complex I activity is a reaction of order 1 which is represented by an exponential curve. However, the activity can be assessed by taking the initial slope (for 1 min). Complex I activity is determined by calculating the difference of the activities in the presence and absence of rotenone (20 µg/ml). The molar extinction coefficient of NADH at 340 nm is: ε (NADH) = 6220 $M^{-1}cm^{-1}$.

3.6
NADH Cytochrome c Reductase (Complex I + III) (Chretien et al. 1990)

3.6.1
Principle

NADH cytochrome c reductase activity is measured by using NADH as the substrate and oxidized cytochrome c as the electron acceptor. Two enzymatic activities participate in this activity:

1. Cytochrome b_5 oxydo-reductase of the outer membrane, the activity of which is insensitive to rotenone.
2. The succession of the respiratory chain complexes I and III; this activity is inhibited by rotenone and antimycin.

The total activity is measured by observing the reduction of cytochrome c at 550 nm.

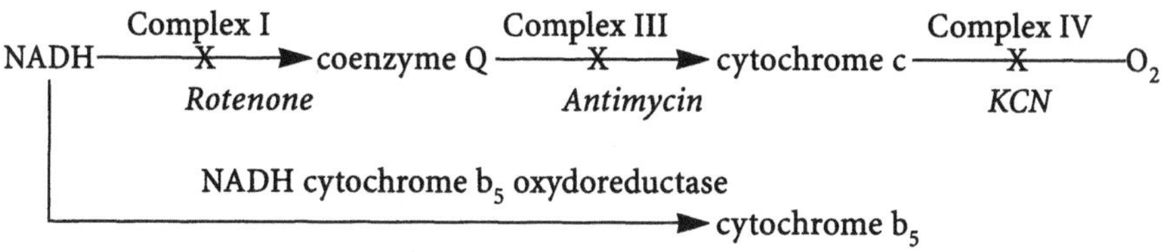

3.6.2
Procedure

The reactional medium includes: 25 mM KH_2PO_4, pH = 7.4, 1 mM KCN, 10 µM EDTA, 0.1 mM oxidized cytochrome c and 15–30 µl homogenate (75–150 µg proteins) or 15–30 µg mitochondrial proteins.

The absorbance is measured after 5 min at 37 °C. The reaction is started by 0.1 mM NADH.

3.6.3
Measurement of Activity

The enzymatic activity of complexes I and III is calculated by the difference of activities in the presence or absence of rotenone, or with or without antimycin, for 1 min. The molar extinction coefficient of cytochrome c at 550 nm is: ε (cyto c) = 18500 $M^{-1}cm^{-1}$.

3.7
Succinate Dehydrogenase (Birch-Machin et al. 1989)

3.7.1
Principle

Succinate deshydrogenase activity, a component of complex II, is measured at 600 nm by observing the reduction of 2,6-dichlorophenol-indophenol (DCIP) in the presence of phenazine methosulfate (PMS). The electron donor is the succinate.

3.7.2
Procedure

The reactional medium includes: 50 mM KH_2PO_4, pH = 7.4, triton X-100 1‰, 3 mM KCN, rotenone (20 μg/ml). Homogenate (10–20 μl) is added (50–100 μg of proteins) or 10–20 μg of mitochondrial proteins and incubated for 10 min at 37 °C. The medium is complemented with 0.18 mM DCIP and 1.3 mM PMS. The reaction is started by 20 mM succinate.

3.7.3
Measurement of Activity

The activity is measured at 600 nm at 37 °C for 1 min. The molar extinction coefficient is: ε (DCIP) = 21000 $M^{-1}cm^{-1}$.

3.8
Succinate Cytochrome c Reductase (Complex II + III)

3.8.1
Principle

The activity of complex (II + III) is measured by observing the reduction of cytochrome c at 550 nm according to the following scheme:

$$\text{Succinate} \xrightarrow[\text{Complex II}]{} \text{coenzyme Q} \xrightarrow[\text{Complex III}]{} \text{oxidized cytochrome c} \xrightarrow[\text{Complex IV}]{\textit{KCN} \; \times} O_2$$

3.8.2
Procedure

The reactional medium includes: 25 mM KH_2PO_4, pH 7.4, 1 mM KCN, 0.1 mM oxidized cytochrome c; then 15 to 30 μl of homogenate (75 to 150 μg of proteins) are added, or 15 to 30 μg of mitochondrial proteins. After stabilization, monitored by absorbance at 37 °C, the reaction is started by addition of 5 mM succinate. Enzymatic activity is measured for 10 min at 550 nm, using the maximal slope at the inflection point with ε (cyto c) = 18500 $M^{-1}cm^{-1}$.

Note that in muscle, the activity of complex (II + III) is weakly sensitive to a complex III deficit (Taylor et al. 1993). This activity is appreciable under these conditions only for a complex III deficit exceeding 40%. Thus, a moderate deficit in complex III will only clearly appear on the assay of enzymatic activity of complex III itself.

3.9
Ubiquinol Cytochrome c Reductase (Complex III)

3.9.1
Principle

Complex III (ubiquinol: cytochrome c oxidoreductase) catalyses the transfer of electrons from reduced ubiquinone (ubiquinol) to cytochrome c. Experimentally, complex I is inhibited by rotenone, complex IV by the KCN; an analogue of ubiquinol, decylubiquinol, is used as complex III substrate. Complex III activity is measured by observing the reduction of cytochrome c at 550 nm. Complex III activity is determined by calculating the difference of activities measured in the absence and presence of antimycin.

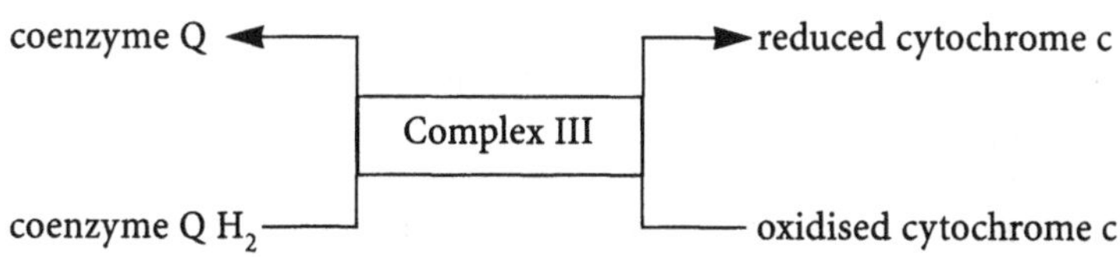

3.9.2
Preparation of Ubiquinol

The preparation of decylubiquinol plus reductor agent is an important step for measuring complex III activity. It is necessary to eliminate any excess of reductor agent which otherwise would interfere with the assay. Several methods have been described in the literature and two are described here. Decylubiquinol (reduced) is identified by its absorption peak at 285 nm and decylubiquinone (oxidized) by a peak at 275 nm.

3.9.2.1
Method 1 (modified from Birch-Machin et al. 1993)

One ml of 6.5 mM decylubiquinone (Sigma D-7911) in ethanol is prepared. The absorption spectrum of a dilution of this solution of decylubiquinone at 275 nm (10 µl/ml ethanol) is recorded. The initial solution (6.5 mM) is reduced by an excess of sodium borohydride (Sigma S-9125). Ten µl concentrated HCl are added and the solution is shaken until it becomes colorless. The absorbance spectrum of reduced CoQ shows a peak at 285 nm. The reduced solution is mixed in 5 ml of buffer (0.1 M KH_2PO_4, 0.25 M sorbitol, pH 7.4) plus 3 ml cyclohexane, vigorously shaken and centrifuged for 5 min at 1000 *g*. The organic phase is then extracted. Several extractions with cyclohexane are performed. After that, all organic phases are evaporated together under a stream of nitrogen. The residue is resuspended in 1 ml cyclohexane and distributed in 50 µl aliquots in Eppendorfs. From one aliquot, the absorbance spectrum is recorded at 285 nm to calculate the decylubiquinol concentration. The other microtubes are then evaporated under a stream of nitrogen and conserved at –80 °C. Before use, a volume of ethanol (approximately 50 µl) is calculated to bring the decylubiquinol concentration of one microtube to 6.5 mM.

3.9.2.2
Method 2 (Rieske 1967)

A concentrated solution of natrium dithionite is prepared (1 g/ml in bidistilled water is heated at 80 °C, until complete dissolution of dithionite). A 15-µl solution of dithionite is added to 88 µl of a solution of 25 mM decylubiquinone in DMSO, shaken on a vortex and put in a water-bath at 37 °C. After 15–20 min, the solution becomes white or slightly yellowish, showing the reduction from decylubiquinone to decylubiquinol. This solution is at 21.4 mM decylubiquinol and can be used as such in the assay of complex III.

3.9.3
Procedure

The reaction medium includes: 35 mM KH_2PO_4 (pH 7.2), 5 mM $MgCl_2$, BSA (2.5 mg/ml), 0.2 mM KCN, 20 µM oxidized cytochrome c, rotenone (10 µg/ml) and 10 to 30 µl of homogenate (50 to 150 µg of proteins) or 15 to 30 µg of mitochondrial proteins. The reaction is stabilized for 5 min at 30 °C. The reaction is started by addition of 20 µM decylubiquinol. The enzymatic activity is recorded by monitoring the reduction of cytochrome c at 550 nm for 2 min.

3.9.4
Measurement of the Activity

The complex III activity is estimated by calculating the difference of the activities measured in the absence and presence of antimycin (10 µg/ml). The molar extinction coefficient of reduced cytochrome c is: ε (cyto c) = 18500 $M^{-1}cm^{-1}$.

3.10
Cytochrome c Oxidase (Complex IV)

3.10.1
Principle

The cytochrome c oxidase activity is estimated by recording the oxidation of reduced cytochrome c at 550 nm. The complex IV activity is determined by the difference of activities measured in the absence and presence of KCN.

$$\text{reduced cytochrome c} + \frac{1}{2}O_2 + 2H^+ \xrightarrow{\text{Complex IV}} \text{oxidized cytochrome c} + H_2O$$

3.10.2
Preparation of Reduced Cytochrome c

As for the complex III assay, the preparation of the reduced substrate (here the reduced cytochrome c) is important. The 1 mM solution of oxidized cytochrome c is diluted to 50 μM (10 ml) in 25 mM KH_2PO_4, pH 7.4. One ml of this solution is reduced by a light excess of natrium dithionite and 1 ml is completely oxidized by an excess of potassium ferricyanure to serve as a reference. The value of the OD ratio (reduced cytochrome c) / (oxidized cytochrome c) at 550 nm is around 0.7. Then, small quantities of the completely reduced solutions are added to the 8 ml of the initial 50 μM oxidized solution until 80–90% reduction is reached (checked spectrophotometrically). This ensures that in the solution there is no excess dithionite which could interfere with the reaction itself.

3.10.3
Procedure

The assay is undertaken from 1 ml of reduced cytochrome c to which 5 to 10 μl of homogenate (25 to 50 μg of proteins) or 5 to 10 μg of mitochondrial proteins are added to start the reaction. The oxidation of the cytochrome c is monitored at 550 nm at 37 °C for 1 min.

3.10.4
Measurement of Activity

The kinetic of cytochrome c is of order 1 under these conditions. However, the kinetic is linear for 1 min providing that the amounts in homogenate or mitochondria are those indicated above. For the calculation of the activity: ε (cyt c) = 18500 $M^{-1}cm^{-1}$.

Table 1. Respiratory chain activities in human muscle homogenate, prepared as described in the text and expressed in nmol min^{-1} mg^{-1} proteins

	Children (quadriceps) (n = 18)	Adults (deltoid) (n = 30)
Citrate synthetase	170 ± 75	197 ± 60
NADH ubiquinone reductase (I)	24 ± 8	38 ± 16
Ubiquinol cytochrome c reductase (III)	84 ± 28	43 ± 20
NADH cytochrome c reductase (I + III)	71 ± 25	70 ± 40
Succinate cytochrome c reductase (II + III)	46 ± 17	79 ± 33
Cytochrome c oxidase (IV)	263 ± 111	226 ± 77
Succinate dehydrogenase (II)	47 ± 15	73 ± 19

3.11 Conclusion

The values obtained in the conditions described above on homogenates from biopsies of child (quadriceps) or adult (deltoid) are summarized in Table 1. Other values are found in Zheng et al. (1990) and Rustin et al. (1994).

4 Polarographic Studies of the Respiratory Chain

Polarographic measurements are useful in controlling the effect of complex deficiencies on oxygen consumption and thus on ATP synthesis. They can also be used to check the coupling between respiration and phosphorylation.

4.1 Principle

Polarographic assays are based on the utilization of an oxygen electrode or Clark electrode (Hansatech). This is constituted by a silver anode and a platinum cathode electrically connected by a solution of saturated or half-saturated KCl and covered by a membrane of teflon permeable to oxygen. The platinum electrode is polarized around −0.8 V with respect to the silver electrode and the current necessary to maintain this difference of potential is, in this range, linearly proportional to the oxygen concentration in the cuvette.

One of the important points with an oxygen electrode is the calibration procedure. This calibration includes the adjustment of 100% (quantity of oxygen dissolved in the buffer) and of zero. The determination of the quantity of oxygen dissolved in the buffer has been described by Hinkle (1995). A value of 480 μM atoms of oxygen (O) or 240 μM O_2, is classically found at 25 °C. At 30 °C, a value close to 450 μM atoms of oxygen is preferred (the concentration in oxygen dissolved in the buffer decreases with the temperature). The adjustment of the zero is carried out with crystals of natrium dithionite ($Na_2S_2O_4$) and makes it possible to adjust the zero in oxygen concentration with the zero of the recorder. Under these conditions, the respiratory rate is expressed in nano-atoms of oxygen/minute/mg of proteins or muscular fibers.

Moreover, a Clark electrode consumes oxygen, and the cuvette and the magnetic stirrer contain oxygen which will be liberated when the oxygen concentration in the buffer decreases. In polarographic assays, particularly when the rate of oxygen consumption being recorded is low, which is often the case when there are deficits in oxidative phosphorylation, it is important to:

1. Close the cuvette (to avoid reoxygenation which might compensate the oxygen consumption which is to be measured).
2. Work between 80 and 100% of oxygen saturation in the buffer in the cuvette, to ensure a low rate of reoxygenation of the buffer.

4.2
Procedure for Isolated Mitochondria

Between 0.25 and 1 mg of mitochondria (the quantity is expressed in mitochondrial proteins) are incubated in 1 ml respiratory buffer at 30 °C including: 0.5 mg/ml BSA, 75 mM mannitol, 25 mM saccharose, 100 mM KCl, 50 µM EDTA, 10 mM Tris/phosphate and 10 mM Tris/HCl pH 7.4. Several respiratory substrates can be used: glutamate (10 mM), pyruvate (10 mM), α cetoglutarate (50 mM) or palmitylcarnitine (20 µM), these four substrates being added in the presence of malate (5 mM), succinate (25 mM) (+ rotenone to avoid the reverse electron pathway), 5 mM ascorbate + 0.5 mM TMPD (tetramethyl-p-phenylenediamine). Three hundred nmol ADP are added for respirations involving complex I, 200 nmol when the respiration is started on succinate, and 100 nmol when the respiration is started with ascorbate + TMPD and involves only complex IV (a coupling is always difficult to see in this case).

The inhibitors are used at the following concentrations: rotenone (0.4 µg/ml), antimycin 5–10 µg/ml), KCN (1 mM), oligomycin 1 µM), atractyloside (100 µM). The uncouplers are CCCP (carbonyl cyanide m-chlorophenylhydrazone) (1 to 2 µM) and 2–4 DNP (2–4 dinitrophenol) (1 mM). Higher concentrations of these uncouplers can have inhibiting effects.

As indicated in Fig. 4, it is possible, in one step to measure the activity of all respiratory chain complexes: with pyruvate or glutamate for example as the first substrate, the activity of complexes I + III + IV is measured. Then with rotenone and succinate added, the activity of complexes II + III + IV is measured. Finally, in the presence of antimycin and ascorbate + TMPD as final respiratory substrate, the complex IV activity is determined. A complex IV deficit is easy to see with this method because it will give weak respirations on all substrates. A complex III deficit should give weak respirations on succinate and on complex I substrates (see Fig. 1). Finally, a complex I deficit will give a weak respiration on pyruvate and glutamate, for example, and a normal respiratory rate on succinate. Similarly, a complex II deficit will give a weak respiration on succinate and normal respirations on the other substrates.

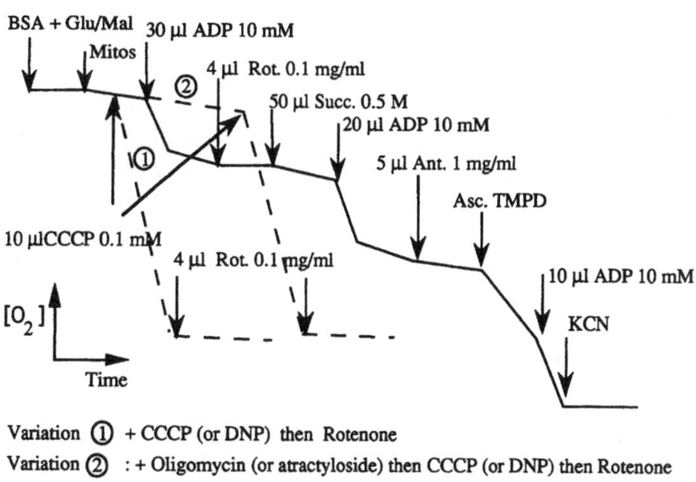

Fig. 4. Polarographic study of isolated mitochondria

Nevertheless, compensations may occur when the activity of several substrates is determined at the same time. We have shown the relative insensitivity of respiratory rate to deficits (up to 50–70%) of different respiratory chain complexes (Letellier et al. 1993, Malgat et al. 1995). For instance, a normal respiration can mask a deficit of 50% in cytochrome c oxidase. In Section 5, we will show that oxygen consumption measurements are a weakly discriminant factor in the analysis of respiratory complex deficiencies. However, the measurement of mitochondrial respiration is an easy way to assess the coupling between phosphorylation and oxygen consumption, revealing uncoupling (as in Luft disease) or phosphorylation deficiencies. It is also the simplest way to display limiting amounts of CoQ and cytochrome c, and deficiencies upstream in the respiratory chain (pyruvate dehydrogenase, α-ketoglutarate dehydrogenase).

4.3 Calculation of Activities

In the study of the mitochondrial consumption of oxygen, several states may be considered according to whether the mitochondria synthesize ATP or not. In the absence of ATP synthesis (no ADP added), the respiration is in state 4, and is characterized by a weak oxygen consumption just necessary to maintain the electrochemical gradient of protons, in compensating the membrane proton leak. In the presence of ADP, the mitochondria synthesize ATP by consuming the proton gradient which is maintained at a weaker value (between –120 and –150 mV) by an acceleration of the respiratory chain: this is state 3.

The rate of respiration at state 4 (absence of phosphorylation) and state 3 (maximal phosphorylation) is expressed in nanoatoms of oxygen/minute/mg of mitochondrial proteins. Normal values in states 3 and 4 can be found in Chretien et al. (1990) and Rustin et al. (1994). The respiratory control: V3/V4 and the ATP/O are calculated as indicated in Fig. 5.

4.4 Protocol of Oxygraphic Determinations of Permeabilized Fibers

The protocol for oxygraphic measurements with permeabilized muscular fibers is similar to that used for isolated mitochondria. The amount of fibers is around 5 to 20 mg (wet weight). A typical recording is shown in Fig. 6. As for the isolated mitochondria, the different respiratory chain complexes can be tested, by acting upon the different substrates and inhibitors. The results obtained are summarized in Table 2. The respiratory buffer used in this case is buffer B used for fiber washing. After measure-

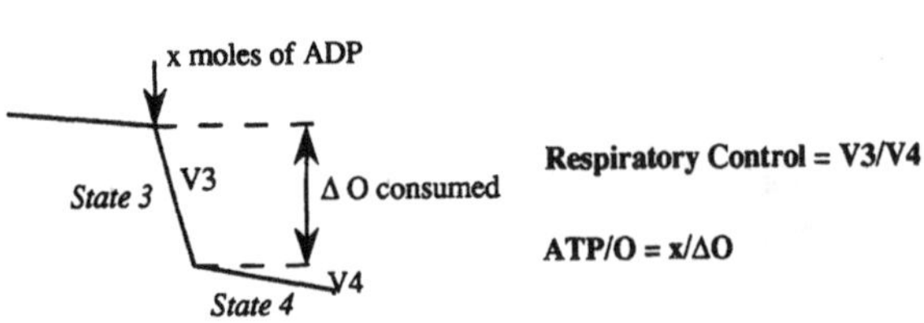

Fig. 5. Calculation of respiratory control and ATP/O

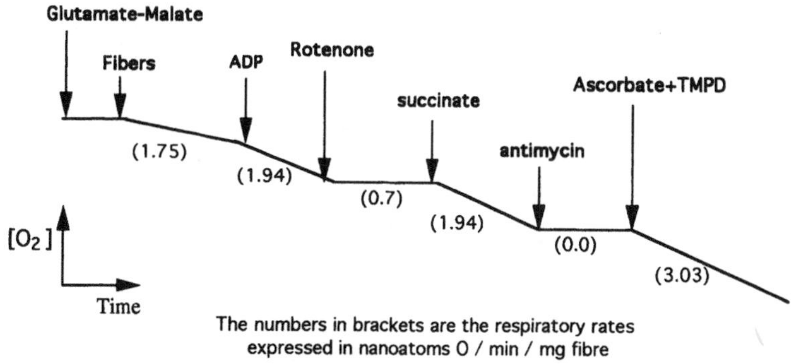

Fig. 6. Polarographic studies of permeabilized fibers

Table 2. Mean volume and standard deviation of respiratory rates of permeabilized fibers on different substrates

	I + III + IV[a]	II + III + IV[b]	IV[c]	RC[d]
Children (n = 18)	1.79 ± 0.45	2.81 ± 0.87	4.11 ± 1.04	2.57 ± 0.17
Adults (n = 30)	1.53 ± 0.44	2.29 ± 0.88	3.76 ± 0.95	3.39 ± 0.88

Rates are expressed in nanoatoms of oxygen per min and per mg of muscular fibers (wet weight).

[a] I + III + IV corresponds to a respiration on pyruvate or glutamate (in the presence of malate).
[b] II + III + IV corresponds to a respiration on succinate (+ rotenone).
[c] IV corresponds to a respiration on ascorbate + TMPD.
[d] RC = respiratory control, is the ratio between the maximal rate of phosphorylating respiration (+ ADP) and the rate with inhibition of phosphorylation in the presence of oligomycin and atractyloside. In this case, succinate (plus rotenone) is used as respiratory substrate.

ment, fibers are dried with blotting paper and weighed. They are conserved for further studies in molecular biology or for measurement of PDH activity.

The same protocol can be used for the measurement of oxygen consumption in cell cultures after permeabilization (fibroblasts, lymphoblasts).

5 Statistical Studies

Analysing results of biochemical studies of muscular biopsies is often difficult and does not clearly indicate the presence of a deficit in one or several complexes of oxidative phosphorylation. These difficulties are due to:

1. The absence of a true control population; control values are in fact determined from the whole set of data collected from a population, a priori suspected of mitochondrial pathology.
2. The small size of the control population, extracted from the whole population, and its unknown distribution, often non-Gaussian, involving a large variability with extreme values.

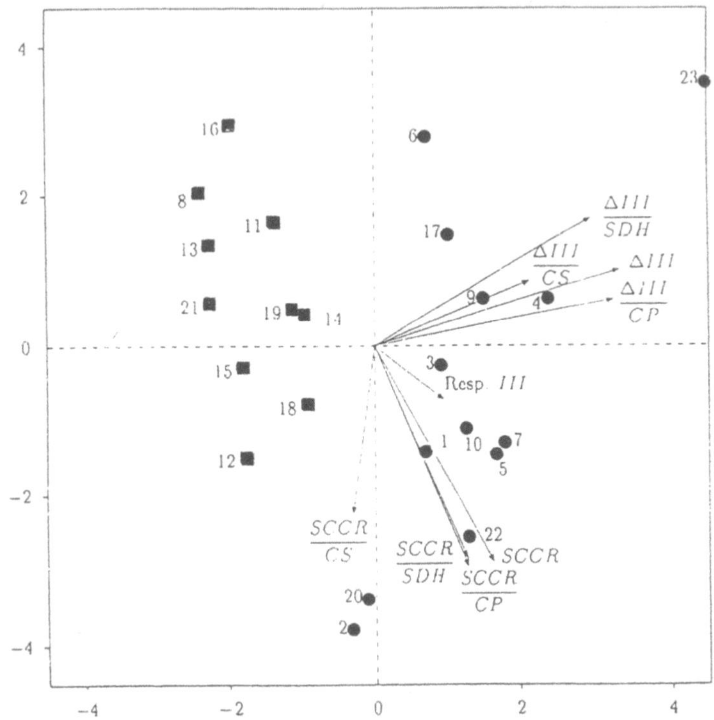

Fig. 7. PCA results for Complex III, ■ Complex III deficiency, ● normal complex III

For these reasons it is difficult to construct a confidence interval using classical methods.

Moreover, the large number of measurements obtained for each biopsy, including about 4 enzymatic and polarographic assays (homogenate of muscle and permeabilized muscular fibers), makes the global interpretation of the results difficult.

For this type of data, principal component analysis (PCA) is used to account for the results for each biopsy. The aim of this method is to reduce the number of measured variables for each complex of the respiratory chain of each patient, but to conserve the maximum amount of information contained in the original variables. From a statistical point of view, PCA involves the search for new independent variables (two in our case, the two first principal components) which are linear combinations of the old variables, with maximal variances, and which retain maximum information. Results concerning each biopsy may be represented in the plane of the first two principal components (main plane, with the first principal component along the abscissa, and the second along the ordinate). The result of such a study is shown in Fig. 7. Each point or square represents a biopsy in the plan of the two principal components. This representation makes it possible to isolate clearly two types of populations, control (circles) and pathological (squares), for complexes I, III and IV of the respiratory chain (the case of complex III is represented in Fig. 7). Pathological values are found in the same left-half plan: consequently, the x-axis discriminates for the deficit. Thus, the more negative the value of the first principal component, the greater is the deficit.

On Fig. 7, the arrows represent the original variables. This representation demonstrates the measured variables that will be the most discriminating for the determination of a deficit. The orthogonal projection of each vector (corresponding to each assay) on the x-axis (axis that discriminates the deficit in our study) gives information on its discriminating power: the higher this projection, the more discriminating the variable for the diagnosis. The analysis of our results shows that the polarographic study involving several complexes is not very discriminating. This is entirely in agreement with the existence of a threshold which makes oxygen consumption only sensitive to a large deficit of a complex (Letellier et al. 1993). Figure 7 also shows the weak discriminating power of the succinate cytochrome c reductase activity for revealing a complex III deficit. This has already been described by Turnbull's group with another approach (R. W. Taylor et al., 1993; see discussion above).

This method of analyzing biochemical results is of great help for the diagnosis of mitochondrial cytopathies.

6 Conclusion

Which Assays to Perform? It is always a problem knowing which assays, polarographic or enzymatic, to do first. We have seen that enzymatic assays are more discriminating than polarographic assays in the case of complex deficiencies. They have to be done in priority. Moreover, it is impossible to do a polarographic study on a frozen biopsy. Nevertheless, the biopsy can be preserved in buffer A for at least 24 h in order to do an enzymatic analysis on a homogenate or polarographic analysis on permeabilized fibers. The polarographic study gives information on the expression of the deficit at the level of the oxygen consumption flux, on defects in coenzyme Q and cytochrome c and on uncoupling of oxidative phosphorylation. Thus, both approaches give different information and are therefore instructive.

Normalization of Results. It is important to clearly establish a deficit. The normalization, compared with the quantity of proteins of a homogenate, seems to provide a correct indication of a deficit (if any). It gives activity of complexes of the respiratory chain with respect to the cellular metabolism in its totality. The normalization with regard to citrate synthase or succinate deshydrogenase also provides an acceptable normalization when these activities are normal; they show the quantity of complexes of the respiratory chain, compared with the mitochondrial metabolism represented by a matricial activity (citrate synthase) or membranous activity (succinate deshydrogenase). These three normalizations will be therefore calculated if possible and compared.

The ratios between the activities of different complexes of respiratory chains are also used to better highlight the deficits (Rustin et al. 1994, Chretien et al. 1994).

Statistical Analysis. The statistical analysis of the results is difficult. Classic methods give very large confidence intervals which sometimes makes diagnosis difficult. PCA takes into account all the results and allows an easier diagnosis.

Permeabilized Fibers. Finally, the method of muscle fiber permeabilization (and of cells in culture) makes it possible to do reliable polarographic studies on minimal quantities of tissue (between 10 and 100 mg). Moreover, this method is rapid and simple to use and takes into account all the mitochondria contained in a muscular fiber, which is not necessarily the case when the mitochondria are isolated, because the mitochondria population may be heterogeneous.

Acknowledgements. The authors thank Drs. F. Sztark and A. Lombès for their critical reading of the manuscript. Mr. Ray Cook for correcting the English and Mrs. I Brault for her highly professional assistance in typing the manuscript. This work was supported by the Association Française contre les Myopathies (AFM), the Université Victor Segalen, Bordeaux 2, INSERM the Région Aquitaine and the "Pôle Médicament Aquitaine" (PMA).

References

Birch-Machin MA, Shepperd IM, Watmough NJ, Sherratt HSA, Barlett K, Darley-Usmar VM, Milligan DWA, Welch RJ, Aynsley-Green A, Turnbull DM (1989) Fatal lactic acidosis in infancy with a defect of complex III of the respiratory chain. Pediatr Res 25 (5):553–559

Birch-Machin MA, Jackson S, Kler RS, Turnbull DM (1993) Study of skeletal muscle mitochondrial dysfunction. Methods Toxicol 2:51–69

Chretien D, Bourgeron T, Rötig A, Munnich A, Rustin P (1990) The measurement of the rotenone-sensitive NADH cytochrome c reductase activity in mitochondria isolated from minute amounts of human skeletal muscle. Biochem Biophys Res Commun 173:26–33

Chretien D, Rustin P, Bourgeron T, Rötig A, Saudubray JM, Munnich A (1994) Reference charts for respiratory chain activities in human tissues. Clin Chim Acta 228:53–70

Chretien D, Pourrier M, Bourgeron T, Sene M, Rötig A, Munnich A, Rustin P (1995) An improved spectrophotometric assay of pyruvate dehydrogenase in lactate dehydrogenase contaminated mitochondrial preparations from human skeletal muscle. Clin Chim Acta 240:129–136

Hinkle PC (1995) Oxygen, proton and phosphate fluxes, and stoichiometries. In: Brown GC, Cooper CE (eds) Bioenergetics, A practical approach. IRL Press, Oxford, pp 1–16

Letellier T, Malgat M, Coquet M, Moretto B, Parrot-Roulaud F, Mazat J-P (1992) Mitochondrial myopathy studies on permeabilized muscle fibres. Pediatr Res 32:17–22

Letellier T, Malgat M, Mazat J-P (1993) Control of oxidative phosphorylation in rat muscle mitochondria: implications for mitochondrial myopathies. Biochim Biophys Acta 1141:58–64

Lowry OH, Rosebrough NJ, Farr AL, Randall RJ (1951) Protein measurement with the Folin phenol reagent. J Biol Chem 193:265–275

Malgat M, Letellier T, Jouaville SL, Mazat J-P (1995) Value of control theory in the study of cellular metabolism – biomedical implications. J Biol Syst 3 (1):165–175

Morgan-Hughes JA, Hayes DJ, Clark JB, Landon DN, Swash M, Strak RJ, Rudge P (1982) Mitochondrial encephalomyopathies. Biochemical studies in two cases revealing defects in the respiratory chain. Brain 105:553–582

Ouhabi R, Boué-Grabot M, Mazat JP (1998) Mitochondrial ATP synthesis in permeabilized cells: assessment of the ATP/O values in situ. Anal Biochem 263:169–175

Rieske JS (1967) Preparation and properties of reduced coenzyme Q-cytochrome c reductase (complex III of the respiratory chain). Methods Enzymol 10:239–245

Rustin P, Chretien D, Bourgeron T, Gérard B, Rötig A, Saudubray JM, Munnich A (1994) Biochemical and molecular investigations in respiratory chain deficiencies. Clin Chim Acta 228:35–51

Shepard D, Garland PB (1969) Citrate synthase from rat liver. Methods Enzymology 13:11–16
Taylor RW, Birch-Machin MA, Barlett K, Turnbull DM (1993) Succinate-cytochrome c reductase: assessment of its value in the investigation of defects of the respiratory chain. Biochim Biophys Acta 1181:261–265
Trounce IA, Kim YL, Jun AS, Wallace DC (1996) Assessment of mitochondrial oxidative phosphorylation in patient muscle biopsies, lymphoblasts, and transmitochondrial cell lines. Methods Enzymol 264:484–509
Veksler VI, Kuznetsov AV, Sharov VG, Kapelko VI, Saks VA (1997) Mitochondrial respiratory parameters in cardiac tissue: a novel method of assessment by using saponin-skinned fibres. Biochim Biophys Acta 892:191–196
Zheng X, Shoffner JM, Voljavec AS, Wallace DC (1990) Evaluation of procedures for assaying oxidative phosphorylation enzyme activities in mitochondrial myopathy muscle biopsies. Biochim Biophys Acta 1019:1–10

CHAPTER 27

Mitochondrial DNA Analysis

27

P. Reynier[1], Y. Malthièry[1], and P. Lestienne[2]

Contents

1 Introduction

The molecular analysis of the mitochondrial genome calls for the same methodologies as those used in the exploration of the nuclear genome. However, the singularity of mitochondrial genetics implies a different interpretation of the relations between genotype and phenotype.

[1] Laboratoire de Biochimie et Biologie Moléculaire A, Centre Hospitalier Régional Universitaire, 49033 Angers, France
[2] E 99-29 INSERM physiologie mitochondriale, Université Bordeaux 2, 146 rue Léo Saignat, 33076 Bordeaux Cedex, France

2
Strategic Exploration of mtDNA Diseases

2.1
Leber's hereditary Optic Neuropathy

The mitochondrial genome can be searched for all the primary and intermediate mutations (mutations which are associated with the pathology on their own) implicated in this disease. The following have been found: G11778A (50 to 70% of European patients), G3460A (15 to 30%), T14484C (10 to 15%), G15257A (5 to 10%) and T4160C (more rare).

On the other hand, 6 mutations among the 12 secondary mutations which are mainly described (mutations which either, when several of them are found together, are able to bring about the illness, or are able to worsen the pathology provoked by a primary or intermediate mutation) are analysed: T4216C, A4917G, G5244A, G7444A, G13708A and G15812A.

2.2
Syndromes with mtDNA Point Mutations

This type of exploration concerns MERRF (myoclonic epilepsy and ragged red fibers), MELAS (mitochondrial encephalomyopathy, lactic acidosis, and stroke-like episodes), NARP (neurogenic muscle weakness, ataxia, and retinis pigmentosa), Leigh's disease and myopathies and mitochondrial cardiomyopathies (Lestienne 1992). The following genes have been implicated:

1 The leucine tRNA gene (UUR) where more than ten pathogenic mutations have been described (A3243G, T3250C, A3251G, A3252G, C3256T, A3260G, T3271C, T3291C, A3302G, C3303T).
2. The lysine tRNA gene (A83446G and T8356C mutations).
3. Sequencing of the area surrounding the T8993G/C mutation in the ATPase 6 gene.
4. The isoleucine tRNA gene (A4269G and A4317G mutations).
5. Threonine and proline tRNA genes (A15923G, A15924G, and C15990T mutations).

2.3
Syndromes with mtDNA Rearrangements

Chronic progressive external ophthalmoplegia (CPEO), Kearns-Sayre syndrome and Pearson's disease are due to mtDNA deletions and/or duplications (Lestienne 1992). Thus we can look for the presence of these rearrangements by using Southern blot and Long PCR. These two techniques are complementary. Long PCR has the advantage of detecting deletions with very low rates, it is particularly suited to the study of multiple deletions (Reynier and Malthiéry 1995). Southern blot allows a better visualization of the duplications and permits a semiquantitative approach towards the heteroplasmy rate. This technique allows us to confirm the presence of multiple deletions when their proportion is sufficient (Lestienne et al. 1997). The joint use of an mtDNA probe and of a ribosomal nuclear DNA probe allows a quantitative evaluation of the amount of mitochondrial genome in the cell, and therefore reveals the existence of depletions and amplifications of the mtDNA (Lestienne et al. 1997).

An important point to note is that the search for deletions is carried out on muscular DNA, as blood samples are generally poor, or are completely lacking in deleted genomes.

2.4 Mitochondrial Diabetes

The mtDNA rearrangements (deletions and duplications) like a mutation (A3243G) are at the origin of a particular form of diabetes, passed on maternally, frequently associated with a deafness (Ballinger et al. 1992; Van den Ouweland et al. 1992).

2.5 Mitochondrial Deafness

Two point mutations were described in this pathology by maternal transmission (Kogelnik et al. 1998). The A7445G mutation is detected by PCR/restriction and the A1555G mutation is detected by sequencing. Since deafness can be a sign of other mitochondrial pathologies, for example Kearns-Sayre syndrome, an analysis of the mtDNA rearrangements using Southern blot and long PCR is systematically carried out.

3 Materials and Methods

3.1 Preparation of DNA

Specific mitochondrial extraction techniques are well known, however, total cellular DNA is sufficient for the type of exploration that we suggest. This total cellular DNA extraction (200 µl of blood sample) is carried out with the help of a fast extraction kit (30 min) which uses glass fibers which bind specifically to DNA after treatment with a chaotropic agent (guanidine), a detergent (Triton X100) and with proteinase K (High Pure PCR Template Preparation Kit, Boehringer, Mannheim). The DNA bound to the glass fibers is then eluted with water. Ethanol precipitation is not necessary because the DNA obtained is concentrated into a small volume. Total muscle DNA extraction (50 mg) is carried out by the conventional phenol/chloroform method after homogenization.

3.2
PCR

The PCRs which allow the detection of restriction polymorphisms or the sequencing of a specific region are carried out under standard conditions: 1.5 mM $MgCl_2$, 75 mM Tris-HCl (pH 9 at 25 °C), 20 mM $(NH_4)_2SO_4$, 0.01% Tween 20, 50 pmol of each primer, 200 µM of each dNTPs, 100 ng of total DNA and 2 units of GoldStar DNA polymerase (Eurogentec, Belgium), in a final volume of 50 µl. The thermocyclers used are the „minicycler“ from MJ Research (USA) and Stratagene's „robocycler“ (USA). The 30 cycles are as follows: after a preliminary denaturation step of 30 s at 94 °C, a hybridation step of 30 s at 48–60 °C (according to the couple of primers used) and an extension step of 1 min at 72 °C are carried out. The negative and positive controls are systematically associated. The PCR products are separated by electrophoresis in a 2% agarose gel at 100 V for 30 min. The gels are next stained by ethidium bromide (10 µg/ml) and analysed by software analysis (Molecular Analyst, Biorad, USA). The localization of the primers and the size of the PCR products are shown in Tables 1 and 2.

Table 1. mtDNA point mutation detection techniques. Nucleotide positions are given on the mtDNA light strand

Mutation	Primers	PCR product (bp)	Restriction site
3460	D21:3139–3162 R21:3519–3495	380	BsaHI suppressed
4160	D3:4036–4054 R3:5650–5630	1614	Sequencing with D3
11 778	D20:11675–11694 R20:11915–11893	240	MaeIII created SfaNI suppressed
14 484	D30:14464–14483[a] R30:14663–14644	199	Sau3AI suppressed
15 257	D26:15141–15160 R26:15599–15580	458	AccI suppressed
4216	D32:3817–3836 R32:4261–4242	444	NlaIII created
4917	D24:4830–4848 R24:5399–5379	569	BfaI created
5244	D24:4830–4848 R24:5399–5379	569	HpaII suppressed
7444/7445	D28:7363–7383 R28:7648–7627	285	XbaI suppressed
13708	D29:13301–13320 R29:13799–13780	498	BstNI suppressed
15812	D31:15601–15620 R31:16059–16040	458	RsaI suppressed

[a] This primer has a mutated nucleotide (C14482G) allowing the creation of a Sau3AI restriction site when the T14484C mutation is absent

Table 2. Primers used for mtDNA sequencing. Nucleotide positions are given on the mtDNA light strand

Sequencing	PCR primers	PCR products (bp)	Sequencing primers
A1555G	D1:350–368 R1:2300–2282	1950	R27:1701–1683
ARNt leucine (UUR)	D2:2240–2260 R2:4200–4180	1960	D21:3139–3162
ARNt isoleucine	D3:4100–4120 R3:5650–5630	1550	D3:4100–4120
ARNt lysine	D5:7170–7190 R5:8620–8600	1450	D25:8180–8199
T8993G	D6:8274–8294 R6:10080–10062	1806	D23:8905–8924
ARNt threonin and proline	D10:15250–15268 R10:400–382	1719	D31:15601–15620

3.3 Enzymatic Digestions

After ascertaining the efficiency of PCR on agarose gel, the PCR product is purified by a quick method according to the same principle as that used for the preparation of DNA (High Pure PCR Product Purification Kit, Boehringer, Mannheim). The purified PCR product is then digested by a restriction enzyme as defined by the supplier (MaeIII: Boehringer, Mannheim and Biolabs, USA, for the rest of the other restriction enzymes used). The visualisation of the restriction fragments is performed after agarose gel electrophoresis. Each detection system for point mutations by PCR/restriction is detailed in Table 1. The detection of the mutation at nucleotide position 11778 is shown in Fig. 1.

3.4 Sequencing

Sequencing of purified PCR products is carried out by a cyclic technique derived from an enzymatic method from Sanger (Amplicycle Sequencing Kit, Perkin Elmer, USA). One of the two primers having allowed the amplification of the region bearing the mutation is labelled on the 5' end with ^{33}p with polynucleotide kinase (Table 2). A series of 30 cycles of primer extension is carried out in four tubes, each containing one of the four ddNTPs in the presence of Taq DNA polymerase. The products of the reactions are subjected to electrophoresis in an 8% polyacrylamide gel in 8 M urea denaturing conditions at 80 W between 1.5 h and 4 h according to the case. The gel is dried and exposed to autoradiography for 24 h.

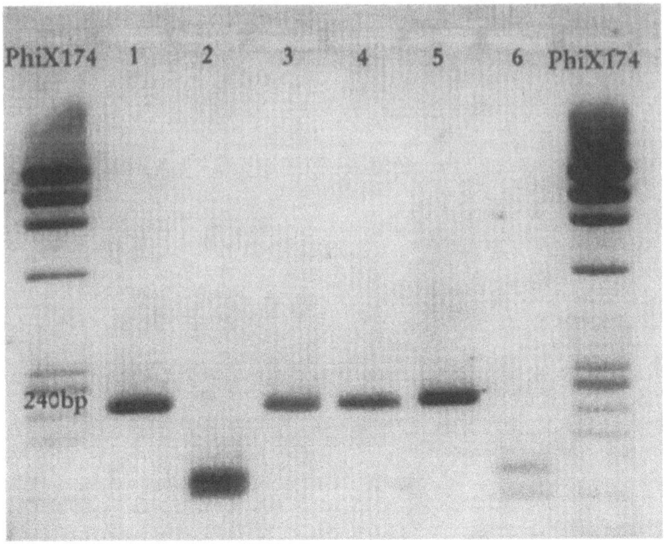

Fig. 1. Electrophoretic migration of PCR/restriction products allowing detection of the 11778 mtDNA mutation. When the DNA is not mutated, the PCR product *(lane 1)* is cut by SfaNI *(lane 2)* and not cut by MaeIII *(lane 3)*. When the DNA is mutated, the PCR product *(lane 4)* is not cut by SfaNI *(lane 5)* and cut by MaeIII *(lane 6)*. The SfaNI and MaeIII digestion products are not exactly the same because SfaNI restriction enzyme cuts the DNA at distance from the restriction site

3.5 Southern Blot

Total cellular DNA (1 to 5 μg) is digested by BamHI (mtDNA cleavage in position 14258) or PvuII (mtDNA cleavage at position 2650) which have in general only one site on mtDNA. Restriction is conducted at 37 °C with an excess of enzyme. Electrophoresis in 0.8% agarose gel is then carried out at 80 V for 4 h. DNA is transferred onto a positively charged Nylon membrane (Boehringer, Mannheim) and fixed by a 30-s exposure to short UV (Stratalinker, Stratagene, USA). The mtDNA probes labelled by digoxigenin are obtained by multi-random priming of an mtDNA matrix (1 μg) prepared by long PCR (Random Primed DNA Labelling Kit, Boehringer, Mannheim). After prehybridization for 2 h at 42 °C in a buffer containing: 5X SSC, 50% formamide, 0.1% sodium-lauroylsarcosine, 0.02% SDS and 2% blocking agent, the Nylon membrane is hybridized overnight at 42 °C with 100 ng of probe. The first washing is carried out at room temperature with a 2X SSC and 0.1% SDS solution. The second washing is carried out at 68 °C, with an SSC 0.1X and SDS 0.1% concentration. Then, the Southern blot is revealed with antibodies labelled by alkaline phosphatase directed against the digoxigenin (DigDNA Labelling and Detection Kit, Boehringer, Mannheim).

3.6
Long PCR

The principle on which Long PCR relies is the joint use of a DNA polymerase with strong processing activity, devoid of 3' → 5' exonuclease activity, and of a low proportion (1/16) of DNA polymerase which does possess this activity (Cheng et al. 1994). The latter polymerase is capable of correcting the induced misincorporations by the predominant polymerase, and therefore allows amplification. Moreover, so as to minimalize the alteration of the template DNA, precautions are taken by reducing the duration of the stages at high temperatures (Reynier and Malthièry 1996). This system allows one to amplify sequences as long as 42 kb, and it also possesses an adequate fidelity (13 times more accurate than the Taq DNA polymerase).

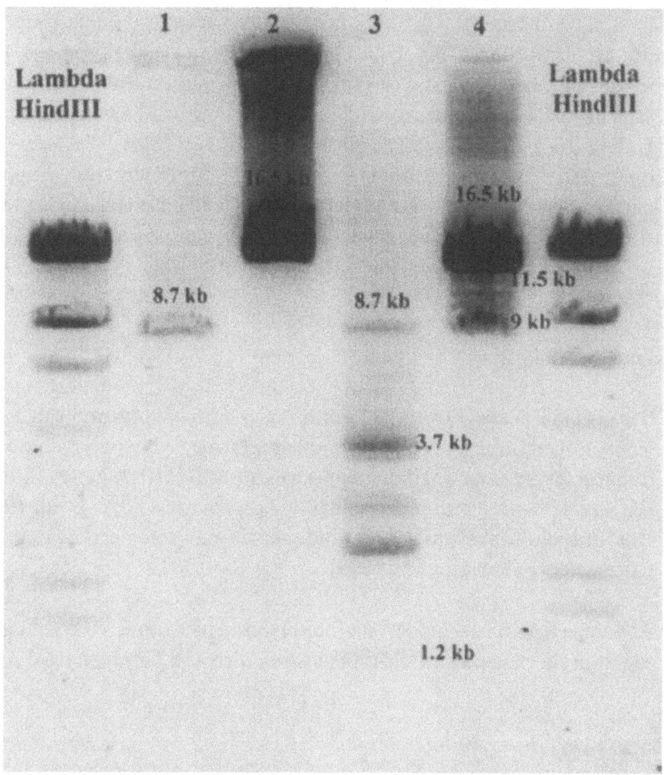

Fig. 2. Electrophoretic migration of long PCR products performed from muscle DNA. In a 20-year-old patient, the PCR using the GoldStar polymerase allows an amplification of 8.7 kb *(lane 1)* and the long PCR using a mixture of two polymerases allows amplification the entire mtDNA *(lane 2)*. In a 70-year-old patient with multiple deletions of mtDNA, the GoldStar PCR *(lane 3)* and the PCR with a mixture of polymerase *(lane 4)* allows amplification of the same fragments associated with multiple deleted fragments. The 1.2 to 3.7 fragments of *lane 3* represent deletions from 7.5 to 5 kb (8.7–1.2 and 8.7–3.7). The 9 to 11.5 kb of lane 4 represent the same deletions of 7.5 to 5 kb (16.5–9 and 16.5–11.5)

Long PCR is carried out from 100 ng of total cellular DNA in a final volume of 50 µl, using 50 mM Tris HCl (pH 9.2 at 25 °C), 14 mM $(NH_4)_2So_4$, 2.25 mM $MgCl_2$, 50 pmol each of primers A1 and A2, and 200 µM of each dNTPs (Reynier and Malthièry 1996). The reaction is conducted as follows: after preliminary denaturation for 1.5 min at 94 °C, 15 first cycles for 20 s at 94 °C and 10 min at 68 °C are carried out. During the following 15 cycles, the duration of the primer extension stage is increased by 15 s/cycle. The thermocycler used is the „Minicycler" (MJ Research, USA). To ensure the success of the long PCR, the technical features of the apparatus must be taken into account (Reynier and Malthièry 1996). The primers A1 (5' CGACATCTGGTTCCTAC-TTCAGG, 16495-16517) and A 2 (5' AACCAGATGTCGGATACAGTTCAC, 16506-16483) are localized within an area conserved from the d-loop, which is outside the zone frequently affected by deletions. Their extremities are adjacent, which allows the amplification of the totality of the mtDNA in a linear form. The PCR products are subjected to electrophoretic migration in a 0.8% agarose gel and are analysed after staining with ethidium bromide to remove (Molecular Analyst, Biorad, USA).

A second type of long PCR is used. This concerns a standard PCR which is carried out under the same experimental conditions as those previously described (Sect 3.2), in the presence of an exo-GoldStar DNA polymerase (Eurogentec, Belgium). This enzyme allows amplification of sequences up to 12 kb (Reynier and Malthièry 1996). Unlike the long PCR which allows quantification of deletions, in this case there is a quantitative amplification bias in favour of the shortest amplified sequences. This amplification bias is used to detect, in an extremely sensitive way, the multiple deletions of mtDNA (Reynier et al. 1994). Some examples of long PCR are shown in Fig. 2.

4 Conclusion

This report deals with the detection of the most common mitochondrial DNA alterations. Nevertheless, evidence of new mutations may be systematically searched for by denaturing gradient gel electrophoresis on the tRNA genes (Sternberg et al. 1998) as well as with new methodological advancements (Chee et al. 1996), bearing in mind that mitochondrial nuclear pseudogenes may interfere in the interpretation of PCR experiments (Parfait et al. 1998).

Acknowledgements. This work is supported by grants from: Le Centre Hospitalier Universitaire d'Angers, la Région Pays de Loire and l'Université d'Angers.

References

Ballinger SW, Shoffner JM, Hedaya EV, Trounce I, Polak MA, Wallace DC (1992) Maternally transmitted diabetes and deafness with a 10.4 kb mitochondrial DNA deletion. Nat Genet 1:11–15

Chee M, Yang R, Hubble E, Berno A, Haung XC, Stern D, Winkler J, Lockar T, Morris MS, Fodor SPA (1996) Accessing genetic information with high-density DNA arrays. Science 274:610–614

Cheng S, Higuchi R, Stoneking M (1994) Complete mitochondrial genome amplification. Nat Genet 7:350–351

Kogelnik AM, Lott MT, Brown MD, Navathe SB, Wallace DC (1998) MITOMAP: a human mitochondrial genome database – 1998 update. Nucleic Acids Res 26(1):112–115

Lestienne P (1992) Mitochondrial DNA mutations in human diseases: a review: Biochimie 74:123–130

Lestienne P, Reynier P, Chrétien MF, Penisson-Besnier I, Malthièry Y, Rohmer V (1997) Oligoasthenospermia associated with multiple mitochondrial DNA rearrangements. Mol Hum Reprod 3:811–814

Parfait A, Rustin P, Munnich A, Rotig A (1998) Coamplification of nuclear pseudogenes and assessment of heteroplasmy of mitochondrial DNA mutations. Biochem Biophys Res Commun 247:57–59

Reynier P, Pellissier JF, Harle JR, Malthièry Y (1994) Multiple deletions of the mitochondrial DNA in polymyalgia rheumatica. Biochem Biophys Res Commun 205:375–380

Reynier P, Malthièry Y (1995) Accumulation of deletions in mtDNA during tissue aging: analysis by long PCR. Biochem Biophys Res Commun 217:59–67

Reynier P, Malthièry Y (1996) PCR longue: progreès récents et application à l'étude des délétions de l'ADN mitochondrial. Médecine/Sciences 12:1011–1016

Sternberg D, Danan C, Lombés A, Laforêt P, Girodon E, Goossens M, Amselem S (1998) Exhaustive scanning approach to screen all mitochondrial tRNA genes for mutations and its application to the investigation of 35 independent patients with mitochondrial disorders. Hum Mol Genet 7:33–42

Van Den Ouweland JMW, Lemkes HHPJ, Ruitenbeek W, Sandkujil LA, de Vijleder MF, Struyvenberg PAA, Van de Kamp JJP, Maassen JA (1992) Mutation in mitochondrial tRNALEU(UUR) gene in a large pedigree with maternally transmitted type II diabetes mellitus and deafness. Nat Genet 1:368–371

Subject Index

A

B

I

K

L

P

Q

R

S

T

Computer to Film: Saladruck, Berlin

Kolibri 360 GmbH
Ferdinand-Jühlke-Straße 7
99095 Erfurt, Deutschland
produktsicherheit@kolibri360.de